Processed Cheese Science and Technology

Woodhead Publishing Series in Food Science, Technology and Nutrition

Processed Cheese Science and Technology

Edited by

Mamdouh El-Bakry

Regulatory Affairs Technologist, Food Supplements & Natural Cosmetics, Martínez Nieto, S.A., Cartagena, Spain

Bhavbhuti M. Mehta

Dairy Chemistry Department, SMC College of Dairy Science, Kamdhenu University, Anand, Gujarat, India

ELSEVIER

WP
WOODHEAD PUBLISHING
An imprint of Elsevier
elsevier.com/books-and-journals

Woodhead Publishing is an imprint of Elsevier
50 Hampshire Street, 5th Floor, Cambridge, MA 02139, United States
The Boulevard, Langford Lane, Kidlington, OX5 1GB, United Kingdom

British Library Cataloguing-in-Publication Data
A catalogue record for this book is available from the British Library

Library of Congress Cataloging-in-Publication Data
A catalog record for this book is available from the Library of Congress

ISBN: 978-0-12-821445-9

For Information on all Woodhead Publishing publications visit our website at https://www.elsevier.com/books-and-journals

Publisher: Nikki P. Levy
Acquisitions Editor: Megan R. Ball
Editorial Project Manager: Billie Jean Fernandez
Production Project Manager: Sojan P. Pazhayattil
Cover Designer: Mark Rogers

Typeset by Aptara, New Delhi, India

Contents

Contributors

Adriana Torres Silva e Alves, Instituto de Tecnologia de Alimentos (ITAL), Centro de Tecnologia de Laticínios (Tecnolat), Campinas, São Paulo, Brazil

Hamadi Attia, University of Sfax, Analysis, Valuation and Food Safety Laboratory (LAVASA), National Engineering School of Sfax (ENIS), Route Soukra, Sfax, Tunisia

Yasser Bouhlal, Corteva Agriscience, Indianapolis, Indiana, United States

Terri D. Boylston, Department of Food Science & Human Nutrition, Iowa State University, Ames, Iowa, United States of America

František Buňka, Department of Food Technology, Faculty of Technology, Tomas Bata University in Zlín, Zlín, Czech Republic; Food Research Laboratory, Department of Logistics, Faculty of Military Leadership, University of Defense, Brno, Czech Republic

Oral Capps Jr., Department of Agricultural Economics, Texas A&M University, College Station, Texas, United States

Michaela Černíková, Department of Food Technology, Faculty of Technology, Tomas Bata University in Zlín, Zlín, Czech Republic

Soumela E. Chatziantoniou, Department of Food Science and Technology, International Hellenic University, Thessaloniki, Greece

Adriano Gomes da Cruz, Instituto Federal de Educação, Ciência e Tecnologia do Rio de Janeiro (IFRJ), Departamento de Alimentos, Brazil

Oznur Cumhur, Bilecik Seyh Edebali University Faculty of Applied Science Deparment of Tourism Management Bilecik, Turkey

Georgia Dimitreli, Department of Food Science and Technology, International Hellenic University, Thessaloniki, Greece

Mamdouh El-Bakry, Regulatory Affairs Technologist, Food Supplements & Natural Cosmetics, Martínez Nieto, S.A., Cartagena, Spain

Zafer Erbay, Adana Alparslan Turkes Science and Technology University, Engineering Faculty, Food Engineering Department, Saricam, Adana, Turkey

Erick Almeida Esmerino, Universidade Federal Fluminense (UFF), Faculdade de Veterinária (UFF)

Imène Felfoul, University of Sfax, Analysis, Valuation and Food Safety Laboratory (LAVASA), National Engineering School of Sfax (ENIS), Route Soukra, Sfax, Tunisia

Mary Ann Ferrer, Department of Food Science, Ontario Agricultural College, University of Guelph, Ontario, Canada

Art Hill, Department of Food Science, Ontario Agricultural College, University of Guelph, Ontario, Canada

Miočinović Jelena, University of Belgrade, Faculty of Agriculture, Department of Animal Source Food Technology, Belgrade, Serbia

Meral Kilic-Akyilmaz, Istanbul Technical University, Department of Food Engineering Istanbul, Turkey

Nurcan Koca, Ege University, Engineering Faculty, Food Engineering Department, Bornova, Izmir, Turkey

Bhavbhuti M. Mehta, Dairy Chemistry Department, SMC College of Dairy Science, Kamdhenu University, Anand, Gujarat, India

Müge Urgu Öztürk, Ege University, Engineering Faculty, Food Engineering Department, Bornova, Izmir, Turkey

Mustafa Ozturk, Sakarya University, Department of Food Engineering Sakarya, Turkey

Gustavo Luís de Paiva Anciens Ramos, Instituto Federal de Educação, Ciência e Tecnologia do Rio de Janeiro (IFRJ), Departamento de Alimentos, Rio de Janeiro, Brazil; Universidade Federal Fluminense (UFF), Faculdade de Veterinária, Niterói,Rio de Janeiro, Brazil

Richardos N. Salek, Department of Food Technology, Faculty of Technology, Tomas Bata University in Zlín, Zlín, Czech Republic

Leila Maria Spadoti, Instituto de Tecnologia de Alimentos (ITAL), Centro de Tecnologia de Laticínios (Tecnolat), Campinas, São Paulo, Brazil

Apostolos S. Thomareis, Department of Food Science and Technology, International Hellenic University, Thessaloniki, Greece

Patrícia Blumer Zacarchenco, Instituto de Tecnologia de Alimentos (ITAL), Centro de Tecnologia de Laticínios (Tecnolat), Campinas, São Paulo, Brazil

Miloradović Zorana, University of Belgrade, Faculty of Agriculture, Department of Animal Source Food Technology, Belgrade, Serbia

Overview of processed cheese and its products

1

Mamdouh El-Bakry[a] *and Bhavbhuti M. Mehta*[b]
[a] Regulatory Affairs Technologist, Food Supplements & Natural Cosmetics, Martínez Nieto, S.A., Cartagena, Spain, [b] Dairy Chemistry Department, SMC College of Dairy Science, Kamdhenu University, Anand, Gujarat, India

1.1 Introduction

Processed cheese manufacture and consumption are constantly growing, especially in recent years, due to a various factors and reasons (Ferrão et al., 2016; Guinee et al., 2004; Mounsey and O'Riordan 2008; Tamime, 2011). The increasingly high consumer demand for ready/prepared or convenience foods could be the first reason. In these foods, processed cheese is used as an alternative to natural cheeses, for example, cheddar-like or mozzarella-like processed cheese in hamburgers, pizzas, pies, lasagne, and cordon bleu products, where cheese texture and melting properties as well as flavor are of crucial importance. The second main reason could be that processed cheese allows the food manufacturer to tailor the product to any specific requirements through the altering of the formulation and processing parameters. This is not the case when natural cheeses are manufactured, as there are many factors that cannot be controlled by the manufacturer. In addition, processed cheese has a very high cost-effectiveness, unlike natural cheeses, which arises from the ease of its manufacture and the replacement of expensive milk ingredients, for example, milk fat or milk proteins, by cheaper vegetable blend products.

Processed cheese has been extensively studied, as various research works have been performed from different perspectives. For instance, processed cheese was manufactured using various combinations of dairy and nondairy ingredients (Zehren and Nusbaum, 2000; Salek et al., 2017). These ingredients may include natural cheeses or casein and vegetable proteins, as a protein source. The lipid phase in processed cheese can include a very wide range of dairy and vegetable fats, such as native or hydrogenated oils (Caric and Kalab, 1987; Ramel and Marangoni, 2017). In addition, carbohydrates such as polysaccharides, native and/ or modified starches, and inulin have been used in the manufacture of processed cheese (Arimi et al., 2010). Most interestingly, the microwave expansion of processed cheese has been deeply investigated, especially the study of the mechanism and factors affecting that expansion. The effect of further ingredients on cheese properties have been studied in research, such as the so-called emulsifying salts (ES), salt, stabilizers, coloring agents, flavors, and flavor enhancers (Fu et al., 2018). These are highly affecting all properties of the processed cheese. As an example, the main role of salt is its influence on the microbial growth and accordingly, the shelf life of the resulting cheese product can be altered (Guinee and Fox, 2004; Taormina, 2010). Cheese functional properties have been also

Processed Cheese Science and Technology: Ingredients, Manufacture, Functionality, Quality, and Regulations.
DOI: https://doi.org/10.1016/B978-0-12-821445-9.00006-6

studied. These include mainly texture and melting properties that can be described by many terms (Bourne, 2002; Joshi et al., 2004). These properties are directly related to the microstructure of processed cheese, that is influenced by various parameters of processing, during the manufacture, as well as the composition (El-Bakry and Sheehan, 2014; Mehta, 2018; Kapoor and Metzger, 2008; Zehren and Nusbaum, 2000).

This chapter is divided into several sections. It starts with the history, definition, and classification of processed cheese. This is followed by a general description of the principles and key ingredients used in the manufacture. The formulations and equipment used are also discussed. A general outline of the textural, melting, and microstructural properties of the cheese is also described. The chapter ends up with the approaches to develop healthy processed cheese, as well as its advantages and applications.

1.2 History of processed cheese

Cheese has been present in the human diet for many years, with recent literature suggesting early cheese-making practices date as far back as 5200 BC (Zehren and Nusbaum, 2000; Salque et al., 2013). In general, the current world production of cheese, which is exceeding 19×10^6 tonnes per annum and which represents over 35% of the total milk production, has increased at an average annual rate of about 4% over the past decades. In addition, the worldwide cheese consumption has increased consistently over many years, where cheese is the principal growth product within the sector of dairy products (Kapoor and Metzger, 2008). There are many reasons for the increased consumption of cheese, including a positive dietary image, convenience, and flexibility in use and the great diversity of flavors and textures. Cheese can be regarded as the quintessential convenience food: it can be used as a major component of a meal, as a dessert, as a component of other foods, or as a food ingredient; it can be consumed without preparation or subjected to various cooking processes. Processed cheese products (PCPs) represent more than 14% of worldwide cheese production, and thus surpass the production of most varieties of cheeses.

Although the high protein, fat, and calcium content of most cheeses, which is a proven nutritional energy source, issues arising over its low stability lead to the initial production of processed cheese (Guinee et al., 2004). This type of cheese is a natural cheese derivative with a much higher stability, where there is usually a reduced need for its refrigerated storage. Processed cheese, which was first manufactured and introduced in the food market about a century ago, that is, in the early 20th century, has many different types that are being manufactured in different countries (Fox et al., 2000). PCPs were mainly developed in order to extend the shelf-life as well as to standardize the properties, such as texture and flavor, of the cheese.

The demand of processed cheese by the consumers is in steady rise, whereas the success is due to a combination of various factors (Talbot-Walsha et al., 2018). Processed cheese, in comparison to the majority of other types of cheeses, is much more economic, commercially available in a very wide variety of flavors, composition, and functional properties, for example, texture, meltability, and flowability, and much more versatile in handling. This is due to the fact that processed cheese has many applications and uses in the food catering sector, especially this kind of cheese has a fairly long shelf life (Guinee et al., 2004).

In USA, recent studies have found that 70% of all households purchased processed cheese for everyday consumption (IDDBA, 2016). This might be a result of having a large variety of PCPs that is currently available on the market. In developing country, where there is a large existing consumer base and a high potential for market growth, it has been predicted that the dairy sector will experience a continued demand of processed cheese, in the near future.

1.3 Definition of processed cheese

All types of processed cheese, in general, can be defined and characterized as a complex, multicomponent dairy-system, described as a stable oil-in-water emulsion. Processed cheese, or sometimes referred to as process cheese as the term widely used in literature for such cheese products, unlike natural cheeses, is not made directly from milk or dehydrated whole or skimmed milk, but rather from various combinations (Zehren and Nusbaum, 2000; Tamine 2011; Fox et al., 2000). The combinations used include dairy and/or nondairy ingredients. These ingredients can mainly include natural raw cheeses, of various degrees of maturity, or cheese curd, casein, caseinates or vegetable proteins, a protein source. The types of cheese predominantly used in different world areas vary. The cheese used is mainly Cheddar, Dutch-type, Swiss-type, Mozzarella, or white brined cheese. For instances, in Britain, USA, Canada, Australia, and New Zealand, the main raw material for processed cheese production is usually cheddar and mozzarella cheeses. In countries around the Mediterranean area and Middle East, white brined cheeses represent the most consumed cheese varieties and are therefore widely used as the main raw material, that is, cheese ingredient, for the production of processed cheese (Salek et al., 2017).

As a fat source, processed cheese may contain milk fat, butter oil, and/ or vegetable oil (Guinee et al., 2004). In addition, carbohydrates, such as polysaccharides, native and/or modified starches, and inulin have been used in the manufacture of processed cheese (Arimi et al., 2010; Duggan et al., 2008). Further ingredients used in the manufacture of processed cheese are the so-called ES, salt, stabilizers, coloring agents, flavors and flavor enhancers (Fu et al., 2018). Processed cheese could be also defined as a viscoelastic dairy-based gel that is considered as a stable oil-in-water emulsion (Hanaei et al., 2015; Lee et al., 2003).

Processed cheese is considered as a non-standard cheese product that is used as an alternative for natural cheese due to the cost efficiency, feasibility in production, versatility in applications and usefulness (Tamime, 2011; Vernon, 1972; Meyer, 1973; Berger et al., 1989; Pellegrino, 1996). Most importantly, the shelf life of processed cheese is much longer than in the case of natural cheeses. In addition, processed cheese, as it can be produced in a wide range of flavors, shapes, and consistencies, is used in many applications in different processed foods, mainly as alternatives for natural cheeses.

1.4 Classification and types of processed cheese

There are several classifications and types of processed cheese. For instance, this cheese can be classified into cheese-based and plant based processed cheese, where natural cheese or ingredients of plant origin are used in the manufacture, respectively. In addition, cheese analogues can be also considered as a main category of processed cheese. However, in general, processed cheese includes mainly two categories, PCPs, and analogue cheese products (ACPs) (Fox et al., 2000; Guinee et al., 2004). The classification varies from country to country and depending on the regulations and legislation on processed cheese (see Chapter 15).

1.4.1 Processed cheese products

PCPs are manufactured mainly by blending ingredients of dairy origin in the presence of the necessary thermal and mechanical energies to form a homogeneous cheese-like mass (Fox et al., 1996). Processed cheese may be further sub-divided into several categories, the number of which varies from country to country and which depends on the main ingredients used. For instance, in the United States, the Code of Federal Regulations (CFR) defined five types of PCP (CFR, 2007), which include pasteurized blended cheese, pasteurized processed cheese, pasteurized processed cheese foods, pasteurized processed cheese spread, and pasteurized cheese spread. The criteria for classification include permitted ingredients and compositional parameters, mainly moisture, fat, and pH. These criteria, including the ingredients used as well as the compositional specifications, are summarized in Table 1.1. In the UK, there are two categories of PCP, namely processed cheese and cheese spreads, as specified by the Cheese and Cream Regulations (2005). Cheese spreads only differ from processed cheese in that milk products other than natural cheese can be used, whereas only natural cheese can be used in processed cheese.

1.4.2 Analogue cheese products

ACPs are cheese-like products, but in which the milk fat has been partly or completely replaced by other nondairy fats (Codex Standard, 1995). This kind of cheese is produced by blending various oils/fats, proteins (usually rennet casein), flavors, and other ingredients with water into a smooth homogeneous cheese-like blend with the aid of heat, shearing forces and ES (Tamime, 2011.). ACP does not necessarily contain natural cheese; however, it can be added at a low level to enhance the cheese flavor or to comply with a particular consumer preference. These cheese analogues can be categorized as either dairy, part dairy, or nondairy analogues, depending on the origin of the fat and protein used (Fox et al., 2000; Guinee, 2004). For instance, ACP, which is produced worldwide in a wide range, may be also classified into three broad categories: analogue cheese, filled cheeses, and tofu-based products. An example of ACP is imitation cheese, which has been extensively studied in various research studies

Table 1.1 Ingredients and composition specifications of different categories of processed cheese products. Adapted from the US Code of Federal Regulations (CFR, 2007).

Processed cheese category	Permitted ingredients	Moisture (%)	Fat (%)	pH
Pasteurized blended cheese	Cheese; cream, milk fat; water; salt; food-grade colors; flavors other than those which simulate the flavor of natural cheese; sorbic acid $\leq$ 0.2%; sodium propionates $\leq$ 0.3%	$\leq$43	–	–
Pasteurized processed cheese	Similar to pasteurized blended cheese but with the following optional ingredients: emulsifying salts (e.g., sodium/potassium phosphates and citrates) $\leq$3%, food-grade organic acids at levels such that the cheese pH is not less than 5.3	$\leq$43	–	Not less than 5.3
Pasteurized processed cheese foods	Similar to pasteurized processed cheese but with the following optional dairy ingredients: milk, buttermilk, whey proteins	$\leq$44	$\geq$ 23%	Not less than 5.0
Pasteurized processed cheese spread	Similar to pasteurized processed cheese but with extra optional food-grade hydrocolloids <0.8%, for example, guar gum, gelatin, and sweeteners, such as sugar, glucose	40–60	$\geq$ 20%	Not less than 4.0
Pasteurized cheese spread	Similar to pasteurized processed cheese spread, except than emulsifying salts are not allowed	40–61	$\geq$ 20%	Not less than 4.1

(Arimi et al, 2010; El-Bakry et al., 2010; Noronha et al, 2008a, Ramel and Marangoni, 2018). Imitation cheese is a term usually applied to products manufactured by mixing water, various edible oil/fats, protein (usually, but not necessarily, rennet casein) and other ingredients in the presence of so-called ES to obtain a homogeneous product having similar properties to natural cheese. This is generally achieved with the aid of a certain required level of heating and mechanical agitation (Guinee, 2003). Imitation cheese differs from PCP in that the protein ingredients are dry ingredients like casein powder rather than natural cheese (Meyer, 1973). An example of ingredients used in ACP are presented in Table 1.2.

Table 1.2 Ingredients used in the manufacture of imitation cheese. Modified from Guinee 2002a; Guinee et al., 2004; Arimi et al., 2010.

Ingredient	Function	Examples
Water	Acts as a medium for dissolving the salts added during manufacture, and as a plasticser causing matrix flexibility (assist in texture and melting properties)	–
Proteins	Give desired composition, semi-hard texture with good flow and stretch characteristics on heating. Assist in the physicochemical stability. Vegetable proteins, used in low levels, to contribute to low cost manufacture.	Caseinates, whey proteins, soy protein
Fats	Give desired composition and texture. Milk fat, if used, contributes to flavor.	Milk fat, vegetable oils
Starches	Substitution of protein or fat	Maize, rice or potato starches
Emulsifying salts	Assist in the protein hydration and fat emulsification, development of texture and melting properties	Phosphates and citrates
Hydrocolloids	Enhance product stability and texture	Guar gum, carageenans
Acidifying Agents	pH control	Lactic, acetic or citric acids
Colours	Impart desired color	Annatto, artificial colors
Preservatives	Inhibit mould growth and prolong shelf life	Nisin, sorbates, propionates, NaCl
Minerals and Vitamins	Improve nutritional value	Iron, vitamin A
Flavourings	Contribute to a "dairy" or "cheesy" desired flavor	EMCs, starter distillate
Flavour enhancers	Emphasize and strengthen the flavor	NaCl, yeast extracts
Sweetening agents	Contribute to the sweetness	Sucrose, dextrose

1.5 General principles of manufacture and key ingredients

The manufacturing protocols involve the application of heat and mechanical shear to the formulated blend, in a reaction container, followed by cooling and cold storage (Zehren and Nusbaum, 2000; Tamime, 2011). The most important part of the process is heating for an appropriate time with continuous stirring, known as the emulsifying process, until a homogeneous mass is formed. The manufactured product, that is, the processed cheese end-product, may be generally described as a cold-set gelled oil-in-water emulsion, which is stabilized by a hydrated protein phase. Processed cheese undergoes relatively very minor or almost no changes during refrigerated storage, in comparison to natural cheeses which generally undergo significant transformations during the post-manufacture ripening period (Fox et al., 2017; Tamime et al., 1990).

1.5.1 Key ingredients

In general, the ingredients used in the manufacture of processed cheese are water, proteins, fats or oils, salt, ES as well as other ingredients such as starch and additives that are used in small amounts, for example, flavoring and coloring agents, vitamins and minerals (Tables 1.1 and 1.2).

1.5.1.1 Proteins

Protein is arguably the most crucial ingredient in the manufacture of processed cheese. This component forms the continuous hydrated matrix of the cheese (Zahren and Nusbaum, 2000; Tamime, 2011). In addition, the protein stabilises the oil-in-water emulsion by reducing the interfacial tension at the aqueous phase-oil droplet interface and by increasing the viscosity of the aqueous phase thereby decreasing the frequency of collisions between oil droplets (Shimp, 1985; Ennis et al., 2000). The sources of protein that have been used in the manufacture of processed cheese include natural cheeses (Tamime, 2011), caseinate (Cavalier-Salou and Cheftel, 1991), whey protein (Mleko and Foegeding, 2000), vegetable proteins and rennet casein (Yang and Taranto, 1982; Cavalier-Salou and Cheftel, 1991; Mounsey and O'Riordan, 2008; Hennelly et al., 2006; Montesinos-Herrero, 2006; Noronha, 2007; Arimi et al., 2010), with the latter being most widely used.

1.5.1.1.1 Natural cheeses

Natural cheeses are used as a main ingredient in the manufacture of processed cheese since more than four decades ago (Corredig et al., 2019; Hyde et al., 2010; Jolly and Kosikowoski, 1975; Moran et al., 1993). The most commonly used natural cheeses are white fresh cheeses and UF-retentate.

1.5.1.1.2 Rennet casein

Rennet casein is obtained by proteolytic destabilisation of the casein micelles in milk by using chymosin or chymosin-like proteinases, which cleave the Phe105–Met106 bond of κ-casein (Southward and Walker, 1980; Ennis et al., 2000). The renneted casein micelles form a curd, which is cooked at ~60°C to increase the firmness and to inactivate the coagulant, is subsequently washed and dried. Rennet casein consists of para-casein micelles, which are native micelles lacking their stabilising layer of caseino-macropeptide (κ-casein-(106–169) peptide). In this respect, rennet casein curd is considered to be similar to cheese curd. Indeed in some instances, it is marketed as dehydrated skimmed milk curd.

Rennet casein is the most commonly used protein source and regarded as the most desirable protein for processed cheese manufacture, due to the significant contribution of the protein to the structure and the functionality of the cheese products. (Aimutis, 1995). However, late-lactation rennet casein has to be avoided in the manufacture of processed cheese, as it was reported that this protein contains high levels of plasmin, a trypsin-like enzyme (fibrinolysin E.C.3.4.21.7), (Farkye and Fox, 1990; O'Malley et al., 2000). Proteolysis has been reported in processed cheese, during the post-manufacture storage, as a result of the presence of plasmin, which is a primary proteolytic agent found in late lactation rennet casein, where α-casein is most susceptible to proteolysis.

Rennet casein is a protein which is insoluble in water due to the calcium-mediated protein-protein interactions. Therefore, adequate casein hydration can only be achieved by the addition of ES, such as sodium phosphates and citrates, in the presence of the necessary agitation and heating to temperatures up to ~ 80°C. The insoluble calcium para-casein is converted to the soluble sodium para-caseinate, which binds water and subsequently fat, as the solubilised casein acts as emulsifier causing the stabilisation of the oil-in-water emulsion by reducing the tension at the aqueous phase-oil droplet interface (Aimutis, 1995; Ennis et al., 2000; Guinee et al., 2004; El-Bakry et al., 2011a).

1.5.1.1.3 Acid casein

Acid casein is defined as the product obtained after isoelectric precipitation (pH 4.6) of the caseins from milk followed by a washing and a drying procedure (Ennis et al., 2000). A distinction can be made between lactic acid casein and mineral acid casein. The former type of casein is obtained from acidification by lactic acid produced by a starter culture. Mineral acid casein is obtained from acidification by mineral acids, such as HCl or H_2SO_4. Both types of acid caseins can contain some residual plasmin and plasminogen, which are associated with the casein fraction. Accordingly, the proteolytic activity of plasmin can affect the stability of products containing this type of casein. These forms of casein can also be used as a protein source in the manufacture of processed cheeses, but because of its low solubility, this protein has to firstly be partially converted into sodium or calcium caseinate using alkali like NaOH or $Ca(OH)_2$ in order to develop a functionality in the cheese obtained (Aimutis, 1995; O'Malley et al., 2000). However, acid casein is not the desired protein source used in cheese manufacture due to the presence of plasmin which may adversely affect the functionality and also, due to its unsatisfactory organoleptic properties, for example, soapy or musty flavor (Zehren and Nusbaum, 2000).

1.5.1.1.4 Caseinates

Caseinates are obtained by addition of alkali to acid casein. The increase in pH will dissolve the casein and, depending on the cation of the alkalising agent, different types of caseinate are obtained, that is, sodium, potassium, calcium, and ammonium caseinates (Dalgleish, 1997). The functionality of calcium caseinate is quite different compared to that of sodium-, potassium-, and ammonium-caseinates, due to the particular interaction of calcium ions with the phosphoserine residues of the casein (Aimutis, 1995). Dissolved calcium caseinate is a milky colloidal dispersion due to calcium-induced formation of micelle-like casein aggregates. Sodium, potassium and ammonium caseinates form nearly clear solutions; in most cases slight opalescence is observed due to the association of the caseins. The functional and organoleptic properties of cheeses manufactured with caseinates can be negatively affected for the same reasons as mentioned for acid casein.

1.5.1.1.5 Whey proteins

These proteins represent approximately 20% of the protein fraction found in milk and are neither rennet nor acid coagulable; thus remain in the whey generated during the manufacture of natural cheeses or casein-making processes. There has been limited work on the usefulness of whey protein in processed cheese (Savello et al., 1989; Gupta and Reuter, 1992; Mleko and Foegeding, 2000). In general, whey proteins exhibit emulsifying abilities inferior to casein and caseinates, but they are less sensitive to changes in pH (Britten et al., 1994). In previous studies, rennet casein was partially replaced by whey proteins without highly altering the functional properties of processed cheese, at very low replacement levels of up to 3.75% (Murphy, 1999). In addition, the incorporation of salty whey in processed cheese has been studied (Kapoor and Metzger, 2004). This approach does not have a significant adverse effect on the quality and properties of processed cheese. Salty whey can be effectively used as an ingredient in various types of processed cheese, where it was indicated that up to 9.5% of salty whey can be included in the formulation of the cheese. The main advantage of using whey and/ or whey proteins, as an ingredient in processed cheese, is their content of most highly nutritionally and nutraceutically valuable milk components, such as lactoferrin, lactoperoxidase, and immunoglobulins (Jelen, 2004). Furthermore, the use of whey appears to be an attractive alternative to other ingredients that could also significantly reduce the need and costs for the disposal of whey, which is considered as a dairy waste product.

1.5.1.1.6 Vegetable proteins

Vegetable proteins such as soya bean, rice protein, wheat gluten and peanut protein have all been successfully used to partially replace the milk proteins used in the manufacture of processed cheese (Chen et al., 1979; Brander et al., 1985; Guirguis et al., 1985; Santos et al., 1989; Kim et al., 1992; Pereira et al., 1992; Murphy, 1999).

Difficulties involved in the use of these proteins in cheese manufacture, include their different molecular and functional properties compared to those of caseins. For instance, soy proteins, compared to milk proteins, are much larger in molecular size and possess a much more complex quaternary structure, and, unlike caseins, they are not phosphoproteins. Using soy proteins in processed cheese resulted in an excessive hardness and poor melting properties, compared to cheeses made with milk proteins (Kim et al., 1992). To improve functionalities of cheeses containing vegetable proteins, various modifications with proteases have been studied in the aim of increasing the solubility and emulsifying capacity of these proteins (Kim et al., 1992; Kamata et al., 1992; 1993). For instance, treatments of soy protein isolates with alcalase and trypsin were used to improve and modify textural properties of the resulting processed cheeses.

1.5.1.2 Fats

The lipid phase in processed cheese can include a wide range of vegetable fats such as native or partially hydrogenated peanut, palm kernel, cotton seed, coconut, corn, rapeseed, sunflower and soy bean oil (Caric and Kalab, 1987). In addition to the nutritional considerations, the popularity of vegetable oils has been driven by their greater cost-efficiency in comparison to milk fat (Kong-Chan et al., 1991; Guinee et al., 2004). The use of vegetable fats can give the processed cheese a consistency that makes it more suitable for certain applications (Shaw, 1984; Brander et al., 1985). In addition, the use of various types of milk fat may highly affect the texture and flavor (Jinjarak et al. 2006; Schieberle et al. 1993). For instance, whey butter has a lower penetration force than sweet butter, whereas there is also a significant difference that has been observed in the flavor of the end-product. In other recent studies (Cerníková *et al.* 2017, Ramel and Marangoni 2017), it has been shown that most of the distinctive properties of the added dairy fats are not observed during the process of manufacture, where the polymorphism of milk fat does not affect the textural properties of the cheese. Regarding the use of vegetable fats or oils, using soybean oil as a sole source of fat resulted in a hard and adhesive cheese, but a low cohesiveness and springiness, while the opposite effect was observed when a combination of soybean oil and butterfat was used in manufacture (Lobato-Calleros et al., 1997). Cheeses formulated with different proportions of tributyrin and soybean oil showed significant variations in texture. The hardness and cohesiveness of the cheese increased on increasing the proportion of the tributyrin in the fat used. Chen et al. (1979) reported that processed cheeses containing peanut oil had stretchy characteristics on heating compared to cheeses containing milk fat; this was attributed to the substantially lower melting point of peanut oil than that of milk fat. In general, a high fat content in the final product contributes to a soft texture and a high flow on melting (Berger et al., 1989; Zehren and Nusbaum, 2000). It has also been demonstrated that the role of fat in texture, particularly on hardness, depends on the physicochemical properties of the fat used (Marshall, 1991). When replacing a native vegetable oil with its hydrogenated equivalent in a cheese formulation, hardness was found to increase and the meltability properties were adversely affected (Zhou and Mulvancy, 1998). Low-calorie fatty substances have been used in processed cheese manufacture as a source of lipids (Kong-Chan et al., 1991). These substances are non-digestible or partially digestible or lower in calories per se, in comparison to vegetable or animal edible fats. It was suggested that the most preferred low-calorie fatty materials for incorporation into the cheese products is replacement of animal and vegetable fats with polyol fatty acid polyesters, sugar alcohols and polyglycerol fatty acid polyesters. For instance, vegetable fats were used in the manufacture of processed cheese in previous studies (Arimi et al., 2010; El-Bakry et al., 2010). The fats used were a mixture of rapeseed oil and hydrogenated palm oil with an approximate ratio of 1:2, respectively. The fat content in the final product was up to 25% in the full-fat cheese. However, a range of half-fat and fat free cheeses have been successfully produced using resistant starch as fat replacers. Processed cheeses containing a fat content comparable to full-fat cheeses, led to the softening and enhanced palatability of the obtained cheese product.

1.5.1.3 Emulsifying salts

ES in processed cheese manufacture are salts that, with the aid of the applied heat and shear, facilitate the hydration of the protein, for example, casein, caseinate or natural cheese, and its conversion to an active emulsifying agent which allows for the subsequent incorporation of fat to form the final homogeneous cheese product (Cavalier-Salou and Cheftel, 1991; Zehren and Nusbaum, 2000). The role of ES involves mainly the chelation and removal of calcium from the rennet casein leading to the absorption of water by the protein (Ennis et al., 2000). The ES blends are also usually selected to adjust the pH, which leads to a better sequestration of calcium by the sodium phosphate ES, as well as a greater negative charge on the protein used as ingredient and accordingly to a higher degree of protein hydration.

There is a wide range of ES used, but the most common ES are the sodium citrates and orthophosphates (Kapoor and Metzger, 2008; El-Bakry et al., 2011a,b). Since ES is the main focus in this research, its role and effects on processed cheese manufacture and functionality will be separately discussed in detail in Chapter 6.

1.5.1.4 Sodium chloride, sorbic acid, and citric acid

Salt or sodium chloride (NaCl) affects cheese pH, which could in turn affect processed cheese manufacture and postmanufacture properties. In general, during natural cheese making, it has been suggested that adding NaCl also promotes calcium solubilization from paracasein (Kindstedt et al., 1992; El-Bakry, 2012), thus displacing calcium from the protein matrix. Increased NaCl content has been reported to cause increased hardness and decreased melting (Olson, 1982; Schroeder et al., 1988). Another main role of NaCl, in natural and processed cheeses, is its influence on the microbial growth and shelf life (Guinee and Fox, 2004; Taormina, 2010). The role of NaCl in processed cheese will be further discussed in Chapter 7.

Sorbic acid and citric acid affect mainly the pH of the processed cheese, where they are considered also as preservative and acidulant, respectively. Therefore, they could have a substantial impact on the properties of processed cheese (see Chapter 5). Sorbic acid is a microbiological preservative agent, which functions mainly in preventing mould growth as well as other microbial growth, by causing dysfunction of the dehydrogenase reaction. This will allow for the manufacture of a microbiologically safe product and an increase the shelf life of processed cheese (Berger et al, 1989; Zehren and Nusbaum, 2000). The antimicrobial action of sorbic acid is pH dependant and it has been reported to be most effective at pH values from 4 to 6.5. In addition, this preservative is a metabolisable food additive, which is converted into calories in a manner similar to the fatty acid caproic in the human body, with no toxic effect (Liewen and Marthy, 1985; Zehren and Nusbaum, 2000).

Citric acid is an acidulant which is added at the end of the manufacture of processed cheese to lower and standardize the pH of the final product. The pH during manufacture of processed cheese is normally relatively raised; to allow for the adequate protein hydration and subsequent emulsification. The addition of citric acid at the end of manufacture, when the protein hydration and emulsification processes are complete, is designed to reduce the pH to allow for a better product stability and shelf life. A pH value of ~ 6 is optimal for processed cheese functional properties, that is, good melting and stretching cheese, and microbial stability (Aimutis, 1995). At lower pH values such as ~5, electrostatic repulsions of the protein are weakened which can result in an excessively hard cheese due to reduced water binding capacity. On the other hand, at higher pH values, greater than 6.5, the cheese products may become excessively soft and microbiologically unsafe (Shimp, 1985).

1.6 Formulations used in the manufacture

The formulations used for the manufacture of processed cheese involve the selection of the correct levels and types of ingredients to give a final product with the desired composition and optimal functional properties such as texture, flavor, and rheological properties. Each manufacturer or researcher has their own specific cheese recipe depending on the ingredients used, desired functional properties and aim of the research. All possible ingredients and their contents used in the manufacture of processed cheese are discussed in detail in this book (see Chapter 2, 3, and 4).

1.7 Cheese processing equipment

The processing equipment can be classified into size-reducing machines, blenders, cookers, cooling, homogenization, and packaging equipment (see Chapter 8).

The main equipment is the so-called cooker or mixer used for the manufacture of the processed cheese. On an industrial scale, the batch and continuous cooking procedures are the two methods for manufacturing (Meyer, 1973; Zehren and Nusbaum, 2000). The batch cookers are producing about 100 to 900 kg of cheese products. One of the most common batch cookers is the Bepex/Rietz cooker, which is a twin-screw direct steam injection cooker (Thomas, 1977). During the process of cheese manufacturing, these screws are rotating in opposite directions with a speed of either 90 or 180 rpm. This mixing system allows for a circular motion and thus an optimum agitation of the cheese mass. The continuous cookers are more advantageous in industry than the batch cookers, due to the fact that the former offer an uninterrupted cost-efficient production on a larger scale (Bosy et al., 1978). A continuous in-line procedure is based on introducing natural cheese ingredients and appropriate ES into a pressurized cooking zone, where direct steam injection is utilized for a certain time. The cheese mass is delivered from this zone to a mixing zone, followed by a final low-pressure cooling zone where deaeration and cooling of the final product takes places.

In scientific research studies on processed cheese, pilot scale or lab scale batch cookers have been used in the manufacture (Chen and Liu, 2012; Kapoor and Metzger, 2004; Arimi et al, 2010; El-Bakry et al., 2010). The Blentech is one of the most common types of cookers used to manufacture processed cheese on a pilot scale. The cooker produces batch sizes ranging from a minimum of 4 up to a maximum of 25 kg (Mounsey and O'Riordan, 2008; Pereira et al., 2001; Glenn et al., 2003;

Shirashoji et al., 2006; 2010; Sharma et al., 2016). The cooker is equipped with nozzles for direct steam injection, a control panel for changing the speed and direction of rotation of the augers and a double discharge for easy collection of the molten cheese mass at the end of manufacture. The mixing chamber is double jacketed so that the temperature of the product can be maintained or adjusted indirectly using steam or hot water. The mixing chamber of the Blentech has twin single ribbon screws rotating towards each other at an equal speeds ranging from 20–200 rpm. The mixing action is completely homogenous and symmetric (Glenn et al., 2003). Due to the shape of the twin screws, while rotating, they cause the cheese mass to also rotate while at the same time conveying the cheese mass around the mixing bowl perpendicularly to the direction of auger rotation in a counter clockwise direction. Another example is the Stephan pilot scale cooker that has been used in the manufacture of processed cheese producing usually batches of about 2 kg (Noronha et al., 2008a; Everard et al., 2006). The mixer has a centrally located wave-cut single-blade rotating at a speed ranging from 100 to 1500 rpm, which centrally shears and chops up the entire cheese mass. The single-blade cooker also had a peripheral scraping paddle, which can be operated at speeds of 75 to 150 rpm to ensure that any product on the sides of the cooker is continuously folded into the central mixing zone.

Extruders have also been reported to be an effective means of manufacturing processed cheese (Cavalier-Salou et al., 1990). Cheeses manufactured in an extruder had similar functional properties to those manufactured in a Blentech cooker. Results showed that the extruded cheese displayed adequate casein hydration but with a lower degree of fat emulsification, that is, under emulsification. Recently, the Farinograph has been used to produce processed cheese on a lab scale, where the batch sizes were ranging from 50 up to 800 g (El-Bakry et al., 2010; Noronha et al., 2008b; Inayat et al., 2014). The properties of the cheese were also studied using this mixer. The Faringraph is a standard dough testing machine approved by the American Association of Cereal Chemists (AACC, 1972), is a low-shear-rate, batch mixer commonly used for flour testing with two nonintermeshing, asymmetrical sigma mixing blades operating at a differential speed ratio of 3:2, with the fast (right) blade turning counter clockwise and the slow (left) blade turning clockwise. The mixing action is extremely nonhomogeneous and the distribution of shear rates depends upon the time-relative motion of the blades (AACC, 1972; Connelly and Kokini, 2006). The instrument also offers the possibility of measurements of the torque-time profile during mixing. Thus a direct proportional relation between the viscosity of the cheese and the measured torque is possible.

1.8 Functional properties of processed cheese

Functional properties or functionality can be generally defined as the behaviour of the cheese during all stages of its manufacture, preparation, and consumption (Guinee, 2002b). Cheese functionality includes a very broad range of properties. The functional properties may be categorized into mainly texture and melting related properties (see Chapter 6 and 9).

1.8.1 Texture related properties

Cheese texture related properties can be described by many terms, and the assessment of these properties is typically done through three different categories of tests, that is, empirical, fundamental, and imitative tests (Bourne, 2002; Joshi et al., 2004).

1.8.1.1 Empirical textural tests

The idea behind these tests is to measure undefined parameters that relate to some aspects of texture in practical terms. The first empirical tests were performed by hand by pressing into the surface to judge firmness and elasticity (Caffyn and Baron, 1947). Simple instruments followed, such as the Cherry-Burrell Curd Tension Meter, which measures the resistance to the passage of a wire through the cheese (Emmons and Price, 1959). Another instrument, the penetrometer, produces force measurements as a needle is pushed into the cheese (Prentice, 1987). Other tests have been performed using a ball compressor, in which a hemisphere presses into the surface of the sample of cheese and the depth of compression and amount of recovery with time is noted (Cox and Baron, 1955). However, these tests are not always satisfactory for research studies due to the other arbitrary nature of the test conditions

1.8.1.2 Fundamental textural tests

Fundamental tests measure well defined physical parameters, that is, rheological properties. Specimens used are of a specific shape and are deformed in a specific manner, and therefore, the results can be analysed systematically (Collyer and Clegg, 1988). Cheeses represent classical viscoelastic behavior, that is, somewhere between the response of a viscous and an elastic material. Thus, in different fundamental rheological tests, several parameters are measured and others are calculated from them. In the dynamic test, a small sinusoidal stress is applied to the cheese sample and the resulting strain is measured (Gunasekaran and Ak, 2000). Two viscoelastic parameters are measured, the storage (G') and loss (G') modulus, which are related to the elastic and viscous properties of the material under study, respectively. Another commonly used viscoelastic parameter is the loss tangent (tan δ = G"/G'), which denotes relative contributions of viscous and elastic moduli in a viscoelastic behaviour. The possibility of examining the effect of heating temperatures on these three above-explained parameters also provides link to melting properties.

1.8.1.3 Imitative textural tests

These tests include the examination of the texture by imitating the chewing of the food stuff, that is, conditions that the cheese is directly subjected to during chewing are mimicked (Voisey, 1976). The first instrument to examine texture in this way was the tenderometer (Volodkevich, 1938; Procter et al., 1955). A wedge designed to imitate teeth on the upper jaw was brought down toward a second wedge, exerting a biting and squeezing action on the sample held between the wedges. The second attempt, to perform imitative texture tests, was by using the general food texturometer (Friedman et al., 1963). This device compresses a bite-sized sample to 25% of its original height, releases, and repeats, with a variable speed thus imitating jaw movements. Accordingly, a force-time curve is produced from which a texture profile analysis (TPA) can be derived. However, due to the variable speed applied, the force-time curve cannot be converted into force-distance curve. Several years later, the Instron Universal Testing Machine was adapted for TPA studies (Bourne, 1976; 1978), and the first studies on cheese using this machine were reported. The speed of compression applied was made constant during the test, which allows the obtaining of both force-time and force-distance curves. Accordingly, the work done during the compression can be calculated by the numerical integration of the latter curve.

In cheeses, the TPA test consists of compressing a sample twice, typically to 70%–80% of its original height between two rigid plates (Bourne, 1978; 2002). Various instrumental parameters are generated. These parameters provide objective and comparative information on rheological properties if obtained under standardized conditions. The typical TPA curve can lead to the extraction of seven textural parameters – five directly measured and two calculated parameters:

1. Fracturability (brittleness) is defined as the force at the first significant break in the curve (Bourne, 1978)
2. Hardness is defined as the peak force during the first compression cycle (Meullenet et al., 1997)
3. Cohesiveness is defined as the ratio of the positive force area during the second compression to that during the first compression (A2/A1), (Szczesniak, 1963; Meullenet et al., 1997)
4. Adhesiveness is defined as the negative force area for the first bite, representing the work necessary to pull the compressing plate away from the sample (Bourne, 1978)
5. Springiness (elasticity) is defined as the height that the food recovers during the time that elapses between the end of the first compression cycle and the start of the second compression cycle (Meullenet et al., 1997)
6. Gumminess is defined as the product of hardness x cohesiveness (Hort and le Grys, 2001)
7. Chewiness is defined as the product of gumminess x springiness (hardness x cohesiveness x springiness) (Bourne, 1978; Hort and le Grys, 2001)

1.8.2 Melting related properties

Several empirical methods have been developed to assess this melting property. These include the Schreiber and Arnott test, in which the change in diameter or height of a cylindrical cheese sample on heating was measured (Kosikowski, 1982; Arnott et al., 1957, respectively). Another test was developed based on measuring the distance of cheese flow in a glass tube after heating for a certain temperature/time (Olson and Price, 1958; Oberg et al., 1991). In comparison to the Schreiber and Arnott tests, this method avoids film formation on the cheese surface and controls the direction of flow during melting. Melting properties include different attributes; meltability, heat-induced flowability, stretchability, and oiling-off (Table 1.3).

Dynamic rheology during heating can be applied to measure temperature induced changes of the rheological properties of the cheese (Nolan et al., 1989;

Table 1.3 Key functional properties of processed cheese. Modified from Guinee et al., 2004.

Melting Characteristic	Description	Related rheological or empirical properties
Meltability	Tendency of the cheese to soften on heating	Change in G', G", and tan δ on heating
Heat-induced flowability	Tendency of the heated cheese to spread unrestrictedly during heating	Increase in length, height, or diameter on heating
Stretchability	Tendency of the heated cheese to form strings when extended uniaxially	Length of strings of heated cheese at failure
Oiling off	Tendency of the cheese to exude oil on heating	Quantity of free oil extractable

Mounsey and O'Riordan, 2008; Arimi et al., 2010; Zehren and Nusbaum, 2000). The dynamic oscillatory measurements were performed at a frequency of 1 Hz and an applied shear stress of 100 Pa. These measurements were also performed at a strain of below 1%, which has been established to be within the linear viscoelastic range of processed cheese. The temperature of the sample was increased from ~15–20 up to 80–100°C using a Peltier heating element.

The parameters measured included G', G", and tan δ. Adding to that the crossover temperature (T_c), that is, when tan $\delta = 1$, may also be determined.

1.9 Microstructure of processed cheese

The microstructure of processed cheese is directly influenced by various parameters of processing, during the manufacture of the cheese product, as well as the composition/formula used (El-Bakry and Sheehan, 2014; Mehta, 2018; Kapoor and Metzger, 2008; Zehren and Nusbaum, 2000, see Chapter 12). In the micrographs of processed cheese, the following can be observed. A fat phase, in the form of fat globules, which highly vary in size from 0.1 to 30 μm, is dispersed a protein matrix. In general, processed cheese can be described as a stable oil-in-water emulsion. The microstructure of processed cheese highly resembles the one of natural cheese, where some differences have to be taken into account (Caric and Kalab, 1987; Mehta, 2018). This was proven using various microscopy techniques, that have been extensively used for the study of the microstructural properties. Fig. 1.1 represents microstructure of natural and processed cheese.

Processing parameters, such as mainly the shear rate, processing temperature, and processing time have a major role in controlling the emulsion formation and thus the microstructure of the cheese matrix (Noronha et al., 2008; Mehta, 2018). The processing time, which is the time for which the shear and heat was applied in the manufacture of processed cheese, is of utmost importance in determining the microstructure of the resulting cheeses (El-Bakry et al., 2010; Glenn and Daubert, 2003; Kapoor et al., 2004).

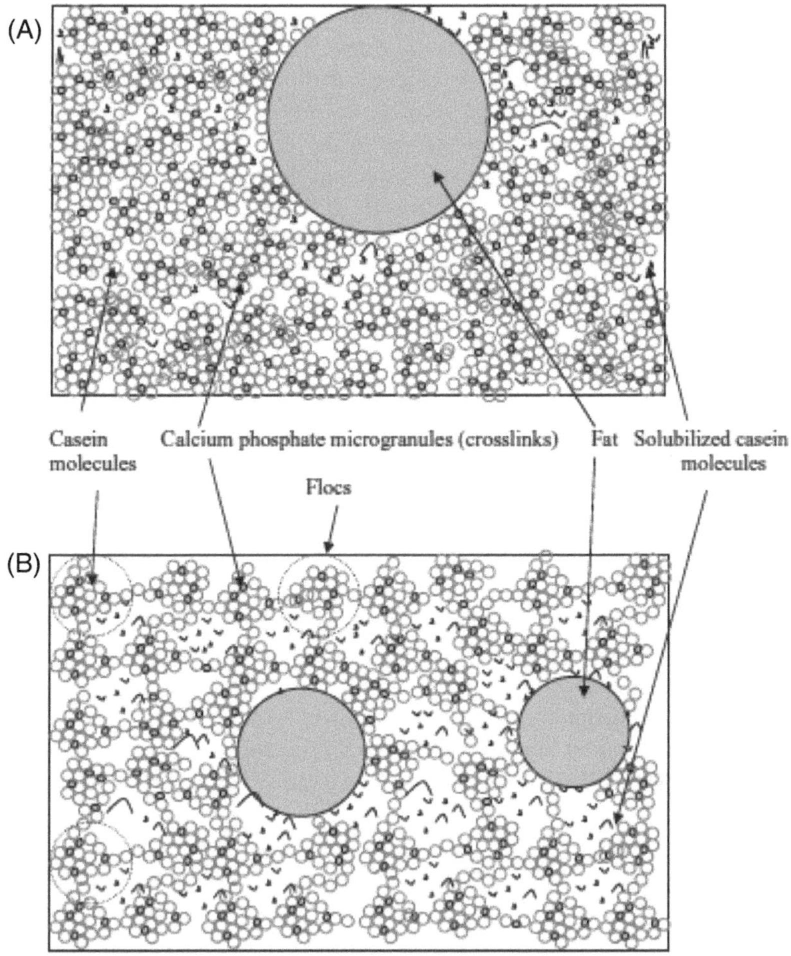

Figure 1.1 Schematic microstructure of (A) natural cheese and (B) processed cheese.
Kapoor and Metzger (2008): Copyright © 2008 by Wiley. Reprinted/reproduced with permission.
Source: Kapoor R., Metzger L.E. (2008). Process cheese: scientific and technological aspects—a review. Compr. Rev. Food Sci. Food Saf., 7:194–214.

The composition and formula used highly and directly affect the microstructure of the resulting cheese (Bachmann, 2001; Tamime, 2011; Zehren and Nusbaum, 2000). These include the variation in the moisture content and the protein, fat, salt and ES types and concentrations. For instance, the ES functions in providing the cheese with its unique microstructure and functional properties (El-Bakry et al., 2011a,b; Meyer, 1973; Berger et al., 1989; Zahren and Nusbaum, 2000). Processed cheese manufactured using dipotassium phosphate and tripotassium citrate was shown to have lower firmness when compared to cheese manufactured using disodium phosphate and trisodium citrate.

In addition, the use of phosphate salts instead of citrate ones, in the manufacture, resulted in a processed cheese with much lower melting properties. Mistry et al. (2006) studied the microstructure of pasteurized process cheese manufactured from vacuum condensed and ultrafiltered milk using cryoscanning electron microscopy. More compact protein masses were observed in cheeses made from concentrated milk whereas relatively large pores were found in control cheese. The highly fused dense protein was observed in ultrafiltered cheeses, which showed the highest resistance to compression as compared to control. The authors have concluded that use of concentrated milks for Cheddar cheese-making influences Process cheese functionality and structure.

1.10 Approaches to develop healthier processed cheeses

Processed cheeses contribute high levels of fats and sodium to the diet. Consumers become health conscious. They are well aware about noncommunicable chronic diseases like obesity and hypertension associated with more consumption of fats and sodium in diet. Foods with low-fat and low-sodium content are available in market and continuously new products with similar properties are introducing into the market (Ferrão et al., 2016). Consumers want low fat/sodium food products without affecting the quality of the products and this is challenging task. Food industry are changing and modifying their food formulations to develop healthier and functional food products. Some of the alterations include limiting use of fats, sodium salts, trans-fatty acids, sucrose, and artificial additives. Use of prebiotic compounds in processed cheese is high demanding by consumers.

In reduced-/low-fat PCPs, the fat content can be replaced by water, protein or other compounds (e.g., gums and stabilizers) not normally found in cheese (Banks, 2004). A fat level between 10% and 24% is found in reduced-fat processed cheese (Lee et al., 2004). The natural cheese contributes more fat percent while some minor sources like dried cream and butteroil contribute less fat percent to the final products. Therefore, use of lower-fat containing natural cheese for processed cheese production is one of the best ways to manufacture a reduced fat processed cheese (Hauerlandová et al., 2014). But use of low-fat containing natural cheese has some disadvantages. The reduction of fat content in processed cheese result in increase in graininess, stickiness, chewiness, acidic and bitter properties with decrease in creaminess and overall sensory acceptability in processed cheese. Researcher has tried to obtained a product with acceptable quality using different fat replacers (i.e., fat substitutes/fat mimetics) at various levels in a PC formulation. Fat substitutes are mainly lipid-based macromolecules whereas fat mimetics are normally carbohydrate-based compounds or protein macromolecules. Various ingredients such milk, eggs, inulin, starches, dextrins, polydextrose, gums, cellulose, maltodextrins, and synthetic fats may be used alone or in combination to produce low-fat cheeses (Ferrão et al., 2016).

Processed cheese contributes in daily sodium intake. The ES together with the NaCl added to the curd contribute sodium content in processed cheese. In processed cheese one of the important ingredients is ES, such as citrate, sodium phosphate, or polyphosphate. The ES provide desired consistency to the product (Chen and Liu, 2012). The application of ES is related to dispersion, hydration, swelling of the proteins, as well as emulsification and stabilization of fat (Hoffmann et al., 2012). Replacement of 25%–27% NaCl by potassium chloride maintains salty taste to the product (Gomes et al., 2011; Chavhan et al., 2015). However, salt reduction in processed cheese affect melting and firmness of the product. The ES containing potassium as replacement of sodium show changes in some characteristics of the processed cheese such as an increased fracture, higher firmness, lower melting capacity and protein solubility (Guinee and O'Kennedy, 2012).

Prebiotic ingredients can also be used to improve functional properties of processed cheese. Prebiotic ingredients defeat potential pathogens and promote intestinal health. Prebiotics as nondigestible carbohydrates and some of them are oligosaccharide, oligofructose, inulin, fructan, stachyose, raffinose, and lactitol. Among them, inulin is widely used. Inulin creates a defensive effect toward various *Lactobacillus* and *Bifidobacterium* species in processed cheese spread. The effect of textural and melting properties of the processed cheese spread have been studied at different levels (0%, 4%, 6%, and 8 %) of inulin addition (Giri et al., 2014). Authors observed that as the levels of inulin addition were increased, the stickiness and work of adhesion decreased whereas the firmness and work of shear of cheese spread increased. Recently, Luana et al. (2020) studied processed cheese with inulin and microencapsulated chia oil, The formulation with inulin had a 36% fat reduction, with no sensorial differences compared to the standard formulation was observed.

1.11 Advantages and applications of processed cheese

Processed cheese is considered as a non-standard cheese product that is used as an alternative for natural cheese due to the cost efficiency, feasibility in production, versatility in applications, and usefulness (Vernon, 1972; Meyer, 1973; Berger et al., 1989; Pellegrino, 1996; Tamime 2011). Most importantly, the shelf life of processed cheese is much longer than in the case of natural cheeses. In addition, processed cheese is used in many applications in different processed foods, mainly as alternatives for natural cheeses (Fox et al., 2000; Guinee et al., 2004).

The advantages of processed cheese, in comparison to natural cheeses, can be summarized as follows:

1. The relatively lower production costs compared to natural cheeses. This is due to the use of some ingredients of vegetable origin like starch and vegetable fats instead of milk fat or proteins. In addition, processed cheeses do not need ripening and the capital costs of the equipment for their manufacture are lower compared to the costs involved in natural cheeses manufacture.
2. Simplicity of manufacture from readily available raw material, eliminating seasonally and supply variations and problems.
3. The manufacture process allows the manufacturer to tailor the product to meet specific requirements in terms of product composition and functional properties (such as texture and

meltability). Certain nutritional attributes can be achieved, for example, vitamin-enriched, low salts products (Keane and Glaeser, 1990). It was recently reported that it is possible to manufacture processed cheese with high functional fibres and low-fat content (Noronha et al., 2007; Arimi et al., 2010). In addition, cheese flavor ingredients, for example, enzyme modified cheeses (EMCs), can be added to processed cheese which greatly improve possibilities for the development of different varieties of processed cheeses. For example, low-moisture processed cheese of mozzarella or cheddar, that is, processed mozzarella cheese or processed cheddar cheese where the corresponding natural cheese is used as a main ingredient, have the advantages of being cheaper and more easily manufactured than natural cheese. In addition, the functional properties of processed cheese could be tailor-made for specific applications. The developments in EMC technology could make it possible to improve and diversify the various flavors of PCPs.

There are a lot of applications of processed cheese (Tamime, 2011; Oliveira et al., 2016), which can be summarized as follows. Processed cheese has been used increasingly as an ingredient in food applications due in particular to the rapid growth of ready meals and processed foods sector (Bachmann, 2001). As an example, the principal applications of these cheese products are as toppings in pizzas and pies, where in this case mozzarella-like processed cheese is used. Cheddar-type processed cheeses are also very common and are predominantly used in sandwich-like products such as cheeseburgers. In addition, processed cheese is used in processed meat products and combined cheese-filled coextruded products (Guinee et al., 2004). Other applications include salads, cheese dips, tacos, enchiladas and lasagne, pasta, etc.

1.12 Conclusion

The world production of cheese, of which processed cheese is form, represents more than 35 % of the total milk production, which is annually increasing. Processed cheese is a complex multicomponent dairy system, which can be described as being a stable cold-set gelled oil-in-water emulsion. This cheese is not made directly from milk, unlike natural cheeses, but rather from various combinations of dairy and/or non-dairy ingredients. These include sources of protein and fat, salt, sorbic acid, citric acid, and ES, where there are numerous formulas used in the manufacture. In general, processed cheese includes two principal categories, PCPs and ACPs. The manufacturing protocols involve the application of heat and mechanical shear to the formulated blend, in a reaction container, well-known as cheese cooker or mixer, followed by cooling and cold storage. There are many cheese processing equipment used, which can be classified into size-reducing machines, blenders, cookers, homogenizers, and cooling and packaging equipment. The cheese end-product is mainly characterised based on its functionality, that is, textural and melting properties, as well as its microstructure. Approaches to develop healthy cheese, containing prebiotics and low sodium/ fat, were extensively studied, as this kind of cheese can be easily tailored to meet specific requirements. This counts as one of the several advantages of processed cheese, in comparison to natural cheese. In addition, processed cheese has relatively low cost and diverse applications.

References

AACC, 1972. In: Locken, L., Loska, S., Shuey, W. (Eds.), The Farinograph Handbook. AACC, St. Paul, Minnesota, USA.

Aimutis, A.W., 1995. Dairy protein usage in processed and imitation cheeses. Food Technol. Eur. 2, 30–34.

Arimi, J.M., Duggan, E., O'Sullivan, M., Lyng, J.G., O'Riordan, E.D., 2010. Effect of moisture content and water mobility on microwave expansion of imitation cheese. Food Chem. 121, 509–516.

Arnott, D.R., Morris, H.A., Combs, W.B., 1957. Effect of certain chemical factors on the melting quality of process cheese. J. Dairy Sci. 40, 957–963.

Bachmann, H., 2001. Cheese analogues: a review. Int. Dairy J. 11, 505–515.

Banks, J.M., 2004. The technology of low-fat cheese manufacture. Int. J. Dairy Technol. 57, 199–207.

Berger, W., Klostermeyer, H., Merkenich, K., Uhlmann, G., 1989. Processed cheese manufacture, a JOHA guide. BK Ladenburg GmbH, Wurzburg, Germany.

Bosy, G., Edwards, E.N., Hoffbeck, W.M., Silvio, P.A., Warwick, L.A., 1978. Method for Manufacture of Process Cheese, 112. U.S. Patent 4, p. 131.

Bourne, M.C., 2002. Food Texture and Viscosity: Concept and Measurement, second ed. Academic Press, New York.

Bourne, M.C., et al., 1976. Interpretation of force curves from instrumental texture measurements. In: deMan, J.M., et al. (Eds.), Rheology and Texture in Food Quality. AVI Publisher Co, Westpot, CT, pp. 244–274.

Bourne, M.C., 1978. Texture profile analysis. Food Technol. 32, 62–66.

Brander, R. W., Raap, T. A., & Gessler, J. F. (1985). Soy milk containing cheese analog. United States Patent, 4556569 (1–4).

Britten, M., Giroux, H.J., Jean, Y., Rodrigue, N., 1994. Composite blends from heat denatured and undenatured whey protein: emulsifying properties. Int. Dairy J. 4, 25–36.

Caffyn, J.E., Baron, M., 1947. The scientific control of cheese making. Dairyman 64, 345–351.

Caric, M., Kaláb, M., 1987. Processed cheese products. In: Fox, P.F. (Ed.), Cheese: Chemistry, Physics and Microbiology. Elsevier Applied Science, London.

Cavalier-Salou, C., Cheftel, J.C., 1991. Emulsifying salts influence on characteristics of cheese analogs from calcium caseinate. J. Food Sci. 56 (1542-1547), 1551.

Cavalier-Salou, C., Queguiner, C., Cheftel, J.C., et al., 1990. Preparation of cheese analogues by extrusion cooking. In: Zeuthen, P., et al. (Eds.), Processing and Quality of Foods: High Temperature Short Time Processing. Elsevier Applied Science Publishers, London, UK, p. 373.

Černíková, M., Nebesářová, J., Salek, R.N., Řiháčková, L., Buňka, F., 2017. Microstructure and textural and viscoelastic properties of model processed cheese with different dry matter and fat in dry matter content. J. Dairy Sci. 100 (6), 4300–4307.

Chavhan, G.B., Kanawjia, S.K., Khetra, Y., Puri, R., 2015. Effect of potassium-based emulsifying salts on sensory, textural, and functional attributes of low-sodium processed mozzarella cheese. Dairy Sci. Technol. 95, 265–278.

Chen, L., Liu, H., 2012. Effect of emulsifying salts on the physicochemical properties of processed cheese made from Mozzarella. J. Dairy Sci. 95, 4823–4830.

Chen, S.L., Wan, P.J., Lusas, E.W., Rhee, K.C., 1979. Utilization of peanut protein and oil in cheese analogs. Food Technol. 33, 88–93.

Cheese and Cream Regulations. (2005). SI 2005/4529, HMSO, London, UK.

Code of Federal Regulations (CFR), (2007). Article 133: Cheese and related products. In Food and Drugs –Title 21, Section 169. US Government Printing Office, Washington, USA.
Codex Standard (1995). Codex general standard for food additives (192–195).
Connelly, R., Kokini, J.L., 2006. 3D numerical simulation of the flow of viscous Newtonian and shear thinning fluids in a twin sigma blade mixer. Adv. Polym. Technol. 25 (3), 182–194.
Collyer, A.A., Clegg, D.W., 1988. Rheological Measurement. Elsevier Applied Science, London, UK.
Corredig, M., Nair, P.K., Li, Y., Eshpari, H., Zhao, Z., 2019. Invited review: Understanding the behavior of caseins in milk concentrates. J. Dairy Sci. 102 (6), 4772–4782.
Cox, C.P., Baron, M., 1955. A variability study in firmness in cheese using the ball-compressor test. J. Dairy Res. 22, 386–390.
Dalgleish, D.G., 1997. Structure-function relationships of caseins. In: Damodaran, S., Paraf, A. (Eds.), Food Proteins and their Application. Marcel Dekker, New York, USA, pp. 199–223.
El-Bakry, M., 2012. Salt in cheese: a review. Curr. Res. Dairy Sci. 4, 1–5.
El-Bakry, M., Duggan, E., O'Riordan, E.D., O'Sullivan, M., 2010. Small scale imitation cheese manufacture using a farinograph. LWT - Food Sci. Technol., 43, pp. 1079–1087.
El-Bakry, M., Duggan, E., O'Riordan, E.D., O'Sullivan, M, 2011a. Casein hydration and fat emulsification during manufacture of imitation cheese, and effects of emulsifying salts reduction. J. Food Eng. 103, 179–187.
El-Bakry, M., Duggan, E., O'Riordan, E.D., O'Sullivan, M, 2011b. Effect of chelating salt type on casein hydration and fat emulsification during manufacture and functionality of imitation cheese. J. Food Eng. 102, 145–153.
El-Bakry, M., Sheehan, J., 2014. Analysing cheese microstructure: a review of recent developments. J. Food Eng. 125, 84–96.
Emmons, D.B., Price, W.V., 1959. A curd firmness test for cottage cheese. J. Dairy Sci. 42, 553–556.
Ennis, M.P., ODowd, J.J., Thornton, A., Mulvihill, D.M., 2000. The effect of varying the calcium-sequestering salt cation on the hydration behaviour of rennet caseins in a simple model system. Int. J. Dairy Technol. 53, 41–44.
Everard, C.D., Fagan, C.C., O'Donnell, C.P., O'Callaghan, D.J., Lyng, J.G, 2006. Dielectric properties of process cheese from 0.3 to 3 GHz. J. Food Eng. 75, 415–422.
Farkye, N.Y., Fox, P.F., 1990. Observation of plasmin activity in cheese. J. Dairy Res. 57, 413–418.
FDA, 2008. 21 CFR, Part 101.61. Nutrient Content Claims for the Sodium Content of Foods. Food and Drug Administration, Department of Health and Human Services, Washington, DC.
Ferrão, L.L., Silva, E.B., Silva, E.B., Silva, H.L.A., Silva, R., Mollakhalili, N., Granato, D., Freitas, M.Q., Silva, M.C., Raices, R.S.L, Padilha, M.C., Zacarchenco, P.B., Barbosa, M.I.M.J., Mortazavian, A.M., Cruz, A.G., 2016. Strategies to develop healthier processed cheeses: Reduction of sodium and fat contents and use of prebiotics. Food Res. Int. 86, 93–102.
Fox, P.F., Guinee, T.P., Cogan, T.M., McSweeney, P.L.H., 2000. Fundamentals of Cheese Science. Springer, New York, pp. 587–590.
Fox, P.F., Guinee, T.P., Cogan, T.M., McSweeney, P.L.H., 2017. Fresh cheese products: Principals of manufacture and overview of different varieties. In: Fox, P.F. (Ed), Fundamentals of Cheese Science. Springer, New York, pp. 543–588.
Fox, P.F., O'Connor, T.P., McSweeney, P.L.H., Guinee, T.P., O'Brien, N.M, 1996. Cheese: physical, biochemical and nutritional aspects. In: Taylor, S.L. (Ed.), Advances in Food and Nutrition Research, 39. Academic Press, London, UK, pp. 163–328.

Friedman, H.H., Whitney, J.E., Szczesniak, A.S., 1963. The texturometer, a new instrument for objective texture measurement. J. Food Sci. 28, 390–396.
FSAI. (2005). Salt and health: review of the scientific evidence and recommendations for public policy in Ireland. www.fsai.ie. Accessed on April 1, 2020.
Fu, W., Watanabe, Y., Satoh, H., Inoue, K., Moriguchi, N., Fusa, K., Yanagisawa, Y., Mutoh, T., Nakamura, T., 2018. Effects of emulsifying conditions on creaming effect, mechanical properties and microstructure of processed cheese using a rapid visco-analyzer. Bioscience, Biotechnology, and Biochemistry, 82 (3), 476–483.
Gianferri, R., Maioli, M., Delfini, M., Brosio, E., 2007. A low-resolution and high-resolution nuclear resonance integrated approach to investigate the physical structure and metabolic profile of Mozzarella di Bufula Campana cheese. Int. Dairy J. 17, 167–176.
Ginty, F., 1998. The effect of dietary sodium intake on biochemical markers of bone metabolism in young women. Br. J. Nutr. 79, 343–350.
Giri, A., Kanawjia, S.K., Khetra, Y., 2014. Textural and melting properties of processed cheese spread as affected by incorporation of different inulin levels. Food Bioprocess Technol. 7, 1533–1540.
Glenn, T.A., Daubert, C.R., 2003. A mixer viscometry approach for blending devices. J. Food Process Eng. 26, 1–16.
Glenn, T.A., Daubert, C.R., Farkas, B.E, 2003. A statistical analysis of creaming variables impacting process cheese melt quality. J. Food Qual. 26, 299–321.
Glenn, T.A., Keener, K.M., Daubert, C.R., 2000. A mixer viscometry approach to use vane tools as steady shear rheological attachments. Appl. Rheolol. 10, 80–89.
GNPD, 2008. Global New Product Database. Mintel Group, Chicago, IL www.gnpd.com.
Gomes, A.P., Cruz, A.G., Cadena, R.S., Celeghini, R.M.S., Faria, J.A.F., Bolini, H.M.A., Granato, D., 2011. Manufacture of low-sodium Minas fresh cheese: effect of the partial replacement of sodium chloride with potassium chloride. J. Dairy Sci. 94, 2701–2706.
Guinee, T.P., et al., 2003. Cheese analogues. In: Roginski, H., et al. (Eds.), Encyclopedia of Dairy Science, 1. Academic Press, Amsterdam, Netherlands, pp. 428–434.
Guinee, T.P., 2004. Salting and the role of salt in cheese. Int. J. Dairy Technol. 57 (2), 99–109.
Guinee, T.P., et al., 2002a. Cheese analogues. In: Roginski, H., et al. (Eds.), Encyclopedia of Dairy Sciences. Academic Press, London, UK, pp. 428–434.
Guinee, T.P., 2002b. The functionality of cheese as an ingredient – a review. Aust. J. Dairy Technol. 57, 79–91.
Guinee, T.P, Carić, M., Kaláb, M., 2004. Pasteurized processed cheese and substitute/imitation cheese products. Cheese: Chemistry, Physics and Microbiology. Elsevier Applied Science, London, New York.
Guinee, T.P., Fox, P.F., et al., 2004. Salt in cheese: physical, chemical and biological aspects. In: Fox, P.F., et al. (Eds.), Cheese - Chemistry, Physics and Microbiology, 1. Academic Press, Amsterdam, Netherlands, pp. 207–259.
Guinee, T.P., O'Kennedy, B.T, 2012. Reducing the level of added disodium phosphate alters the chemical and physical properties of processed cheese. Dairy Sci. Technol. 92, 469–486.
Guirguis, A.H., Abdel Baky, A.A., El-Neshawy, A.A., El-Shafy, N.M, 1985. Peanut curd in the manufacture of processed cheese-like spread. Dairy Ind. Int. 50, 37–41.
Gunasekaran, S., Ak, M.M., 2000. Dynamic oscillatory shear testing of foods - selected applications. Trends Food Sci. Technol. 11, 115–127.
Gupta, V.K., Reuter, H., 1992. Processed cheese foods with added whey-protein concentrates. Lait 72, 201–212.
Hanaei, F., Cuvelier, G., Sieffermann, J.M., 2015. Consumer texture descriptions of a set of processed cheese. Food Quality and Preference, 40, 316–325.

Hauerlandová, I., Lorencová, E., Buňka, F., Navrátil, J., Janečková, K., Buňková, L., 2014. The influence of fat and monoacylglycerols on growth of spore-forming bacteria in processed cheese. Int. J. Food Microbiol. 37–43, 182–183.
Hennelly, P.J., Dunne, P.G., O'Sullivan, M, O'Riordan, E.D, 2006. Textural, rheological and microstructural rheological properties of imitation cheese containing inulin. J. Food Eng. 75, 388–395.
Hoffmann, W., Gärtner, J., Lück, K., Johannsen, N., Maurer, A., 2012. Effect of emulsifying salts containing potassium on the quality of block-type processed cheese. Int. Dairy J. 25, 66–72.
Hort, J., Le Grys, G., 2001. Developments in the textural and rheological properties of UK cheddar cheese during ripening. Int. Dairy J. 11, 475–481.
Hyde, M., Carruthers, L., McGeown, J., Trecker, G., 2010. Methods to produce cheese products and processed cheese bases. Canadian Int. Pro. Off. 364–393.
IDDBA. (2016). What's in store 2016: section and chapter highlights (30 ed.).
Inayat, S., Ayaz, M., Pasha, T., Abdullah, M., Waheed, A., 2014. Evaluation of Commercially Available Mozzarella Cheese for its melt and stretching behaviour through Farinograph-E. The J. of Ani. Pla. Sci. 24 (3), 770–773.
Jelen, P., et al., 2004. Whey processing: utilization and products. In: Roginski, H., et al. (Eds.), Encyclopedia of Dairy Sciences. Academic Press, London, UK, pp. 2739–2745.
Jinjarak, S., Olabi, A., Jiménez-Flores, R., Walker, J.H., 2006. Sensory, functional, and analytical comparisons of whey butter with other butters. J. Dairy Sci. 89 (7), 2428–2440.
Jolly, R.C., Kosikowski, F.V., 1975. A new blue cheese food material from ultrafiltrated skim milk and microbial enzyme-mold spore reacted fat. J. Dairy Sci. 58 (9), 1272–1275.
Joshi, N.S., Jhala, R.P., Muthukumarappan, K., Acharya, M.R., Mistry, V.V., 2004. Textural and rheological properties of processed cheese. Int. J. Food Prop. 7, 519–530.
Kamata, Y., Chiba, K., Yamauchi, F., Yamada, M., 1992. Selection of commercial enzymes for soymilk-curd production by limited proteolysis with immobilized enzyme reactor. J. Japanese Soc. Food Sci. Technol. 39, 102–105.
Kamata, Y., Sanpei, S., Yamauchi, F., Yamada, M., 1993. An immobilized protease reactor for cheese analog production from soymilk. Nippon Shokuhin Kogyo Gakkaishi 40, 296–298.
Kapoor, R., Lehtola, P., Metzger, L.E., 2004. Comparison of pilot-scale and rapid visco analyzer process cheese manufacture. J. Dairy Sci. 87, 2813–2821.
Kapoor, R., Metzger, L.E., 2008. Process cheese: scientific and technological aspects - a review. Compr. Rev. Food Sci. Food Saf. 7, 194–214.
Keane, M., Glaeser, H., 1990. Imitation products. Dairy Ind. Int. 55, 30–35.
Kim, S.Y., Park, P.S.W., Rhee, K.C., 1992. Textural properties of cheese analogs containing proteolytic enzyme-modified soy protein isolates. J. Am. Oil Chem. Soc. 69, 755–759.
Kindstedt, P.S., Kiely, L.J., Gilmore., J.A, 1992. Variation in composition and functional properties within brine-salted Mozzarella cheese. J. Dairy Sci. 75, 2913–2921.
Kong-Chan, J. L. Y., Hellyer, J. A., & Tafuri, M. W. (1991). Simulated cheese products with reduced animal fat and calories. United States Patent, 5.061.503 (1–14).
Kosikowski, F., 1982. Process cheese and related products. Cheese and Fermentd Milk Foods. Kosikowski and Associations, Brooktondale, N.Y., USA, pp. 405–406.
Lee, S.K., Buwalda, R.J., Euston, S.R., Foegeding, E.A., McKenna, B., 2003. Changes in the rheology and microstructure of processed cheese during cooking. LWT - Food Sci. Technol. 36, 339–345.
Lee, S.K., Anema, S., Klostermeyer, H., 2004. The influence of moisture content on the rheological properties of processed cheese spreads. Int. J. Food Sci. Technol. 39, 763–771.

Liewen, M.B., Marthy, E.H, 1985. Growth and inhibition of microorganisms in the presence of sorbic acid: a review. J. Food Prot. 48, 364–375.
Lobato-Calleros, C., Vernon-Carter, E.J., Guerrero-Legarreta, I., Soriano-Santos, J., Escalona-Buendia, H., 1997. Use of fat blends in cheese analogs: influence on sensory and instrumental characteristics. J. Texture Stud. 28, 619–632.
Luana, G.C., Ivan, J.B.J., Raissa, V., Juliane, M, Christian, O., Vandré, B.B., Luciane, M.C, 2020. Processed cheese with inulin and microencapsulated chia oil (Salvia hispanica). Food Biosci. 37, 100731.
Marshall, R.J., 1991. Combined instrumental and sensory measurement of the role of fat in food texture. Food Qual. Pref. 2, 117–124.
Mehta, B.M., 2018. Microstructure of Cheese Products. In: ElBakry, M., Sanchez, A., Mehta, B.M. (Eds.), Microstructure of Dairy Products. John Wiley & Sons Ltd, West Sussex, UK, pp. 145–180.
Meyer, A., 1973. Processed Cheese Manufacture. Food Trade Press Ltd, London, UK.
Meullenet, J., Carpenter, J.A., Lyon, B.G., Lyon, C.E., 1997. Bi-cyclical instrument for assessing texture profile parameters and its relationship to sensory evaluation of texture. J. Texture Stud. 28, 101–118.
Mistry, V.V., Hassan, A.N., Acharya, M.R., 2006. Microstructure of pasteurized process cheese manufactured from vacuum condensed and ultrafiltered milk. Lait 86, 453–459.
Mleko, S., Foegeding, E.A., 2000. Physical properties of rennet casein gels and processed cheese analogs containing whey proteins. Milchwissenschaft 55, 513–516.
Montesinos-Herrero, C., Cottell, D.C., O'Riordan, E.D., O'Sullivan, M, 2006. Partial replacement of fat by functional fibre in imitation cheese: Effects on rheology and microstructure. Int. Dairy J. 16, 910–919.
Moran, J.W, Posdal, J.R, Trecker, G.W., 1993. Manufacture of high-solids pre-cheese for conversion into natural cheese. United States Patent 5262183.
Mounsey, J.S., O'Riordan, E.D., 2008. Characteristics of imitation cheese containing native or modified rice starches. Food Hydrocoll. 22, 1160–1169.
Murphy, K.J., 1999. Properties of imitation cheese containing non-casein proteins M.Agr.Sc. thesis. University College, Dublin, Ireland.
Nolan, E.J., Holsinger, V.H., Shieh, J.J., 1989. Dynamic rheological properties of natural and imitation Mozzarella cheese. J. Texture Stud. 20, 179–189.
Noronha, N., O'Riordan, E.D, O'Sullivan, M., 2007. Replacement of fat with functional fibre in imitation cheese. Int. Dairy J. 17, 1073–1082.
Noronha, N., O'Riordan, E.D, O'Sullivan, M., 2008a. Influence of processing parameters on the texture and microstructure of imitation cheese. Eur. Food Res. Technol. 226, 385–393.
Noronha, N., Duggan, E., Ziegler, G.R., O'Riordan, E.D., O'Sullivan, M., 2008b. Investigation of imitation cheese matrix development using light microscopy and NMR relaxometry. Int. Dairy J. 18, 641–648.
Oberg, C.J., Merrill, R.K., Brown, R.J., Richardson, G.H., 1991. Effects of milk-clotting enzymes on physical properties of Mozzarella cheese. J. Dairy Sci. 75, 669–675.
Oliveira, R.B., Margalho, R.P., Nascimento J.S., Costa, L.E.O., Portela J.B., Cruz, A.G., Sant'Ana, A.S., 2016. Processed cheese contamination by spore-forming bacteria: A review of sources, routes, fate during processing and control. Trends Food Sci. Technol. 57, 11–19.
Olson, N.F., 1982. Salt affects cheese characteristics. Dairy Field 72–73.
Olson, N.F., Price, W.V., 1958. A melting test for pasteurized process cheese spreads. J. Dairy Sci. 41, 389–392.
O'Malley, A.M., Mulvihill, D.M., Singh, T.K., 2000. Proteolysis in rennet casein-based cheese analogues. Int. Dairy J. 10, 743–753.

Pellegrino, L., Resmini, P., De Noni, I., Masotti, F., 1996. Sensitive determination of lysinoalanine for distinguishing natural from imitation Mozzarella cheese. J. Dairy Sci. 79, 725–734.

Pereira, R.B., Antunes, L.A.F., dos Santos Ferreira de Silva, R.S., 1992. Development and characterization of a cheese analogue containing aqueous soya extract (soya milk), whey and cow milk. Archivo de Biológica y Técnica 35, 99–115.

Pereira, R.B., Bennett, R.J., Hemar, Y., Camplanella, O.H., 2001. Rheological and microstructural characteristics of model processed cheese analogues. J. Texture Stud. 32, 349–373.

Prentice, J.H., 1987. Cheese rheology. In: Fox, P.F. (Ed.), Cheese: Chemistry, Physics and Microbiology. Elsevier Applied Science, London, UK, pp. 299–314.

Proctor, B., Davison, G., Malecki, G., Welch, M., 1955. A recording strain-gage denture tenderometer for foods. Instrumental evaluation and initial tests. Food Tech. 9, 471–477.

Ramel, P.R., Marangoni, A.G., 2017. Characterization of the polymorphism of milk fat within processed cheese products. Food Str. 12, 15–25.

Ramel, P.R., Marangoni, A.G., 2018. Processed cheese as a polymer matrix composite: A particle toolkit for the replacement of milk fat with canola oil in processed cheese. Food Res. Int. 107, 110–118.

Salek, R.N., Černíková, M., Pachlová, V., Bubelová, Z., Konecná, V., Buňka, F. Properties of spreadable processed Mozzarella cheese with divergent compositions of emulsifying salts in relation to the applied cheese storage period. LWT - Food Sci. Technol. 77, 30–38.

Salque, M., Bogucki, P.I., Pyzel, J., Sobkowiak-Tabaka, I., Grygiel, R., Szmyt, M., Evershed, R.P., 2013. Earliest evidence for cheese making in the sixth millennium bc in northern Europe. Nature 493 (7433), 522–525.

Santos, B.L., Resurreccion, A.V.A., Garcia, V.V., 1989. Quality characteristics and consumer acceptance of a peanut-based imitation cheese spread. J. Food Sci. 54, 468–471.

Savello, P.A., Ernstrom, C.A., Kalab, M., 1989. Microstructure and meltability of model process cheese made with rennet and acid casein. J. Dairy Sci. 72, 1–11.

Schieberle, P., Gassenmeier, K., Guth, H., Sen, A., Grosch, W., 1993. Character impact odour compounds of different kinds of butter. LWT - Food Sci. Tech. 26 (4), 347–356.

Schroeder, C.L., Bodyfeh, E.W., Wyatt, C.J., McDaniel, M.R., 1988. Reduction of sodium chloride in cheddar cheese: effect on sensory, microbiological, and chemical properties. J. Dairy Sci. 71, 2010–2020.

Sharma, P., Munro, P.A., Dessev, T.T., Wiles, P.G., Buwalda, R.J., 2016. Effect of shear work input on steady shear rheology and melt functionality of model Mozzarella cheeses. Food Hydrocoll. 54, 266–277.

Shaw, M., 1984. Cheese substitutes: threat or opportunity? J. Soc. Dairy Technol. 37, 27–31.

Shimp, L.A., 1985. Process cheese principles. Food Technol. 39, 63–69.

Shirashoji, N., Jaeggi, J.J., Lucey, J.A., 2006. Effect of trisodium citrate concentration and cooking time on the physicochemical properties of pasteurized process cheese. J. Dairy Sci. 89, 15–28.

Shirashoji, N., Jaeggi, J.J., Lucey, J.A., 2010. Effect of sodium hexametaphosphate concentration and cooking time on the physicochemical properties of pasteurized process cheese. J. Dairy Sci. 93, 2827–2837.

Southward, C.R., Walker, N.J., 1980. The manufacture and industrial use of casein. N.Z. J. Dairy Sci. Technol. 15, 201–217.

Szczesniak, A.S., 1963. Classification of textural characteristics. J. Food Sci. 28, 385–389.

Talbot-Walsh, G., Kannar, D., Selomulya, C., 2018. A review on technological parameters and recent advances in the fortification of processed cheese. Trends in Food Science and Technology, 81, 193–202.

Tamime, A.Y., 2011. Processed cheese and analogues: an overview. Tamime, Processed Cheese and Analogues. Wiley Blackwell.

Tamime, A.Y., Kalab, M., Davies, G., Younis, M.E., 1990. Microstructure and firmness of processed cheese manufactured from cheddar cheese and skim milk powder cheese base. Food Struct. 9, 23–37.

Taormina, P.J., 2010. Implications of salt and sodium reduction on microbial food safety. Crit. Rev. Food Sci. Nutr. 50 (3), 209–227.

Thomas, M.A., 1977. The Processed Cheese Industry. Department of Agriculture, New South Wales (Sydney), Australia.

Vernon, R.H., 1972. Non-dairy cheese - a unique reality. Food Product Development 6, 22–26.

Voisey, P.W., et al., 1976. Instrumental measurements of food texture. In: deMan, J.M., et al. (Eds.), Rheology and Texture in Food Quality. AVI Publisher Co, Westpot, CT, pp. 79–141.

Volodkevich, N.N., 1938. Apparatus for measurements of chewing resistance or tenderness of foodstuffs. J. Food Sci. 3, 221–238.

Yang, C.S.T., Taranto, M.V., 1982. Textural properties of Mozzarella cheese analogs manufactured from soybeans. J. Food Sci. 47, 906–910.

Zehren, V.L., Nusbaum, D.D., 2000. Process Cheese. Cheese Reporter Publishing, Co, Inc, Madison, Wisconsin, USA.

Zhou, N., Mulvaney, S.J., 1998. The effect of milk fat, the ratio of casein to water, and temperature on the viscoelastic properties of rennet casein gels. J. Dairy Sci. 81, 2561–2571.

Dairy ingredients in processed cheese and cheese spread

2

Art Hill and Mary Ann Ferrer
Department of Food Science, Ontario Agricultural College, University of Guelph, Ontario, Canada

2.1 Introduction

A 1937 research bulletin (Sommer and Templeton, 1939) from the Agricultural Experiment Station, University of Wisconsin, lists some 100 U.S. patents associated with processed cheese beginning with a patent by J.L Kraft in June 1916 (U. S. Pat. 1,186,524). According to reviews by Price and Bush (1974) "The objectives were to produce the sensory characteristics approximating natural cheese, to extend keeping quality, and to deliver it in convenient form to please the consumer." While early processed cheese formulations sourced all solids from cheese, various patents and publications on the use of emulsifying salts and other sources of nonfat milk solids appeared during the 1920s (Price and Bush, 1974). The roles of nonfat milk components have continued to expand and evolve as reviewed in (Guinee, 2009; Kapoor and Metzger, 2008a), and also in a more recent review of processed cheese technology (Guinee and Kilcawley, 2017). In addition to cheese varieties and noncheese dairy ingredients, cheese bases can be considered a third category of dairy ingredients introduced in the 1970s (Ernstrom et al., 1980). Cheese bases are distinct from other protein ingredients in that they may contain coagulants, lactic cultures, other acidulants and fat.

This chapter focuses on the properties and roles of milk components and dairy ingredients, including milk fat components and ingredients, and protein concentrates, including cheese bases. The use of cheese varieties in processed cheese is discussed in Chapter 3. As we begin, we also need to clarify some terminology. Processed cheese, processed cheese foods, processed cheese spreads, and processed cheese products have different legal definitions in different jurisdictions, and those regulations have important implications for cheese product names, formulations, processing, and packaging as discussed in Chapter 15. However, for most purposes in this chapter, we can discuss the roles of dairy-based ingredients without reference to particular processed cheese formulations. Accordingly, in this chapter we use processed cheese products (PCP) as a general term for all processed cheese products unless otherwise specified.

The ultimate goal of PCP manufacture is to create a processed cheese formula that will result in a specific functionality in the end product. For example, a mozzarella cheese analog (MCA) must possess a certain capacity to melt, flow, and stretch on baking while being able to be shredded before cooking (Kindstedt, 1999). Therefore, manufacturers control the chemical composition and physical–chemical properties of cheese analogues and PCP by selecting ingredients and processing parameters to obtain the targeted functionality. Availability and cost of natural cheese (type and age),

Processed Cheese Science and Technology: Ingredients, Manufacture, Functionality, Quality, and Regulations.
DOI: https://doi.org/10.1016/B978-0-12-821445-9.00012-1

rework, and other ingredients change from day to day. Accordingly, PCP manufacturers must understand the functionality of the different ingredients and be able to reformulate within certain bounds to optimize ingredient cost efficiency. Further, because dairy ingredients in PCP are mainly used as substitutes for natural cheese, it is important to understand the physical-chemical properties of natural cheese when used in PCP (Chapter 3), as well as the functional properties of the different ingredients used to replace or extend it.

Use of dairy ingredients in processed cheese is limited by their individual functionalities and by regulations, which are different from country to country. PCP regulations where reviewed by Hickey (2011) and in Chapter 15 of this text. For some particular products and jurisdictions the only dairy ingredient allowed is cheese. Others limit noncheese dairy ingredients by the amount of lactose they provide. For example, in Spain, lactose is limited to not more than 6g/100 g of the finished product (Gobierno de Espana, Ministerio de la Presidencia, 2006). Similar limitations apply for countries like Hungary and the Czech Republic. Other jurisdictions like the US, also allow the use of some dairy fat sources up to 5g/100 g if the product is called "processed cheese," while use of other dairy ingredients is allowed if the product is called processed cheese foods or processed cheese spreads. Therefore, manufacturers must review local regulations.

Formulation is usually carried out using computer programs to calculate variable ingredients and composition parameters based on prescribed or fixed parameters such as a predetermined ratio of rennet casein to cheese protein. Typical composition parameters include levels of protein, fat, lactose, moisture, intact casein, and calcium to casein ratios (Tamime, 2011). However, functional properties of those ingredients also need to be considered, such as pH, water binding capacity, wettability, emulsifying capacity, etc. As discussed in Chapter 8, dairy ingredients are added directly to the cooker together with the natural cheese, water, and emulsifiers or they can be preblended to ensure homogeneity of the blend. Preblending allows ingredient interactions and resulting physical-chemical changes to occur at lower temperatures, which may have important impacts on the resulting PCP. Milk protein ingredients, in particular, replace young cheese in the formula to preserve the appropriate balance of intact casein and hydrolyzed casein. The specific type of hydrolysis may also need to be considered. For example, Brickley et al. (2007) reported that proteolysis of α_{s1}-casein, relative to hydrolysis of β-casein, had greater impact on softening Cheddar cheese.

The use of milk protein ingredients in PCPs is limited by their protein profile, solubility, pH, and mineral composition, which differ according to the manufacturing process of each ingredient. While most dairy protein ingredients can be transformed into powders, some have better functionality in liquid and/or concentrated forms. Fig. 2.1 is a flow chart showing differentiation of the principal dairy ingredients from milk and/or whey to intermediates and to powders.

2.2 Milk composition and properties

Milk is a mix of water, proteins, fats, sugars, minerals, and vitamins; all of which bring certain functionalities to milk and the dairy products created from them. Typical levels

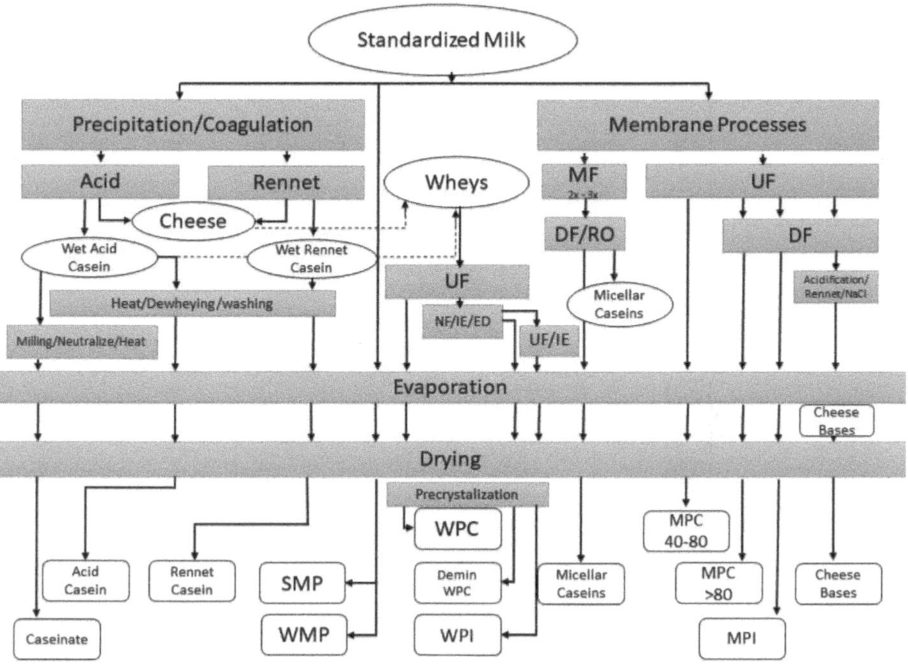

Figure 2.1 Differentiation of milk to produce intermediate and dried dairy ingredients.

Table 2.1 Typical proximate analysis of the common dairy species (% w/w).

	Cow	Dairy sheep	Water buffalo	Goat	Camel
Fat	3.9	7.2	7.4	4.5	2.9–5.4
Total protein[a]	3.3	4.6	3.8	3.2	3–3.9
Casein	2.6	3.9	3.2	2.6	1.9–2.3
Whey[a]	0.7	0.7	0.6	0.6	0.7–1
Lactose	4.5	4.8	4.8	4.3	3.3–5.8
Ash	0.7	0.9	0.8	0.8	0.6–1
Total solids	12.5	17.5	16.83	12.8	12–14

Reproduced with permission from Amigo and Fontecha (2011), Farah (2011), Ramos et al. (2015), Sindhu and Arora (2011), Wong et al. (1988).

[a] Estimated as crude protein, namely, N*6.38.

of principal components in milk of dairy cow breeds and other dairy species are summarized in Table 2.1. Milk can be defined as a solution, a dispersion and an emulsion all at once. Table 2.2 summarizes the properties of the principal structural features of milk, namely, fat globules, casein micelles, globular proteins, and lipoprotein particles. Taken together these components represent a vast range of functionalities. External conditions, such as pH, temperature or processing can change parts of this structure creating an even wider range of functionalities. Some physical properties of milk and

Table 2.2 Properties of the main structural components of milk.

	Milk			
		Plasma		
			Serum	
	Fat globules	Casein micelles	Globular proteins	Lipoproteins particles
Main component	Fat	Casein, water, salts	Serum proteins	Lipids, proteins
To be considered as	Emulsion	Fine dispersion	Colloidal solution	Colloidal dispersion
Content (% dry matter)	4	2.8	0.6	0.01
Volume fraction	0.04	0.1	0.006	10^{-4}
Particle diameter	0.1–10 um	20–300 nm	3–6 nm	10 nm
Number per mL	10^{10}	10^{14}	10^{17}	10^{14}
Surface area (cm^2/mL milk)	700	40,000	50,000	100
Density (20C; kg/m^3)	920	1100	1300	1100
Visible with	Microscope	Ultramicroscope		Electron microscope
Separable with	Milk separator/microfiltration	High–speed centrifuge, ultrafiltration	Ultrafiltration	Ultrafiltration
Diffusion rate (mm in 1 h)	0	0.1–0.3	0.6	0.4
Isoelectric pH	~3.8	~4.6	4–5	~4

Reproduced with permission from Walstra et al. (2005).
Note: Numerical values are approximate averages.

Table 2.3 Some physical properties of milk as affected by processing.

Parameter	Comments
pH @20C	Increases as Na+ and Cl- increase, as inorganic P decreases. Precipitation of calcium phosphate by warming or concentration (e.g., UF) lowers pH.
Titratable acidity	Useful to follow acidification of milk.
Buffering capacity (BC)	Main buffering components are soluble phosphate, colloidal calcium phosphate (CCP), citrate, bicarbonate, and casein. Heat treatments shift buffering maxima to lower pH. Increasing concentration increase the BC.
Redox Potential (Eh)	Inverse to pH; if pH decreases Eh increases. Heating and bacterial activity lowers the Eh.
Density	Affected mainly by composition and temperature, and slightly by processing. Temperature affect is mainly due to changing proportions of solid/liquid fat.
Viscosity mP/s	Influenced mainly by casein micelles, secondly by fat content and much less by lactose and whey proteins. pH influences viscosity. At the same solids content P/F is proportional to Viscosity. Cooling increases viscosity while warming decreases viscosity. Concentration increases viscosity at degree depending on the processing method.
Surface tension	Decreases with increased temperature, increases as fat content increases (up to 4% fat) and increases with homogenization.

Reproduced with permission from McCarthy 2002 and 2009.

related information are summarized in Table 2.3. For more depth and breadth several texts and reviews are available. See for examples (Fox et al., 2015; Walstra et al., 2005; Wong et al., 1988).

2.3 Caseins

2.3.1 Casein micelle composition and structure

Milk proteins are the principal structural elements in PCP as they are in all cheese. The principal proteins of interest are the caseins and the quantitatively major whey proteins. Since the introduction of membrane processes, beginning in the 1960s, milk proteins are available in a growing number of forms. Table 2.4 lists some properties of the principal milk proteins. See also recent reviews (Crowley et al., 2018; Horne, 2019).

Caseins are empirically defined as that group of milk proteins which are precipitated from milk by acidification to pH 4.6. Caseins account for about 74%–80% of milk crude protein (total nitrogen x 6.38) and include four principle fractions. Those fractions are distributed approximately as: 38% α_{s1}–casein (>90% variant B in most breeds), 10% α_{s2}–casein, 36% β–casein (mainly A variant), and 13% κ–casein (mostly variants A and B) (Eigel et al., 1984). Levels of water soluble C–terminal β–casein fragments

Table 2.4 Some properties of milk proteins.

	α_{s1}-casein B	α_{s2}-casein A	β-casein A	κ-casein A	Para-κ-casein[a]	Macro-peptide	ß-LG B	α-LB B	Serum albumin
Concentration g/L	10	2.6	9.3	3.3	2.13	1.17	3.2	1.2	0.4
Molecular weight KD	23.614	25.230	23.983	19.023a	12.269	6.754	18.283	14.176	66.267
Amino acids/molecule	199	207	209	169	105	64	162	123	583
Cystein/molecule	0	2	0	2	2	0	5	8	35
S–S linkages	0	1	0	–	1	0	2	4	17
Phosphate/molecule	8	11	5	1	0	1	0	0	0
HΦ (kJ/amino acid)	4.9	4.7	5.6	5	5.5	4.5	5.1	4.26	4.5
Charge (mol%)	34	36	23	21			30	28	34
Net charge/amino acid	–0.1	–0.07	–0.06	–0.02	0.04	–0.13	–0.04	–0.02	–0.02
Distribution of charge	Uneven	Uneven	Very uneven	Very uneven			Even	Even	
Isoelectric pH	4.46	4.4	5.13	3.7		<3.8	5.2	4.3	4.8
Association tendency	Strong	Strong	([b])	Strong	Strong	Very weak	Strong (dimers)	Weak	Strong
Ca^{2+}	++	++	+	–	–	–	–	([c])	[d]–
T_D @ pH 6.5 (°C)							75.9	61	71.9
T_D @ pH 3.5 (°C)							81.9	58.6	73.5

Reproduced with permission from Walstra et al. (2005), Berger et al. (2002), Le et al. (2017), Azuma et al. (1991), Shi et al. (2008), Hirayama (1990), Bernal and Jelen (1985), Guo and Wang (2016).

[a] Excluding carbohydrate residues

[b] Poor below 5°C; strong with micelle formation at 37°C

[c] Strongly binds 1 mol Ca2+ per mole

Td = midpoint of thermal denaturation transition

[d] Td of WPC is 76.9°C at pH 6.5 and 88°C at pH 3.5

(γ–caseins) in milk serum are increased by activity of the endogenous milk protease, plasmin (Eigel et al., 1984).

Caseins have minimal tertiary structure but are assembled into particles called micelles (Table 2.2). The average hydrodynamic diameter of casein micelles is about 200 nm (De Kruif and Holt, 2003). Phosphoserine groups of α_{s1}–, α_{s2}–, and β–caseins bind Ca strongly (Table 2.4), which helps stabilize the micelle. Other salts, sometimes called calcium nanoclusters, mostly calcium phosphate with a small amount of citrate, also help stabilize micelle structure. Further stability is provided by uneven distribution of charged and hydrophobic groups, particularly for β–casein. Because of its hydrophobicity, β–casein reversibly dissociates from the micelle at lower temperatures (around 5°C or lower) where hydrophobic interactions are weak. This creates the possibility of using cold microfiltration (MF) to prepare β–casein rich and β–casein depleted milk protein concentrates (MPC) as reported by van Hekken and Holsinger (2000).

A most important stabilizing factor is stearic hindrance provided by the strongly amphiphilic κ–casein which is mostly near or on the surface of the micelles. It is well known that enzymatic cleavage of κ–casein between residues 105 and 106, forming para–κ–casein and a caseinomacropeptide (CMP), is the biochemical phase of rennet coagulation. At native micelle concentration, when cleavage is about 90% complete, aggregation of casein micelles begins, due mainly to ionic salt bridges (especially calcium salts) and hydrophobic interactions among micelles (Walstra and Van Vliet, 1986). The CMP portion is often referred to as glycomacropeptide, but in the milk of most dairy cow breeds only about 30% of κ–casein is glycosylated (Eigel et al., 1984).

Caseins can be added to PCP in a wide variety of formats including rennet and acid caseins, several caseinates (mainly sodium and calcium caseinates), skimmed and whole milk powders, ultrafiltered MPCs and MPIs (milk protein isolates), micellar casein concentrates, and coprecipitates. Composition data for various dairy powders are presented in Table 2.5. For all casein based ingredients the state of association of the micelles and the pH and ionic environments are critical to their functionality. For example, Kap*oor* et al, (2007) studied the effect of pH at draining and the amount of salt added to Cheddar cheese on the properties of PCP made from it. They concluded that Cheddar cheese with increased Ca and P effected by higher pH at draining, and increased salt–to–moisture percentage, increased hardness as measured by texture profile analysis (TPA) and decreased the meltability of the PCP.

Data mainly collected from: Evans et al., 2009; Huppertz and Gazi, 2017; Huppertz and Patel, 2016; Kilara and Vaghela, 2018; Kosikowski, 1979; Olson and Aryana, 2009. Minimum and maximum values include standards from Codex Alimentarius and American Dairy Products Institute (ADPI, 2020).

2.3.2 Rennet and acid caseins

Isolation of caseins is accomplished by renneting and then stirring at (50–55°C) followed by liquid–solid centrifugation or separation on a vibrating sieve. The resulting curd is washed, pressed, milled and dried, usually by drum drying. The highly insoluble resulting product is called rennet casein. A similar process but initiated by isoelectric precipitation of caseins from diluted milk results in a product called acid casein (See

Table 2.5 Composition of some dairy powders (%w/w).

Product	Protein	Fat	Lactose	Ash	Moisture	pH
Whole milk powder	25–28	25–29	36–39	6–7	2–4	6.7
Skim milk powder	34–37	0.5–1.0	48–52	7–8	3–5	6.7
MPC42	42	1	45.7	7.8	3.5	6.3–6.8
MPC70	70	1.4	17	7.2	4.4	6.3–6.8
MPC85	85	1.8	1.2	7.4	4.6	6.3–6.8
Caseins						
Acid casein	84–90	0–1	0–1	1–2	8–12	4.6–5.2
Sodium caseinate	90–94	0–1	0–1	3–6	3–5	6.5–8.0
Calcium caseinate	90–94	0–1	0–1	3–5	3–5	6.5–8.0
Rennet casein	80–85	0–1	0–1	6–9	8–12	7.0–7.5
Whey						
Sweet WP	11–15	0–2	63–75	8–10	3–5	5.2–6.4
Acid WP	11–14	0–2	61–70	10–13	3–5	4.4–4.8
Demineralized WP	11–15	0–2	70–80	1–7	3–5	6.2–7.0
Lactose reduced WP	18–24	1–4	52–58	11–22	3–4	
WPC35	34–36	3–5	48–52	6–8	3–5	6.0–6.7
WPC60	60–62	4–8	25–30	4–6	3–5	6.0–6.7
WPI	90–92	<1	<1	2–3	4–5	6.0–6.7
Permeate						
Milk permeate	3–5	0–1.0	76–88	8–11	3–4.5	5.5–6.6
Whey permeate	2–7	0–1.0	76–85	8–11	3–4.5	5.5–6.6
Mineral powder						
Mineral Powder	7	1	11	70	6	6.4–7.5

MPC, milk protein concentrate; WP, whey powder; WPC, whey protein concentrate; WPI, whey protein isolate.
Reproduced with permission from Evans et al. (2009), Huppertz and Pattel (2016), Huppertz and Gazi (2017), Kilara and Vaghlea (2018), Kosikowski (1979), Olson and Aryana (2009), Sikand et al. (2011).
Minimum and maximums include standards from Codex Alimentarius and American Dairy Products Institute (ADPI).

Fig. 2.1). Acid casein is produced by direct acidification with mineral or organic acids, like hydrochloric or acetic acid, or by fermentation with lactic acid bacteria (Golding, 2016). The main difference is reduced lactose in acid casein produced by fermentation. In both cases, processing can include ion exchange or concentration before or after precipitation or gelation. Composition of caseins and caseinate powders are presented in Table 2.5. Commercial rennet and acid caseins have lower protein content (80%–90%) relative to commercial caseinates (90%–95%), mainly due to differences in moisture retention during drying, 8%–12% versus 3%–5%, respectively.

Rennet casein, prepared at pH 6.8–7.5, retains the calcium–phosphate paracaseinate structure, similar to rennet cheese curd, while the CMP, which originally contributed to the stabilization and water binding capacity of the casein micelle, remains in the whey. In contrast, acid casein retains the κ–casein primary structure intact, but the dissociation of colloidal calcium phosphates due to acidification leads to disintegration of the casein micelle. This confers acid casein with greater hydration and water binding capacity when the pH is raised back to that of natural milk (Modler, 1985; Ruegg and Blanc, 1976; Solowiej et al., 2008; Sood et al., 1979).

PCP and cheese analogs have been manufactured successfully using rennet casein (Leko and Foegeding, 2000; Miralles et al., 2006; O'Malley et al., 2000; Shah et al., 2010) and acid casein (Savello et al., 1989; Solowiej, 2012). Rennet casein is most frequently, but not exclusively, used because it has better flavor and functional properties. It is particularly preferred for use in Mozzarella cheese analogs (MCA) because of its ability to stretch and melt (Ennis and Mulvihill, 1999; Rahul and Dhanraj, 2015). Both acid and rennet caseins may improve the shredability of mozzarella cheese (Jana, 2001). Also, while rennet casein contributes to the stringiness and stretchability of MCA, Berta et al. (2016) found that MCA made with rennet casein showed lower stringiness and viscosity after melting than natural mozzarella cheese. Solowiej et al. (2008) found that when compared with rennet casein, acid casein (10%–13% of formula) produced MCA with better melting properties.

PCPs are formed by a protein matrix entrapping water, fat, and other components. Milk proteins, especially caseins and denatured whey proteins, have the ability to adsorb rapidly at the oil–water interface in PCP. This emulsification capacity of dairy proteins is essential to create adequate texture, properly incorporate the target amount of fat (including reduced fat applications), and prevent defects, such as overcreaming, coalescence, and oiling off. The capacity of milk proteins to emulsify is affected by pH and ionic strength.

Caseins owe their emulsifying properties to their "disordered" structure containing both hydrophobic and hydrophilic amino-acids compartmentalized in different regions of their structures, which can interact with the oil interface and extend into the water phase, a concept known as the "train–loop–tail" model (Dickinson, 1998; Singh, 2011). In contrast, whey proteins naturally present more rigid secondary and tertiary structures that allow them to "hide" hydrophobic regions from the water phase, thus limiting their emulsifying capacity. However, conditions, such as temperature, ionic strength, and pH change these structures, expose hydrophobic regions, and enhance emulsifying and water holding capacity of WP (Sarkar and Singh, 2016; Singh and Ye, 2020).

Also of importance when using acid or rennet casein is the type and amount of emulsifying salts, since one of the main roles of the emulsifying salts is to remove colloidal calcium phosphate, which is already depleted in acid casein. For example, adding oxalate, as a calcium binding agent, increased the meltability of rennet casein samples using phosphate as emulsifiers, but reduced the emulsification of the fat droplets. In general, rennet caseins provide better meltability to PCP, but to a degree that depends, among other factors, on the type of emulsifier used. Savello et al. (1989) showed that rennet casein conferred better meltabilty than acid casein to model PCP when using trisodium citrate as emulsifying salt, but they also showed that conditioning acid casein with NaOH before cooking and adjusting the final pH of all treatments after cooking, increased the meltability of PCP made with acid casein. Similarly, Solowiej et al. (2008) observed that meltability of processed cheese analogs produced with only acid casein increased with increasing pH values (5.0–7.0). Further, Savello et al. (1989) found that acid casein produced good melting response when used with disodium phosphate (DSP) or trisodium phosphate (TSP) as emulsifiers. Further,

Kwak et al. (2002) reported that certain casein hydrolysates could be used to replace ES.

2.3.3 Caseinates

Wet acid casein curd or reconstituted rennet casein powders can be dissolved in alkali until they reach pH 6.8–7.5. These neutralized caseins can be spray dried to produce sodium, calcium, potassium, or ammonium caseinates. As opposed to the close structure of native casein micelles in native milk, caseins in sodium and potassium caseinates are arranged in an open flexible structure while calcium caseinates form "pseudo" micelles stabilized by intermolecular links of calcium with phosphoserine residues (Tamime, 2009). This structure gives calcium caseinate solutions a white, milky appearance while sodium, potassium and ammonium caseinates form clear or slightly opalescent solutions. Caseinates are soluble; in particular, sodium caseinate is the most soluble and the most widely used caseinate. Better water binding capacity of caseinates, relative to acid casein is associated with higher viscosity and better emulsification properties.

In general, caseinates are known to improve consistency of PCP, but functional properties of commercial caseinates vary greatly, so selection for use in PCP has to be based on individual caseinate functionality (Bastier et al. 1993). Some examples follow.

- Calcium and magnesium caseinates have significantly lower solubility index, water absorption capacity and apparent viscosity than sodium and potassium caseinates (Bastier et al., 1993; Huppertz and Patel, 2016; Mulvihill and Murphy, 1991).
- Complete substitution of cheese solids nonfat with calcium caseinate will decrease spreadability, while partial substitution may increase meltability as shown by Gouda and El–Zayat (1988). This may be related to how caseinates influence the creaming effect.
- Dhanraj et al, (2017) reported that 10% substitution of rennet casein with WPC produced satisfactory Mozzarella cheese analog with respect to shredability and performance in baking tests.
- Emulsions formed in natural mozzarella cheese versus partial (2%) substitution of natural mozzarella with calcium caseinate, show that fat globules were heterogeneously dispersed and some were coalesced (Tunick et al., 1989).
- Emulsification properties of caseins depend to a great extent on the structure of the micelles; restructuring due to calcium depletion results in different emulsification capacities. For example, treatment with tri–sodium citrate formed smaller casein aggregates with better emulsifying capacities, but also caused over creaming and flocculation (Lazzaro et al, 2017).

In summary, emulsification and water holding properties of caseins and caseinates are affected by environmental factors, such as pH, ionic strength, presence of salts and particularly calcium. All these factors must be considered while selecting or exchanging, for example, replacing sodium caseinate with calcium caseinate. While various workers have studied the use of high levels of substitution of caseinates for cheese solids in PCP, in practice caseinates (and also WPCs) are typically added at levels in the range of 5–7 g/100 g (Tamime, 2011).

2.3.4 Micellar caseins

Micellar casein concentrates (MCC) are obtained using MF with a pore size of 0.1 μm to separate casein micelles from whey proteins. This allows preparation of caseins without chemical alterations to micelle composition and structure that result from acidification or renneting (Grossbier, 2016). MF in commercial practice can be used to increase the mass ratio of casein to whey proteins from the native level of 82:18 up to 95:5 (Hammam and Metzger, 2018). MF to produce MCC also produces "whey" that is very close to milk serum in terms of native composition and native protein structures. MCC whey is sometimes called purified whey concentrate (Maubois, 2002) because microorganisms are retained with the MCC rather than the whey. This at least partly accounts for the limited shelf life of wet MCC; some workers have attempted to extend refrigerated shelf life of wet MCC by using MF to reduce microorganisms in the milk prior to MF concentration of casein micelles (Amelia and Barbano, 2013). Purified whey, also called ideal whey, is further discussed in a subsequent subsection on whey protein concentrates and isolates.

MCCs, also known as native phosphocaseinate, are viscous MF retentates with excellent gelling properties. MCC solutions at different ranges have shown half the renneting time and double the gel strength relative to reconstituted skimmed milk at similar protein concentrations (Pierre et al., 1992). MCCs will also gel in the absence of rennet at low temperatures depending on protein concentration. At 17% protein, MCC will gel at 4°C and 20% MCC will gel at 24°C (Hussain et al., 2011). Gelation in the absence of rennet is reversible after rewarming to around 50°C (Amelia and Barbano, 2013). This allows for further processing of the concentrate, which can be spray dried to create a powder rich in caseins and colloidal minerals.

High casein micelle concentration results in high buffering capacity and an abundance of colloidal calcium phosphate that has to be dealt with during PCP manufacturing. Another limitation compared with SMP, MPC and WPC, is that powdered MCC is difficult to reconstitute (Grossbier, 2016; Schokker et al., 2011). MCC has a relatively short wetting time but is slow to disperse (hours not minutes) and has a high insolubility index. For example, the insolubility index of SMP is maximum 1.2 mL (ADPI, 1971), while the MCC insolubility index can be as high as 15 mL. Solubility (Nasser et al., 2017) and reconstitution–ability (Schokker et al., 2011) of MCC decrease during storage and are related to browning and fragmentation time (amount of time to achieve maximum amount of particles during hydration) (Nasser et al., 2017). Accordingly, liquid MCC, frozen or refrigerated, is preferred because it retains the original functionality of casein micelles.

Hussain et al. (2011) found that using NaCl in the dispersion media (>3% NaCl) at room temperature increases the wettability time but decreases the dispersion time of the powder from about 6 h to 3 h. It is important to note that with NaCl concentrations less than 3%, wettability times increased. In addition, adding 1.5% $CaCl_2$ almost halved the dispersion time, but the effect stayed constant for higher concentrations (2.25% to 12%) of $CaCl_2$. When extrapolating this data to a complex system, such as PCP, where heating, high shear, and multiple ingredients are involved, manufacturers need to consider that even small changes in the formulation might create substantial changes

Table 2.6 Composition (g/L) of powders from sweet whey, acid whey, heat–acid precipitated whey, and Greek yogurt whey.

Components	Sweet whey[a]	Acid whey[a]	Paneer whey[b]	Greek Yogurt whey[c]
Total solids	63.0–70.0	63.0–70.0	61	55.5
Lactose	46.0–52.0	44.0–46.0	50	3.2
Protein	6.0–10.0	6.0–8.0	3	2.4
Fat	5.0	4.0	1.3	
Lactic acid	2.0	6.4	nil	5.5
Ash	5.0	8.0	6.0	7.3
Calcium	0.4–0.6	1.2–1.6		1.3
Phosphate	1.0–3.0	2.0–4.5		1.8
pH	6.0–6.6	4.0 – 5.0		4.5

[a] Yadav et al., 2015.
[b] Kumar et al., 2018.
[c] Nishanthi et al., 2017.

in the dispersion of the MCC, and consequently in its functionality. Sensitivity of MCC to mineral salts may create opportunities to develop ingredients with different functionalities. For example, structure and functionality of MCC can be altered by adding calcium sequestrants (Guinee et al., 2013) or mineral salts (Schuck et al., 2002) to the skimmed milk before MF.

Until now there are few reports on using MCC in PCPs. However, one study reported that substitution of 83% of Cheddar cheese with a mixture of acid curd made from MCC and powdered MCC produced an acceptable Cheddar PCP in the absence of added emulsifying salts. The ideal proportion of proteins coming from acid curd to MCC was 2:1 (Hammam and Metzger, 2019).

2.4 Whey based ingredients

2.4.1 Types of whey

Whey is the liquid remaining after removing caseins during cheese making or during manufacture of caseins and caseinates. It is similar in composition to milk serum except that whey from rennet coagulated cheese contains the CMP portion of κ–casein. See Tables 2.6 and 2.7 and the following general descriptions of several types of whey.

Sweet whey is separated from rennet-coagulated milk, at pH $>$ 6.0, typically 6.1–6.5. Sweet whey proteins include the CMP portion of κ–casein. Fat content of sweet whey is typically 2–5 g/L depending on the initial fat content of the standardized cheese milk and cheese making conditions. Normally, centrifugal separation is used to reduce the whey fat content to $<$1g/L. The whey cream can be used as sweet cream in many applications (Sodini et al., 2006; Morin et al., 2007).

Acid whey is the byproduct of: (1) casein preparation via isoelectric precipitation with acidification to pH 4.6 via lactic acid fermentation or by addition of mineral or organic acids, (2) whey of pH near 4.6 removed from Greek yoghurt, or (3) cheese

Table 2.7 Whey protein and nitrogen distribution.

	% of true protein		Nitrogen distribution[a]	
Component	A	B	Nx6.38 (g/L)	% of N
β–lactoglobulin	45–57	48–58	3.3	37
α–lactalbumin	15–25	13–19	1.4	15
Serum albumin	3.5–10	6	0.4	4
Immunoglobulins	4–10	8–12	0.5	5
Caseinomacropeptide	15–20	12–20	1.2	13
γ–caseins			0.7	8
NPN			1.6	18
Total			9.0	100

A: Kumar et al., 2018.
B: Guo and Wang, 2016.
[a] Calculated from various sources, including Crowley et al. (2018), DePeters and Ferguson (1992).

making that employs acid coagulation at or near the pI of casein micelles in the pH range of 4.4–5.0. The higher pH may be used for some lactic cheese varieties that employ small amounts of rennet to improve curd elasticity. Acid whey composition is similar to sweet whey with the important differences that: (1) its pH is much lower (4.4–5.0), (2) lactose is reduced from 4.5%–5% to 3.8%–4.3% and replaced with lactic acid, except when acidification is by addition of organic or mineral acids, (3) Protein content, in some cases, is reduced by heat treatment applied before fermentation as is the case with the Greek yogurt whey in Table 2.6, and (4) it contains higher levels of calcium and calcium phosphate salts due to their solubilization from the micelles and colloidal calcium phosphates. These salts precipitate and cause membrane fouling when whey is warmed for UF treatment. This effect can be reduced by addition of citrate to chelate Ca and Mg (Abd El–Salam et al., 2009). Lactate in acid whey can also be removed by nanofiltration (Chandrapala et al., 2016a,b). Vitamin and mineral contents in acid and sweet wheys were reported by Glass and Hedrick (1977).

Heat-acid whey is the byproduct of varieties manufactured by addition of organic acids to heated milk or blends of milk and whey. As an example, the composition of paneer whey is given in Table 2.6. The whey protein is reduced due to coprecipitation of denatured whey proteins with caseins.

Salty whey results from cheese varieties like Domiati that are made from salted milk, and dry salted cheese like Cheddar, Cheshire, and American varieties. Blaschek et al. (2007) reported that, relative to whey collected at draining, α–lactalbumin (α–LA) and β–lactoglobulin (β–LG) were reduced in whey collected after salting of commercial Cheddar, American Mozzarella, and Colby cheeses, but lactoferrin was increased from 1 g/L to as high as 15 g/L. Other workers have characterized salty whey and salty whey proteins with a view to finding suitable end uses (Nishanthi et al., 2017; Talebi et al., 2019). Kapoor and Metzger (2004) observed no differences between control PCP and PCP formulated with salty whey to replace added moisture and salt.

Purified whey. Multistep MF of milk to reduce bacterial load and then to create MCC leads to the production of whey with minimal processing. This whey can be treated with UF membranes of 10,000 Daltons and diafiltered to produce a highly purified and native whey protein isolate described as "ideal whey" (Maubois, 2002) or as "soluble protein concentrate" (SPC) (Evans et al., 2010). The main differences between WPC and SPC are: (1) WPC contains up to 8% fat (see Table 2.5) versus less than 0.5% in SPC, (2) WPC has more Ca, depending on pH at draining, (3) fermentation and other treatments reduce pH and create various volatiles in WPC that are absent from SPC, and (4) WPC includes CMP (absent in SPC) and a higher amount of NPN. It was reported that CMP reduces gelling and foaming of WPC (Abd El-Salam *et al.*, 2009; Veith and Reynolds, 2004).

2.4.2 Whey proteins structure and properties

Whey proteins constitute about 20 - 23% of total milk proteins corresponding to the historic and still often cited casein number of 78% (Van Slyke and Price, 1949; Coggins, 1991). Whey protein composition varies substantially depending on original milk composition and diverse procedures used for making cheese and caseins. There are also inconsistencies around reporting of CMP, γ-caseins, and true versus crude protein. To help clarify, the distribution of the principal proteins in sweet whey, as reported by two authors, is shown in Table 2.7, along with estimated values for the nitrogen distribution of sweet whey including the CMP, γ-caseins, and NPN. Minor proteins not listed in Table 2.7 include lactoferrin, lactoperoxidase, lysozyme, and various proteins associated with the fat globule membrane. See Table 2.4 and recent reviews of whey proteins (FAO/WHO, 2011; Edwards and Jameson, 2014; Kelly, 2018; Kilara and Vaghela, 2018).

Whey proteins are mainly compactly folded and hydrophobic globular proteins that are denatured by heat. β–LG represents almost 50% (by weight) of all whey proteins and its behavior dominates the behavior of whey proteins as a group, especially during thermal processing (Crowley et al., 2018). That behavior depends strongly on pH, ionic strength and temperature. At high concentrations and under certain conditions, β-LG forms gels, including cold-set gels (Ju and Kilara, 2008).

α-LA is a low-molecular-weight and tightly folded protein, which tends not to associate except at low ionic strength. At low pH (about 4), α-LA releases one calcium ion per molecule, which causes partial unfolding into a molten globular state. This makes the protein susceptible to further denaturation by heating followed by aggregation and precipitation. At pH above 4, α-LA is relatively stable to thermal aggregation and precipitation. However, calorimetry reveals conformational changes starting at temperatures as low as about 45°C (Bernal and Jelen, 1985) and the T_D value for α-LA at pH 6.5 is 61°C. See Table 2.4. In this case thermostability observed after heat treatment of milk should be interpreted in terms of renaturation and not of high temperatures of denaturation. In practice this means that complete thermal aggregation of α-LA with itself and other proteins requires time.

2.4.3 Whey protein concentrates and isolates

Decades ago, almost all whey was considered a waste, an environmental hazard, or used as animal feed or fertilizer (Kosikowski, 1979; Price, 2018). Currently advanced processing options allow preparation of value added whey products, particularly whey protein concentrates (WPC) and whey protein isolates (WPI) (Tsakali et al., 2010). These processes capitalize on the most valuable component of whey, which is its protein content. Depending on the presence or absence of CMP and other processing parameters, the true protein in whey is about 6–8 g/L.

The simplest whey protein concentrate is whey powder which is prepared by evaporation, lactose crystallization, and drying of sweet or acid whey. Evaporation aims to concentrate whey to 42%–60% of total solids, while crystallization of lactose aims to convert most of the lactose to its crystalline α–hydrate form. This makes the lactose much less hygroscopic and reduces caking, which is common for high lactose powders. Protein content in both acid and sweet whey powders is 11%–15% (Table 2.5). Relative to WPC, whey powder has the advantage that it includes all whey solids; there is no residual BOD. BOD associated with whey is 35,000–55,000 mg O_2/L (Tsakali et al., 2010).

Separation and concentration of whey proteins from whey can be achieved by UF membranes followed by dialysis, if necessary, and spray drying, producing WPC with 35% to 80% protein, and WPI with around 90% protein. During manufacture of WPC and WPI, preheating at temperatures between 68–72°C is used to precipitate calcium phosphate and avoid membrane fouling during processing. Whey proteins have different functionalities; especially important to understand, when formulating processed cheese, is their solubility over a wide range of pH (Pelegrine and Gasparetto, 2005), water binding capacity (De Wit, 1988; Abd El–Salam et al., 2009), and limited emulsification capacity (De Wit et al., 1986). Manufacturing variables lead to different functionalities, so use of different whey protein ingredients has to be carefully tested.

The solubility of whey proteins is highly dependent on pH and temperature (De Wit et al., 1986; Lee et al., 1992; Pelegrine and Gasparetto, 2005). Solubility is high near neutral pH and pH around 3.5, and minimum at pH 4.5 at all temperatures. Solubility of WPC is the lowest at pH 4.5 and 60°C (Pelegrine and Gasparetto, 2005). This fact may be important when preblending ingredients for PCP, for example blending WPC and cheese with pH around 5. Lee et al. (1992) reported that solubility of WPI was highest at 25°C versus 55 and 65°C.

The emulsification capacity of whey proteins, like solubility, is pH and temperature dependant (Lee et al., 1992), and is described as poor or very good according to different authors. These differing observations may be explained by the reactive behaviour of WP; when denatured within the food system containing the fat, whey proteins unfold exposing hydrophobic sites that interact with the fat phase and promote emulsification (Modler, 1985). Gupta and Reuter (1992) found that substituting up to 20% of young cheese solids in PCP with WPC did not create differences in the free oil released during heating. However, using SEM, Hill and Smith (1992) showed that in cheese spread, made with acid–heat precipitated whey proteins, the fat globules were not embedded in the protein matrix as it was in spreads made with Cheddar cheese. Whey protein products are known to have a higher concentration of phospholipids in their fat fraction than other protein ingredients. These phospholipids may contribute to the emulsifying capacity of WPC (Young, 1999). An attempt to replace ES with polymerized WP in PCP made with acid casein (Sołowiej et al., 2014) resulted in a reduction in disodium phosphate content from 0.8% to 0.4% while keeping the rheological properties of the PCP.

WPC and WPI can be used to confer flow resistance in certain PCP applications or increase viscosity in other applications (Gupta and Reuter, 1993; Mleko and Foegeding, 2001; Mounsey and O'Riordan, 2008; Savello et al., 1989; Szafrańska and Sołowiej, 2020). The increased viscosity results from the water binding capacity of whey proteins. Increased viscosity is desired when low cost dairy ingredients are required as bulking agents or a source of proteins. Typical usage rate of WPC in PCP is around 5–7 g 100 g−1 of blend; with higher amounts reported to create issues with the flavor and structure of the product (Caric and Kalab, 1999, Guinee and Kilcawley, 2017). Gupta and Reuter (1992, 1993) reported that WPC created a sandy and chalky texture independent of the emulsifier used or the blending method. The defect was only prevented by the use of trisodium citrate at 3% as an emulsifier.

Hill and Smith (1992) found that whey protein curds made by heat–acid precipitation from reconstituted WPC35, produced a chalky mouthfeel in PCP that was not present in PCP made by direct addition of reconstituted whey powders. This may be because whey proteins in the reconstituted whey powder would be native and, therefore, during heat treatment, rather than forming larger particles, they would form small aggregates with each other and associate with κ–casein on the surface of the micelles or in the serum (more on this in the next subsection). In another study, Kaminarides and Stachtiaris (2000) were able to substitute up to 10% of Greek cheese in a cheese analog with up to 5% of whey solids from UF sweet whey concentrate without significantly decreasing the organoleptic characteristics of the analogs; however, hardness, adhesiveness, gumminess, elasticity and chewiness all decreased as the level of substitution increased. This is consistent with other reports. Schulz (1976) used whey products to reduce meltability of PCP made with natural cheese or with acid and rennet curds. Similarly, Savello et al. (1989) concluded that in most cases meltability is reduced as the amount of whey protein, denatured or not, is increased.

Another noteworthy effect of adding whey protein powders (and also milk powders and concentrates) is that high levels of lactose in processed cheese may cause browning during cooking, which can be controlled by selection of ingredients to minimize lactose, lower cooking temperatures, and rapid cooling of the PCP (Bley et al., 1985; Kombila–Moundounga and Lacroix, 1991). Too much lactose may also result in sandy texture due to crystallization of lactose (Patocka and Jelen, 1988).

2.4.4 Denatured vs native whey proteins

One of the most important chemical interactions associated with milk processing is the thermal associations among whey proteins and caseins, mainly, but also involving other milk components. When milk is heated near or above the T_D of β-LG (>70°C) the principal whey proteins denature, including BSA and IgG (See Table 2.4) (Bernal and Jelen, 1985). As noted in an earlier section, β-LG comprises about 50% of whey proteins and tends to dominate thermal interactions among WPs and also disulfide, and perhaps hydrophobic, associations of β-LG with κ-casein on the micelle surface and also in the serum. The relative distribution of WP- κ-casein complexes between the micelles and the serum phase depends on the pH of heating. Heating milk at its natural pH (~6.7) transfers about a third of the serum WPs to the casein micelles, heating at pH 6.3 increases this proportion to nearly 75% of the total serum proteins, but heating at pH 6.8–7.1 moves almost all of the WP from the micelles to the serum phase in complexation with κ–casein (Kethireddipalli et al., 2011; Kethireddipalli and Hill, 2015).

The α–LA readily unfolds but requires considerable time to form stable aggregates with other whey proteins and with κ–casein (Hill, 1989). The CMP and γ–caseins are not heat sensitive and do not precipitate with heating. Doing the arithmetic from Table 2.7, this means that about 60% of sweet whey N or 75% of true whey proteins are heat labile. It's not surprising, then, that thermal interactions among whey proteins and caseins also occur before and during manufacture of PCP and make interesting contributions to PCP quality. Some examples follow.

Native whey proteins form a denser more uniform matrix in PCP. Increased firmness caused by incorporation of whey proteins in PCP made with acid casein was reported by Solowiej et al. (2008). Similarly, Gupta and Reuter (1993) found increased firmness and decreased meltability when substituting young Cheddar with WP. Later, Lee et al. (2013) studied the incorporation of native or pre–denatured WP into PCP made with rennet caseins. When native WPs were included, cooking between 90–85°C caused denaturation of around 55% of the WP, which was able to interact with other components in the matrix forming a denser, more uniform network as shown by transmission electron microscopy (TEM). Microstructural examination of selected cheeses indicated that, when pre–denatured whey proteins were used, the whey proteins were incorporated as large aggregates dispersed within the cheese matrix. These aggregates did not contribute to, and may disrupt the structure of the cheese. Predenatured WP also produced softer cheese which melted more than those prepared with native whey proteins.

WP products affect firmness of PCP depending on pH of processing. WP products tend to increase firmness of PCP but the effects are complicated by interaction effects of pH during pre–heating for denaturation and while processing the PCP. Solowiej et al. (2008), reported that 2% substitution of acid casein with WP products substantially increased firmness. However, firming caused by WP products differed according to pH. Lee and Anema (2009) obtained similar results for PCP made with rennet casein and WP comprising 20% of the total protein. Mixes were heated (80–90°C, total of 10 min) at 10 levels of pH ranging from pH 5.6–7.6. After processing, the pH of the final product was adjusted to pH 5.7. All samples cooked at pH 6.7 or higher were 85%–95% denatured. A plot of firmness versus pH shows maximum firmness at pH 6.7, and mixes cooked at pH near 5.6 and 7.5 were not different than controls without added WP. It's not clear why WPs denatured at pH > 6.7, apparently, did not interact with and strengthen the rennet casein network. Nevertheless, these results show the importance of managing processing parameters and ingredients to the final texture of the PCP.

Denatured and native WPs decrease meltability of PCP. It is well known that meltability of cheeses made from higher heated milk is reduced due to interaction of denatured whey proteins with casein micelles. It is therefore, not surprising that WPs tend to reduce meltability of PCP, often along with increased firmness. For example,

Lee and Anema (2009) and Savello et al. (1989) reported that WP reduced meltability in PCP made with both rennet and acid casein, but did not affect emulsification. Solowiej et al. (2014) reported that (partial) substitution of rennet or acid caseins with native or denatured WPs decreases the meltability of PCP.

Predenatured WPIs may replace ES in processed cheese. Nastaj et al. (2019) studied the effects of 2%–6% of WPI versus 2% –6% of predenatured WPI (denatured at 80°C for 50 min) with 10% acid casein as the main structural protein. They reported that the polymerized WPI formed more hydrophobic surfaces with unevenly distributed fat droplets relative to emulsions formed with WPI. The same research group (Sołowiej et al., 2020) explored the use of small amounts of polymerized WPI (denatured at 80°C for 50 min) to partially replace ES (disodium phosphate) in PCP made with 11%–13% acid caseins. WPI in the range of 0%–0.8% was balanced with 0.8%–0% ES. The pH was adjusted to 5.8 before cooking, based on favorable results for hardness, cohesiveness, meltability, and spreadability, the authors concluded that denatured WPI could be used to replace or partially replace ES in processed cheeses.

2.4.5 Microparticulated whey proteins

Dynamic high pressure shearing can be used to create microparticulated whey proteins (MWP). MWP are aggregated and smooth colloidal particles ranging in diameter from 0.5 to 4 μm (Ipsen, 2017), similar to fat globules in size. These small particles are not perceived in the mouth as individual particles, but rather as a creamy, smooth fluid (Lucca and Tepper, 1994), giving MWP the ability to mimic fat in several applications. MWP can also coat the mouth and taste buds in a way similar to fats, an effect known to reduce the perception of some undesirable flavors such as bitter and astringent in low fat foods (Lucca and Tepper, 1994).

The first MWP was patented by the NutraSweet Company in 1988 and distributed under the trade name, Simplesse®, but it was not until 2004 when the patent expired that the supply for MWP started to expand, with brands such as Nutrilac® and Dairy–Lo™ entering the market. MWP are made from WPC (35%–55% proteins) treated with or without heating; higher temperatures create bigger particles (Sánchez–Obando et al., 2020). Dissanayake et al. (2012) found that microfluidization of WP at low pH with heating produced smaller particles with better colloidal stability and improved solubility, both, of which, could have important effects on the general functionality of the powders. Microfluidization at low pH also improved emulsification compared with microfluidization at neutral pH. MWP formed weaker heat gels and denatured MWP formed firmer cold acid gels than non microparticulated WP. On the other side, Dissanayake and Vasiljevic (2009) found that, during microfluidization at neutral pH, heating decreased solubility but microfludization increased it, while the emulsification properties were enhanced by both microfluidization and heating.

The process of microparticulation makes thiol groups in WPs less available to interact with other whey proteins and caseins, which may result in better texture without sandy or gritty texture; the smaller particles give better mouthfeel and creamy texture (Havea, 2009). Microparticulation of WP also enhances the heat stability of WP–stabilized emulsions, and, depending on the extent of denaturation, it allows addition of more WP to emulsions without causing flocculation (Çakir-Fuller, 2015). MPW may contain other nondairy ingredients depending on the functionality required. For example, Tobin et al. (2010) studied how inulin could replace lactose in MWP to produce MWP that will increase viscosity and decrease sedimentation, while also decreasing the caloric content of the end product.

One of the early applications for MWP was as a fat replacer in Cheddar cheese (Stankey et al., 2017) and yogurt (Torres et al., 2011) with satisfactory results. MWP with a high degree of denaturation (>80%) and particle diameter within 1–1.5 μm has been used to substitute up to 5% of the fat in a cheese spread and 40% fat in petit–Swiss without affecting the sensory attributes. While the viscoelastic properties of petit–Swiss were unaffected by the fat substitution, substituting 19% of the fat of the cheese spread did cause a decrease in the gel strength (Sánchez–Obando et al., 2020).

2.5 Whole milk protein ingredients

This section considers sources of whole milk proteins for use in PCP. Whole milk means ingredients, such as skim milk powder, UF separated MPC, or coprecipitates with casein:WP ratio similar to native milk proteins.

2.5.1 Skimmed milk powder

Whole milk and skimmed milk powders (SMPs) are made by pasteurizing standardized skimmed or whole milk, evaporating it to around 45% DM, homogenizing and cooling/storing (if needed) and heating before (most frequently) spray drying. See Fig. 2.1 and Table 2.5. The resulting powder can be agglomerated to obtain better dispersion properties. Whole milk powder and SMP retain the protein, lactose and minerals in similar proportions as in the original milk plasma. SMP is frequently used in PCP (10–12 g 100 g^{-1} PCF) because of its versatility, and capacity to improve the stability of processed cheese and contribute to creaming and spreadability (Gouda and Abou El-Nour, 2003). However, its use in PCP is limited by its high lactose content (50%), which causes browning and crystallization during processing and storage of PCP. In addition, use at levels of 7%–10% was reported to create crumbliness and lack of cohesiveness (Gouda and Abou El-Nour, 2003).

SMP is used as a cheap source of solids and protein instead of more costly natural cheese. SMP contains intact caseins as opposed to partially hydrolyzed caseins in natural cheese (especially old cheese), a lower casein:whey protein ratio, and a different colloidal calcium content than cheese, all factors known to affect the properties of PCP. The significance of intact casein in PCP is explained in detail in chapter 3. When compared with other protein ingredients (MPC, UF-Cheese Base), SMP used in PCP (cheese sauce) showed higher oil separation index, lower viscosity and lower sensory score, with a sweet flavor probably because of the higher lactose content (Saad et al., 2016). Relative to SMP, whole milk powder is more difficult to produce, less stable in storage, more difficult to reconstitute, and generally less convenient to use in PCP. Interestingly, full fat camel milk powder was reported to produce good sensorial characteristics when used to replace up to 15% Cheddar cheese in PCP (cheese sauce) but with decreased meltability when the substitution level achieved 15% (compared with 10%) (Desouky et al., 2019).

2.5.2 Buttermilk powder

Buttermilk is usually obtained as a by–product of butter manufacturing from sweet cream. In much lower quantities it can also be obtained from churning cultured cream or whey cream. It's composition is similar to skimmed milk except that it contains much of the surface active compounds (SAC) from the milk fat globule membrane (MFGM), which is rich in phospholipids and glycoproteins (Sodini et al., 2006; Zanabria and Corredig, 2011), The concentration of phospholipids on the dry basis in buttermilk is reported to be four, five and ten times that of skimmed milk, whole milk and cream, respectively (Christie et al., 1987; Rombaut and Dewettinck, 2006) and up to 7 times the phospholipids of whole milk (Zanabria and Corredig, 2011). Buttermilk is commonly spray dried and used in applications similar to SMP.

Functionality of buttermilk depends on how it was processed (Zanabria and Corredig, 2011). In particular, heat treatment increases interactions between MFGM and whey proteins at least partly due to exposure of sulfhydryl groups (Sharma et al., 2012a). There has also been an attempt to isolate SAC components, for example, using microfiltration (Morin et al., 2007). Other workers reported that buttermilk can be used to increase the heat stability of recombined milks (O'Connell and Fox, 2000; Singh, 1990), and further that buttermilk added to cheese milk increases water holding capacity of the cheese (Govindasamy–Lucey et al., 2007; Raval and Mistry, 1999).

Buttermilk is recognized as a good emulsifier, at least in some applications, due to SAC released from the fat globule membrane as described by Vanderghem et al. (2010). However, when a hot cheese emulsion was prepared with sodium caseinate or buttermilk as emulsifiers, sodium caseinate improved fat emulsification and emulsion stability, while buttermilk decreased fat particle size but did not control creaming and precipitation; however, blending equal parts of butter milk and sodium caseinate achieved short term emulsion stability similar to the control emulsified with disodium hydrogen phosphate (Kelimu et al., 2017). This suggests the possibility of using buttermilk–protein blends to replace ES.

Sodini et al. (2006) compared the emulsification capacity of powdered sweet, cultured, and whey buttermilk to each other and to powdered sweet whey and skim milk. All three buttermilk powders demonstrated better emulsification properties than skim milk or whey powders. But most interesting was the performance of whey butter milk powder in two respects. Firstly, emulsification capacity of whey buttermilk was independent of pH, while solubility and emulsification capacity of sweet and cultured butter milk decreased at lower pH. Secondly, whey buttermilk demonstrated the highest emulsification properties and lowest foaming capacity. The authors suggested this may be due in part to a much lower ratio of protein to phospholipids in the whey buttermilk.

Similarly, another study used buttermilk to replace up to 2% of ES. They reported that the level of free oil did not change, but the meltability of PCP decreased with the use of buttermilk. They also found that buttermilk created a better dispersion of fat globules in PCP independent of ES (Raval and Mistry, 1999). Found that addition of concentrated buttermilk (up to 30% of the amount of natural cheese) caused lower meltability of the PCP but also decreased firmness (penetration force). Sensory evaluation of the different treatments did not show significant differences within the control and different levels of buttermilk, but the samples with higher amounts showed higher overall preference scores. Addition of Buttermilk to milk during pizza cheese manufacturing also decreased meltability (Govindasamy–Lucey et al., 2007)

Buttermilk seems to contribute to the prevention of fatty acid oxidation and lipolytic rancidity during storage of PCP; samples prepared with heated buttermilk concentrate presented lower POV (meq of peroxide per Kg fat) and acid degree values (meq/100 g fat) than the samples without buttermilk addition (Doosh et al., 2014).

2.5.3 Milk protein concentrates

MPCs are manufactured by ultrafiltration (UF) and, sometimes, diafiltration of pasteurized skimmed milk followed by spray–drying to secure shelf stability (Mistry and Hassan, 1991). This process retains the original casein to protein ratio of the original milk (80:20) (Rollema and Muir, 2009). Different levels of concentration by membrane processes allow for the production of MPC ranging from 40% to 85% protein on a dry weight basis, while further concentration is possible through diafiltration to obtain MPC with more than 80% of protein and milk protein isolates (MPI) with a minimum of 89.5% protein (ADPI, 2020). As the protein concentration increases, the levels of lactose and soluble minerals decrease, while some minerals (mainly colloidal calcium phosphate) are retained in the casein micelles (Ferrer et al., 2011). Diafiltration can be used to further reduce lactose and minerals (Ferrer et al., 2014). Therefore, MPCs with higher protein concentrations present different functionalities even when reconstituted to similar protein levels (Ferrer et al., 2008). MPI is rarely used in PCP because it's more expensive'

One of the most important functional properties of MPC powders is solubility (Anema et al., 2006). Low solubility is primarily due to the close proximity of protein molecules and the hydrophobic nature of the casein protein constituents (Crowley et al., 2015; De Kruif et al., 2012; Horne, 2006; Mimouni et al., 2010). Several different formulation and technological approaches have been investigated to improve rehydration of MPCs. For examples:

- Ultrasonication of liquid MPC
- Calcium depletion using ion exchange and addition of calcium chelating agents
- Replacing calcium with sodium during UF (Meena et al., 2019; Sikand et al., 2013). Solubility is affected by the ionic composition of the serum phase during reconstitution (Crowley et al., 2015, Lin et al., 2018, Schuck et al., 2002).
- Static high pressure treatments (Augustin and Udabage, 2007).
- Pre–acidification before UF (Liu et al., 2019),
- Adjustment of pH. The dissolution of high–protein powders, especially casein–based powders, is sensitive to pH (Eshpari et al., 2014). Solubility of MPI increases in alkaline environments (pH 8.4) (Wu et al., 2019).

- Optimize storage conditions. Solubility of caseins (not WP) in MPC decreases during storage as a function of temperature (Anema et al., 2006b; Havea, 2006; Fang et al., 2011; Mimouni et al., 2010).

The heat stability of MPCs is known to be related to the level of protein concentration. When 3.5% protein dispersions of MPC were heated at 140°C and pH <6.8, MPC80 and 70 had lower heat stability than MPC50 (Crowley et al., 2014). Processing parameters also affect the functional behavior of MPC (Corredig et al., 2019).

Regarding functionality in PCPs, several properties of MPC are worth mentioning. (1) The surface coverage of emulsions made with MPC–85 is approximately 10x greater than those prepared using sodium caseinate or whey protein concentrate. Also emulsions made with the large casein aggregates found in MPC85 and skim milk powder are more stable to creaming than those made with caseinate. It is believed that the protein aggregates in MPC may be too large to cause significant depletion flocculation, or alternatively, that they trigger a high degree of depletion flocculation and thus form very viscous emulsions, which will also reduce creaming (Kelly, 2011). (2) MPCs are the only dairy powder containing the same casein to whey proteins ratio as skimmed milk, but with considerably lower amounts of lactose, allowing manufacturers to include higher amounts of protein without the deleterious effects of excessive lactose such as browning and crystallization. (3) As the protein in MPC increases so does the calcium. (4) Replacement of rennet casein with MPC85 increased firmness and improved sliceability (resistance to cutting), but decreased the meltability of cheese spread analogue (Abou-El-Nour et al., 1998). And (5) Souza et al. (2014) prepared processed cheese with acid casein and MPC, WPC or a mixture of them; decreasing the pH from 6.2 to 5.6 either before or after cooking decreased the firmness of the PCP. WPC, relative to MPC, always produced the firmest cheese.

2.5.4 Coprecipitates

Co–precipitates, similar to total milk proteinate (Morr, 1985) are manufactured by precipitating caseins and whey proteins together using high temperatures (>60°C, mostly around 90°C) followed by acidification to low pH (4.6–5.9) with or without added calcium. Heat induces the denaturation of whey proteins and formation of some primary aggregates of WPs with themselves and with caseins. Acidification contributes to precipitation of caseins, denatured WPs and their complexes. Recovery of proteins and functionality of coprecipitates depend on temperature, pH, ionic strength, and the presence of calcium to contribute to aggregation and precipitation. The coprecipitate is then washed and dried in a similar fashion to acid caseins (Rollema and Muir, 2009). In general, coprecipitates are known for their low solubility and limited functionality, and therefore, they are not so available in the market (Kelly, 2011). Their use is mainly in bakery, meat and cereal based foods, as well as yogurt. Coprecipitates can have good surface active properties and heat stability (Bohoua-Guichard et al., 1997). And, emulsifying properties of heat-induced protein complexes formed by β-LG and κ-casein are better than the emulsifying capacity of either protein heated individually (Bohoua-Guichard et al., 1997; Grufferty, 1990).

When used in PCP, coprecipitates increase firmness and decrease meltability (Abou–El–Nour et al., 1996). PCP made with 8% co–precipitate at pH 9.5 had similar rheological properties and meltability to PCP made with acid caseins, but as the manufacturing pH of the coprecipitate decreased, the meltability also decreased (Mounsey et al., 2007). The levels of whey proteins in the co–precipitates increased as the pH of cooking increased from 3 to 7.5. Levels of coprecipitate up to 5% can be used in PCP without inducing browning (Thomas, 1969). When added to PCP made with natural cheese, 3% and 5% of coprecipitates improved the body of the PCP compared with samples with no or 1% of the co–precipitate; similar results were obtained with low or high calcium coprecipitates, but melting decreased, especially at the 5% level. Addition of coprecipitates reduces the pH of the PCP by up to 0.5 units (Thomas, 1970).

According to Thomas and Hyde (1972) ES in PCP could be reduced from 3.0% to 2.0%–2.5%, if calcium fortified coprecipitate was added at a level of 2%–3% to the processed cheese formulation. However, high levels (>3%) reduced the flowability of the PCP, especially when a high proportion of young cheese was used.

2.6 Quality of dairy protein powders

Utilization of dairy ingredients in liquid forms has the advantage of convenience; liquid ingredients can be pumped and measured precisely with volumetric meters. However, liquid ingredients typically have short shelf lives and are difficult and expensive to move over great distances. Accordingly, because of their long shelf–life and lower moisture content, protein powders are usually preferred as ingredients, including ingredients for PCP. The American Dairy Products Institute developed standards for different dairy protein powders (ADPI, 2020). International standards and definitions for different milk derived powders can be found at (FAO/WHO, 2011). Although these standards specify that only milk and milk–derived fractions can be used as raw materials for dairy derived powders, the standard also lists other additives such as stabilizers, emulsifiers, acidifiers, anticaking agents and antioxidants. This section considers some quality aspects of some dairy ingredients with particular focus on protein powders that may be used in PCP.

2.6.1 Composition

Table 2.5 shows proximate analysis of powdered dairy ingredients that are commonly used or may be used in PCPs. Data was obtained from different sources and aims to be representative of products in the market. Protein content is frequently reported as crude protein, which includes non–protein nitrogen (e.g., urea). True protein in milk is the total nitrogen minus the nonprotein nitrogen (NPN) (NPN is about 5.5% of total nitrogen) multiplied by the commonly used coefficient of 6.38 g protein per g nitrogen. Also, moisture in dairy protein powders is low, generally between 2–5% moisture; the exception is caseinates which can have moisture between 8%–12% (see Table 2.5). Consequently, composition in dairy powders is frequently reported on a dry basis. So, when formulating with protein powders, manufacturers must ensure that the protein and other nutrient calculations are performed with comparable data. Note also reviews of dairy protein ingredients (Guinee, 2009; Jensen and Nielsen, 1982; Tamime, 2009).

2.6.2 Microbial quality of ingredients

A detailed review of microbial quality of ingredients is beyond the scope of this chapter. Here we mention only two microbial concerns of particular interest to the use of milk powders in PCP. First, milk powders are known to contain endospores of thermophilic bacteria, particularly *Clostridium spp.* (Hill and Smythe, 2012) which cause spongy texture, gas, and bad aroma in processed cheese. Spores of *Clostridium spp.* can survive heating at 80°C for 30 min. Germination and growth of some *Clostridium spp.* in PCP can be reduced or prevented by inclusion of longchain polyphosphates as part of the ES blend (Loessner et al., 1997). Secondly, it is well known that *Salmonella spp* survive for extended periods of time in milk powders and various other dried products. See reviews (Podolak et al., 2010; Finn et al., 2013)

2.6.3 Whey protein nitrogen index

There is some level of heat treatment involved in the manufacturing of dairy powders. Heating not only helps to control the microbial quality but also imparts functionality during and after processing (Tamime, 2009). Therefore, skimmed milk powders are classified according to the heat treatment they receive; this classification is based on the denaturation of whey proteins caused by heat and measured by two different indices, the whey protein nitrogen index (WPNI) and the casein number (CN). WPNI measures the denatured whey proteins in the reconstituted SMP (Sharma et al., 2012b). Based on their WPNI, SMPs are classified as low heat <6; medium heat 4.5–5.9; medium high heat 1.5–4.5, and high heat <1.5 mg undenatured WPN/g powder (Sharma et al., 2012b), each of them with different functionalities. For other protein powders, the casein number, calculated as the proportion of nitrogen precipitated at pH 4.6 (in relation to total protein) can range within 80–82 in good quality milk while fully denatured milk has values in excess of 82. Because this value is calculated on the basis of the total proteins, the casein number can be used in many dairy powders with the understanding that casein number in this context simply means the percentage of total proteins that precipitate at pH 4.6.

2.6.4 Protease activity

Also of importance is the potential for residual plasmin, since this enzyme tends to associate with the caseins (Kelly and Fox, 2016). Caseins and caseinates, as well as whey and whey products can carry plasmin activity and this residual activity might affect the quality of PCP (Ismail and Nielsen, 2011; O'Malley et al., 2000). Li et al. (2020) found that Cheddar cheese made with MCC powders was affected by residual plasmin in the powder. In addition, residual rennet activity in rennet casein may induce physical-chemical changes during storage of PCP.

2.6.5 Nitrates and nitrites

An important consideration, when using dairy powders as ingredients, is the potential to contribute to the nitrate content of the product being manufactured. Although nitrate and nitrites are not allowed as dairy ingredients (FAO/WHO, 2011; USDA, 2018), they can be found naturally in milk (Genualdi et al., 2020) in very low levels (<1 mg/Kg) and also in dairy powders (Lakritz and Pensabene, 1981; Manning et al., 1968; Nygren, 1982). Dehydration of milk can produce nitrates and nitrites (Kelly et al., 1989), while nitrosamines can be produced as early as during pasteurization in the manufacturing process (Lakritz and Pensabene, 1981). During dehydration, N-nitrosodimethylamine (NDMA) forms mainly during indirect heating (Scanlan et al., 1994). According to (Indyk, 2011), concentrations of nitrate can go up to 50 mg/Kg in whey powders and up to 102 mg/kg in milk powders, while Yeh et al. (2013) found up to 312 mg/Kg of nitrates in certain milk powders. These amounts can add considerably to the amounts of nitrites and nitrate present in cheese; in Canada, for example, some cheese may contain, by regulation, up to 50 mg/kg of residual nitrites (Canada Minister of Justice 2020).

2.6.6 Storage stability

The relationship between solubility and other functional properties of dairy powders is not fully understood; Castro-Morel (2002) found no correlation between protein content, solubility, viscosity and emulsification capacity of commercial dairy protein powders with a wide range of protein content (MPCs 41.5%–90% protein). When comparing MPCs with similar high protein concentration (82%–86%), the authors observed a correlation among protein, pH, viscosity and solubility. However, it is recognized that crystallization of lactose during storage leads to caking and insolubility while fat crystallization can lead to agglomeration resulting in impaired functionality of dairy powders. The presence of oxygen and moisture leads to accelerated oxidation and browning during storage. Both lipid oxidation and Maillard reactions can lead to protein–protein interactions which are known to decrease the emulsifying capacity of protein powders (Thomas et al., 2004).

2.6.7 Solubility and reconstitution

For use in PCPs, the most important functional properties for protein ingredients, in general terms, are solubility, water holding capacity, gelling and emulsification capacity. Solubility or lack of sediment after rehydration, is a necessary primary functionality because protein particles need to be dispersed and/or solubilized to be able to function properly (Augustin and Udabage, 2007; Havea, 2009; Jimenez-Flores and Kosikowski, 1986), especially since denaturation of whey proteins is one of the main causes of poor solubility. Rehydration of powders can be divided into four steps: (1) wetting, water absorption, (2) sinking, (3) dispersing into separated discreet particles, and (4) dissolution, or separation of molecules in the water (Freudig et al., 1999). Proper dissolution, usually measured as solubility index, is the ultimate goal of a functional dairy powder. Therefore, when choosing dairy protein powders, manufacturers need to understand the solubility of the different ingredients and interpret the data provided by the ingredient manufacturer. A standard method for solubility determination has been developed by a joint committee from ISO/IDF (2005), this standard replaces that of the American Dry Milk Institute (ADPI, 2020)

2.6.8 Sensory quality

The sensory characteristics of dairy protein powders vary even within products of similar composition, showing that raw materials, processing conditions, and storage conditions all influence color, aroma and flavor of final products. Bleaching liquid whey proteins modifies the sensory profile of the whey and the flavor profile is transferred into the powder during drying (Croissant, et al., 2009). Perhaps the most important color issue in dairy powders is Maillard browning, which in turn is associated with aromas and flavors. Smith et al. (2016) found that burnt sugar flavor followed the course of Maillard reactions during storage of powders of MPC, MPI, caseins, and MCC.

Other workers have also reported Off–flavors in powdered dairy ingredients and some are carried forward into final products such as PCP (Drake et al., 1999; Evans et al., 2010; Wright et al., 2011). For example, WPC and SPC are characterized by sweet, aromatic cooked flavors, with SPC lacking the diacetyl flavour observed in WPC. Also, WPC is more likely to present cardboard flavor (Carunchia et al., 2005; Croissant et al., 2009; Evans et al., 2009 and 2010; Karagul et al., 2003), which has also been observed in MPCs and MCCs, especially in the powders versus the wet concentrates (Carter et al., 2016).

Another example is casein containing ingredients such as MPCs, MCIs, MCCs and acid and rennet caseins, which present sweet aromatic, cooked milk aromas, but also brothy, tortilla, soapy and fatty flavors (Smith et al., 2016). In general, acid and rennet caseins have higher aroma intensities, especially wet-dog/animal aroma (Karagul et al., 2003) . Whey protein products carry flavors derived from the cheese they came from, for example, diacetyl coming from fermentation by starter cultures. Other aromas in whey protein products come from lipid oxidation (Carunchia et al., 2005; Drake et al., 2014; Morr and Ha, 1993; Wright et al., 2006; Wright et al., 2009) and, to a lesser degree, protein oxidation; for example, degradation of sulfur–containing amino acids can result in dimethyl disulfide and dimethyl trisulfide, which are associated with garlic, cabbage or potato aromas (Wright et al., 2006).

2.7 Milk fat

2.7.1 Milk fat structure and composition

Milk fat is the most diverse of all natural fats with somewhere between 250 (Walstra et al., 2005) and 400 (Lindmark–Månsson, 2008) different fatty acids. The composition of milk fat varies with diet, season, and stage of lactation (Taylor and MacGibbon, 2011). Almost all milk fat is in the form of fat globules, which are 0.1 to 10 μm in size (Table 2.2) and much less dense than other components in milk. This makes milk fat easily separable by churning, centrifugation and MF. Gravity creaming can be used to separate larger globules comprising about 40% of the total fat in raw milk. The size of the fat globule is important, not only for the stability of the fat phase but because the size of the fat globule relates to its composition (Couvreur et al., 2007; Fauquant et al., 2005). The globule is stabilized in the aqueous plasma by a membrane of surface active materials (SAC).

The MFGM accounts for 2% to 6% of the total mass of the fat globules (Singh, 2006) and protects them from coalescence. The MFGM contains organized and complex layers of proteins, phospholipids, sphingolipids (up to 43% of the lipids in the membrane), glycoproteins, cholesterol, enzymes and other components. Conventionally, the MFGM is described as a tri–layer structure of thickness between 10–50 nm with the interior layer (facing the core of the globule) formed by proteins and polar lipids and an outer bilayer of polar lipids (Lopez, 2011). The quantitative distribution of the MFGM components, varies with the size of the fat globule. Also, the size and composition of the milk fat globule is important for the structure and characteristics of rennet and milk gels and therefore the quality of cheese (Michalski et al., 2002; Martini et al., 2016). When milk fat is separated and/or processed, its functionality changes.

The interior of the fat globules is mostly TAG which are 98% of the fat in milk. Approximately 2/3 of fatty acids in milk are saturated, with palmitic, myristic and stearic accounting for more than 50% of all fatty acids. Oleic acid (monounsaturated) represents 24% of the fatty acids and linoleic and linolenic acids are the main polyunsaturated fatty acids (Keenan and Mather, 2009). Milk also contains conjugated linoleic acid (CLA), which is associated with several health benefits (Spitsberg, 2005). In fact, the milk and meat of ruminants is the only significant dietary source of CLA. The content of CLA and the proportion of its isomers in milk depends on cow's age and breed, diet and season. Thermal processing may also affect CLA composition. Shantha et al. (1992) reported that CLA concentration in several PCPs ranged from 3.2 to 8.9 mg/g fat, and further that CLA increased during heating of PCP under atmospheric conditions, but not while processing under nitrogen. They also observed that addition of WPC increased CLA formation. Similarly, Zhang et al. (2006) observed increased CLA in both processed cheese food and processed cheese spread processed at 90°C but not at 75°C.

The organization of the fatty acids in the TAG molecule is of particular interest; short chain fatty acids like butyric, caproic and caprylic are located mainly in position sn-3, palmitic (C16) and medium chain (C12 and C14) saturated fatty acids are preferentially located in position sn-2 and long chain saturated and unsaturated tend to be located at positions sn-1 and sn-3. Long and unsaturated fatty acids each have preferential positions in the TAG molecule. Crystallization and melting behavior of milk fat is influenced by fatty acid composition, fatty acid positions within the TAG, polymorphism, and heating and cooling during processing (Lopez, 2011; Ramel and Marangoni, 2017).

2.7.2 Butter and butter oil

Butter is a water–in–oil emulsion where the discontinuous aqueous phase is droplets of milk serum, and the continuous oil phase is a mixture of globular (native) fat, semi–continuous networks of fat crystals, and liquid fat (Wright et al., 2011). Butter oil for practical purposes is 100% milk fat. The technologies of butter and butter oil manufacture are explained in many texts, for example, the Dairy Processing Handbook (Anonymous, 2015) and the Dairy Education E–Book (Goff, 2020). Briefly, butter is normally made by continuous flotation churning of cream and sometimes whey cream. Flotation churning means that cream (35%–55% fat) is cold aged (12–15°C) for optimum fat crystallization and agitated in the presence of air, which associates with fat globule membranes and helps destabilize the milk fat globules. Alternatively, butter can be made from plastic cream, which is warmed sweet cream concentrated to 80% fat by centrifugation. The 80% oil-in-water emulsion (cream) is inverted to water–in–oil (butter) by cooling. As the plastic cream is cooled, the oil begins to crystallize and ruptures the membranes of the densely packed fat globules, effectively inverting the emulsion. Surface active compounds (SAC) associated with the MFG membrane are retained in butter made by inversion of plastic cream, but are mostly lost in the butter milk during conventional flotation churning. A third not commonly used process for butter making is to blend butter oil (also known as AMF for anhydrous milk fat) with salt and water in a manner similar to margarine manufacture. Butter made from AMF contains no fat globule membrane materials.

Sweet cream butter (salted or unsalted), which is the most abundant in the market, has a higher pH and shorter shelf life than cultured butter, which is made from cream fermented with heterofermentative lactic bacteria resulting in an intense diacetyl flavor. Sweet whey typically contains 2–5 g/L of fat, which is normally removed by centrifugation to produce whey cream. Whey cream and whey cream products, such as whey butter can potentially be used in PCP. In the last two decades, whey butter, produced from whey cream has been looking for a place in the market due to the high amounts of whey being processed around the world. Bulk butter is usually packed in 25 kg polyethylene–lined cardboard cartons for refrigerated or frozen storage and distribution.

The Codex Alimentarius, and most standards around the world define butter as the product composed of at least 80% of fat all coming from milk (FAO/WHO, 2018a) while butter oil and AMF must have minimum fat contents of 99.6 and 99.8% respectively (FAO/WHO, 2018b). For the manufacture of butter oil, the butter is melted and centrifuged, then vacuum dried to obtain fat values higher than 99.5%, and, finally, cooled before packaging (FAO/WHO, 2018b). During the process oxygen is removed, which together with packing in nitrogen filled bulk drums (mainly 200 L) extends the shelf life of the product by protecting it from oxidative rancidity. Butter oil can also be manufactured from cream, by use of a hermetic solids–ejecting separator which destabilizes the fat globules and separates the residual serum (Goff, 2020).

The quality of milk fat ingredients, especially lipolytic and oxidative rancidity must be considered. It is known, for example, that heating accelerates oxidation of fats. Van Ruth et al. (2008) found that the peroxide value (POV) and free fatty acid (FFA) levels resulting from heating were similar for butter and butter oil. The Codex Alimentarius (FAO/WHO, 2018a) recommends maximum free fatty acids (FFA) of 0.3% and 0.4% (of free oleic acid) and a maximum POV of 0.3 and 0.6 milliequivalents of oxygen kg^{-1} fat for AMF/ABO and MF/Butter oil, respectively. Because iron and copper act as catalysts for lipid oxidation resulting in fishy/metallic or oily flavor, the same code recommends a maximum of 0.05 mg kg^{-1} of copper and 0.2 mg kg^{-1} of iron in milk fats of good quality.

2.7.3 Milk fat in PCP

When comparing dairy fat products, Jinjarak et al., (2006) found that whey butter had lower penetration force (softer) than sweet butter, but panelists were not able to distinguish the firmness of the different butters (sweet, cultured and whey) with the exception of a higher porosity of whey butter. Contrary to texture, the flavor differences between sweet, cultured, and whey butter are distinct. However, in general, studies have shown that texture and flavor profiles, and color are more influenced by processing parameters than by type of dairy fat (sweet butter, cultured butter, whey butter, butter oil or AMF) (Jinjarak et al., 2006; Schieberle et al., 1993). With respect to flavor profiles, this may be due in part to release of volatiles caused by heating and agitation during processing. Regarding color, PCP color is normally standardized with one or more carotenoid colorants, such as β-Apo-8-carotenal.

In regards to texture differences affected by different fat sources, an important consideration is that high shear and high temperature processing disrupts the natural globular structure and creates new globules of smaller sizes stabilized by the protein network (Purna et al., 2006). The size and stability of the newly formed globules depend on several factors, including large molecule emulsifiers like proteins (e.g., caseins, denatured proteins), fatty acid composition, added emulsifying salts, shear, temperature and viscosity of the dispersed phase. As emulsification improves, the surface area of the now smaller globules increases, and, being surrounded by a new protein "membrane," the fat globules participate in protein to protein interactions, thus contributing to the structure and texture of the protein network. In other words, many of the distinctive properties of the added dairy fats are erased during PCP processing. For example, Cerníková et al., (2017) found that substituting almost all the fat in natural cheese (used as raw material) with butter had no effect on the viscoelastic properties of model PCP. In this vein it's interesting that Ramel and Marangoni (2017) who studied polymorphism in PCP, concluded that although milk fat because of its complexity is rarely found in the most stable β polymorph, this form does occur in processed cheese. However, they also observed that polymorphism does not affect rheological properties of PCP.

2.7.4 Low fat PCP

Decreasing fat in PCP leads to excessive firm, rubbery textures with reduced meltability and other textural defects (Drake et al., 1999; Swenson et al., 2000) and decreased consumer acceptance (Muir et al., 1997). In general terms, several strategies are used to minimize these defects.

Substitution of natural cheese with low fat natural cheese is often the first option used to reduce fat in PCP (Caric and Kalab, 1999). However, manufacturing low fat cheeses is challenging and requires important changes in the processing parameters which result in secondary effects on the quality of PCP made from it (Banks, 2004; Johnson et al., 2009b).

Reducing fat from sources other than cheese, depending on the ingredients originally used in the PCP, may achieve up to 25% reduction. More than 25% fat reduction typically requires the use of reduced fat natural cheese (Johnson et al., 2009b).

Replace protein with water. Fat has a smaller influence on the firmness and meltability of processed cheese than protein. Accordingly, increasing moisture while keeping the ratio of fat to protein constant, substantially decreases firmness of PCP (Lee et al., 2004; Pereira et al., 2001 and 2002; Ramos et al., 2015). In contrast, lowering the fat to protein ratio produces a denser protein network resulting in increased firmness and decreased meltability (El–Assar et al., 2019; Guinee and O'Callaghan, 2013; Lee, 2015a). While commercial PCP may contain between 40 to 60% of moisture, low fat PCP may contain up to 73% moisture (Lee et al., 2004).

Use a nondairy ingredient. El–Shibmy et al., (2007) were able to reduce fat in PCP by 50% by replacing Ras cheese with skimmed rennet curd plus rice powder while maintaining sensory scores similar to the full fat PCP; substitution with acid curd produced lower sensory scores.

2.8 Lactose and other solids

Lactose is the major solid (and the only carbohydrate) in milk. It's a reducing disaccharide with relatively low solubility (about 200 g L^{-1} at 20°C). The combination of free glucose and galactose resulting from lactase activity, is sweeter than lactose, a fact that needs to be considered when formulating lactose reduced or free products.

With the advent of membrane processing, which made it possible to separate lactose from milk proteins, ultrafiltered milk and whey permeate replaced whey as the primary BOD yielding waste stream in the dairy industry. It is, therefore important to find profitable uses for lactose. As listed in Table 2.5, lactose containing products include milk and whey powders, milk and whey protein concentrates, powdered milk, and UF whey permeate. Lactose in these forms is mainly used as a filler. Essentially, further processors select milk or whey protein concentrates with protein to lactose ratios suitable for particular products. Pure lactose does have some value in further processing, for example, in pharmaceuticals. The global value of the combined markets for lactose, pharmaceutical lactose and permeate powder is estimated to be US $1.9 billion in 2020 (Affertsholt and Pedersen, 2017). Of course, lactose is also critically important as the substrate for lactic acid bacteria in all fermented dairy products.

The main limitation of adding lactose to PCP is that it causes nonenzymatic browning during cooking and storage of PCP (Bley et al., 1985; Kombila–Moundounga and Lacroix, 1991; Thomas, 1969), which leads to color defects, but also undesirable flavours and aromas. High pH (>5.7) and high processing and storage temperatures contribute to browning. Excess lactose may also crystalize during storage

(Kombila–Moundounga and Lacroix, 1991), which together with crystals that may form from salts and amino acids can result in sandy texture. Hydrolysis of lactose can prevent such crystallization; when lactose was enzymatically hydrolysed prior to manufacturing of a whey–buttermilk spread, reduction of lactose to levels under 12.5% eliminated crystallization (Patocka and Jelen, 1988). Note that: (1) the critical value with respect to crystallization is the concentration of lactose as a percentage of the moisture in the PCP, (2) PCP is often stored under refrigeration after opening or may be exposed to lower temperature during transport, and (3) solubility decreases in the presence of salts (Choscz et al., 2019). An industry rule of thumb is lactose as a percentage of PCP moisture should not exceed 17. That is, for a PCP containing 44% moisture lactose must not exceed 7.5% (approximately depending on composition) or 10% for a PCP with 60% moisture (Kapoor and Metzger, 2008b).

Lactose in PCP is mainly added in the form of protein ingredients, especially MPC and WPC. It can also be added directly, but either way it's added as a filler. Model PCP containing 1% lactose had reduced firmness relative to controls; the same reduction was found for galactose and saccharose (Macků et al., 2009), suggesting that the effect will be the same if the sugar is coming from cultured (e.g., cheese) or uncultured (e.g., rennet casein) sources.

Regarding other solids, milk mineral powders, consisting mainly of calcium phosphate salts are available (See Table 2.5). However, it seems likely their use in PCP would tend to impair emulsification, although we did not find any literature reports on this.

2.9 Cheese bases

A cheese base (CB), also called precheese, is a product obtained by ultrafiltration (UF) and diafiltration (DF) of milk standardized to obtain the targeted protein and fat concentration of a certain cheese. The goal of using cheese bases in processed cheese is to reduce the cost of using natural cheese and increase yield efficiency by incorporating all the whey proteins into the cheese. The proximate chemical composition of a CB must be similar to that of the natural cheese being created or replaced in a PCP (Maubois and Mocquot, 1975). The use of CB in Cheddar processed cheese has been reported by (Kycia, et al., 2006; Maubois and Mocquot, 1975; Rubin and Bjerre, 1983; Tamime et al., 1990). Cheese bases made from the milk of other dairy species such as sheep milk have also been explored. Some examples of cheese bases used in PCP are in Table 2.8.

2.9.1 Cheese bases versus conventional cheese

Preparation of CB by ultrafiltration or evaporation retains whey proteins representing up to 20% of the total proteins and increased cheese yields between 16%–20% (Maubois and Mocquot, 1975; Rubin and Bjerre, 1983).Whey proteins confer flow resistance and decrease stretchability of cheese upon heating, but these effects depend on the level of denaturation of the whey proteins. Other advantages are that liquid

Table 2.8 Examples of cheese bases used in processed cheese products.

Product	Effect	Reference
Block Type processed Cheese from CB (4.5% to 42.5%) made using reconstituted ultrafiltered and diafiltered SMP. AMF as fat source.	Cheese base produced: firmer PCP and changes in the microstructure with some crumbliness and graininess on some of the treatments (high CB + enzymes). α_{s1}–caseins in CB PCP were mostly intact in contrast with those of the Cheddar cheese	Younis and Davies (1991), Tamime et al. (1990), Abdou et al. (1991 a,b'), Younis et al. (1991)
A "blue cheese food" made with skimmed milk CB, butter or cococunt oil, lipase and mold spores.	A food product with the flavour of blue cheese that can be used as a flavouring ingredient for other food products	Jolly and Kosikowski (1975)
Mozzarella style PC made with plastic cream and a cheese base made with ultrafiltered skim milk or diafiltered and acidified skim milk .	Cheese base made with UF slowed down fermentation and pH decrease and had poor stretchablity. Diafiltration combined with acidification produced CB that resulted in better stretchability and sensory profile. Yield was good (8.4 Kg of skim milk per Kg of PC).	Covacevich and Kosikowski (1978)
Cheese base by UF and DF, acidified or not to obtain proper lactose/buffer capacity ratios, evaporated to achieve Cheddar cheese composition.	Milk yield increased 16%–18% compared with natural Cheddar cheese yield. When used to replace young Cheddar (80% base/20% old Cheddar) in PC and PCF, cheese base produced firmer PC and good flavor in both PC and PCF.	Ernstrom (1980)
A solid matrix (CB) made out of milk powders (SMP, MPC), NaCl, fat and water, preservatives, and acidifiers	The patent claims that cheese base can be used for processed cheese (up to 20% cheese base)	Hyde et al. (2005)
A cheese base made by UF (approx. 2X) and DF of skim milk, fermentation and adjusting ionic strength with NaCl to prevent coagulation during further concentration by evaporation	Increased yield of 13%–16%. Adjusted buffer capacity/lactose. The cheese base did not coagulate during processing, all WP was retained, and the macro composition of the CB and natural cheese were similar.	Jameson and Sutherland (1982)

(*continued on next page*)

Table 2.8 Examples of cheese bases used in processed cheese products—cont'd

Product	Effect	Reference
Whole milk concentrated by UF fermented with yogurt starter	Cheese base can replace up to 30% of the process cheese ingredients while retaining physicochemical, sensory, and textural characteristics. Higher amounts of CB increased firmness of PCP. Conclusion is UF and fermented CB can replace young cheese in PCP.	Kycia et al. (2006)
A patent for cheese base prepared with milk (skimmed or full fat) ultrafiltered, acidified, diafiltered and then evaporated	Patent claims that CB can substitute up to 80% of cheese in PCP	Rubin and Bjerre (1983)
A patent claiming to produce cheese base without diafiltration, acidifying before or after UF with different acids	Patent claims to produe good PC (slices or spread) with up to 100% cheese substitution by CB. Citric and lactic acid worked better as acidifiers	Simbuerger et al. (1997)

CB can be stored frozen and therefore shelf life can be extended considerably. Also, there is no need to chop or grate, so CB may be easier to handle than conventional cheese (CC). Another interesting property is that while heating milk at native casein concentration prevents subsequent rennet coagulation, rennetability of caseins is restored by acidification, addition of $CaCl_2$, or concentration by any means such as UF, MF, evaporation, or reverse osmosis. This suggests that CB could be treated at high temperatures (>110°C) to destroy microorganisms, including spore forming bacteria (Maubois et al., 1975).

The buffer capacity of milk increases during UF due to concentration of proteins (especially caseins), calcium phosphate salts associated with the caseins, and colloidal calcium phosphates. Accordingly, lactic acid fermentation or acidification with organic acids may be carried out before or during UF to reduce colloidal calcium and magnesium phosphates, and prevent excessively increased buffer capacity (Brule et al., 1974b). If the buffering capacity is not controlled, more lactic acid is required to achieve the target pH. For example, (Brule et al., 1974b) determined the amount of lactic acid required to achieve the target pH in milk ultrafiltrates. Fermentation with selected cultures decreases lactose and colloidal calcium, favors a more pronounced flavor and desirable texture, and increases the ratio of water soluble nitrogen to total nitrogen (WSN/TN) (El–Garhi et al., 2018). Brickley et al. (2008) found that acidification of the cheese base to pH 5.2 to 5.4 produced good melting properties while those at pH 5 had significantly lower meltability and increased firmness. Poor control of the fermentation, amount of lactose, and colloidal calcium phosphate may lead to cheese bases with flavor defects (too flat or too acidic) and improper texture like sandiness, firmness, and crumbliness. The lactose to buffer capacity ratio can also be controlled by the combination of ultrafiltration and diafiltration (Corredig et al., 2019; Jameson and Sutherland, 1982).

After membrane filtration and fermentation, vacuum evaporation can be used to remove excess moisture to reach composition similar to natural cheese (Ernstrom et al., 1980; Jameson and Sutherland, 1994; Moran et al., 1989 and 1993; Tamime et al., 1990). Vacuum evaporation can also be used in preparation for spray drying to produce CB powders. This is useful where long term storage or long distance transport is needed such as export to countries where fluid milk availability is limited (Mistry and Maubois, 2017).

2.9.2 Cheese bases in processed cheese products

The earliest attempts to produce processed cheese with a cheese base were in the 1970s (Covacevich and Kosikowski, 1978; Jolly and Kosikowski, 1975). Sood and Kosikowski (1979) were able to produce processed cheese with acceptable flavor and texture by replacing up to 60% of aged Cheddar cheese with an enzyme–treated cheese base containing 30 percent solids. Moran *et al.,* (1993) patented cheese bases made by fermenting UF milk retentate to pH 4.8–5.6, adding a nonclotting amount of coagulating enzyme, and evaporating to minimum solids of 55%. Small amounts of sodium chloride were used to prevent coagulation of the retentates. Jameson and Sutherland (1982) claimed that other salts like potassium chloride, buffered phosphate, citrates and mixtures of such salts can be used to control coagulation through ionic strength. The addition of NaCl (0.5%–0.9%) also contributes to the release of colloidal calcium through ionization of casein phosphoserine groups (Brule et al., 1974a,b) and the release of soluble proteins from the micelles (Corredig et al., 2019).

The use of powders for manufacture of CB has been explored. Hyde et al. (2010) patented a precheese made from milk powders mixed with sodium chloride, milk fat, water and preservatives or acids which after cooling forms a solid matter that can be consumed or ground to be used in processed cheese. The authors claim that this mass has the texture and consistency of processed cheese. In addition, Simbuerger et al. (1997) patented a cheese base made by acidification to pH 3.8–4.6 of a UF retentate containing up to 36.6% dry matter. The authors claim a process cheese can be obtained substituting cheese with up to 100% of this base. The same authors claimed that using other milk derived powder to increase the dry matter and fat content can lead to the production of good processed cheese spreads. The use of powders to manufacture CB can be useful in countries with low availability of fresh milk.

The use of evaporation alone to prepare CB (as opposed to UF cheese base) is limited because of the high levels of lactose and ions in the resulting base. Evaporation also produces a CB with poor melting properties (Acharya and Mistry, 2004, 2005). The soluble caseins and calcium in the serum phase shift to the colloidal phase as a function of concentration (Corredig et al., 2019) as opposed to what occurs during diafiltration (Ferrer et al., 2014).

When substituting natural cheese with cheese bases, the age of the natural cheese being replaced needs to be considered carefully. As natural cheese ages, hydrolysis of caseins decreases the amount of intact caseins; the result is that un–melted firmness of the PCP decreases (Černíková et al., 2017; Piska and Štětina, 2004; Purna et al., 2006; Templeton and Sommer, 1930). This observation suggests that cheese bases can substitute for younger cheese in processed cheese formulas. Ernstrom et al., 1980 used CB to substitute up to 80% of the natural cheese (80% CB to 20% aged cheese) producing a satisfactory process cheese food, and a good flavored but excessively firm processed cheese. Other workers, Ganguli (1991) and Kycia et al. (2006), respectively, report successful replacement of up to 80 and 70 g 100 g^{-1} of CC with CB.

2.10 Summary and new directions

In most jurisdictions, regulations governing permitted ingredients, within the broad category of PCPs, include formulations with little or no requirement for natural cheese. This makes it possible for processors to select from a growing diversity of dairy ingredients to create PCP formulae that optimize cost efficiency within quality (flavor, texture, appearance) and other objectives such as waste and BOD reduction. This suggests several lines of continuing development.

Native or minimally processed ingredients have potential to reduce waste and preserve native functionality of milk components. A good example is separation of whey proteins prior to lactic fermentation to produce Greek yogurt or any of the lactic cheeses such as cottage cheese or cream cheese. This would avoid the ongoing challenge of recovering high quality whey proteins from acid whey. Another example is micellar casein concentrates (MCC).

Dairy byproducts can potentially be better utilized in PCP. Kapoor and Metzger (2004) observed no differences between control PCP and PCP formulated with salty whey to replace added moisture and salt. Another example is whey cream or whey butter. Buttermilk including whey butter milk is interesting because its emulsification capacity is often compromised by the high temperatures (>80°C) used to pasteurize cream before butter manufacture (Corredig and Dalgleish, 1998). However, in PCP other ingredients could be used to balance emulsification capacity.

Clean labels is another driver for dairy ingredient development. Various workers have experimented with dairy ingredients with potential to replace all or part of ES in PCP. For examples: (1) blends of buttermilk and sodium caseinate (Kelimu et al., 2017), (2) denatured whey proteins (Nastaj et al., 2019; Sołowiej et al., 2020), (3) MCC (Hamman and Metzger, 2019), (4) hydrolyzed caseins (Kwak et al., 2002), (5) calcium reduced ingredients (Smith and Rivera, 2008), and (6) calcium fortified coprecipitates (Thomas, 1969).

Recombinant milk proteins will likely become a substantial commercial reality in the dairy ingredient marketplace during the next decade. It seems likely, pending cost and regulatory considerations, that recombinant milk proteins in the form of caseinates will find their way into PCP formulae. It is known that β–casein spontaneously forms micelles, but at this time the possibility of recombining several recombinant caseins into micelles with functionalities similar to native casein micelles is speculative.

References

Abd El–Salam, M., El–Shibiny, S., Salem, A., 2009. Factors affecting the functional properties of whey protein products: A review. Food Rev. Int. 25 (3), 251–270.

Abdou, S.M., Younis, M.F., Tamime, A.Y., DavieS, G., Hunter, E.A, Dawood, A.H., 1991a. Production of processed cheese using cheddar cheese and cheese base. Milchwissenschaft 46, 566–569.

Abdou, S.M., Younis, M.F., Tamime, A.Y., Davies, G., Hunter, E.A., Dawood, A.H., 1991b. production of processed cheese using Cheddar cheese and cheese base. 4. microbiological and organoleptic qualities. Milchwissenschaft 46, 645–664.

Abou–El–Nour, A.M., Schenrer, G.J., Omar, M.M., Buchheim, W., 1996. Physico–chemical and rheological properties of block–type processed cheese analogue made from rennet casein and total milk protein. Milchwissenschaft 51 (12), 684–687.

Acharya, M.R., Mistry, V.V., 2004. Comparison of effect of vacuum–condensed and ultrafiltered milk on cheddar cheese. J. Dairy Sci 87 (12), 4004–4012.

Acharya, M.R., Mistry, V.V., 2005. Effect of vacuum–condensed or ultrafiltered milk on pasteurized process cheese. J. Dairy Sci. 88 (9), 3037–3043.

ADPI, 2020. Dairy Products. American Dairy Products Institute https://www.adpi.org/DairyProducts/tabid/62/Default.aspx.

Affertsholt, T., Pedersen, D., 2017. Whey & lactose: market fueled by high protein, The world of food ingredients, 26–28. https://www.3abc.dk/wp–content/uploads/2017/06/Whey–and–Lactose–Market–Fueled–by–High–Protein.pdf. (Accessed 15 July 2020).

Amelia, I., Barbano, D.M., 2013. Production of an 18% protein liquid micellar casein concentrate with a long refrigerated shelf life. J. Dairy Sci. 96 (5), 3340–3349.

Amigo, L., Fontecha, J., 2011. Milk: Goat milk. In: Fucquay, J.W., Fox, P.F., McSeewney, P.H.L. (Eds.), Encyclopedia of Dairy Sciences. Elsevier, pp. 484–493.

Anema, S.G., Pinder, D.N., Hunter, R.J., Hemar, Y., 2006. Effects of storage temperature on the solubility of milk protein concentrate (MPC85). Food Hydrocoll. 20 (2–3 Spec. Iss.), 386–393.

Anonymous, 2015. Dairy processing handbook. Revision 1. Tetra Pak® Processing Systems, Lund, Sweden. https://dairyprocessinghandbook.tetrapak.com/.

Augustin, M.A., Udabage, P., 2007. Influence of processing on functionality of milk and dairy proteins. Adv. Food. Nutr. Res. 53 (07), 1–38.

Azuma, N., Nara, K., Kanno, C., 1991. Enzymic modification of αsl–casein with peptidylarginine deiminase: preparation of less acid–coagulable and less calcium–sensitive casein. J. Dairy Res. 58 (4), 421–429.

Banks, J.M., 2004. The technology of low–fat cheese manufacture. Int. J. Dairy Technol. 57 (4), 199–207.

Bastier, P., Dumay, E., Cheftel, J.C., 1993. Physico–chemical and functional properties of commercial caseinates. LWT – Food Sci. Techno. 26 (6), 529–537.

Berger, W., Klostermeyer, H., Merkinich, K., Uhimann, G., 2002. Processed Cheese Manufacture: A JOHA(R) Guide. Ed. H. Klostermeyer. Ladenburg, Germany: BK Giulini Chemie.

Bernal, V., Jelen, P., 1985. Thermal stability of whey proteins – a calorimetric study. J. Dairy Sci. 68 (11), 2847–2852.

Berta, M., Muskens, E., Schuster, E., Stading, M., 2016. Rheology of natural and imitation mozzarella cheese at conditions relevant to pizza baking. Int. Dairy J. 57, 34–38.

Blaschek, K.M., Wendoff, W.L., Rankin, S.A., 2007. Survey of salty and sweet whey composition from various cheese plants in Wisconsin. J. Dairy Sci. 90 (4), 2029–2034.

Bley, M.E., Johnson, M.E., Olson, N.F., 1985. Factors affecting nonenzymatic browning of process cheese. J. Dairy Sci. 68 (3), 555–561.

Bohoua–Guichard, L., Haque, Z., Gnakri, D., Kamenan, A., 1997. Effect of the relative proportion of κ–casein to β–lactoglobulin on food functionality of their complex. Sciences de Aliments 17 (6), 671–678.

Brickley, C.A., Auty, M.A.E., Piraino, P., McSweeney, P.L.H., 2007. The effect of natural cheddar cheese ripening on the functional and textural properties of the processed cheese manufactured therefrom. J. Food Sci. 72 (9), C483–C490.

Brickley, C.A., Govindasamy–Lucey, S., Jaeggi, J.J., Johnson, M.E., McSweeney, P.L.H., Lucey, J.A, 2008. Influence of emulsifying salts on the textural properties of nonfat process cheese made from direct acid cheese bases. J. Dairy Sci. 91 (1), 39–48.

Brule, G., Maubois, J., 1974b. Étude de la Teneur en Eléments Minéraux des Produits Obtenus Lors de l'Ultrafiltration du Lait sur Membrane. Le Lait 54 (539–540), 600–615.

Brule, G., Maubous, J.L., Vandeweghe, J., Fauqant, J., Goudendranche, H., 1974a. Ultilisation de l'ultrafiltration sur d emembrane pour la fabrication de fromages de type pate fraiches. Lait 54, 600–615.

Çakır–Fuller, E., 2015. Enhanced heat stability of high protein emulsion systems provided by microparticulated whey proteins. Food Hydrocoll. 47, 41–50.

Canada Minister of Justice, 2020. Food and drug regulations. Part B. Division 16. Table XI, Part 1. Food Additives That May Be Used as Class I Preservatives. https://laws–lois.justice.gc.ca/eng/regulations/C.R.C.,_c._870/page–74.html#h–573289. Accessed November 9, 2021.

Caric, M., Kalab, M., 1999. In Fox, P.F. (ed.) Cheese: Chemistry, Physics and Microbiology. 2nd edn. vol. 2, Aspen, Gaithersburg, pp.467–505.

Carr, A., Golding, M., 2016. Functional milk proteins. Production and utilization: casein–based ingredients. In: McSweeney, P.L.H., O'Mahony, S.A. (Eds.), Advanced Dairy Chemistry, vol. 1 Proteins, 4th edn, 1. Springer Publishers, NY, New York, pp. 35–66.

Carter, B., Patel, H., Barbano, D.M., Drake, M., 2016. 0710 Characterization of flavor and functional properties of liquid and dried WPC 80, WPI, MPC 85, and micellar casein concentrates. J. Anim. Sci. 94 (s5), 340.

Carunchia–Whetstine, M.E., Croissant, A.E., Drake, M.A., 2005. Characterization of dried whey protein concentrate and isolate flavor. J. Dairy Sci. 88 (11), 3826–3839.

Castro–Morel, W.H., 2002. Basic functionality of commercial milk protein concentrates. Milchwissenschaft 57, 367–370.

Černíková, M., Nebesářová, J., Salek, R.N., Řiháčková, L., Buňka, F., 2017. Microstructure and textural and viscoelastic properties of model processed cheese with different dry matter and fat in dry matter content. J. Dairy Sci. 100 (6), 4300–4307.

Chandrapala, J., Chen, G.Q., Kezia, K., Bowman, E.G., Vasiljevic, T., Kentish, S.E., 2016a. Removal of lactate from acid whey using nanofiltration. J. Food Eng. 177, 59–64.

Chandrapala, J., Duke, M.C., Gray, S.R., Weeks, M., Palmer, M., Vasiljevic, T., 2016b. Nanofiltration and nanodiafiltration of acid whey as a function of pH and temperature. Sep. Purif. Technol. 160, 18–27.

Choscz, C., Held, C., Eder, C., Sadowski, G., Briesen, H., 2019. Measurement and modeling of lactose solubility in aqueous electrolyte solutions. Ind. Eng. Chem. Res. 58 (45), 20797–20805.

Christie, W.W., Noble, R.C., Davies, G., 1987. Phospholipids in milk and dairy products. J. Soc. Dairy Technol 40, 10–12.

Coggins, J.S., 1991. Predicting cheddar cheese yield in an individual plant: Van Slyke revisited. J. Dairy Sci. 74 (2), 359–368.

Corredig, M., Nair, P.K., Li, Y., Eshpari, H., Zhao, Z., 2019. Invited review: understanding the behavior of caseins in milk concentrates. J. Dairy Sci. 102 (6), 4772–4782.

Couvreur, S., Hurtaud, C., Marnet, P.G., Faverdin, P., Peyraud, J.L., 2007. Composition of milk fat from cows selected for milk fat globule size and offered either fresh pasture or a corn silage–based diet. J. Dairy Sci. 90 (1), 392–403.

Covacevich, H.R., Kosikowski, F.V., 1978. Mozzarella and cheddar cheese manufacture by ultrafiltration principles. J. Dairy Sci. 61 (6), 701–709.

Croissant, A.E., Kang, E.J., Campbell, R.E., Bastian, E., Drake, M.A., 2009. The effect of bleaching agent on the flavor of liquid whey and whey protein concentrate. J. Dairy Sci. 92 (12), 5917–5927.

Crowley, S.V., Desautel, B., Gazi, I., Kelly, A.L., Huppertz, T., O'Mahony, J.A., 2015. Re-hydration characteristics of milk protein concentrate powders. J. Food Eng. 149, 105–113.

Crowley, S.V., Megemont, M., Gazi, I., Kelly, A., Huppertz, T., O'Mahony, J.A., 2014. Heat stability of reconstituted milk protein concentrate powders. Int. Dairy J. 37 (2), 104–110.

Crowley, S.V., O'Mahony, J.A., Fox, P.F., 2018. The proteins of milk. In: Van Benzel, N. (Ed.), Achieving Sustainable Production of Milk, 1. Burleigh Dodds Science Publishing, London, pp. 3–62.

De Kruif, C.G., Holt, C., 2003. Casein micelle structure, functions and interactions. In: Fox, P.F., McSweeney, P.L.H. (Eds.), Advanced Dairy Chemistry—1 Proteins. Springer US, Boston, MA, pp. 233–276.

De Kruif, C.G., Huppertz, T., Urban, V.S., Petukhov, A.V., 2012. Casein micelles and their internal structure. Adv. Colloid Interface Sci. 171–172, 36–52.

De Wit, J., Klarenbeek, G., Adamse, M., 1986. Evaluation of functional properties of whey protein concentrates and whey protein isolates. 2. Effects of processing history and composition. Neth. Milk Dairy J. 40 (1), 41–56.

De Wit, J.N., 1988. Empirical observations and thermodynamical considerations on water-binding by whey proteins in food products. J. Food Sci. 53 (5), 1553–1559.

DePeters, E.J., Ferguson, J.D, 1992. Nonprotein nitrogen and protein distribution in the milk of cows. J. Dairy Sci. 75 (11), 3192–3209.

Desouky, M.M., Salama, H.H., El–Sayed, S.M., 2019. The effects of camel milk powder on the stability and quality properties of processed cheese sauce. Acta Sci. Pol. Technol. Aliment. 18 (4), 349–359.

Dhanraj, P., Jana, A., Modha, H., Aparnathi, K.D., 2017. Influence of using a blend of rennet casein and whey protein concentrate as protein source on the quality of Mozzarella cheese analogue. J. Food Sci. Technol. 54 (3), 822–831.

Dickinson, E., 1998. Proteins at interfaces and in emulsions; stability, rheology and interactions. J. Chem. Soc. 94 (12), 1657–1669.

Dissanayake, M., Liyanaarachchi, S., Vasiljevic, T., 2012. Functional properties of whey proteins microparticulated at low pH. J. Dairy Sci. 95 (4), 1667–1679.

Dissanayake, M., Vasiljevic, T., 2009. Functional properties of whey proteins affected by heat treatment and hydrodynamic high–pressure shearing. J. Dairy Sci. 92 (4), 1387–1397.

Doosh, K., Alhusyne, L.A., Almosawi, B.N., 2014. Utilization of concentrated buttermilk in functional processed cheese manufacturing and studying some of its physicochemica properties. Pak. J. Nutr. 13 (1), 33–37.

Drake, M.A., Miracle, R.E., Wright, J.M., 2014. Sensory properties of dairy proteins. In: Thompson, A., Boland, M., Singh, H. (Eds.), Milk Proteins: From Expression to Food. 2nd edn. Elsevier, San Diego, California, pp. 473–492.

Drake, M.A., Truong, V.D., Daubert, C.R., 1999. Rheological and sensory properties of reduced–fat processed cheeses containing lecithin. J. Food Sci. 64 (4), 744–747.

Edwards, P.B., Creamer, L.K., Jameson, G.B., 2014. Structure and stability of whey proteins. In: Thompson, A., Boland, M., Singh, H. (Eds.), Milk Proteins: From Expression to Food. 2nd edn. Elsevier, Sand Diego, CA, pp. 202–242.

Eigel, W.N., Butler, J.E., Ernstrom, C.A., Farrell, H.M., Harwalker, V.R., Jenness, R., Whitney, R.M., 1984. Nomenclature of proteins of cow's milk: fifth revision. J. Dairy Sci. 67 (8), 1599–1631.

El–Assar, M., Abou–Dawood, S.A., Sakr, S.S., Younis, N.M., 2019. Low–fat processed cheese spread with added inulin: Its physicochemical, rheological and sensory characteristics. Int. J. Dairy Sci. 14 (1), 12–20.

El–Garhi, H.E.M., El–Aidie, S.A.M., Rashid, N.A., Hayee, Z.A., 2018. Quality improvement of spreadable processed cheese made from ultrafiltered milk retentates using commercial starter cultures. Food Sci. Technol. Int. 24 (6), 465–475.

El–Nour, A., A.M., S., J., G., Bucheim, W., 1998. Use of rennet casein and milk protein concentrates in the production of spread type processed cheese analogue. Milchwissenschaft 53 (12), 685–690.

El–Shibmy, S., Shenana, M.E., El–Nagar, G.F., Abdou, S.M., 2007. Preparation and properties of low fat processed cheese spreads. Int. J. Dairy Sci. 2 (1), 13–22.

Ennis, M.P., Mulvihill, D.M., 1999. Compositional characteristics of rennet caseins and hydration characteristics of the caseins in a model system as indicators of performance in Mozzarella cheese analogue manufacture. Food Hydrocoll. 13 (4), 325–337.

Ernstrom, C.A., Sutherland, B.J., Jameson, G.W., 1980. Cheese base for processing. A high yield product from whole milk by ultrafiltration. J. Dairy Sci. 63 (2), 228–234.

Eshpari, H., Tong, P.S., Corredig, M., 2014. Changes in the physical properties, solubility, and heat stability of milk protein concentrates prepared from partially acidified milk. J. Dairy Sci. 97 (12), 7394–7401.

Gobierno de Espana, Ministerio de la Presidencia, 2006. Real Decreto 1113/2006, del 29 de Septiembre, por el que se aprueban las normas de calidad para quesos y quesos fundidos. Boletin Oficial del Estado 236, 34717–34720.

Evans, J., Zulewska, J., Newbold, M., Drake, M.A., Barbano, M., 2009. Comparison of composition, sensory, and volatile components of thirty–four percent whey protein and milk serum protein concentrates. J. Dairy Sci. 92 (10), 4773–4791.

Evans, J., Zulewska, J., Newbold, M., Drake, M.A., Barbano, M., 2010. Comparison of composition and sensory properties of 80% whey protein and milk serum protein concentrates. J. Dairy Sci. 93 (5), 1824–1843.

Fang, Y., Selomulya, C., Ainsworth, S., Palmer, M., Chen, X.D., 2011. On quantifying the dissolution behaviour of milk protein concentrate. Food Hydrocoll. 25 (3), 503–510.

FAO/WHO, 2011. Milk and milk products. Joint FAO/WHO Codex Alimentarius Commission http://www.fao.org/3/i2085e/i2085e00.pdf.

FAO/WHO, 2018a. Standard for butter. Joint FAO/WHO Codex Alimentarius Commission CXS 279–1971. http://www.fao.org/fao–who–codexalimentarius/sh–proxy/es/?lnk=1&url=https%253A%252F%252Fworkspace.fao.org%252Fsites%252Fcodex%252FStandards%252FCXS%2B279–1971%252FCXS_279e.pdf .

FAO/WHO, 2018b. Standard for Milkfat Products. Joint FAO/WHO Codex Alimentarius Commission CXS 280–1973. http://www.fao.org/fao–who–codexalimentarius/sh–proxy/en/?lnk=1&url=https%253A%252F%252Fworkspace.fao.org%252Fsites%252Fcodex%252FStandards%252FCXS%2B280–1973%252FCXS_280e.pdf .

Farah, Z., 2011. Milk: Camel milk. In: Fucquay, J.W., Fox, P.F., McSeewney, P.H.L. (Eds.), Encyclopedia of Dairy Sciences. 2nd edn. Elsevier, San Diego, CA, pp. 512–517.

Fauquant, C., Briard, V., Leconte, N., Michalski, M.C., 2005. Differently sized native milk fat globules separated by microfiltration: Fatty acid composition of the milk fat globule membrane and triglyceride core. Eur. J. Lipid Sci. Technol. 107 (2), 80–86.

Ferrer, M., Alexander, M., Corredig, M., 2014. Changes in the physico–chemical properties of casein micelles during ultrafiltration combined with diafiltration. LWT – Food Sci. Technol. 59 (1), 173–180.

Ferrer, M.A., Alexander, M., Corredig, M., 2011. Does ultrafiltration have a lasting effect on the physico–chemical properties of the casein micelles? Dairy Sci. Technol. 91 (2), 151–170.

Ferrer, M.A., Hill, A.R., Corredig, M., 2008. Rheological properties of rennet gels containing milk protein concentrates. J. Dairy Sci. 91 (3), 959–969.

Finn, S., Condell, O., McClure, P., Amézquita, A., Fanning, S., 2013. Mechanisms of survival, responses and sources of Salmonella in low–moisture environments. Front. Microbiol. 4, 331.

Fox, P.F., Uniacke–Lowe, T., McSweeney, P.L.H., O'Mahony, J.A., 2015. Physical properties of milk. Dairy Chemistry and Biochemistry, 2nd edn. Springer, Heidelberg, Germany, pp. 321–343.

Freudig, B., Hogekamp, S., Schubert, H., 1999. Dispersion of powders in liquids in a stirred vessel. Chem. Eng. Process. 38 (4–6), 525–532.

Ganguli, N.C., 1991. Cheese base—a base for new foods in its own right. Indian Dairyman 43, 449–451.

Genualdi, S., Matos, M.P.V., Mangrum, J.B., Dejager, L., 2020. Investigation into the concentrations and sources of nitrates and nitrites in milk and plant–based powders. J. Agric. Food Chem. 68 (6), 1725–1730.

Glass, L., Hedrick, T.I, 1977. Nutritional composition of sweet– and acid–type dry wheys. II. Vitamin, mineral, and calorie contents. J. Dairy Sci. 60 (2), 190–196.

Goff, D., 2020. The Dairy Education eBook Series. Food Science. https://www.uoguelph.ca/foodscience/industry/dairy–education–ebook–series.

Gouda, A., Abou El–Nour, A., 2003. Cheeses, processed cheese. In: Caballero, B. (Ed.), Encyclopedia of Food Sciences and Nutrition. Academic Press, Oxford, pp. 1108–1115.

Gouda, A., El–Zayat, A.I., 1988. Use of calcium caseinate in processed cheese spreads making. Egypt J. Dairy Sci. (13) 115–119.

Govindasamy–Lucey, S., Lin, T., Jaeggi, J.J., Martinelli, C.J., Johnson, M.E., Lucey, J.A., 2007. Effect of type of concentrated sweet cream buttermilk on the manufacture, yield, and functionality of pizza cheese. J. Dairy Sci. 90 (6), 2675–2688.

Grossbier, D., 2016. Manufacture, concentration, and functionality of micellar casein concentrate. Thesis and Dissertations. South Dakota State University Paper 993.

Grufferty, M.B., Mulvihill, D.M., 1990. Hydration related properties of protein isolates prepared from heated milks. J. Soc. Dairy Technol. 43 (4), 99–103.

Guinee, T.P., 2009. The role of dairy ingredients in processed cheese products. In: Corredig, M. (Ed.), Dairy–Derived Ingredients: Food and Nutraceutical Uses. Woodhead Publishing, Cambridge, England, pp. 507–538.

Guinee, T.P., Kilcawley, K.N., 2017. Pasteurized processed and imitation cheese products. In: McSweeney, P.L.H., Fox, P.F., Cotter, P.D., Everret, D.W. (Eds.), Cheese Chemistry, Physics & Microbiology, 2. Elsevier, London, pp. 1133–1184.
Guinee, T.P., O'Callaghan, D.J., 2013. Effect of increasing the protein–to–fat ratio and reducing fat content on the chemical and physical properties of processed cheese product. J. Dairy Sci. 96 (11), 6830–6839.
Guinee, T.P., O'Kennedy, B., Kelly, P., Mounsey, J., 2013. Process for manufacturing cheese using milk protein concentrate powder. European Patent Application EP 2 647 293 A1. https://patents.google.com/patent/EP2647293A1/en. Accessed November 9, 2021.
Guo, M., Wang, G., 2016. Whey protein polymerisation and its applications in environmentally safe adhesives. Int. J. Dairy Technol. 69 (4), 481–488.
Gupta, V.K., Reuter, H., 1992. Processed cheese foods with added whey protein concentrates. Le Lait 72 (2), 201–212.
Gupta, V.K., Reuter, H., 1993. Firmness and melting quality of processed cheese foods with added whey protein concentrates. Le Lait 73 (4), 381–388.
Hammam, A.R.A., Metzger, L.E., 2019. Manufacturing of pro-cess cheese without emulsifying salt using acid curd. American Dairy Science Association, 81–82. Cincinnati, Ohio. Retrieved from https://www.adsa.org/Meetings/2019-Annual-Meeting.
Hammam, A.R.A., Metzger, L.E., 2018. Production and storage stability of liquid micellar casein concentrate. J. Dairy Sci. 101, 8.
Havea, A.J., Baldwin, A.J., Carr, A.J., 2009. Specialised and Novel Powders. In: Tamime, A.Y. (Ed.), Dairy Powders and Concentrated Products. Wiley–Blackwell, Ayr, UK, pp. 268–293.
Havea, P., 2006. Protein interactions in milk protein concentrate powders. Int. Dairy J. 16 (5), 415–422.
Havea, P., Baldwin, A.J., Carr, A.J., 2009. Specialised and Novel Powders. In: Tamime, A.Y. (Ed.), Dairy Powders and Concentrated Products. Wiley–Blackwell, Ayr, UK, pp. 268–293.
Hickey, M., 2011. Current legislation on processed cheese and related products. In: Tamine, A.Y. (Ed.). Processed Cheese and Analogues. Ayr, UK, pp. 25–80.
Hickey, M., 2017. Current legislation in relation to cheese. In: McSweeney, P.L.H., Fox, P.F., Cotter, P.D., Everret, D.W. (Eds.), Cheese Chemistry, Physics & Microbiology, vol. 1. Elsevier, London, pp. 757–780.
Hill, A.R., 1989. The β–lactoglobulin–χ–casein complex. Can. Inst. Food Sci. Technol. J. 22 (2), 120–123.
Hill, A.R., Smith, S., 1992. Texture and ultrastructure of process cheese spreads made from heat precipitated proteins. Milchwissenschaft 74 (2), 69–74.
Hill, B.M., Smythe, B.W., 2012. Endospores of thermophilic bacteria in ingredient milk powders and their significance to the manufacture of sterilized milk products: an industrial perspective. Food Rev. Int. 28 (3), 299–312.
Hirayama, K., Akashi, S., Furuya, M., Fukuhara, K., 1990. Rapid confirmation and revision of the primary structure of bovine serum albumin by ESIMS and frit–FAB LC/MS. Biochem. Biophys. Res. Commun. 173 (2), 639–646.
Horne, D.S., 2006. Casein micelle structure: models and muddles. Curr. Opin. Coll. Interface Sci. 11 (2–3), 148–153.
Horne, D.S., 2019. Casein micelle structure and stability. In: Thompson, A., Boland, M., Singh, H (Eds.), Milk Proteins from Expression to Food. Elsevier, San Diego, CA, pp. 133–162.
Huppertz, T., Gazi, I., 2017. Caseins and caseinates. In: Huppertz, T., Patel, H. (Eds.), Ingredients From Milk For Use In Food And Non–Food Products: From Commodity To Value–Added Ingredients, 2016. Burleigh Dodds Science Publishing, Cambridge, UK, pp. 121–144.

Hussain, R., Gaiani, C., Aberkane, L., Scher, J., 2011. Characterization of high–milk–protein powders upon rehydration under various salt concentrations. J. Dairy Sci. 94 (1), 14–23.

Hyde, M., Carruthers, L., McGeown, J., Trecker, G., 2010. Methods to produce cheese products and processed cheese bases. Canadian Intelectual Property Office 2, 364–393.

Hyde, M.A., Carruthers, L., Mcgeown, J.J., Trecker, G.W., 2005. Method for preparing cheese products and process cheese bases. Australian Patent Office 200197128 B2. http://pericles.ipaustralia.gov.au/ols/auspat/applicationDetails.do?applicationNo=2001097128. Accessed November 9, 2021.

Indyk, H.E., Woollard, D.C., 2011. Nitrates and nitrites. In: Fucquay, J.W., Fox, P.F., McSeewney, P.H.L. (Eds.), Encyclopedia of Dairy Sciences. Elsevier, San Diego, CA, pp. 906–911.

Ipsen, R., 2017. Microparticulated whey proteins for improving dairy product texture. Int. Dairy J. 67, 73–79.

Ismail, B., Nielsen, S., 2011. Enzymes indigenous to milk: plasmin system in milk. In: Fucquay, J.W., Fox, P.F., McSeewney, P.H.L. (Eds.), Encyclopedia of Dairy Sciences. 2nd edn. Elsevier, San Diego, CA, pp. 309–313.

ISO/IDF, 2005. Dried milk and dried milk products — Determination of insolubility index, ISO8156:2005. https://www.iso.org/obp/ui/#iso:std:iso:8156:ed–2:v1:en. Accessed July 29, 2020.

Jameson, G., Sutherland, S. 1994. Process for producing cheese containing substantially all the casen and whey proteins in milk. World Intelectual Property Organization Patent WO 82/01806.

Jameson, G.W., Sutherland, B., 1982. Process and apparatus for making cheese or cheese base and a novel intermediate product of said process. World Intelectual Property Organization. WO 82/01806.

Jana, A.H., Upadhyay, K.Y., 2001. Development of a formulation and process standardization for Mozzarella cheese analogue. Indian J. Dairy Sci. 54, 1–7.

Jensen, G.K., Nielsen, P., 1982. Milk powder and recombination of milk and milk products. J. Dairy Res. 49 (3), 515–544.

Jimenez–Flores, R., Kosikowski, F.V., 1986. Properties of ultrafiltered skim milk retentate powders. J. Dairy Sci. 69 (2), 329–339.

Jinjarak, S., Olabi, A., Jiménez–Flores, R., Walker, J.H., 2006. Sensory, functional, and analytical comparisons of whey butter with other butters. J. Dairy Sci. 89 (7), 2428–2440.

Johnson, M.E., Kapoor, R., McMahon, D., McCoy, D.R., Narasimmon, R.G., 2009. Reduction of sodium and fat levels in natural and processed cheeses: scientific and technological aspects. Compr. Rev. Food Sci. Food Saf. 8 (3), 252–268.

Jolly, R.C., Kosikowski, F.V, 1975. a new blue cheese food material from ultrafiltrated skim milk and microbial enzyme–mold spore reacted fat. J. Dairy Sci. 58 (9), 1272–1275.

Ju, Z., Kilara, A., 2008. textural properties of cold–set gels induced from heat–denatured whey protein isolates. J. Food Sci. 63 (2), 288–292.

Kaminarides, S., Stachtiaris, S., 2000. Production of processed cheese using kasseri cheese and processed cheese analogues incorporating whey protein concentrate and soybean oil. Int. J. Dairy Technol. 53 (2), 69–74.

Kapoor, R., Metzger, L.E., 2004. Evaluation of salt whey as an ingredient in processed cheese. J. Dairy Sci. 87 (5), 1143–1150.

Kapoor, R., Metzger, L.E., 2008. Process cheese: scientific and technological aspects – a review. Compr. Rev. Food Sci. Food Saf. 7 (2), 194–214.

Kapoor, R., Metzger, L.E., Biswas, A.C., Muthukummarappan, K., 2007. Effect of natural cheese characteristics on process cheese properties. J. Dairy Sci. 90 (4), 1625–1634.

Karagül–Yüceer, Y., Vlahovich, K.N., Drake, M.A., Cadwallader, K.R., 2003. Characteristic aroma components of rennet casein. J. Agric. Food Chem. 51 (23), 6797–6801.
Keenan, T.W., Mather, I.H., 2009. Intracellular origin of milk fat globules and the nature of the milk fat globule membrane. Advanced Dairy Chemistry: Lipids, Third edn. 2. Springer US, Boston, MA, pp. 137–171.
Kelimu, A., Silva, F., D., G., X., I., R., H., A., B., 2017. Effects of different dairy ingredients on the rheological behaviour and stability of hot cheese emulsions. Int. Dairy J. 71, 35–42.
Kelly, P., 2018. Whey protein ingredient applications. In: Deeth, H., Bansal, N. (Eds.), Whey Proteins: From Milk to Medicine. Academic Press, London, UK, pp. 335–375.
Kelly, A.L., Fox, P.F., 2016. Manufacture and properties of dairy powders. In: Mc-Sweeney, P.L.H, O'Mahony, J.A. (Eds.), Advanced Dairy Chemistry–1: Proteins. Fourth edn. Springer, New York, NY, pp. 1–33.
Kelly, P.M., 2011. Milk protein products: Membrane–based fractionation. In: Fucquay, J.W., Fox, P.F., McSeewney, P.H.L. (Eds.), Encyclopedia of Dairy Sciences: 2nd Ed., San Diego, CA: Academic Press, pp. 864–872.
Kelly, P.M., Gray, J.I., Slatterty, J., 1989. Direct low–nox gas combustion heating of a spray drier during milk powder manufacture. J. Soc. Dairy Technol. 42 (1), 14–18.
Kethireddipalli, P., Hill, A.R, 2015. Rennet coagulation and cheesemaking properties of thermally processed milk: overview and recent developments. J. Agric. Food Chem. 63 (43), 9389–9403.
Kethireddipalli, P., Hill, A.R., Dalgleish, D.G, 2011. Interaction between casein micelles and whey protein/κ–casein complexes during renneting of heat–treated reconstituted skim milk powder and casein micelle/serum mixtures. J. Agric. Food Chem. 59 (4), 1442–1448.
Kilara, A., Vaghela, M.N., 2018. Whey proteins. In: Yada, R. (Ed.), Proteins in Food Processing. Second edn. Woodhead Publishing, Cambridge, England, pp. 93–116.
Kindstedt, P.S., 1999. Mozzarella and pizza cheese. In: Fox, P.F. (Ed.), Cheese: Chemistry, Physics and Microbiology, 2. Aspen, Gaithersburg, pp. 337–362.
Kombila–Moundounga, E., Lacroix, C., 1991. Effet des combinaisons de chlorure de sodium, de lactose et de glycérol sur les caractéristiques rhéologiques et la couleur des fromages fondus à tartiner. Can. Inst. Food Sci. Technol. J. 24 (5), 239–251.
Kosikowski, F.V., 1979. Whey utilization and whey products. J. Dairy Sci. 62 (7), 1149–1160.
Kumar, R., Chauhan, S.K., Shinde, G., Subramanian, V., Nadanasabapathi, S., 2018. Whey proteins: a potential ingredient for food industry– a review. Asian J. Dairy Food Res. 37 (4), 283–290.
Kwak, H.S., Choi, S.S., Ahn, J., Lee, S.W., 2002. Casein hydrolysate fractions act as emulsifiers in process cheese. J. Food Sci. 67 (2), 821–825.
Kycia, K., Pluta, A., Zmarlicki, S., 2006. Physicochemical and textural properties of processed cheese spreads made with the addition of cheese base obtained from UF milk retentates. Polish J. Food Nutr. Sci. 15 (56), 115–118.
Lakritz, L., Pensabene, J.W., 1981. Survey of fluid and nonfat dry milks for n–nitrosamines. J. Dairy Sci. 64 (3), 371–374.
Lazzaro, F., Saint–Jalmes, A., Violleau, F., Lopez, C., Gaucher–Delmas, M., Madec, M.N., Beaucher, E., Gaucheron, F., 2017. Gradual disaggregation of the casein micelle improves its emulsifying capacity and decreases the stability of dairy emulsions. Food Hydrocoll. 63, 189–200.
Le, T.T., Deeth, H.C., Larsen, L.B., 2017. Proteomics of major bovine milk proteins: Novel insights. Int. Dairy J. 67, 2–15.

Lee, S.K., Anema, S., Klostermeyer, H., 2004. The influence of moisture content on the rheological properties of processed cheese spreads. Int. J. Food Sci. Technol. 39 (7), 763–771.

Lee, S.K., Anema, S.G., 2009. The effect of the pH at cooking on the properties of processed cheese spreads containing whey proteins. Food Chem. 115 (4), 1373–1380.

Lee, S.K., Huss, M., Klostermeyer, H., Anema, S.G., 2013. The effect of pre–denatured whey proteins on the textural and micro–structural properties of model processed cheese spreads. Int. Dairy J. 32 (2), 79–88.

Lee, S.K., Klostermeyer, H., Anema, S.G., 2015. Effect of fat and protein–in–water concentrations on the properties of model processed cheese. Int. Dairy J. 50, 15–23.

Lee, S.–Y., Morr, C.V., Ha, Y.E., 1992. Structural and functional properties of caseinate and whey protein isolate as affected by temperature and pH. J. Food Sci. 57 (5), 1210–1229.

Leko, S., Foegeding, E.A., 2000. Physical properties of rennet casein gels and processed cheese analogs containing whey proteins. Milchwissenschaft 55, 513–516.

Li, B., Waldron, D.S., Tobin, J.T., Subhir, S., Kelly, A.L., McSweeney, P.L.H., 2020. Evaluation of production of Cheddar cheese from micellar casein concentrate. Int. Dairy J. 107, 1–10.

Lin, Y., Kelly, A.L., O'Mahony, J.A., Guinee, T.P., 2018. Effects of milk heat treatment and solvent composition on physicochemical and selected functional characteristics of milk protein concentrate. J. Dairy Sci. 101 (8), 6799–6813.

Lindmark–Månsson, H., 2008. Fatty acids in bovine milk fat. Food Nutr. Res. 52, 1–3.

Liu, D., Zhang, J., Yang, T., Liu, X., Hemar, Y., Regenstein, J.M., Zhou, P., 2019. Effects of skim milk pre–acidification and retentate pH–restoration on spray–drying performance, physico–chemical and functional properties of milk protein concentrates. Food Chem. 272, 539–548.

Loessner, M.J., Maier, S.K., Schiwek, P., Scherer, S., 1997. Long–chain polyphosphates inhibit growth of Clostridium tyrobutyricum in processed cheese spreads. J. Food Prot. 60 (5), 493–498.

Lopez, C., 2011. Milk fat globules enveloped by their biological membrane: Unique colloidal assemblies with a specific composition and structure. Curr. Opin. Colloid Interface Sci. 16 (5), 391–404.

Lucca, P.A., Tepper, B.J., 1994. Fat replacers and the functionality of fat in foods. Trends Food Sci. Technol. 5 (1), 12–19.

Ma, X., James, B., Zhang, L., Emanuelsson–Patterson, E.A.C., 2013. Correlating mozzarella cheese properties to its production processes and microstructure quantification. J. Food Eng. 115 (2), 154–163.

Macků, I., Buňka, F., Voldánová, B., Pavlínek, V., 2009. Effect of addition of selected solid cosolutes on viscoelastic properties of model processed cheese containing pectin. Food Hydrocoll. 23 (8), 2078–2084.

Manning, P.B., Coulter, S.T., Jenness, R., 1968. Determination of nitrate and nitrite in milk and dry milk products. J. Dairy Sci. 51 (11), 1725–1730.

Martini, M., Salari, F., Altomonte, I., 2016. The macrostructure of milk lipids: the fat globules. Crit. Rev. Food Sci. Nutr. 56 (7), 1209–1221.

Maubois, J.L., 2002. A tool for a new approach in dairy technology. Aust. J. Dairy Technol. 57 (2), 92–96.

Maubois, J.L., Germain, C., Vassal, L.J., 1975. United States Patent 3 914 435. https://patents.google.com/patent/US3914435A/en

Maubois, J.L., Mocquot, G., 1975. Application of membrane ultrafiltration to preparation of various types of cheese. J. Dairy Sci. 58 (7), 1001–1007.

McCarthy, O.J., 2002. Milk: Physical and physico–chemical properties of milk. In: Fucquay, J.W., Fox, P.F., McSeewney, P.H.L. (Eds.), Encyclopedia of Dairy Sciences. Elsevier, pp. 467–477 http://doi.org/10.1016/B978–0–12–374407–4.00311–3.
Meena, G.S., Singh, A.K., Gupta, V.K., Borad, S.G., Arora, S., Tomar, S.K., 2019. Alteration in physicochemical, functional, rheological and reconstitution properties of milk protein concentrate powder by pH, homogenization and diafiltration. J. Food Sci. Technol. 56 (3), 1622–1630.
Michalski, M.C., Cariou, R., Michel, F., Garnier, C., 2002. Native vs. damaged milk fat globules: membrane properties affect the viscoelasticity of milk gels. J. Dairy Sci. 85 (10), 2451–2461.
Mimouni, A., Deeth, H.C., Whittaker, A.K., Gidley, M.J., Bhandari, B.R., 2010. Investigation of the microstructure of milk protein concentrate powders during rehydration: alterations during storage. J. Dairy Sci. 93 (2), 463–472.
Miralles, B., Krause, I., Ramos, M., Amigo, L., 2006. Comparison of capillary electrophoresis and isoelectric focusing for analysis of casein/caseinate addition in processed cheeses. Int. Dairy J. 16, 1448–1453.
Mistry, V.V., Hassan, H.N., 1991. Delactosed, high milk protein powder. 1. Manufacture and composition. J. Dairy Sci. 74 (4), 1163–1169.
Mistry, V.V., Maubois, J.L., 2017. Application of membrane separation technology to cheese production. In: McSweeney, P.L.H., Fox, P.F., Cotter, P.D., Everret, D.W. (Eds.), Cheese Chemistry, Physics & Microbiology, Fourth edn. 2. Elsevier, London, pp. 677–697.
Mleko, S., Foegeding, E., 2001. Incorporation of polymerized whey proteins into processed cheese analogs. Milchwissenschaft 56, 612–615.
Modler, H.W., 1985. Functional properties of nonfat dairy ingredients — a review. modification of lactose and products containing whey proteins. J. Dairy Sci. 68 (9), 2206–2214.
Moran, J.W., Posdal, J.R, Trecker, G.W., 1989. Manufacture of curd and cheese from mlk retentate. United States Patent 4820530. https://patents.google.com/patent/US4820530A/en
Moran, J.W, Posdal, J.R, Trecker, G.W., 1993. Manufacture of high–solids pre–cheese for conversion into natural cheese. United States Patent 5262183. https://patents.google.com/patent/US5262183A/en
Morin, P., Britten, M., Jiménez–Flores, R., Pouliot, Y., 2007. Microfiltration of buttermilk and washed cream buttermilk for concentration of milk fat globule membrane components. J. Dairy Sci. 90 (5), 2132–2140.
Morr, C., 1985. Functionality of heated milk proteins in dairy and related foods. J. Dairy Sci. 68, 2773–2781.
Morr, C.V., Ha, E., 1993. Whey–protein concentrates and isolates – processing and functional–properties. Crit. Rev. Food Sci. Nutr. 33 (3), 431–476.
Mounsey, J.S., O'Kennedy, B.T., Kelly, P.M., 2007. Effect of the aggregation state of whey protein–based ingredients on processed cheese functionality. Milchwissenschaft 62, 44–47.
Mounsey, J.S., O'Riordan, E.D., 2008. Modification of imitation cheese structure and rheology using pre–gelatinised starches. Eur. Food Res. Technol. 226 (5), 1039–1046.
Muir, D.D., Williams, S.A.R., Tamime, A.Y., Shenana, M.E., 1997. Comparison of the sensory profiles of regular and reduced–fat commercial processed cheese spreads. Int. J. Food Sci. Technol. 32 (4), 279–287.
Mulvihill, D.M., Murphy, P.C., 1991. Surface active and emulsifying properties of caseins/caseinates as influenced by state of aggregation. Int. Dairy J. 1 (1), 13–37.
Nasser, S., Moreau, A., Jeantet, R., Hédoux, A., Delaplace, G., 2017. Influence of storage conditions on the functional properties of micellar casein powder. Food Bioprod. Process. 106, 181–192.

Nastaj, M., Terpiłowski, K., Sołowiej, B.G., 2019. The effect of native and polymerised whey protein isolate addition on surface and microstructural properties of processed cheeses and their meltability determined by Turbiscan. Int. J. Food Sci. Technol. 55, 2179–2187.

Nishanthi, M., Vasiljevic, T., Chandrapala, J., 2017. Properties of whey proteins obtained from different whey streams. Int. Dairy J. 66, 76–83.

Nygren, O., 1982. Determination of nitrate content in whey and milk powders at a dairy after an occupational accident with nitric acid. Scand. J. Work Environ. Health 8 (3), 227–230.

O'Connell, J.E., Fox, P.F., 2000. Heat stability of buttermilk. J. Dairy Sci. 83 (8), 1728–1732.

Olson, D., Aryana, K.J., 2009. Ingredients in Dairy Products. In: Chandan, R.C. (Ed.), Dairy Processing and Quality Assurance. John Wiley & Sons, West Sussex, UK, pp. 189–202.

O'Malley, A.M., Mulvihill, D.M., Kumar Singh, T., 2000. Proteolysis in rennet casein–based cheese analogues. Int. Dairy J. 10 (11), 743–753.

Patocka, J., Jelen, P., 1988. Enzymatic lactose hydrolysis for prevention of lactose crystallization in a whey spread. J. Food Sci. 53 (5), 1370–1372.

Pelegrine, D.H.G., Gasparetto, C.A., 2005. Whey proteins solubility as function of temperature and pH. LWT – Food Sci. Technol. 38 (1), 77–80.

Pereira, R.B., Bennett, R.J., Hemar, Y., 2001. Rheological and microstructural characteristics of model processed cheese analogues. J. Texture Stud. 32, 349–373.

Pereira, R.B., Bennett, R.J., Mcmath, K.L., Luckman, M.S., 2002. In–hand sensory evaluation of textural characteristics in model processed cheese analogues 33(4) 255–268.

Pierre, A., Fauquant, J., Le Graet, Y., Piot, M., 1992. Native micellar casein separation through cross flow membrane microfiltration. Lait 72 (5), 461.

Piska, I., Štětina, J., 2004. Influence of cheese ripening and rate of cooling of the processed cheese mixture on rheological properties of processed cheese. J. Food Eng. 61 (4), 551–555.

Podolak, R., Enache, E., Stone, W., Black, D.G., Elliott, P.H., 2010. Sources and risk factors for contamination, survival, persistence, and heat resistance of Salmonella in low–moisture foods. J. Food Prot. 73 (10), 1919–1936.

Price, J., 2018. History of the development and application of whey protein products. In: Deeth, N., Bansal, H. (Eds.), Whey Proteins: From Milk to Medicine. Academic Press, San Diego, CA, pp. 51–95.

Price, W.V., Bush, M.G., 1974. The process cheese industry in the United States: a review. I. Industrial growth and problems. J. Milk Food Technol. 37 (3), 135–152.

Purna, S.K.G., Pollard, A., Metzger, L.E., 2006. Effect of formulation and manufacturing parameters on process cheese food functionality – I. Trisodium citrate. J. Dairy Sci. 89 (7), 2386–2396.

Rahul, S., Dhanraj, P., 2015. Influence of rennet casein levels on the chemical, baking and sensory quality of mozzarella cheese analogue. J. Dairy, Veterinary & Animal Res. 2 (3), 99–105.

Ramel, P.R., Marangoni, A.G., 2017. Characterization of the polymorphism of milk fat within processed cheese products. Food Struct. 12, 15–25.

Ramos, O.L., Pereira, R.N., Rodrigues, R.M., Teixeira, J.A., Vicente, A.A., Malcata, F.X., 2015. Whey and whey powders: production and uses. In: Caballeros, B., Finglas, P., Toldra, F. (Eds.), Encyclopedia of Food and Health. Elsevier Ltd, Amsterdam, Netherlands, pp. 498–505.

Raval, D.M., Mistry, V.V., 1999. Application of ultrafiltered sweet buttermilk in the manufacture of reduced fat process cheese. J. Dairy Sci. 82 (11), 2334–2343.

Rollema, H.S., Muir, D., 2009. Casein and related products. In: Tamime, A.Y. (Ed.), Dairy Powders and Concentrated Products. Wiley–Blackwell, Ayr, UK, pp. 235–255.

Rombaut, R., Dewettinck, K., 2006. Properties, analysis and purification of milk polar lipids. Int. Dairy J. 16, 1362–1373.
Rubin, J., Bjerre, P., 1983. Process for preparing cheese–base . United States Patent 4401679. https://patents.google.com/patent/US4401679A/en.
Rüegg, M., Blanc, B., 1976. Effect of ph on water vapor sorption by caseins. J. Dairy Sci. 59 (6), 1019–1024.
Saad, S.A., El–Mahdi, L.D., Awad, R.A., Hassan, Z.M.R., 2016. Impact of different food protein sources in processed cheese sauces manufacture. Int. J. Dairy Sci. 11 (2), 52–60.
Sánchez–Obando, J.D., Cabrera–Trujillo, M.A., Olivares–Tenorio, M.L., Klotz, B., 2020. Use of optimized microparticulated whey protein in the process of reduced–fat spread and petit–suisse cheeses. LWT – Food Sci. Technol. 120, 108933 2019.
Sarkar, A., Singh, H., 2016. Emulsions and foams stabilised by milk proteins BT. In: McSweeney, P.L.H, O'Mahony, J.A. (Eds.), Advanced Dairy Chemistry–1: Proteins. Springer, New York, pp. 133–153.
Savello, P.A., Ernstrom, C.A., Kalab, M., 1989. Microstructure and meltability of model process cheese made with rennet and acid casein. J. Dairy Sci. 72 (1), 1–11.
Scanlan, R.A., Barbour, J.F., Bodyfelt, F.W., Libbey, L.M., 1994. N– Nitrosodimethylamine in nonfat dry milk. Symposium Series: Nitrosamines and Related N–Nitroso Compounds: Chemistry and Biochemistry ACS. ACS Publications, Washington, DC, 34–31.
Schieberle, P., Gassenmeier, K., Guth, H., Sen, A., Grosch, W., 1993. Character impact odour compounds of different kinds of butter. LWT – Food Sci. Technol. 26 (4), 347–356.
Schokker, E.P., Church, J.S., Mata, J.P., Gilbert, E.P., Puvanenthiran, A., Udabage, P., 2011. Reconstitution properties of micellar casein powder: Effects of composition and storage. Int. Dairy J. 21 (11), 877–886.
Schuck, P., Davenel, A., Mariette, F., Briard, V., Méjean, S., Piot, M., 2002. Rehydration of casein powders: Effects of added mineral salts and salt addition methods on water transfer. Int. Dairy J. 12 (1), 51–57.
Schulz, M.E., 1976. Preparation of melt–resistant process cheese. US patent 3962483. https://patents.google.com/patent/US3962483A/en
Shah, R., Jana, A.H., Aparnathi, K.D., Prajapati, P.S., 2010. Process standardization for rennet casein based mozzarella cheese analogue. J. Food Sci. Technol. 47 (5), 574–578.
Shantha, N.C., Decker, E.A., Ustunol, Z., 1992. Conjugated linoleic acid concentration in processed cheese. J. Am. Oil Chem. Soc. 69 (5), 425–428.
Sharma, A., Jana, A.H., Chavan, R.S., 2012. Functionality of milk powders and milk–based powders for end use applications–a review. Compr. Rev. Food Sci. Food Saf. 11 (5), 518–528.
Shi, Q., Zhou, Y., Sun, Y., 2008. Influence of pH and ionic strength on the steric mass–action model parameters around the isoelectric point of protein. Biotechnol. Prog. 21 (2), 516–523.
Sikand, V., Tong, P.S., Walker, J., 2013. Effect of adding salt during the diafiltration step of milk protein concentrate powder manufacture on mineral and soluble protein composition. Dairy Sci. Technol. 93 (4–5), 401–413.
Simbuerger, S., Wolfschoon, A.F., Kempter, K., Rose, M., Marder, U. 1997. Acidified ultrafiltration concentrates as raw material for the production of processed cheese. European Patent Office 0 755 630 A1. http://scholar.google.com/scholar?hl=en&btnG=Search&q=intitle:Designated+Extension+States+:#0.
Sindhu, J.S., Arora, S., 2011. Milk: Buffalo milk. In: Fucquay, J.W., Fox, P.F., McSeewney, P.H.L. (Eds.), Encyclopedia of Dairy Sciences. Second edn. Academic Press, San Diego, CA, pp. 503–511.

Singh, H., 2006. The milk fat globule membrane–A biophysical system for food applications. Curr. Opin. Colloid Interface Sci. 11 (2–3), 154–163.
Singh, H., 2011. Aspects of milk–protein–stabilised emulsions. Food Hydrocoll. 25 (8), 1938–1944.
Singh, H., Tokley, R.P., 1990. Effects of preheat treatments and buttermilk addition on the seasonal variations in the heat stability of recombined evaporated milk and reconstituted concentrated milk. Aust. J. Dairy Technol. 45 (1), 10–16.
Singh, H., Ye, A., 2020. Interactions and functionality of milk proteins in food emulsions. In: Thompson, A., Boland, M., Singh, H (Eds.), Milk Proteins From Expression to Food. Academic Press, San Diego, CA, pp. 321–345.
Smith, G.F., Rivera, E., 2008. Processed cheese without emulsifying salts. US Patent Office US9532584B2. https://patents.google.com/patent/US9532584B2/en.
Smith, T.J., Campbell, R.E., Jo, Y., Drake, M.A., 2016. Flavor and stability of milk proteins. J. Dairy Sci. 99 (6), 4325–4346.
Sodini, I., Morin, P., Olabi, A., Jiménez–Flores, R., 2006. Compositional and functional properties of buttermilk: a comparison between sweet, sour, and whey buttermilk. J. Dairy Sci. 89 (2), 525–536.
Solowiej, B., 2012. Textural, rheological and melting properties of acid casein reduced–fat processed cheese analogues. Milchwissenschaft 67 (1), 9–13.
Sołowiej, B., Cheung, I.W.Y., Li–Chan, E.C.Y., 2014. Texture, rheology and meltability of processed cheese analogues prepared using rennet or acid casein with or without added whey proteins. Int. Dairy J. 37 (2), 87–94.
Solowiej, B., Mleko, S., Gustaw, W., 2008. Physicochemical properties of acid casein processed cheese analogs obtained with different whey products. Milchwissenschaft 63 (3), 299–302.
Sołowiej, B.G., Nastaj, M., Szafrańska, J.O., Muszyński, S., Gustaw, W., Tomczyńska–Mleko, M., Mleko, S., 2020. Effect of emulsifying salts replacement with polymerised whey protein isolate on textural, rheological and melting properties of acid casein model processed cheeses. Int. Dairy J. 105, 104695.
Sommer, H.H., Templeton, H.L., 1939. Research bulletin 137. The Making of Processed Cheese. Wisconsin Agricultural experimental Station, Madison.
Sood, S.M., Gaind, D.K., Dewan, R.K., 1979. Corrrelation between micelle solvation and calcium content. N.Z. J. Dairy Sci. Technol. 14 (1), 32–34.
Sood, V.K., Kosikowski, F.V., 1979. Process cheddar cheese from plain and enzyme treated retentates. J. Dairy Sci. 62 (11), 1713–1718.
Souza, A.B., Carlos Gonçalves Costa Júnior, L., Tuler Perrone, Í., Stephani, R., De Freitas Almeida, D., 2014. parâmetros de textura em queijos processados: influência da utilização de concentrados proteicos de leite e de Soro. Revista do Instituto de Laticínios Cândido Tostes 69 (3), 181.
Spitsberg, V.L., 2005. Invited review: bovine milk fat globule membrane as a potential nutraceutical. J. Dairy Sci. 88 (7), 2289–2294.
Stankey, J.A., Lu, Y., Abdalla, A., Govindasamy–Lucey, S., Jaeggi, J.J., Mikkelsen, B., Pedersen, K.T., Andersen, C.B., 2017. Low–fat cheddar cheese made using microparticulated whey proteins: effect on yield and cheese quality. Int. J. Dairy Technol. 70 (4), 481–491.
Swenson, B.J., Wendorff, W.L., Lindsay, R.C., 2000. Effects of ingredients on the functionality of fat–free process cheese spreads. J. Food Sci. 65 (5), 822–825.
Szafrańska, J.O., Sołowiej, B.G., 2020. Effect of different fibres on texture, rheological and sensory properties of acid casein processed cheese sauces. Int. J. Food Sci. Technol. 55 (5), 1971–1979.

Talebi, S., Kee, E., Chen, G.Q., Bathurst, K., Kentish, S.E., 2019. Utilisation of salty whey ultrafiltration permeate with electrodialysis. Int. Dairy J. 99, 104549.

Tamime, A., Kalab, M., Davies, G., Younis, M., 1990. Microstructure and firmness of processed cheese manufactured from cheddar cheese and skim milk powder cheese base. Food Microstruct. 9 (1), 23–37.

Tamime, A.Y., 2009. Dairy Powders and Concentrated Products. In: Tamime, A.Y. Wiley–Blackwell. Ayr, UK.

Tamime, A.Y., 2011. Processed Cheese and Analogues: An Overview. In: Processed Cheese and Analogues (pp. 1–24). Wiley-Blackwell. https://doi.org/10.1002/9781444341850.ch1.

Taylor, M.W., MacGibbon, A.K.H., 2011. Milk lipids: fatty acids. In: Fucquay, J.W., Fox, P.F., McSeewney, P.H.L. (Eds.), Encyclopedia of Dairy Sciences. Elsevier, pp. 655–659.

Templeton, H.L., Sommer, H.H., 1930. Some observations on processed cheese. J. Dairy Sci. 13 (3), 203–220.

Thomas, M.A., 1969a. Browning reaction in cheddar cheese. Aust. J. Dairy Technol. 24 (4), 185–189.

Thomas, M.A., 1970. Use of calcium co–precipitates in processed cheese. Aust. J. Dairy Technol. 25 (1), 23–26.

Thomas, M.A., Hyde, K.A., 1972. The manufacture of processed cheese— a working manual. Bulletin D40. Division of Dairy. New South Wales Department of Agriculture, Sydney.

Thomas, M.E.C., Scher, J., Desobry–Banon, S., Desobry, S., 2004. Milk powders ageing: effect on physical and functional properties. Crit. Rev. Food Sci. Nutr. 44 (5), 297–322.

Tobin, J.T., Fitzsimons, S.M., Kelly, A.L., Kelly, P.M., Auty, M.A.E., Fenelon, M.A., 2010. Microparticulation of mixtures of whey protein and inulin. Int. J. Dairy Technol. 63 (1), 32–40.

Torres, I.C., Janhøj, T., Mikkelsen, B.Ø., Ipsen, R., 2011. Effect of microparticulated whey protein with varying content of denatured protein on the rheological and sensory characteristics of low–fat yoghurt. Int. Dairy J. 21 (9), 645–655.

Tsakali, E., Petrotos, K., D'Alessandr, A.G., Goulas, P., 2010. A review on whey composition and the methods used for its utilization for food and pharmaceutical products. In: 6th International Conference on Simulation and Modelling in the Food and Bio–Industry 2010 (FOODSIM 2010), 195–201.

Tunick, M.H., Basch, J.J., Maleeff, B.E., Flanagan, J.F., Holsinger, V.H., 1989. Characterization of natural and imitation mozzarella cheeses by differential scanning calorimetry. J. Dairy Sci. 72 (8), 1976–1980.

USDA, 2018. 21 Code of Federal Regulations, Part 172. Sodium Nitrite Vol. 3.e. https://www.accessdata.fda.gov/scripts/cdrh/cfdocs/cfcfr/CFRSearch.cfm?fr=172.175. Accessed July 25, 2020.

van Hekken, D.L., Holsinger, V.H., 2000. Use of cold microfiltration to produce unique beta–casein enriched milk gels. Le Lait 80 (1), 69–76.

Van Ruth, S.M., Koot, A., Akkermans, W., Araghipour, N., Rozijn, M., Baltussen, M., Wisthaler, A., Märk, T.D., Frankhuizen, R., 2008. Butter and butter oil classification by PTR–MS. Eur. Food Res. Technol. 227 (1), 307–317. https://doi.org/10.1007/s00217-007-0724-7.

Van Slyke, L.L., Price, V., 1949. Cheese, Second edn. Orange Judd Publ. Co., New York.

Vanderghem, C., Bodson, P., Danthine, S., Paquot, M., Deroanne, C., Blecker, C., 2010. Milk fat globule membrane and buttermilks: from composition to valorization. Biotechnol. Agron. Soc. Environ. 14 (3), 485–500.

Veith, P.D., Reynolds, E.C., 2004. Production of a high gel strength whey protein concentrate from cheese whey. J. Dairy Sci. 87 (4), 831–840.

Walstra, P., van Vliet, T., 1986. The physical chemistry of curd making. Neth. Milk Dairy J. 40 (2–3), 241–259.

Walstra, P., Wouters, JT.M., Guerts, T.J., 2005. Dairy Science and Technology, second ed., CRC Taylor & Francis, Boca Raton.

Wong, N.P., Jenness, R., Keeney, M., Marth, E., 1988. Fundamentals of Dairy Chemistry, Third edn Van Nostrand Reinhold Company Inc., New York, NY.

Wright, A.J., Marangoni, A.G., Hartel, R.W., 2011. Milk Lipids: rheological properties and their modification. In: Fucquay, J.W., Fox, P.F., McSeewney, P.H.L. (Eds.), Encyclopedia of Dairy Sciences. Elsevier, pp. 704–710.

Wright, B.J., Zevchak, S.E., Wright, J.M., Drake, M.A., 2009. The impact of agglomeration and storage on flavor and flavor stability of whey protein concentrate 80% and whey protein isolate. J. Food Sci. 74 (1), 17–29.

Wright, J.M., Whetstine, C., M.E., M., R.E., D., M., A., 2006. Characterization of a cabbage off–flavor in whey protein isolate. J. Food Sci. 71 (2), C86–C90.

Wu, S., Fitzpatrick, J., Cronin, K., Miao, S., 2019. The effect of pH on the wetting and dissolution of milk protein isolate powder. J. Food Eng. 240 (2018), 114–119.

Yadav, J.S.S., Bezawada, J., Elharche, S., Yan, S., Tyagi, R.D., Surampalli, R.Y., 2015. Cheese whey: a potential resource to transform into bioprotein, functional/nutritional proteins and bioactive peptides. Biotechnol. Adv. 33 (6), 756–774.

Yeh, T.S., Liao, S.F., Kuo, C.Y., Hwang, W.I., 2013. Investigation of the nitrate and nitrite contents in milk and milk powder in Taiwan. J. Food Drug Anal. 21 (1), 73–79.

Young, S., 1999. Whey Products in Cold Pack and Pasteurized Processed Cheese Products — Classifications. US Dairy Export Council, pp. 1–8.

Younis, M.F., Tamime, A.Y., Davies, G., Hunter, E.A., El–Hady, S.M., 1991. Production of processed cheese using Cheddar cheese and cheese base. 5. Rheological properties. Milchwissenschaft 46, 701–705.

Younis, M.F., Davies, G., 1991. Production of processed cheese using Cheddar cheese and cheese base. 2. production of a cheese base from skim milk powder. Milchwissenschaft 46 (8), 495–498.

Zanabria Eyzaguirre, R., Corredig, M., 2011. Milk lipids: buttermilk and milk fat globule membrane fractions. In: Fucquay, J.W., Fox, P.F., McSeewney, P.H.L. (Eds.), Encyclopedia of Dairy Sciences. Second edn. Academic Press, San Diego, CA, pp. 691–697.

Zhang, W., Kakuda, Y., Hill, A., 2006. Conjugated linoleic acid (CLA) in processed cheese. Milchwissenschaft (62) 174–178.

The use of different cheese sources in processed cheese

3

Richardos N. Salek[a], *František Buňka*[a,b] *and Michaela Černíková*[a]
[a] Department of Food Technology, Faculty of Technology, Tomas Bata University in Zlín, Zlín, Czech Republic, [b] Food Research Laboratory, Department of Logistics, Faculty of Military Leadership, University of Defense, Brno, Czech Republic

3.1 General aspects in processed cheese technology

3.1.1 Historical background

Natural cheese varieties are known and appreciated as a valuable food component since ancient Roman times (more than 2000 years ago). However, processed cheese could be characterized as a "new-born" dairy product, originating in the 20th century. Moreover, the idea of processed cheese was though based on natural cheese (Tamine, 2011). The production of processed cheese started in Europe and might be dated near to the early 1890s. It may be recommended, that the concept of processed cheese originated from the search of new strategies for a more sophisticated manufacture technique of cheese, in order to extend the shelf-life of natural cheese, bypassing traditional preservation methods; such as air-drying or smoking. Likewise, another possible reason could be the desire to develop a new dairy product characterized of milder taste and more stable structure (Tamine, 2007; 2011). According to Guinee et al. (2004) the idea of processed cheese stemmed from a typical Swiss traditional dish called "Fondue," for which natural cheese is melted (together with application of heat) in the presence of wine, in which tartrate from the wine provided an emulsifying effect. However, Jan Hendrikzoon was the first who established a thermal treatment for canned Gouda-cheese in 1899 (Kammerlehner, 2012). The Swiss researchers Walter Gerber and Fritz Stettler closely followed traditional fondue preparation and, after several experiments with tartaric acid, added sodium citrate prior to heat-processing of the shredded raw material to achieve a stable but modified final product (Tamine, 2011). Industrial production of processed cheese began in Europe and the USA between 1910 and 1920. The first processed cheese was developed by the Swiss scientists mentioned earlier; Walter Gerber and Fritz Stettler in Thun, Switzerland (1911). At this certain attempt, natural Emmentaler cheese was applied to produce a heat-treated cheese (known as "Schachtelkäse") by the addition of sodium citrate as an emulsifying salt. About the same period in the USA (1917), independently, James Lewis Kraft, an entrepreneur, inventor and industrialist started working with pieces of Cheddar cheese, citrates and monophosphates. Hence, in 1916, he published a patent describing the method of heating cheese and its emulsification in the presence of alkaline salts. However, these early attempts were of limited success in order to manufacture a

Processed Cheese Science and Technology: Ingredients, Manufacture, Functionality, Quality, and Regulations.
DOI: https://doi.org/10.1016/B978-0-12-821445-9.00010-8

good-quality processed cheese. The process became widespread by the 1930s, due to the appearance of polyphosphate emulsifying salts on the world market (Fox, 2000; Hui, 2005; Tamine, 2007).

3.1.2 Processed cheese – an overview

Processed cheese is a multicomponent dairy complex system; described as stable oil-in-water emulsion (Chen and Liu, 2012; Hanaei et al., 2015; Lee et al., 2004). In a physicochemical point of view, processed cheese could be characterized as a dispersion of fat droplets in a concentrated, gelled protein network matrix (Hasenhuetl and Hartel, 2008). Therefore, the multilateralism of the above-mentioned system (processed cheese) derives from the fact that it can contain plethora of interacting components and relatively high amount of water content (Marchesseau et al., 1997). Hence, processed cheese matrix is formed by blending shredded natural cheese (of different types/varieties and maturity degrees) in the presence of appropriate emulsifying salts (mainly sodium salts of phosphates, polyphosphates and/or citrates), under partial vacuum, constant stirring and upon heat treatment; resulting in the constitution of a homogeneous and smooth mass with desired functional and organoleptic properties (Anema and Klostermeyer, 2004; Sádlíková et al., 2010). Nevertheless, the application of heat during the processing can inactivate the starter culture microorganisms and other bacteria, including also enzymes, all present in natural cheese; resulting in the shelf-life extension of the final developed product (Tamine, 2007; 2011). However, there is no specific European Union legislation on processed cheese. For a more complete oversight on the legislation of processed cheese and related products within the European Union, it is necessary to look at a selection of Member States. In addition, processed cheese according to international standards is grouped by the following characteristics: composition, water content and consistency. According to these characteristics, exist three main categories (mainly in the USA): processed cheese blocks, processed cheese foods, processed cheese spreads. Moreover, additional subcategories are processed cheese slices and smoked processed cheese (Tamine, 2007; 2011).

Furthermore, emulsifying salts (sometimes also called emulsifying agents) are more commonly known in the dairy industry as "melting" salts and are of crucial importance during processed cheese manufacture. These salts are ionic compounds made up of monovalent cations (sodium, potassium) and polyvalent anions (phosphates, polyphosphates and/or citrates) (Chen and Liu, 2012; Buňka et al., 2014). The most commonly applied emulsifying salts are sodium citrates, sodium hydrogen monophosphates, diphosphates and polyphosphates. Nowadays, emulsifying salts are rarely used as individual compounds, whereas they are applied rather in the form of phosphate and phosphate-citrate blends (binary, ternary or even quaternary mixtures) (Chen and Liu, 2012; Hasenhuetl and Hartel, 2008). However, emulsifying salts cannot be characterized as "*true*" emulsifiers (low molecular-weight surfactants) since they do not indicate surface-activity. In the narrow sense, emulsifying salts cannot be used for the preparation of oil-in-water (O/W) or water-in-oil (W/O) emulsions, whereas they play important role in modifying the emulsifying activity of the present surface-active proteins (mainly caseins from the applied natural cheese as the main raw material) within the processed cheese matrix. (Carić et al., 1985; Lucey, 2002; Zehren and Nusbaum, 2000). In addition, caseins have the ability of binding calcium, which has the effect of reducing their solubility and thus enhancing their emulsifying properties. Furthermore, emulsifying salts have a higher affinity for calcium than do the caseins, and thus, are able to improve the solubility and emulsifying ability of the caseins. In general, two types of emulsifying salts exist; those that bind calcium relatively "weak" and those that bind calcium more "strongly." "Weak" emulsifying salts have a modest effect on the emulsifying properties of the caseins, leading to the formation of a soft cheese with relatively large fat droplets. On the other hand, "strong" emulsifying salts give a greater improvement in the emulsifying capacity and result in a firmer cheese with smaller droplets of fat (Hasenhuetl and Hartel, 2008). According to the European Union legislation (Commission Regulation No. 1333/2008) the maximum permitted level of emulsifying salts (expressed as P_2O_5) in processed cheese production is 20,000 mg/kg (or mg/L), as appropriate. In general, depending on the nature of the applied salt, are added in an amount from 1%–3 % (w/w) (Tamine, 2007; 2011). The essential role of emulsifying salts is to solubilize calcium paracaseinate, sequester calcium and thus dispersing the present proteins. Calcium in the calcium-paracaseinate complex of natural cheese is removed by the ion-exchange properties of the emulsifying salts; solubilizing the paracaseinate, usually as sodium paracaseinate (Fox, 2000; Hui, 2005). Moreover, pH adjustment and calcium sequestering are among the main functions of emulsifying salts during processed cheese production. They possess the ability to sequester calcium from the casein matrix; by exchanging sodium ions, resulting in the conversion of insoluble calcium paracaseinate into soluble sodium paracaseinate (Chen and Liu, 2012; Kapoor et al., 2007; Salek et al., 2015). Within the matrix sodium paracaseinate acts as an emulsifier stabilizing the oil-in-water interface. The control and stabilization (upward adjustment) of the pH level and an influence on the formation of the final product structure after cooling are some of the additional properties of emulsifying salts (Guinee et al., 2004;, Dimitreli and Thomareis, 2009; El-Bakry et al., 2011). The application of the correct blend of emulsifying salts increases (due to their buffering capacity) the pH from typical values of ~4.6–5.5 in the natural cheese to values ranging from 5.6 to 6.2 in the spread-type processed cheese. Furthermore, the increase in pH extends the calcium-sequestering ability of the emulsifying salts and the negative charge on the paracaseinate. The dispersed hydrated paracaseinate contributes to the emulsification of the free fat by coating the surfaces of the dispersed free fat globules, resulting in the formation of an artificial (recombined) membrane. Additionally, the high water-binding capacity of the paracaseinate enhances the formed emulsion stability as it leads to high viscosity of the aqueous phase and thus a reduction in the collision frequency of the emulsified particles is observed. However, not all emulsifying salts have the same calcium ion-exchange ability (affinity). The phosphate ion-exchange ability increases with increasing P_2O_5 content in the following order: monophosphate $<$ citrates $\approx$ diphosphate $<$ triphosphate $<$ polyphosphate (Guinee et al., 2004; Shirashoji et al., 2006; Buňka et al., 2013). El-Bakry et al. (2011) and Mizuno and Lucey (2005) stated that trisodium citrate presents better calcium chelating ability and casein peptization properties than do sodium monophosphates and diphosphates. Moreover, on a mole for mole basis, phosphates have a higher calcium chelating ability than citrates (Dimitreli and Thomareis, 2009). The ion-exchange ability increases with the extending length of the polyphosphate chain; it results in better casein dispersion and better fat emulsification and water stabilization, which leads to better crosslinking of the matrix network in the final product (Shirashoji et al., 2010; Kaliappan and Lucey, 2011; Molins, 1991). The absence of emulsifying salts (during heating and shearing of the applied mixture of various ingredients), could lead to the formation of an undesired inhomogeneous mass (Guinee et al., 2004; Buňka et al., 2014). Concretely, by the heating (to temperatures applied in processed cheese production) and stirring of the used natural cheese together with the other ingredients of the formula (in the absence of emulsifying salts); would result in the membrane destruction of the emulsified fat globules, leading to their clustering into larger units. Moreover, the combined effect of low pH and high processing temperatures application would cause aggregation and contraction of the casein molecules, resulting in water release followed by the separation of hydrophilic and hydrophobic phases (a defect known as oiling-off) (Guinee et al., 2004). Furthermore, another important property of emulsifying salts is their bacteriostatic effects. Manufacture of processed cheese normally involves processing temperatures (70–95°C) which are lower than that used for sterilization. Hence, processed cheese might contain viable microbial spores (especially of *Clostridium* genus) originating mainly from the applied raw materials. Nevertheless, spores germination during storage often can lead to serious technological problems (or defects) such as blowing of the packaging material (when cans are used), protein putrefaction and off-flavors. The bacterial spoilage, in general, can be eliminated by preservatives addition. Hence, some emulsifying salts possess noteworthy bacteriostatic properties. Concretely, polyphosphates inhibit many microorganisms (*Staphylococcus aureus, Bacillus subtilis, Clostridium sporogenes* and some *Salmonella* spp.). Moreover, monophosphates and diphosphates have been found to inhibit the growth of dangerous for public health *Clostridium botulinum*; depending on the levels of moisture, NaCl and pH of the processed cheese. However, on the contrary, citrates do not possess bacteriostatic properties, and may be even degraded by bacteria and thus reducing the shelf-life of the final processed cheese product (Buňková et al., 2008; Fox et al., 2000; Tanaka et al., 1986).

The natural cheese varieties which are predominantly applied in various countries of the world vary (including mainly Cheddar, Dutch-type, Swiss-type, Mozzarella-type (or other pasta filata cheeses) and white brined cheeses). In general, in the English-speaking countries (e.g., Britain, USA, Canada, Australia, and New Zealand) the basic natural cheese as the main raw material for processed cheese manufacture is commonly Cheddar and Mozzarella cheeses. On the other hand, in regions around the Mediterranean basin, Balkan, the Near and Middle East, white brined cheeses belong among the most preferable and consumed natural cheese types and are utilized as the basic ingredient for the production of processed cheese and related products (Černíková et al., 2017; Moatsou et al., 2011; Salek et al., 2020). For the past years several studies (Table 3.1) have been performed in which divergent natural cheese varieties were implemented for the production of processed cheese and similar products. In particular, Adbel-Hamid et al. (2000); Awad et al. (2002) used Ras cheese, Adhikari et al. (2009); Biswas et al. (2015); Brickley et al. (2007); Guinee and O'Kennedy (2009);

Table 3.1 An overview of published scientific papers examining properties of processed cheese manufactured from different natural cheeses as the main raw material.

Natural cheese type	Authors
Ras cheese	Abdel-Hamid et al. (2000)
	Awad et al. (2002)
Cheddar cheese	Adhikari et al. (2009)
	Biswas et al. (2015)
	Brickley et al. (2007)
	Fagan et al. (2007)
	Fu and Nakamura (2020)
	Fu et al. (2018)
	Guinee and O'Callaghan (2013)
	Guinee and O'Kennedy (2009; 2012)
	Hassan et al. (2007)
	Hoffmann et al. (2012)
	Janevski et al. (2012)
	Kommineni et al. (2012)
	Rafiq and Ghosh (2017)
	Salek et al. (2020)
	Shirashoji et al. (2010; 2016; 2006)
Mozzarella type	Chavhan et al. (2015)
	Chen and Liu (2012)
	Salek et al. (2017)
	Seth and Bejwa (2015)
Requeijão cremoso cheese	Alves et al. (2007)
	Belsito et al. (2017)
	Cunha and Viotto (2010)
	Cunha et al. (2012)
	Ferrão et al. (2018)
	Nogueira et al. (2018)
	Torres et al. (2017)
Edam cheese	Buňka et al. (2004; 2013)
	Černíková et al. (2017)
	Hauerlandová et al. (2014)
	Macků et al. (2009)
	Nagyová et al. (2014)
	Piska and Štětina (2004)
	Přikryl et al. (2018)
	Sádlíková et al. (2010)
	Salek et al. (2016; 2017; 2020)
	Schädle et al. (2020)
	Weiserová et al. (2011)
Gouda cheese	Dimitreli and Thomareis, (2004; 2007; 2008)
	Fu and Nakamura (2020)
	Fu et al. (2018)
	Mozuraityte et al. (2019)
	Stangieski et al. (2019)
	Weiss et al. (2018)

(continued on next page)

Table 3.1 An overview of published scientific papers examining properties of processed cheese manufactured from different natural cheeses as the main raw material—cont'd

Natural cheese type	Authors
Myzithra-type whey cheese	Chatziantoniou et al. (2015, 2019) Kontou et al. (2019)
Halloum cheese	Hanna (1999)
Kasseri/Kashar (pasta-filata) cheese	Kaminarides and Stachtiaris (2000) Topcu et al. (2020)
Feta cheese	Ghods Rohani and Rashidi (2019) Kontou, Dimitreli and Raphaelides (2019)
Swiss-type cheese	Salek et al. (2016)
Goat cheese	Burgos et al. (2020)
White cheese	Yılmaz et al. (2011) Salek et al. (2020)

2012; Rafiq and Ghosh (2017), Salek et al. (2020); Hassan et al. (2007); Fu and Nakamura (2020); Fu et al. (2018); Janevski et al. (2012); Shirashoji et al. (2010; 2016; 2006); Guinee and O'Callaghan (2013); Kommineni et al. (2012); Fagan et al. (2007); Hoffmann et al. (2012) applied Cheddar cheese, Chavhan et al. (2015); Seth and Bejwa (2015); Salek et al. (2017); Chen and Liu (2012) used Mozzarella type cheese, Cunha and Viotto (2010); Cunha et al. (2012); Nogueira et al. (2018); Ferrão et al. (2018); Torres et al. (2017); Belsito et al. (2017); Alves et al. (2007) used Requeijão cremoso cheese, Černíková et al. (2017); Salek et al. (2016; 2017; 2020); Hauerlandová et al. (2014); Nagyová et al. (2014); Macků et al. (2009); Buňka et al. (2004; 2013); Weiserová et al. (2011); Schädle et al. (2020); Přikryl et al. (2018); Sádlíková et al. (2010); Piska and Štětina (2004) utilized Edam cheese, Stangieski et al. (2019); Weiss et al. (2018); Dimitreli and Thomareis (2004; 2007; 2008); Fu and Nakamura (2020); Fu et al. (2018); Mozuraityte et al. (2019) applied in their studies Gouda cheese, Chatziantoniou et al. (2015, 2019); Kontou et al. (2019) used a Myzithra-type whey cheese, Hanna (1999) applied Halloum cheese, Kaminarides and Stachtiaris (2000); Topcu et al. (2020) utilized in their work Kasseri/Kashar (pasta-filata) cheese, Ghods Rohani and Rashidi (2019), Kontou et al. (2019) used Feta cheese, Salek et al. (2016) used Swiss-type cheese, Burgos et al. (2020) used goat cheese, Yılmaz et al. (2011) applied white cheese.

3.2 Basic aspects in natural cheese production technology

Cheese-making is a rather simple process in itself, but it involves complex chemical and physical phenomena. It is essentially a concentration process, beginning with the coagulation of the main milk protein, casein and then proceeding with manufacturing steps designed to control the chemistry of the casein molecules. The physical or rheological characteristics of cheese are governed mainly by interactions between casein molecules (Johnson and Lucey, 2006). Moreover, the factors that can influence the above-mentioned interactions are the following: pH; dissolution of colloidal calcium phosphate; proteolysis; temperature; cheese composition (in particular, casein content and distribution of moisture and fat). Additionally, each factor can be considered independently, it must also be considered in context with all the other factors. In turn, these physical and chemical parameters "dictate" the manufacturing process. The key elements in producing a cheese with desired properties are: (1) milk composition (because this, in part, determines cheese composition) and (2) the rate and extent of acid development during manufacture (because this influences the loss of moisture, the extent of dissolution of colloidal calcium phosphate and the lowest pH obtainable in the cheese, all key factors in deciding the texture of the finished cheese) (Tamine, 2007; 2011).

3.2.1 Selection of milk

The chemical composition of milk (especially; the content of fat, protein, calcium and pH value), that will be applied in cheese-making process has a strong impact on the composition and sensory properties of the final product. Moreover, the milk's chemical composition is influenced by several factors; including the species, breed, individuality, nutritional status, health, and lactation stage of the producing mammal (animal). The milk should be free of chemical taints and free fatty acids, which could lead to undesirable off-flavors in cheese, and free of antibiotics, which could also cause bacterial cultures inhibition. Furthermore, the milk should be of good microbiological quality, as contaminating bacteria would concentrate in cheese curd and may cause potential quality defects, or even worse, public health problems (Guinee et al., 2004; Fox, 2000).

3.2.2 Standardization of milk

In order to produce a cheese of consistent composition, the first condition is necessary to start with milk of consistent composition. However, to avoid the problem of producing a cheese of different characteristics is critical to maintain milk in consistent composition. Most countries have laws governing two aspects of the composition of cheese; minimum fat in dry matter and maximum moisture levels. Nevertheless, not all cheese varieties have legislation standards. The latter serve to ensure a certain degree of composition continuity for a given variety, even though produced by different manufacturers, and to protect both the consumer and the manufacturing process. These setting standards also can vary from country to country. The composition of milk which will be subjected to cheese production can be modified (standardized) by adding milk solids (condensed milk, skim milk, milk powder, ultrafiltration retentate). The ratio of casein to fat determines cheese composition in terms of the amount of fat in the total solids portion of cheese. The total amount of casein and fat determines the yield potential of milk. In addition, during the coagulation of milk by rennet and the subsequent processing of the formed coagulum, calcium ions play major role. A common practise in cheese making is the addition of calcium in the form of $CaCl_2$ to cheese milk. Another critical factor affecting the whole process of cheese production is the pH of the milk. The pH is inadvertently adjusted by the addition of starter cultures, which decreases the pH value of the milk immediately by about 0.1 units (Fox, 2000; Tamine 2011; Zerfiridis, 2001).

3.2.3 Heat treatment of milk

The reason that selected thermal treatment is applied to cheese milk is mainly to ensure the microbiological quality and hence, to avoid public health problems and inconstant end-products. The fermentation of the milk itself will often inhibit the growth of undesirable bacteria (pathogenic/spoilage), whereas in order to avoid disadvantageous bacterial enzymatic activities, it is preferred that the contaminating flora should be removed. The inactivation is commonly performed by heat treatment (pasteurization) of milk (at a minimum temperature of 72°C for at least 15 s, or any combination of time and temperature in order to achieve an equivalent effect) using plate heat-exchangers. Pasteurization of milk is based on a 9-log destruction of Coxiella burnetti. Nevertheless, some spores in particular may survive the thermal treatment and might cause problems during subsequent ripening of cheese, particularly spores of Clostridium tyrobutyricum. The named spores during cheese maturation grow and produce H_2 and CO_2, which may cause "blowing" of the cheese as well as butyric acid and other fermentation products resulting in unpleasant flavor or defects (Hui et al., 2012; Johnson and Lucey, 2006). During traditional cheesemaking all cheese was produced from raw milk, until the 1940s, when the pasteurization process of milk became widespread. However, even nowadays, significant amounts of unique varieties of cheese are made via using raw milk (Swiss Emmental, Gruyère, Parmigiano-Reggiano). Cheeses made from unpasteurized milk shall be held for 60 days at a temperature not less than 1.7°C (Codex Alimentarius, 2009). It is though that pathogens will die out during this storage period because of acidic conditions in cheese and due to the growth of non-starter lactic acid bacteria (Marth and Steele, 2001). The need of milk with longer shelf-life resulted from the development of the dairy industry, as the production of cheese significantly grew and larger amounts of milk were stored for longer periods. In addition, mainly four alternatives technological operations for lessening the number of microorganisms in milk exist; treatment with H_2O_2, activation of the lactoperoxidase-H_2O_2-thiocyanate system, bactofugation and microfiltration (Fox, 2000). After heat treatment milk is cooled to the temperature conductive for optimal starter activity into specially designed vessels, named vats.

3.2.4 Clotting (gelation) of milk – formation of the cheese curd

At this particular stage of cheese making the milk is converted into cheese curd; a process involving three principal operations; acidification, coagulation (precipitation) and dehydration. After the standardization, thermal treatment or other proper applied treatment of milk have been successfully performed, milk is transferred into vats (of various shapes, constructions, and volumes), where the clotting will take place. Acidification is achieved by the addition of lactic acid bacteria (biochemical acidification) or minority by the direct acidification (chemical) using an acid (usually lactic or hydrochloric) or an acidogen (usually gluconic acid-δ-lactone). When biological acidification is applied the production of lactic acid is a product of lactose fermentation by lactic acid bacteria. Moreover, direct acidification method is commercially applied in the manufacture of Cottage cheese and Feta-type cheese from ultrafiltered milk and Mozzarella cheese. Furthermore, direct acidification has the advantage of a more controllable process than biochemical acidification. However, the addition of bacteria has key functions during cheese ripening. The ultimate pH of the curd for most hard cheese varieties lies in the range of 5.0 to 5.8, while for soft cheese varieties it is within the range of 4.3–5.0 (Aguilera and Stanley, 2000). Coagulation is the essential step in the manufacture of all cheese varieties and is in close relation to caseins of milk in order to form a gel which will enclose the present fat. Three ways exist by which the caseins can be coagulated and the cheese variety to be produced dictates the method employed. Concretely, coagulation may be achieved by: (1) limited proteolysis using selected proteinases (rennet), (2) acidification to pH 4.6, and (3) acidification to a pH level higher than 4.6 ($\approx$ 5.2) in combination with heating to 90°C, (iv) precipitation of the caseins by 40 % (v/v) ethanol solution at pH 6.7 or at lower concentrations if the pH value is reduced. The majority of cheese varieties (representing about 75 % of the total production) are rennet-coagulated. In this case the added coagulating enzyme, destabilizes the casein micelles colloidal suspension and thus makes them aggregate to form a gel. Moreover, the first coagulants were derived from calf stomachs (called rennets), whereas today there are many sources for coagulants, including plants and fungi. Additionally, the characteristic rennet enzymes are chymosine and bovine pepsin, which can be added either as purified standardized enzyme or as an extract of the fourth stomach of calves. The optimal temperature for rennet activity is within 29 – 33°C; normally renneting takes 20–30 min. The direct acidification method is mainly used for cottage, cream cheese, fromage frais and some varieties of Quesco blanco, when low pH is used to cause the caseins to form a clot. In this case, the destabilization process is purely physical. The lessen pH value reduces the repulsive chargé between the casein micelles to a point at which they aggregate and start forming a gel or clot. The third coagulation method is a combination of acid and high temperature to cause precipitation of both casein and whey proteins to form a clot. These cheese varieties are called acid-heat-coagulated cheeses and are of relatively minor importance. They are produced from whey or a blend of whey and skim milk. In this cheese category the most known members are Riccota and related varieties, Queso Blanco, Anari and Manouri cheeses (Fox, 2000; Hui et al., 2012; Law and Tamine, 2010). Regardless of the coagulation method applied, milk fat is surrounded by the present casein as the coagulum forms and is trapped together with serum (whey). The serum contains water-soluble components (lactose, whey proteins, and minerals). The subsequent processing steps are used to remove the serum coagulum. However, these steps differ, and are based on the method that the casein was coagulated (Law and Tamine, 2010).

3.2.5 Cutting of the formed cheese-curd

Renneting process is followed by the cutting of the curd. This is a very important and effective operation for whey release from the coagulated casein. By the moment that the coagulum will reach the required level of firmness, it is then very carefully cut and separated into small pieces (called cheese grains), by special equipment supplied with knife blades or wires. This cutting tool can be designed in different ways. Internally, the casein molecules are rearranging and "tightening", and thus the curd immediately begins to shrink and expel whey. This process is known as "syneresis" and results in the squeezing of the whey from the casein network (Zerfiridis, 2001).

3.2.6 Stirring, heating (cooking or scalding) and whey draining (dewheying)

After the cutting is performed, the curd must be handled gently, so that the formed cheese grains are not broken apart, leading to fat loss in the whey. Stirring is applied to assist the process of "syneresis," in a way that the mechanical effect (of stirring) will cause the collision of the grains, and the resulted pressure will cause the whey to be pressed out of the grains. The current process proceeds for 1 to 2 h. The main purpose of this phase is to regulate the pH and the moisture content of the final cheese. The increase of the final temperature inhibits the growth of the added starter culture, resulting in a higher pH of the cheese. Heating to temperature above 40°C, is typically known as cooking and heating to temperatures beyond 44°C, is called scalding. Moreover, the velocity of the heating influences whey expel. The cooking of semi hard cheeses made with mesophilic starter cultures (Cheddar, Gouda, Edam) is usually performed at a temperature range 34 – 40°C; depending on the fat content. Furthermore, scalding for hard cheese varieties made with thermophilic cultures (Emmental, Gruyère, Comté, Parmigiano Reggiano, Grana Padano) is performed at 45–56°C. The final temperature also has important effect on the later forming curd. The lower the temperature at which the cheese is formed, the more difficult it is to achieve proper pressing and get the cheese grains to stick together. The process can take about 1 h (for many cheese varieties), although it can vary from 30 to 90 min (Fox, 2000; Hui et al., 2012; Law and Tamine, 2010). The texture, color and flavor of the cheese are affected by the method applied for the separation of the whey from the curd formed. The curd can be placed directly in the molds (in the manufacture of soft cheese) or it can be given further treatment before final molding (in the manufacture of most semi hard or hard cheese); depending on the cheese variety to be produced. Whey may be withdrawn directly from the cheese vat. Another possible way to withdraw whey is; by pumping the curd/whey mixture across vibrating or sloping sieves, so that the whey passes through the sieve and the curd grains are separated from the whey (Chandan et al., 2009; Hui et al., 2012; Law and Tamine, 2010).

3.2.7 Salting of cheese

Salting is a vital part of the cheese making process. Salt (NaCl) has very important roles in flavor enhancement and development, in the control of microbiology, the final cheese pH; giving a balanced, pleasant taste, with desired consistency product. Salt increases the osmotic pressure of the aqueous phase, leading to the dehydration of bacterial cells and killing them or inactivating their growth and thus extending the shelf-life of the product. Moreover, for many cheese varieties, salt also affects the moisture content decrease. Concretely, salt affects the syneresis of the curd, resulting in whey expulsion and thus in reduction of the moisture of cheese, which also affects the activity of microorganisms and enzymes. There are two components to the salting process; the application of salt (salting) and the subsequent mixing uptake, associated with moisture loss (mellowing). The amount of salt required will vary depending on the type of cheese being manufactured. In general, salt levels in cheese range from ~1.0% in Swiss type cheese, 1.8% in Cheddar, to 3% or more in Feta type cheeses. For most semihard and semisoft varieties, a NaCl concentration of 1.5%–2.0 % is suitable. On the other hand, for Emmental or other large-eyed cheeses, the salt content is lower (e.g. 0.8%–1.4%), in order to enhance the growth of propionic acid bacteria. Salting can be proceeded, mainly by four ways; by adding salt into the curd before molding, by diffusion of salt into cheese after molding, by immersion in brine and by dry salting (Chandan et al., 2009; Fox, 2000; Law and Tamine, 2010).

3.2.8 Ripening/maturation of cheese

Cheese can be described as an essential matrix principally, formed by protein, fat and carbohydrates, containing a wide range of enzymes and microorganisms. Their activities are allowing the changes that convert the young (fresh) cheese into an end-product with desired properties, mainly via proteolysis, lipolysis, and glycolysis. Furthermore, ripening is a highly complex of several phenomena involving changes to the microflora of the cheese, including the death and lysis of the starter cells, development of the adventitious non-starter microflora and in many cheeses growth of a secondary microflora (Guinee et al., 2004). Moreover, cheese ripening can be a slow process ranging from about 3 weeks (Mozzarella) to 2 or more years (parmesan and extra-mature Cheddar cheeses, respectively) (Fox, 2000). During this period, in general, a very complex group of biological, biochemical and chemical reactions occur; resulting in the development of the characteristic flavor compounds and texture modification. Moreover, the rate of ripening is directly related to the moisture content of the cheese; low-moisture cheeses have slow ripening rate (Fox, 2000). Concretely, the reactions occurring in the cheese matric during ripening are categorized into 4 major groups: (1) glycolysis of the residual lactose and lactate catabolism, (2) catabolism of citrate, (3) lipolysis and free fatty acids catabolism, and (4) proteolysis and amino acids catabolism. The metabolism of lactose to lactate is essential in the production of all cheese varieties. Since cheese curd contains a low level of residual lactose (most of lactose is lost in the whey – draining process), the latter is rapidly metabolized to lactate at the early phases of ripening. The final step in the glycolysis process is the conversion of pyruvate to lactate, which is catalyzed by lactate dehydrogenase. Lactate is the end product of glycolysis which converts 1 mol of lactose to 4 mol of lactate with the production of 4 mol of ATP. Lactate is an important substrate for a range of reactions, with positive or negative impact on the final cheese product. Furthermore, in Cheddar and Dutch-type cheeses, L-lactate is produced by *Lactococcus* strains, which can be racemised to DL-lactate by non-starter lactic acid bacteria. The DL-lactate is however, less soluble than L-lactate, resulting in the formation of Ca-D-lactate crystals, which appear as white specks on the surface of the matured cheese. Nevertheless, catabolism of lactate is very important in the production of Swiss-type and mould-ripened cheeses. During the production of Swiss-type cheeses, lactate is catabolized by, for example, *Propionibacterium freundenreichii* subsp. *shermanii* or *Propionibacterium freundenreichii* subsp. *freundenreichii* to propionate, acetate, H_2O, and CO_2. Propionate and acetate contribute to the formation of Swiss-type cheese typical flavor. Hence, the CO_2 formed migrates through the curd to points of "weakness", where it collects and forms the large eyes, characteristic for this type of cheeses (Guinee et al., 2004). Milk fat is essential in the development of flavor of all ripened cheese varieties. Lipids in cheese can undergo hydrolytic or oxidative degradation. However, in most of varieties, lipolysis is rather limited. Major exceptions are Blue (Gorgonzola, Roquefort) and some Italian cheese varieties (Romano and Provolone); in which the fatty acids and/or their degradation products contribute in flavor. Enzymatic activities of lipases of various sources (raw milk, rennet, starter bacteria) contribute to lipolysis during ripening. Concretely, by the lipase enzymes the triacylglycerides of milk can by hydrolyzed into free fatty acids and glycerol. The short chain fatty acids (C4–C10) have a sharp, pungent flavor (Fox, 2000; Hui et al., 2012). Levels of fatty acids vary considerably between varieties. Many internal bacterially ripened varieties (Edam, Swiss-type, and Cheddar) contain low levels of free fatty acids (200–2,000 mg/kg), on the other hand, Blue cheeses have high levels of fatty acids (30,000 mg/kg). Additionally, fatty acids are precursors for the production of other volatile compounds (fatty acid esters; ethyl esters are more common in cheese, thioesters, fatty acid lactones, and methyl ketones) during ripening (Guinee et al., 20041). Proteolysis is the most complex, in most cheese varieties, also the most significant event occurring during ripening of cheese. Additionally, is very important affecting cheese texture; by hydrolyzing the para-casein matrix, which gives cheese its structure (body) and by increasing the water-binding ability of the curd (Guinee et al., 2004). The caseins are gradually by rennet proteinase enzymes and by plasmin (an indigenous milk protease), which yields a number of polypeptides. The latter does not affect the flavor of the cheese, however, some hydrophobic peptides may be astringent or may cause bitter taste (Hui et al., 2012). These first stages of proteolysis are of crucial importance for the transformation of the structure of cheeses; from a rubber-like to a sliceable consistency. Furthermore, the large polypeptides generated by chymosin (which is trapped in the curd) are degraded further by enzymes from the starter bacteria and adventitious populations of non-starter lactobacilii in cheese. Lactobacilii have proteinases that can degrade α_{s1}- and β-caseins. The above mentioned microorganisms also possess intramolecular aminopeptidases, dipeptidases, carboxypeptidases, and endopeptidases, playing a vital role in the production of free amino acids. The latter is characterized as precursors for a range of catabolic reactions which produce many volatile flavor compounds (McSweeney, 2004; McSweeney and Sousa, 2000; Yvon and Rijnen, 2001).

3.3 Manufacture of selected cheese varieties

In the following subchapters will be described the essential steps implemented in the production of Edam, Swiss-type, Mozzarella, Cheddar and White brined cheeses, respectively. As it was previously mentioned the latter cheese varieties belong to those that are widely used in the production of processed cheese or related products.

3.3.1 Edam cheese manufacture

Edam cheese (Edammer kaas – in Dutch) together with Gouda cheese (Goudse kaas – in Dutch) are traditionally the two predominant Dutch-type cheeses, which are mainly produced in the Netherlands (Guinee et al., 2004; Weimer, 2007). Furthermore, Edam cheese originates from the homonymous area in the province of North Holland and derive its name from a city (Edam, Netherlands) which is famous market for this kind of cheese (Zerfiridis, 2001). According to the Codex standard for Edam (Codex Alimentarius, 2009), developed by the Codex Alimentarius, Edam cheese is defined as it follows: Edam is a ripened firm/semi-hard cheese in conformity with the General Standard for Cheese. The body has a near white or ivory through to light yellow or yellow color and a firm-textured (when pressed by thumb) texture, suitable for cutting, with few more or less round rice to pea sized (or mostly up to 10 mm in diameter) gas holes, distributed in a reasonable regular manner throughout the interior of the cheese, but few openings and splits are acceptable. The shape is spherical, of a flat block or of a loaf. The cheese is manufactured and sold with dry rind, which may be coated. Edam of a block or loaf shape is also sold without rind. Edam cheese is characterized of a semi-hard to hard consistency, and a smooth texture having small eyes (about the size of a pea – diameter between 2 and 10 mm) with a flavor intensity that ranges greatly (Codex Alimentarius, 2009; Fox, 2000; McSweeney, 2007; Weimer, 2007; Zerfiridis, 2001). According to the FDA standard of identity for Edam cheese, the latter may be classified as a food prepared by any procedure which produces an end-product (cheese) with a minimum milk fat content of 40% by weight of the solids, and a maximum moisture content of 45% by weight (Weimer, 2007; Zerfiridis, 2001). Edam is a mild cheese with a rubbery texture. Traditionally, it is manufactured in a spherical shape and is coated in a red/yellow wax, or plastic coating, however, other shapes may also be found (rectangular blocks, loaf). Moreover, mature Edam should have a fat in dry-matter content of 42 % (w/w) and consequently the milk for cheese production is standardized to 2.5% (w/w) fat. For the production of Edam cheese a mixture of evening milk and morning milk is used during the traditional cheese-making procedure. The standardized milk (for its fat:protein ratio in order to control the fat content in the dry matter of the cheese) is pasteurized and mixed-strained with mesophilic starter culture (containing citrate fermenting lactococci and *Leuconostoc* spp.) is added at a level 0.3%–0.5 % w/v. Frequently used mesophilic dairy cultures during the production of natural Dutch-type cheese include the representatives of *Lactococcus lactis* subsp. *lactis, Lactococcus lactis* subsp. *lactis* biovar *diacetylactis* and *Lactococcus lactis* subsp. *cremoris* (Flasarová et al., 2016; McSweeney, 2007; Weimer, 2007). Together with the starter culture addition, the milk coagulation process (approximately at 30°C) occurs and is enhanced by the action of renneting enzymes (25–30 mL/100 L of milk), usually chymosin preparation (calf rennet). Additionally, the milk is often bactofugated to reduce the number of spores of butyric acid bacteria. However, the bactofugation technique may be applied when cheese is made from raw milk. Following starter addition, calcium chloride (0.015 %) is added in order to optimize renneting of the low acid milk. In addition, at the same time, sodium or potassium nitrate (0.015 %) or lysozyme can be added to prevent blowing of the cheese through gas production by coliforms or *Clostridium tyrobutyricum*. Moreover, annatto or β-carotene can be added at low levels as coloring agents. Thereafter, the formed curd is cut approximately 30 min after the rennet addition. The curd is then gently stirred and 50% of the whey is removed. Water of 55–60°C is added into the vat to bring the temperature of the curd-whey mixture to 36°C (mild scalding). Washing of the curd reduces the lactose level and prevents the development of excess acidity during cheese ripening. Once the final scald temperature has been achieved, approximately 30% of the whey is removed and the curd is pressed mechanically in the vat to encourage it to fuse together. The curd then is transferred into molds and pressed lightly. Moreover, the pressed curd is immersed in saturated brine (14%–20 % NaCl content w/w, at 12–15°C) for 2–3 days, to obtain a salt content of 2.5%. The cheeses then are stored in ripening cellars (14 -15°C at a relative humidity of 85 – 90 %) for the rind to dry off and harden. Edam cheese must be sold after undergoing ripening of 4 weeks at minimum. The ripening period of Dutch type cheeses can range from some weeks or longer, to sometimes for more than a year. Nevertheless, a typical ripening period for Edam cheese lies within the interval of 6 up to 8 weeks (Aljewicz and Cichosz, 2015; Pachlová et al., 2011; 2012).

3.3.2 Swiss-type cheese manufacture

Swiss-type cheeses are hard or semi hard brine-salted cheeses, of nutty flavor, containing characteristic eyes, resulting from the metabolism of various substances. This cheese category was originally manufactured in the Swiss German: Emmental (Emmen valley) in Switzerland; their precursors were mountain cheesed. However, there is no general internationally recognized definition of Swiss-type cheeses that could differentiate them among other cheese varieties and there is a lack of consensus regarding the cheese varieties to be included in this category (Bisig et al., 2010). Emmental cheese is probably the best-known representative of all Swiss-type cheeses around the world, and is frequently referred to simply as Swiss cheese (it comprises cheese with the desired propionic acid fermentation (Bisig et al., 2010). According to Zerfiridis (2001), the general term Swiss cheese (referring to Emmental) probably dominated due to the fact that Emmental presents the largest size (or shape) and the greatest eyes and in Switzerland is considered at the "king" of all cheeses. The characteristic eyes, a typical sign of the Swiss-type cheeses are formed due to the production of CO_2 which is product of the propionic acid fermentation. Hence, these cheeses are also called "propionic acid fermentation cheeses" (Bisig et al., 2010; Zerfiridis, 2001). Nevertheless, in 2007, the Codex Alimentarius Commission of the Food and Agriculture Organisation (FAO) of the United Nations developed an internationally recognized Codex standard for Emmental as follows: Emmental is a ripened hard cheese, usually manufactured as wheels and blocks of weights from 40 kg or more, adapted to various use conditions. The body of the cheese has an elastic texture, with regular cherry to walnut sized gas holes, or eyes. The activity of propionic acid bacteria is essential to eye formation. Swiss-type cheeses are manufactured by techniques differing from traditional Swiss procedures. Thus, the milk treatment, the starters used, the weight, shape, ripening period and shelf-life are some of the main differences (Bisig et al., 2010, Isolini et al., 2011).

3.3.3 Mozzarella cheese manufacture

Mozzarella cheese originates from the southern regions of Italy and is typical of pasta filata or stretched-curd variety (together with Provolone the most important cheeses of this category). The term pasta filata is derived from an Italian phrase, that literally means "spun paste" or "stretched curd," which refers to a unique plasticization and stretching process; that is shared by all cheeses in this category, giving them their common identity. Pasta filata cheeses enclose a wide range of cheese varieties (Mozzarella, Provolone, Scarmorza, Caciocavallo, Kashkaval), which originated primarily in the northern Mediterranean region (Italy, Greece, the Balkans, Romania and Turkey (Ferrari et al., 2003; Gernigon et al., 2010; Pisano et al., 2016). The term Mozzarella stemmed from the Italian word "mozzare", meaning to tear, and indicative to the stretching process applied during its manufacturing (Hui, 2005). Furthermore, over the last decades, this cheese variety has experienced substantial growth, mainly because of its use as ingredient in a range of foods (including pizzas) and is a high volume cheese supporting food service industries. Mozzarella exhibits desirable functional properties, such as stretching and melting. Moreover, according to the Codex Standard for Mozzarella cheese: Mozzarella is an unripen cheese in conformity with the General Standard for Cheese and the Standard for unripen Cheese Including Fresh Cheese. It is a smooth elastic cheese with a long stranded parallel-orientated fibrous protein structure without evidence of curd granules. The cheese is rindless and may be formed into various shapes. Nowadays, Mozzarella is typically manufactured from cow's milk, whereas during traditional manufacture is made mainly from buffalo (*Bubalus bubalis*) milk [in 1996 gained a Protected Designation of Origin (PDO) recognition under European Union disciplinary], or mixtures of cow and buffalo milk. This cheese is considered a fresh cheese and is characteristic for its typical mild flavor and recognizable rubbery texture. However, Mozzarella cheese presents high values of moisture up to 60 % (generally 45 – 52 % w/w), elevated pH (in the range of 5.1 – 5.3), lack of ripening and maximum of 2 % of salt. The latter "microenviromental" cheese conditions provide adequate surroundings for the survival and growth of pathogenic microorganisms that may pose risks to public health. Hence, due to the above mentioned reason, Mozzarella is made from pasteurized milk (Amagliani et al., 2007; Costas Dias et al., 2012; Gorrasi et al., 2016; Rubio-Texeira, 2006). Typical pasteurization treatments are in the range of 73–74°C for 17–20 s. Such thermal treatments are considered sufficient (assuming of good quality raw milk), to minimize pathogens and lessen spoilage and competitive microflora, as well as inactivating several enzyme systems (Fox et al., 2015). Milk that is used for Mozzarella manufacture is almost always standardized to a specific protein-to-fat or casein-to-fat ratio to produce a cheese of desired fat in dry-matter content (Law and Tamine, 2010). The traditional manufacture of Mozzarella is similar to Cheddar cheese manufacture (in the initial production-phases) as far as the milling stage. However, higher temperatures are used in processing (up to 42°C), thus consequently a thermophilic starter culture (usually of 1.5%–2.0 %) is more essential. This normally consists of a mixture of *Streptococcus thermophilus* alone or in combination with Lactobacillus *delbrueckii* subsp. *bulgaricus* and/or *Lactobacillus helveticus* (De Angelis and Gobbetti, 2004; Guidone et al., 2016; Law and Tamine, 2010). Aditionally, Mozzarella can be manufactured also by using mesophilic (*Lactococcus lactis* subsp. *lactis, Lactococcus lactis* subsp. *cremoris*) lactic acid bacteria, typically used in the manufacture of Cheddar. However, the final product has a blander flavor than the typical acetaldehyde flavor obtained when thermophilic starters are applied. The main role of the added starter culture is to produce lactic acid (in adequate level) to transform the curd into that will plasticize and stretch in hot water. Therefore, the starter culture selection is dependent on the desired rate of acid production during the subsequent steps of cheesemaking (Hui, 2005; Law and Tamine, 2010). These bacteria convert lactose to lactic acid, lowering the pH, promoting whey expulsion from the curd and altering the protein structure. In addition, an alternative method to destabilize the casein micelles and thus cause the curd formation (also used in Mozzarella manufacture), is the direct application of a food-grade acidulant (e.g., acetic, lactic, or citrc acid) (Faccia et al., 2009; Guidone et al., 2016). Thereafter, in the inoculated milk rennet is added. The resultant coagulum is cut gently, approx. into 1.0–1.5 cm cubes. Immediately the curd begins to expel whey and shrink. Subsequently, the curd-mass is left to rest (or to "heal") for 5 min and then is gently stirred for clumping prevention. If cultures have been applied the temperature of the curd is increased to 41°C. The curd is separated from the whey, at the moment when proper level of acidity of the curd is reached (typically pH of 5.9). At this stage, the curd is allowed to mat together or is gently stirred, until is ready for stretching (optimal curd pH within the interval of 5.0–5.3). Additionally, at the pH range mentioned previously, calcium phosphate is becoming more dissociated and a decrease in the net charge on proteins facilitates an increasing degree of hydrophobic interaction between casein molecules. Curd stretched at pH 5.3 is more structured, as indicated by a higher apparent viscosity immediately after manufacture that continues during the ripening process, compared to Mozzarella curd made from curd stretched at pH 5.0 (Rowney et al., 2003a; 2003b; Yun et al., 1993). Before mixing takes place, the curd is usually pre-salted at a rate of up to 2 % (salt to curd ratio), significantly reducing the future brining period (Hui, 2005). Furthermore, the subsequent step is the heating and stretching process of the curd-mass, resulting in a visible alignment of protein fibers within the curd. Concretely during this stage, fat globules form channels between the protein strands and allow protein fibers to align and separate (McMahon and Oberg, 2011; Oberg et al., 2015). In reality, the curd is immersed in hot water (68–74°C), salt brine or whey, where is warmed gradually to 60°C. The heated curd is kneaded by an auger system. Protein fibers align and entrap coalesced fat and moisture, resulting in a plasticizing effect on the curd (Hui, 2005). Thereafter, the hot plasticized curd is forced under pressure into chilled molds, giving the cheese the desired shape. The cheeses then may undergo further cooling and salting (by immersion in cold brine or by direct salting in some cases). However, in practice many variations of the basic Mozzarella manufacture protocol exist, but the underlying principles are common to all procedures (Hui, 2005; Law and Tamine, 2010).

3.3.4 Manufacture of cheddar cheese

Cheddar cheese is one of the most important cheese varieties worldwide and one of the most consumed variety of natural (Paludetti et al., 2020). The Cheddar cheese originates in England (in Somerset) were it has been produced since the middle ages. Moreover, Cheddar cheese probably got its name from the town of Cheddar where traditionally caves were used in order to store the produced cheese during the ripening period. Nowadays, Cheddar cheese is produced not only in the British regions but in most areas of the world. Moreover, Cheddar cheese was traditionally produced as large cylindrical shaped loafs (weighting nearly 30 kg) but nowadays under modern industrial scale process the cheese is usually formed into rectangular blocks in order to promote handling (Walstra et al., 2005). Cheddar cheese can be described as a ripened hard cheese in conformity with the General Standard for Cheese. The body of the cheese has a white or ivory through to light yellow or orange color and a firm and short (when is pressed), smooth and waxy texture. In addition, gas holes are absent (due to the use of homo-fermentative starter cultures), but some openings and splits can be acceptable. The manufactured Cheddar cheese should not possess any surface microflora and shall be available with and without a rind (Walstra et al., 2005). Furthermore, Cheddar is a matured cheese consisting of approximately 55 % (w/w) of moisture, belonging to the category of hard cheeses. The ripening period needed to developing the typical organoleptic properties (mainly in terms of flavor and body characteristics development) is normally from 5 weeks up to 2 years (at 7–15°C), depending on the extend of the maturity level required (Salek et al., 2020). Many factors can influence the consistency of the Cheddar cheese. Hence, among these factors are the production steps of cheddaring and the salting process of the cheese-curd before pressing. The cheddaring process affects the structure of the cheese by making the closed structure of the cheese more enhanced due to the removal of entrapped air from the cheese. The addition of salt to the cheese curd can influence its consistency characteristics. The salt content in a Cheddar cheese should be in the range of 1.75–1.95 % w/w (Walstra et al., 2005). Moreover, the pH (4.9–5.3) is also a factor affecting the texture of Cheddar cheese. Hence, the higher the pH the softer the texture of the cheese will be (Walstra et al., 2005). Cheddar cheese production process involves 2 basic phases (1) the conversion of milk into curd in the first phase, which is generally accomplished over 24 h and (2) and the transformation of the young "green" cheese into a mature cheese in the second phase. Traditionally, the milled-curd method is applied for Cheddar cheese manufacture. The latter method involves the process of "cheddaring" (which involves turning and flipping loaves of warm curd at regular intervals for 1 to 2 h for the development of acid, leading to a fibrous, "chicken breast-like" structure), followed by milling, hooping, and pressing. In the stirred-curd method of manufacture, the curds are continuously stirred after the drainage of the whey. The stirred-curd method is the method of choice in modern and highly mechanized production plants. In addition, acidification could be characterized as one of the most important operations over the production protocol of Cheddar cheese. Starters are added to milk in order to achieve a uniform and predictable rate of acid production. Furthermore, the post-coagulation steps for milled-curd Cheddar are different from those for stirred-curd Cheddar; therefore, the development of acidity and the growth rate of starter bacteria might be different. Starter bacteria also contribute to proteolysis during Cheddar cheese maturation through the action of their cell envelope-associated proteinase and intracellular peptidases (Singh et al., 2003).

3.3.5 Manufacture of white brined cheeses

The East-Mediterranean regions are these where probably white brined cheeses originate (about 8000–9000 years ago). Furthermore, the current cheese variety is produced under various names in different countries of the world and the most well-known representatives are the following: *Feta* (Greece); *Telemea* or *Telemes* (Romania, Greece); *Bjalo salamureno sirene* (Bulgaria); *Mohant* (Slovenia); *Sjenicki, Homoljski, Zlatarski* (Serbia); *Pljevaljski, Polimsko-Vasojevaski, Ulcinjski* (Montenegro); *Travnicki* (Bosnia-Herzegovina); *Beyaz peynir* (Turkey); *Liqvan* (Iran); *Akawi* (Lebanon, the Czech Republic); *Domiati* (Egypt), *Brinza* or *Brynza* (Russia, Ukraine, Slovakia); *Halloumi* (Cyprus). Cheeses matured and preserved in brine (solution consisting of 6-12 % w/w of NaCl), and in most cases these cheeses could be stored without the need for refrigeration temperatures/regimes. Hence, this method of cheese storage (in brine) can influence the biochemical, textural, structural and flavor changes during the phenomenon of cheese ripening. In addition, white brined cheeses are without rind and can vary in consistency properties (from soft to semi-hard type). Traditionally, sheep's, goat's, and buffalo's milk is used for the production of this cheese category type of cheese. Therefore, they retain the white color from the main raw material (milk). Furthermore, some white brined cheeses can be also described by the term "white soft cheese". Moreover, the color of white brined cheeses is white (e.g. porcelain white, marble white, or snow white) when they are made from sheep's, goat's, or buffalo's milk. Nevertheless, their color ranges from off-white to yellowish when they are made from cow's milk. In general, when cow's milk is used and as the consumers expectation is a final product with a white color, in some regions (when is permitted by the legislation laws) some 'decolorants' (e.g., chlorophylls, titanium dioxide) or other treatments are applied in order to cover or eliminate the yellowish color. Moreover, no gas holes or other openings are allowed, with exception for some small mechanical openings. However, the presence of small or large gas holes is a visual defect of the product. Their texture can be described as smooth, soft, and crumbly but in most of the cases the end product is sliceable. Moreover, the shape of white brined cheeses is associated with that of the container in which the product is stored, and usually cheese blocks are rectangular, cubic or in some circumstances cylindrical. The final cheese products are packed in containers (varying in size) which are full-filled with brine solution. Moreover, in some countries (especially in small artisanal units), cheese whey containing 8%–12% salt can be applied as the brining solution. In addition, the increased demand for cheeses matured in brine, particularly for Feta cheese, led to the need for standardized techniques for their manufacture. Feta is the most known and consumed white brined cheese over the world and is registered as a Product of Designation of Origin (PDO) by a European Union Commission Regulation. Moreover, Feta cheese has a salty and slightly acidic and piquant taste, a soft texture, few openings are present, and a white color (Alichanidis and Polychroniadou, 2008;

Al-Nabulsi et al., 2020; Miloradovic et al., 2018). In addition, Feta is produced using ewes' milk or a mixture of ewes' and goats' milk. Furthermore, the latter milk (goat) shall not exceed 30% of the total cheese milk used for the production of Feta cheese. The levels of moisture and fat-in-dry matter content should be 56% (maximum) and 43% (minimum), respectively, and the cheese must be ripened for at least 60 days. According to the Commission Regulation published in 2002 by the European Union, the cheese milk for Feta production must originate only from selected geographic areas (Macedonia, Thraki, Ipiros, Thessalia, Sterea Ellada, Peloponnissos, and Lesvos Island) of Greece (Hayaloglu, 2017).

3.4 The impact of natural cheese maturity degree on processed cheese properties

The basic raw material for the traditional production of processed cheese is natural cheese. Therefore, by the selection of this main raw material, it is possible to influence the consistency of the final processed cheese product. During the selection of the natural cheese, it is necessary to take into account the type of natural cheese (variety), the degree of maturity, its composition (dry matter, fat, protein, and calcium contents, respectively), the pH value and the required properties (organoleptic or functional) of the final product.

Proteolysis is one of the principal biochemical changes during cheese ripening leading to physical, chemical, flavor, and functional changes in natural cheese. Proteolysis occurs when various enzymes hydrolyze proteins in cheese during ripening. The sources of enzyme for proteolysis originate from chymosin and bovine pepsin in rennet residual, indigenous and endogenous heat-stable proteinases in milk, and various proteinases from starter and non-starter bacteria. In other words, during the ripening process of natural cheeses, the phenomenon of proteolysis occurs, where the casein (*para*-casein) component is hydrolyzed into peptides and free amino acids. In particular, proteolysis of casein occurs by the action enzymes involving the residual activity of native milk enzymes, rennet and the exogenous enzymatic system of acidic and non-acidic lactic acid bacteria. The ratio of cleaved and practically non-cleaved (intact) casein is an important criterion influencing the consistency of processed cheeses (Kapoor and Metzger, 2008). The term intact casein in cheese refers to the casein that has not been hydrolyzed during the ripening process, which is used as a non-specific method to measure cheese proteolysis. Intact casein content in the final product has been also correlated to cheese rheological and functional properties. Experimentally, intact casein is determined by the difference between total protein and the pH 4.6-soluble protein using nitrogen-based quantifications. Furthermore, intact casein and total protein contents can provide indirect essential information for processed cheese manufacturing as they both can determine its functional and textual properties (Kapoor and Metzger, 2008). Furthermore, it is generally accepted that the intact casein content of natural cheese is inversely related to the age of the natural cheese. Hence, as a natural cheese is matured, its intact casein content (Kapoor et al., 2007).

Moreover, Guinee et al. (2004) stated that "young," unripe natural cheese containing 70%–95 % of intact casein is suitable for the production of sliceable processed cheese products, whilst medium-ripened cheese with 60%–75% of intact casein is suitable for the manufacture of spreadable processed cheese products. Thus, if an immature raw material with a low degree of proteolysis is used, the processed cheese product obtained is characterized by a stiffer, gummier consistency and an empty taste. On the other hand, the advantage of using this kind of raw material (unripen natural cheese) is the reduction of raw material costs. In addition, by using a ripened natural cheese with a high degree of proteolysis, an easily melting mixture is obtained and the final processed cheese product has a fine and spreadable consistency. During a longer maturing process, plethora of sensory active substances are formed, giving to the developed processed cheese melt a full and distinctive aroma (Guinee et al., 2004; Lee and Klostermeyer, 2001; Swenson et al., 2000). In practice, a mixture of more or less matured natural cheeses is usually applied during the production of processed cheese. If the raw material composition contains older, very mature natural cheeses with excessively hydrolyzed proteins, the addition of intact casein (in the form of young cheeses or curd) is usually included, which will help to form a stable protein matrix. Otherwise, there is a risk of an unstable to liquid-like consistency of the product. This phenomenon would occur due to the low binding of water and its release and the inability to form a stable protein matrix. The effect of increasing the degree of proteolysis of natural cheese (Cheddar natural cheese was the basic raw material) on the textural and viscoelastic properties of processed cheese was demonstrated in a study by Brickley et al. (2007). The degree of proteolysis of Cheddar increased most rapidly in the first 28 days after its production, which was in line with the highest decrease in the firmness of the processed cheese produced. Piska and Štětina (2004) produced processed cheeses from mixtures of natural cheeses with different degrees of proteolysis. It was found that the sample showing the highest stiffness, adhesiveness and gumminess was made from natural cheese with a low proteolysis index (ratio of water-soluble nitrogen at pH 4.6 to total nitrogen). On the contrary, the values of these textural parameters were the lowest for the product manufactured from the raw material with a higher proteolysis index.

Additionally, Bunka et al. (2013) studied the dependence of natural cheese (Edam block; 2, 4, and 8 weeks of maturity) maturity level on selected textural properties (hardness, cohesiveness and relative adhesiveness) of spreadable processed cheese (40%, w/w, dry matter, 50% w/w fat in dry matter contents) during a 30 day storage period. The above-mentioned authors stated that hardness of processed cheese decreased when a raw material with higher level of maturity was applied for the production of the examined samples. Furthermore, with the raising maturity degree of the cheese, the values of processed cheese cohesiveness slightly decreased. On the other hand, the obtained values of relative adhesiveness of the processed cheese samples increased. Furthermore, significant changes in the relative adhesiveness values were reported when cheese with 8-week maturity was used. A possible explanation of this could be due to a shortening length of casein fractions caused by the rising intensity of proteolytic changes during the ripening of the utilized raw material (natural cheese) (Brickley et al., 2007; Piska and Stětina,2004; Wang et al., 2011). The results are in accordance to the findings reported by Brickley et al. (2007) and Guinee et al. (2004). In general, the values processed cheese relative adhesiveness should increase with the rising maturity level of the used natural cheese. Moreover, according to the work of Hladká et al. (2014) model samples of processed cheeses (40% w/w dry matter content and 50% w/w fat in dry matter content; Edam cheese block was applied as the main raw material; 50% w/w dry matter, 30% w/w fat in dry matter) were homogeneous regardless of the maturity degree of cheese (the maturity of the applied raw material was in the range of 1–16 weeks). Moreover, the latter authors developed processed cheese samples without emulsifying salts addition. The highest values of hardness were observed in the products made of cheese with 1–2 weeks of maturity. Moreover, with the increasing degree of cheese maturity (4–12 weeks), the hardness of the tested samples was gradually decreasing. Additionally, the hardness of processed cheese made of natural cheeses with maturity of 12–16 weeks did not differ. Nevertheless, the lowest values of hardness were observed in the samples examined on the production day and this parameter was significantly increasing with the prolonging of the storage period. In addition, other series of studies performed by Salek et al. (2016; 2017). Salek et al. (2020) examined the effect of various natural cheese varieties and their degree of maturity on selected textural and rheological properties of model processed cheese samples. The above-mentioned authors applied Mozzarella-type (42 % w/w dry matter content; 35 % w/w fat in dry matter content; 0, 2, 4 weeks of maturity), Swiss-type (60% w/w dry matter content, 30% w/w fat in dry matter content; 4, 8, 12, and 16 weeks of maturity), Cheddar (dry matter content 62% w/w, fat in dry matter content, 50 % w/w; 4, 8, 12 and 16 weeks of maturity) and White brined cheeses (dry matter content 48% w/w, fat in dry matter content 38 % w/w; 2, 4, 8, 16 and 24 weeks of maturity) respectively, as the basic raw material in the formulation of the resultant processed cheese samples. The hardness of the samples obtained decreased with the rising maturity degree of the natural cheese used (regardless of the applied cheese variety). Moreover, the monitored values of the gel strength and interaction factor decreased with the increasing maturity degree of the cheese used. According to Kapoor and Metzger (2008), the reported decrease in the gel strength values could be probably due to fewer number of interactions within the processed cheese matrix. In general, it could be stated that a more rigid end-product can be expected with higher gel strength values. The intensity of rigidity of the processed cheese samples has an analogous relationship to the intensity of the gel strength; the higher the gel strength of the sample, the more firmer the product that can be expected (Salek et al., 2016; 2017). Moreover, higher values of gel strength were reported for the processed cheese samples produced with Cheddar cheese in comparison to those made from White brined cheese. According to Salek et al. (2020) from the comparison of processed cheese samples manufactured from two different natural cheeses (Cheddar and White brined cheese) of the same maturity level (4-, 8- and 16-week maturity), it can be seen that the White brined processed cheese samples were significantly less hard that those made of Cheddar cheese. Moreover, this important difference in the hardness values development could be due to different chemical composition (including pH, calcium content, NaCl content, residual lactose content and mainly the amount of intact casein) and cheesemaking process of the applied natural cheese (Piska and Štětina, 2004; Purna et al., 2006). As it was mentioned previously, the level of intact casein present in the natural cheese can influence the properties of the resultant processed cheese. Furthermore, the structural

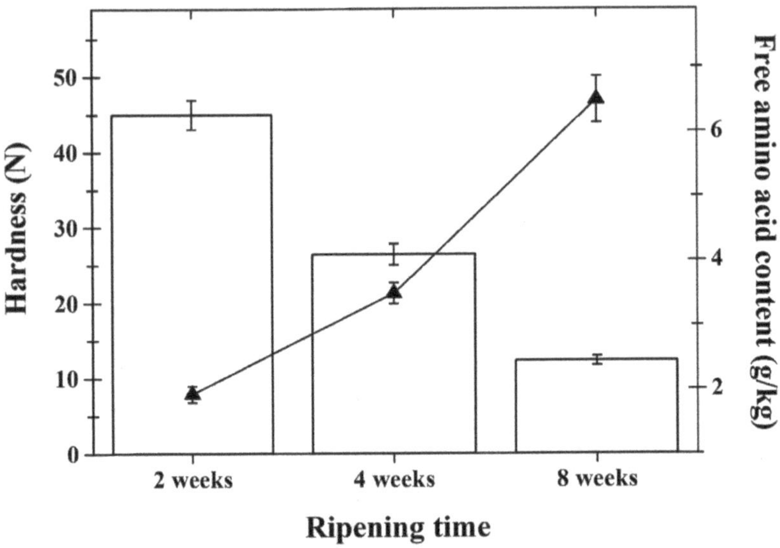

Figure 3.1 Dependence of processed cheese hardness (N; columns) and free amino acid content (g/kg; ▲) in natural cheese (the main raw material for the production of processed cheese) on the ripening time (weeks) of Edam cheese. The results were expressed as mean and standard deviation (bars).

network of processed cheese is mainly formed by intact casein and thus, a high level of intact casein could be a potential indicator for abundant interactions between the present proteins and fat. According to the literature, processed cheese produced from natural cheese with lower intact casein level is soft, whereas processed cheese made from natural cheese with a higher level of intact casein is firm. On the same token, probably the level of intact casein in the utilized White brined cheese was lower in comparison to that in Cheddar cheese, resulting in significant differences in their values of hardness. On the whole, from the reported results of rheological and hardness analyses it could be concluded that the processed cheese gel strength and hardness development follow a contiguous trend which is not affected by the type of applied natural cheese (Figs. 3.1–3.5).

In addition, Burgos et al. (2020) evaluated the effect of the maturity degree of natural goat cheese on the textural, rheological and organoleptic properties of processed cheese products. Furthermore, the developed processed cheeses were formulated using goat cheeses with 10, 20, and 40 days of ripening. In particular, the hardness, adhesiveness and complex modulus (G*) of the tested samples decreased as the maturity degree of the utilized natural cheese (raw material) increased, which is probably associated with the intact casein content. According to the latter authors, increasing the amount of natural goat cheese having hogher amount of intact casein in the formulation can influence the complex modulus of the resultant end product. Finally, a general statement could be reported that as the maturity degree of natural cheese applied in process cheese manufacture increased, the hardness of the resultant processed cheese decreased (Garimella

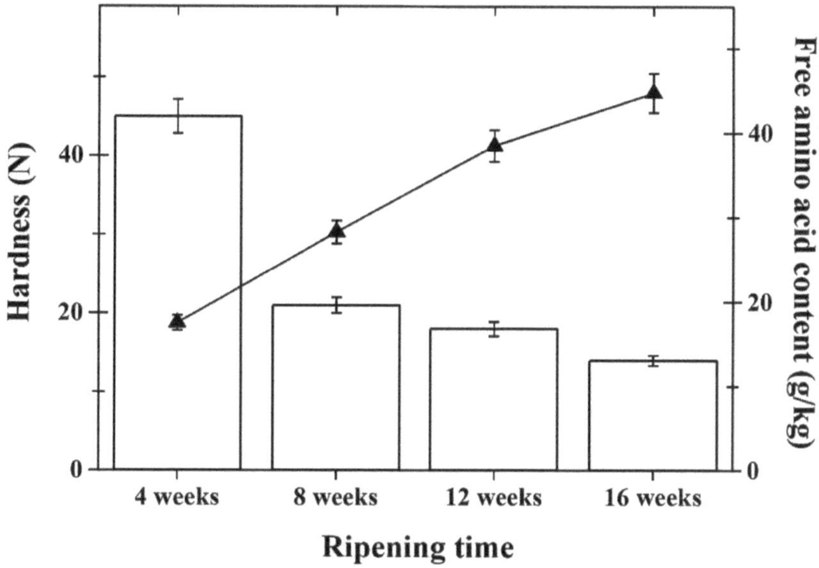

Figure 3.2 Dependence of processed cheese hardness (N; columns) and free amino acid content (g/kg; ▲) in cheese (the main raw material for the production of processed cheese) on the ripening time (weeks) of Swiss-type cheese. The results were expressed as mean and standard deviation (bars).

Purna et al., 2006; Kapoor, Metzger, Biswas and Muthukummarappan, 2007; Piska and Štětina, 2004).

3.5 The effect of natural cheese pH, calcium and phosphorus content and salt-in-moisture content on the functional properties of processed cheese

Active acidity (pH value) has a significant effect on the consistency of processed cheeses. In particular, this parameter is mainly influenced by the type and concentration of emulsifying salts used and the pH value of the applied natural cheese. Moreover, processed cheese with low pH values (in the range of 4.8–5.2) can be expected to have a stiffer consistency than less acidic processed cheeses with higher values of pH (pH > 6.0), which are usually more spreadable under otherwise identical circumstances (Awad et al., 2004). The closer the pH value of the processed cheese is to the value of the isoelectric point of casein (pI ≈ 4.6), the more attractive protein interactions occur, resulting in a stiffer product (Lee et al., 2004). Marchesseau et al. (5) investigated the effect of increasing pH on the rheological properties of processed cheeses with a constant content of emulsifying salts. The result was decreasing values of loss and elastic moduli of elasticity, indicating a decrease in viscosity and elasticity of the tested processed cheese with an increasing pH.

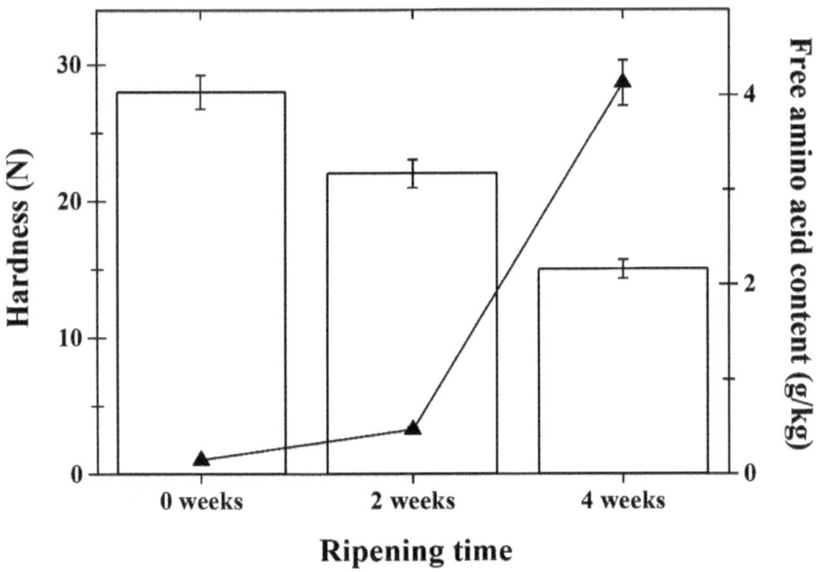

Figure 3.3 Dependence of processed cheese hardness (N; columns) and free amino acid content (g/kg; ▲) in cheese (the main raw material for the production of processed cheese) on the ripening time (weeks) of Mozzarella-type cheese. The results were expressed as mean and standard deviation (bars).

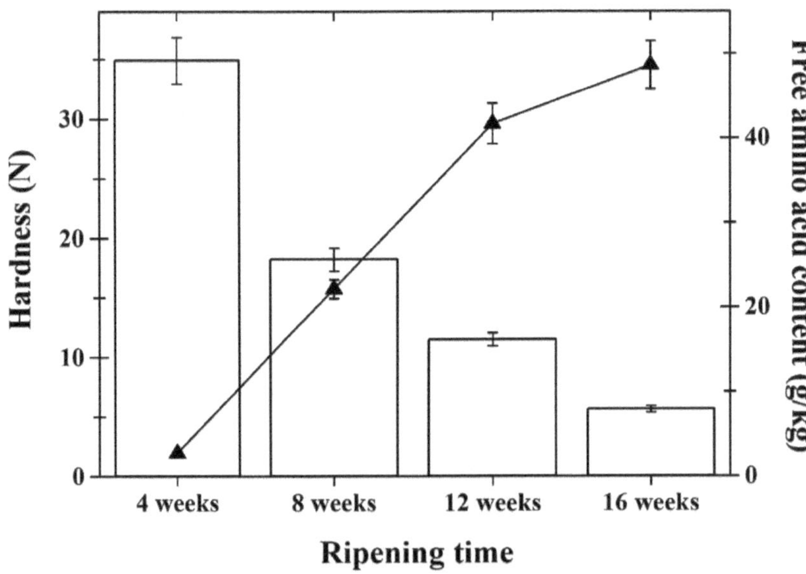

Figure 3.4 Dependence of processed cheese hardness (N; columns) and free amino acid content (g/kg; ▲) in cheese (the main raw material for the production of processed cheese) on the ripening time (weeks) of Cheddar-type cheese. The results were expressed as mean and standard deviation (bars).

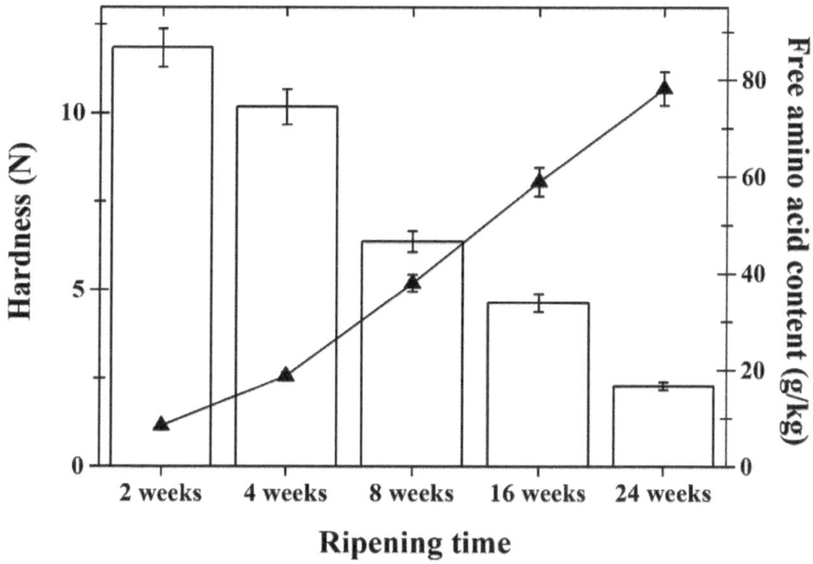

Figure 3.5 Dependence of processed cheese hardness (N; columns) and free amino acid content (g/kg; ▲) in cheese (the main raw material for the production of processed cheese) on the ripening time (weeks) of white brined cheese. The results were expressed as mean and standard deviation (bars).

Moreover, Lee and Klostermeyer (2004) also achieved the same results in processed cheese spreads with reduced fat content (12.0% w/w of fat). In contrast, Swenson et al. (2000) found a decrease in the firmness of processed cheese containing only 0.6 % w/w of fat with a decreasing pH. However, these different results are probably due to the analysis of different types of matrices (e.g., different processing equipment and conditions, chemical composition of the matrix). Moreover, factors such as natural cheese calcium and phosphorus contents and salt-moisture-content can influence the functional properties of processed cheese. According to Kapoor et al. (2007) high calcium and low phosphorus content in natural cheese can influence the processed cheese melting properties, viscous modulus and apparent viscosity. Moreover, the above-mentioned results were in accordance to those reported by Biswas et al. (2008). Moreover, decreased levels of calcium content can affect the rheological properties of processed cheese in such a way that the resultant end product will present lower values in fracture test, strain and hardness. In general, such products are characterized as soft and brittle in texture (Guinee and O'Kennedy, 2009). In addition, the salt-in-moisture content of natural cheese can influence the proteolysis degree. In particular, an increase in the salt-in-moisture content can result in a decreased degree of proteolysis. Thus, the hardness of the resultant processed cheese should be higher, whereas the values of cohesiveness will be lower. Furthermore, according to Biswas et al. (2015) processed cheese produced from Cheddar cheese with high calcium and phosphorus contents and high salt-in-moisture content is harder with lower tendency to flow.

3.6 Conclusion

Processed cheese is one of the youngest dairy products and also due to the complexity of the processed cheese matrix, there is still insufficient relevant information in the literature on the effect of individual factors on the consistency of processed cheeses. The current chapter aimed to summarize the existing knowledge in the field of characteristics of natural cheese for the production of processed cheeses. On the same token, the chapter also dealt with the use of different natural cheeses with different maturity degrees and how this factor for industrial practice affects the consistency of processed cheeses. It has been shown that the choice of a particular natural cheese will significantly affect the consistency of processed cheeses and must be taken into account when "modeling" the raw material composition formula. Furthermore, it has been found that as the maturity degree of natural cheese increases, the hardness of processed cheeses will decrease.

References

Adbel-Hamid, L.B., El-Shabrary, S.A., Awad, R.A., 2000. Chemical properties of processed Ras cheese spreads as affected by emulsifying salt mixtures. J. Food Process. Preserv. 24, 191–198.

Adhikari, K., Cole, A., Grün, I., Heymann, H., Hsieh, F.-H., Huff, H., 2009. Physical and sensory characteristics of processed cheeses manufactured by extrusion technology. J. Sci. Food Agric. 89 (8), 1428–1433.

Aguilera, J.M, Stanley, D.W, 2000. Microstructural Principles of Food Processing and Engineering, 24143, Second Ed SciTech Book News, Switzerland, ISSN:0196-6006.

Alichanidis, E., Polychroniadou, A., 2008. Characteristics of major traditional regional cheese varieties of East-Mediterranean countries: a review. Dairy Sci. Technol. 88 (4-5), 495–510.

Aljewicz, M., Cichoz, G., 2015. Protective effects of Lactobacillus cultures in Dutch-type cheese-like products. LWT-Food Sci. Technol. 63 (1), 52–56 ISSN:0023-6438.

Al-Nabulsi, A.A., Osaili, T.M., AbuNaser, R.A., Olaimat, A.N., Ayyash, M., Al-Holy, M.A., Kadora, K.M., Holley, R.A., 2020. Factors affecting the viability of Staphylococcus aureus and production of enterotoxin during processing and storage of white-brined cheese. J. Dairy Sci. 103 (8), 6869–6881.

Alves, R.M.V., Van Dender, A.G.F., Jaime, S.B.M., Moreno, I., Pereira, B.C., 2007. Effect of light and packages on stability of spreadable processed cheese. Int. Dairy J. 17 (4), 365–373.

Amagliani, G., Giammarini, C., Omiccioli, E., Brandi, G., Magnani, M., 2007. Detection of *Listeria monocytogenes* using a commercial PCR kit and different DNA extraction methods. Food Control 18 (9), 1137–1142.

Awad, R.A., Abdel-Hamid, L.B., El-Shabrawy, S.A., Singh, R.K., 2002. Texture and microstructure of block type processed cheese with formulated emulsifying salt mixtures. LWT-Food Sci. Technol. 35 (1), 54–61.

Awad, R.A., Abdel-Hamid, L.B., El-Shabrawy, S.A., Singh, R.K., 2004. Physical and sensory properties of block processed cheese with formulated emulsifying salt mixtures. Int. J. Food Prop. 7, 429–448.

Belsito, P.C., Ferreira, M.V.S., Cappato, L.P., Cavalcanti, R.N., Vidal, V.A.S., Pimentel, T.C., Esmerino, E.A., Balthazar, C.F., Neto, R.P.C., Tavares, M.I.B., Zacarchenco, P.B., Freitas, M.Q., Silva, M.C., Raices, R.S.L., Pastore, G.M., Pollonio, M.A.R., Cruz, A.G., 2017. Manufacture of Requeijão cremoso processed cheese with galactooligosaccharide. Carbohydr. Polym. 174, 869–875.

Bisig, W., Frohlich-Wyder, M.T, Jakob, E., Wechsler, D., 2010. Comparison between Emmentaler PDO and generic Emmental cheese production in Europe. Aust. J. Dairy Technol. 65 (3), 206–213.

Biswas, A.C., Muthukumarappan, K., Marella, C., Metzger, L.E., 2015. Understanding the role of natural cheese calcium and phosphorous content, residual lactose and salt-in-moisture content on block-type processed cheese functional properties: Cheese hardness and flowability/meltability. Int. J. Dairy Technol. 68 (1), 44–53.

Biswas, A.C., Muthukumarappan, K., Metzger, L.E., 2008. Dynamic rheological properties of process cheese: Effect of Ca and P Content, residual lactose, Salt-to-Moisture ratio and cheese temperature. Int. J. Food Pro. 11 (2), 282–295. https://doi.org/10.1080/10942910701329369.

Brickley, C.A., Auty, M.A.E., Piraino, P., McSweeney, P.L.H., 2007. The effect of natural Cheddar cheese ripening on the functional and textural properties of the processed cheese manufactured therefrom. J. Food Sci. 72 (9), C483–C490.

Buňka, F., Doudová, L., Weiserová, E., Černíková, M., Kuchař, D., Slaviková, S., Nagyová, G., Ponížil, P., Gruber, T., Michálek, J., 2014. The effect of concentration and composition of ternary emulsifying salts on the textural properties of processed cheese spreads. LWT - Food Sci. Technol. 58 (1), 247–255.

Buňka, F., Kříž, O., Veličková, A., Buňková, L., Kráčmar, S., 2009. Effect of acid hydrolysis time on amino acid determination in casein and processed cheeses with different fat content. J. Food Compos. Anal. 22 (3), 224–232.

Buňka, F., Doudová, L., Weiserová, E., Kuchař, D., Ponížil, P., Začalová, D., Nagyová, G., Pachlová, V., Michálek, J., 2013. The effect of ternary emulsifying salt composition and cheese maturity on the textural properties of processed cheese. Int. Dairy J. 29 (1), 1–7.

Buňka, F., Hrabě, J., Kráčmar, S., 2004. The effect of sterilisation on amino acid contents in processed cheese. Int. Dairy J. 14 (9), 829–831.

Buňková, L., Pleva, P., Buňka, F., Valášek, P., Kráčmar, S., 2008. Antibacterial effects of commercially available phosphates on selected microorganisms. Acta Univ. Agric. et Silvic. Mendelianae Brun., 56, pp. 19–24.

Burgos, L., Pece, N., Maldonado, S., 2020. Textural, rheological and sensory properties of spreadable processed goat cheese. Int. J. Food Stud. 9, SI62–S74.

Carić, M., GantarANTAR, M., Kaláb, M., 1985. Effects of emulsifying agents on the microstructure and other characteristics of process cheese — a review. Food Microstruct. 4, 297–312.

Carić, M., Kaláb, M., 1997. Processed cheese products. In: Fox, P.F. (Ed.), Cheese: Chemistry, Physics and Microbiology. Chapman & Hall, London, pp. 467–505 ISBN 0-412-535106.

Černíková, M., Nebesářová, J., Salek, R.N., Řiháčková, L., Buňka, F., 2017. Microstructure and textural and viscoelastic properties of model processed cheese with different dry matter and fat in dry matter content. J. Dairy Sci. 100 (6), 4300–4307.

Chandan, R.C, Kilara, A., Shah, N., 2009. Dairy Processing and Quality Assurance. Wiley-Blackwell, USA. ISBN: 9780813827568.

Chatziantoniou, S.E., Thomareis, A.S., Kontominas, M.G., 2015. Effect of chemical composition on physico-chemical, rheological and sensory properties of spreadable processed whey cheese Eur. Food Res. Technol. 241, 737–748.

Chatziantoniou, S.E., Thomareis, A.S., Kontominas, M.G., 2019. Effect of different stabilizers on rheological properties, fat globule size and sensory attributes of novel spreadable processed whey cheese. Eur. Food Res. Technol. 245 (11), 2401–2412.

Chavhan, G.B., Kanawjia, S.K., Khetra, Y., Puri, R., 2015. Effect of potassium-based emulsifying salts on sensory, textural, and functional attributes of low-sodium processed Mozzarella cheese. Dairy Sci. Technol. 95 (3), 265–278.

Chen, L., Liu, H., 2012. Effect of emulsifying salts on the physicochemical properties of processed cheese made from Mozzarella. J. Dairy Sci. 95 (9), 4823–4830.

Commissiion, 2009. The Codex Alimentarius. CXS 283 - Codex General Standard for Cheese. Codex Alimentarius 1.

Commission, 2009. The Codex Alimentarius. CXS 269 - Codex Standard for Emmental. Codex Alimentarius 1.

Commission, 2009. The Codex Alimentarius. CXS 265 - Codex Standard for Edam. Codex Alimentarius 1.

Costa Dias, M.A, Sant'ana, A.S, Cruz, A.G, Faria, J.D.A.F, Fernandes De Oliveira, C.A, Bona, E., 2012. On the implementation of good manufacturing practices in a small processing unity of Mozzarella cheese in Brazil. Food Control 24, 199–205.

Cunha, C.R., Viotto, W.H., 2010. Casein peptization, functional properties, and sensory acceptance of processed cheese spreads made with different emulsifying salts. J. Food Sci. 75 (1), C113–C120.

da Cunha, C.R., Alcântara, M.R., Viotto, W.H., 2012. Effect of the type of emulsifying salt on microstructure and rheological properties of "Requeijão Cremoso" processed cheese spreads. J. Food Sci. 77 (8), E176–E181.

De Angelis, M., Gobbetti, M., 2004. Environmental stress responses in *Lactobacillus*: a review. Proteomics 4, 106–122.

Dimitreli, G., Thomareis, A.S., 2004. Effect of temperature and chemical composition on processed cheese apparent viscosity. J. Food Eng. 64 (2), 265–271.

Dimitreli, G., Thomareis, A.S., 2007. Texture evaluation of block-type processed cheese as a function of chemical composition and in relation to its apparent viscosity. J. Food Eng. 79 (4), 1364–1373.

Dimitreli, G., Thomareis, A.S., 2008. Effect of chemical composition on the linear viscoelastic properties of spreadable-type processed cheese. J. Food Eng. 84 (3), 368–374.

Dimitreli, G., Thomareis, A.S., 2009. Instrumental textural and viscoelastic properties of processed cheese as affected by emulsifying salts and in relation to its apparent viscosity. Int. J. Food Prop. 12 (1), 261–275. ISSN:1094-2912.

El-Bakry, M., Duggan, E., O'Riordan, E.D, O'Sullivan, M., 2011. Effect of chelating salt type on casein hydration and fat emulsification during manufacture and post-manufacture functionality of imitation cheese. J. Food Eng. 102 (2), 145–153.

Faccia, M., Trani, A., Di Luccia, A., 2009. Short Communication: Relationships between milk quality and acidification in the production of table Mozzarella without starters. J. Dairy Sci. 92 (9), 4211–4217.

Fagan, C.C., Everard, C., O'Donnell, C.P., Downey, G., Sheehan, E.M., Delahunty, C.M., O'Callaghan, D.J., Howard, V., 2007. Prediction of processed cheese instrumental texture and meltability by mid-infrared spectroscopy coupled with chemometric tools. J. Food Eng. 80 (4), 1068–1077.

Ferrão, L.L., Ferreira, M.V.S., Cavalcanti, R.N., Carvalho, A.F.A., Pimentel, T.C., Silva, H.L.A., Silva, R., Esmerino, E.A., Neto, R.P.C., Tavares, M.I.B., Freitas, M.Q., Menezes, J.C.V., Cabral, L.M., Moraes, J., Silva, M.C., Mathias, S.P., Raices, R.S.L., Pastore, G.M.,

Cruz, A.G., 2018. The xylooligosaccharide addition and sodium reduction in requeijão cremoso processed cheese. Food Res. Int. 107, 137–147. https://doi.org/10.1016/j.foodres.2018.02.018.

Ferrão, L.L., Ferreira, M.V.S., Cavalcanti, R.N., Carvalho, A.F.A., Pimentel, T.C., Silva, H.L.A., Silva, R., Esmerino, E.A., Neto, R.P.C., Tavares, M.I.B., Freitas, M.Q., Menezes, J.C.V., Cabral, L.M., Moraes, J., Silva, M.C., Mathias, S.P., Raices, R.S.L., Pastore, G.M., Cruz, A.G., 2018. The xylooligosaccharide addition and sodium reduction in requeijão cremoso processed cheese. Food Research International 107, 137–147. https://doi.org/10.1016/j.foodres.2018.02.018.

Ferrari, E., Gamberi, M., Manzini, R., Pareschi, A., Persona, A., Regattieri, A., 2003. Redesign of the Mozzarella cheese production process through development of a microforming and stretching extruder system. J. Food Eng. 59 (1), 13–23.

Flasarová, R., Pachlová, V., Buňkova, L., Menšíková, A., Georgová, N., Drab, V., Buňka, F., 2016. Biogenic amine production by *Lactococcus Lactis* subsp. *Cremoris* strains in the,model system of Dutch-type cheese. Food Chem. 194, 68–75.

Fox, P.F., 2000. Fundamentals of Cheese Science. Aspen Publishing, Gaithersburg, MD ISBN 978-0-8342-1260-2.

Fox, P.F, McSweeney, P.L.H., 1996. Proteolysis in cheese during ripening. Food Rev. Int 12 (4), 457–509 1996ISSN:8755-9129.

Fox, P.F, Uniacke-Lowe, T., MsSweeney, P.L.H, O'Mahony, J.A., 2015. Dairy Chem. Biochem, Second Edition ISBN: 3319148915.

Fox, P.F., Guinee, T.P., Cogan, T.M., McSweeney, P.L.H., 2000. Fundamentals of Cheese Science. Aspen Publication, Gaithersburg, USA.

Fu, W., Nakamura, T., 2020. Moisture content impact creaming effect and microstructure of processed cheese containing different textural starches. Int. Dairy J. 105.

Fu, W., Watanabe, Y., Inoue, K., Moriguchi, N., Fusa, K., Yanagisawa, Y., Mutoh, T., Nakamura, T., 2018. Effects of pre-cooked cheeses of different emulsifying conditions on mechanical properties and microstructure of processed cheese. Food Chem. 245, 47–52.

Gernigon, G., Schuck, P., Jeantet, R., 2010. Processing of Mozzarella cheese wheys and stretchwaters: a preliminary review. Dairy Sci. Technol. 90 (1), 27–46.

Ghods Rohani, M., Rashidi, H., 2019. Improving the characteristics of spreadable processed cheese using Konjac and Xanthan gums. J. Food Process. Preserv. 43 (12).

Gorrasi, G., Bugatti, V., Tammaro, L., Vertuccio, L., Vigliotta, G., Vittoria, V., 2016. Active Coating for storage of Mozzarella cheese packaged under thermal abuse. Food Control 64, 10–16.

Guidone, A., Zotta, T., Matera, A., Ricciardi, A., De Filippis, F., Ercolini, D., Parente, E., 2016. The microbiota of high-moisture Mozzarella cheese produced with different acidification methods. Int. J. Food Microbiol. 216, 9–17.

Guinee, T.P, Carić, M., Kaláb, M., 2004. Pasteurized processed cheese and substitute/imitation cheese products. Cheese: Chemistry, Physics and Microbiology 2, 349–394.

Guinee, T.P., O'Callaghan, D.J, 2013. Effect of increasing the protein-to-fat ratio and reducing fat content on the chemical and physical properties of processed cheese product. J. Dairy Sci. 96 (11), 6830–6839.

Guinee, T.P., O'Kennedy, B.T, 2009. The effect of calcium content of Cheddar-style cheese on the biochemical and rheological properties of processed cheese. Dairy Sci. Technol. 89 (3-4), 317–333.

Guinee, T.P., O'Kennedy, B.T, 2012. Reducing the level of added disodium phosphate alters the chemical and physical properties of processed cheese. Dairy Sci. Technol. *92* (5), 469–486.

Guinee, T.P., 2011. Pasteurized processed cheese products. Encyclopedia of Dairy Sciences, pp. 805–813 ISBN: 978-0-1237-4407-4.

Hanaei, F., Cuvelier, G., Sieffermann, J.M., 2015. Consumer texture descriptions of a set of processed cheese. Food Qual. Preference 40, 316–325.

Hanna, S.S., 1999. The use of halloum cheese for the production of processed cheese. Int. J. Dairy Technol. 52 (3), 117–118.

Hasenhuettl, G.L, Hartel, R.W., 2008. Food Emulsifiers and their Applications, second ed., pp. 1–426 ISBN: 9780387752839.

Hassan, A.N., Awad, S., Mistry, V.V., 2007. Reduced fat process cheese made from young reduced fat Cheddar cheese manufactured with exopolysaccharide-producing cultures. J. Dairy Sci. 90 (8), 3604–3612.

Hauerlandová, I., Lorencová, E., Buňka, F., Navrátil, J., Janečková, K., Buňková, L., 2014. The influence of fat and monoacylglycerols on growth of spore-forming bacteria in processed cheese. Int. J. Food Microbiol. 182-183, 37–43.

Hayaloglu, A.A., 2017. Cheese varieties ripened under brine. Cheese. Elsevier, Netherlands, pp. 997–1040.

Hoffmann, W., Gärtner, J., Lück, K., Johannsen, N., Maurer, A., 2012. Effect of emulsifying salts containing potassium on the quality of block-type processed cheese. Int. Dairy J. 25 (1), 66–72.

Hui, Y.H, Evranuz, E.Ö, Chandan, R.C., 2012. Handbook of animal-based fermented food and beverage Technology ISBN: 9781439850237.

Hui, Y.H., 2005. Handbook of Food Science, Technology, and Engineering, 148 ISBN: 9781466507876.

Isolini, D., Bachmann, H.P, Butikofer, U., Jakob, E., Fröhlich-Wyder, M.T, 2011. Swiss-type cheeses. Encyclopedia of Dairy Sciences 712–720.

Janevski, O., Hassan, A.N., Metzger, L., 2012. Application of salt whey in process cheese food made from Cheddar cheese containing exopolysaccharides. J. Dairy Sci. 95 (7), 3609–3616.

Johnson, M.E, Lucey, J.A, 2006. Major technological advances and trends in cheese. J. Dairy Sci. 89 (4), 1174–1178.

Kaliappan, S., Lucey, J.A., 2011. Influence of mixtures of calcium-chelating salts on the physicochemical properties of casein micelles. J. Dairy Sci. 94 (9), 4255–4263.

Kaminarides, S., Stachtiaris, S., 2000. Production of processed cheese using kasseri cheese and processed cheese analogues incorporating whey protein concentrate and soybean oil. Int. J. Dairy Technol. 53 (2), 69–74.

Kammerlehner, J.K, 2012. Verlug Reisinger, Kuenstlerpresse W Bode, Frising ISBN 978-00-037724-2.

Kapoor, R., Metzger, L.E, Biswas, A.C, Muthukummarappan, K., 2007. Effect of natural cheese characteristics on process cheese properties. J. Dairy Sci. 90 (4), 1625–1634.

Kapoor, R., Metzger, L.E., 2008. Process cheese: scientific and technological aspects – a review. Food Sci. Food Saf. 7, 194–214.

Kommineni, A., Amamcharla, J., Metzger, L.E., 2012. Effect of xylitol on the functional properties of low-fat process cheese. J. Dairy Sci. 95 (11), 6252–6259.

Kontou, V., Dimitreli, G., Raphaelides, S.N., 2019. Elongational flow studies of processed cheese spreads made from traditional greek cheese varieties. LWT 107, 318–324.

Law, B.A, Tamine, A.Y., 2010. Technology of Cheesemaking ISBN: 9781444323740.

Lee, S.K., Anema, S., Klostermeyer, H., 2004. The influence of moisture content on the rheological properties of processed cheese spreads. Int. J. Food Sci. Technol. 39, 763–771.

Lee, S.K., Klostermeyer, H., 2001. The effect of pH on the rheological properties of reduced-fat model processed cheese spreads. Food Sci. Technol. 34 (5), 288–292.

Lucey, J.A., 2002. Formation and physical properties of milk protein gels. J. Dairy Sci. 85 (2), 281–294.
Macků, I., Buňka, F., Voldánová, B., Pavlínek, V., 2009. Effect of addition of selected solid cosolutes on viscoelastic properties of model processed cheese containing pectin. Food Hydrocoll. 23 (8), 2078–2084.
Marchesseau, S., Gastaldi, E., Lagaude, A., Cuq, J.L., 1997. Influence of PH on protein interactions and microstructure of process cheese. J. Dairy Sci. 80 (8), 1483–1489.
Marth, E.H, Steele, J.L, 2001. Applied Dairy Microbiology, second. CRC Press, Boca Raton ISBN 978-0824705367.
McMahon, D., Oberg, J., 2011. Pasta-filata cheeses: low-moisture part-skim Mozzarella (pizza cheese). Encyclopedia of Dairy Sciences 737–744.
McSweeney, P.L.H, Sousa, M.J., 2000. Biochemical Pathways for the production of flavour compounds in cheeses during ripening: a review. Lait 80 (3), 293–324.
McSweeney, P.L.H., 2004. Biochemistry of cheese ripening. Int. J. Dairy Technol. 57 (2-3), 127–144.
McSweeney, P.L.H. Cheese problems solved. 2007. ISBN 978-1-84569-060-1.
Miloradovic, Z., Smigic, N., Djekic, I., Tomasevic, I., Kljajevic, N., Nedeljkovic, A., Miocinovic, J., 2018. The influence of NaCl concentration of brine and different packaging on goat white brined cheese characteristics. Int. Dairy J. 79, 24–32.
Mizuno, R., Lucey, J.A., 2005. Effects of emulsifying salts on the turbidity and calcium-phosphate–protein interactions in casein micelles. J. Dairy Sci. 88 (9), 3070–3078.
Moatsou, G., Govaris, A., 2011. White brined cheeses: a diachronic exploitation of small ruminants milk in Greece. Small Ruminant Res 101 (1-3), 113–121.
Molins, R.A., 1991. Phosphates in Food. CRC Press, Boca Raton, p. 261 ISBN 0-8493-4588-X.
Mozuraityte, R., Berget, I., Mahdalova, M., Grønsberg, A., Øye, E.R., Greiff, K., 2019. Sodium reduction in processed cheese spreads and the effect on physicochemical properties. Int. Dairy J. 90, 45–55.
Nagyová, G., Buňka, F., Salek, R.N., Černíková, M., Mančík, P., Grůber, T., Kuchař, D., 2014. Use of sodium polyphosphates with different linear lengths in the production of spreadable processed cheese. J. Dairy Sci. 97 (1), 111–122.
Nogueira, E.B., Costa-Lima, B.R.C., Torres, F., Regazone, A.V., Melo, L., Franco, R.M., Cortez, M.A.S. 2018. Effect of potassium-based emulsifying salts on the sensory and physicochemical parameters of low-sodium spreadable processed cheese. Int. J. Dairy Technol. 71 (3), 717–722.
Oberg, E.N, Oberg, C.J, Motawee, M.M, Martini, S., McMahon, D.J., 2015. Increasing stringiness of low-fat Mozzarella string cheese using polysaccharides. J. Dairy Sci. 98 (7), 4243–4254.
O'Sullivan, D.J., McSweeney, P.L.H, Cotter, P.D, Giblin, L., Sheehan, J.J. Compromised. J. Dairy Sci., 99, 4, 2625-2640.
Pachlová, V., Buňka, F., Buňkova, L., Weiserová, E., Budinský, P., Žaludek, M., Kráčmar, S., 2011. The effect of three different ripening/storage conditions on the distribution of selected parameters in individual parts of Dutch-type cheese: cheese ripened/stored under three different conditions. Int. J. Food Sci. Technol. 46 (1), 101–108.
Pachlová, V., Buňka, F., Flasarová, R., Válková, P., Buňková, L., 2012. The effect of elevated temperature on ripening of Dutch type cheese. Food Chem. 132 (4), 1846–1854.
Paludetti, L.F., O'Callaghan, T.F., Sheehan, J.J., Gleeson, D., Kelly, A.L., 2020. Effect of Pseudomonas fluorescens proteases on the quality of Cheddar cheese. J. Dairy Sci. 103 (9), 7865–7878. https://doi.org/10.3168/jds.2019-18043.

Pisano, M.B., Scano, P., Murgia, A., Cosentino, S., Caboni, P., 2016. Metabolomics and microbiological profile of Italian Mozzarella cheese produced with buffalo and cow Milk. Food Chem. 192, 618–624.

Piska, I., Štětina, J., 2004. Influence of cheese ripening and rate of cooling of the processed cheese mixture on rheological properties of processed cheese. J. Food Eng. 61 (4), 551–555.

Přikryl, J., Hájek, T., Švecová, B., Salek, R.N., Černíková, M., Červenka, L., Buňka, F., 2018. Antioxidant properties and textural characteristics of processed cheese spreads enriched with rutin or quercetin: The effect of processing conditions. . LWT - Food Sci. Technol., 87, pp. 266–271.

Purna, S.K.G., Pollard, A., Metzger, L.E. 2006. Effect of formulation and manufacturing parameters on process cheese food functionality—I. trisodium citrate. J. Dairy Sci. 89 (7), 2386–2396.

Rafiq, S.M., Ghosh, B.C., 2017. Effect of potato incorporation on the physico-chemical, textural and sensory properties of processed cheese. J. Food Meas. Charact. 11 (2), 776–780.

Rowney, M.K., Hickey, M.W, Roupas, P., Everett, D.W., 2003. The Effect of homogenization and milk fat fractions on the functionality of Mozzarella cheese. J. Dairy Sci. 86 (3), 712–718.

Rubio-Texeira, M., 2006. Endless versatility in the biotechnological applications of *Kluyveromyces* LAC genes. Biotechnol. Adv. 24 (2), 212–225.

Sádlíková, I., Buňka, F., Budinský, P., Barbora, V., Pavlínek, V., Hoza, I., 2010. The effect of selected phosphate emulsifying salts on viscoelastic properties of processed cheese. LWT - Food Sci. Technol., 43, pp. 1220–1225.

Salek, R.N., Černíková, M., Lorencová, E., Pachlová, V., Kůrová, V., Šenkýřová, J., Buňka, F., 2020. The impact of Cheddar or white brined cheese with various maturity degrees on the processed cheese consistency: a comparative study. Int. Dairy J.

Salek, R.N., Černíková, M., Maděrová, S., Lapčík, L., Buňka, F., 2016. The effect of different composition of ternary mixtures of emulsifying salts on the consistency of processed cheese spreads manufactured from Swiss-type cheese with different degrees of maturity. J. Dairy Sci. 99 (5), 3274–3287.

Salek, R.N., Černíková, M., Pachlová, V., Bubelová, Z., Konečná, V., Buňka, F., 2017. Properties of spreadable processed Mozzarella cheese with divergent compositions of emulsifying salts in relation to the applied cheese storage period. LWT 77, 30–38. https://doi.org/10.1016/j.lwt.2016.11.019.

Salek, R.N., Černíková, M., Nagyová, G., Kuchař, D., Bačová, H., Minarčíková, L., Buňka, F., 2015. The effect of composition of ternary mixtures containing phosphate and citrate emulsifying salts on selected textural properties of spreadable processed cheese. Int. Dairy J. 44, 37–43.

Salek, R.N., Lorencová, E., Míšková, Z., Lazárková, Z., Pachlová, V., Adámek, R., Bezděková, K., Buňka, F., 2020. The impact of Chios mastic gum on textural, rheological and melting properties of spread-type processed cheese during storage. Int. Dairy J. 109, 10475–104759.

Schädle, C.N., Eisner, P., Bader-Mittermaier, S., 2020. The combined effects of different fat replacers and rennet casein on the properties of reduced-fat processed cheese. J. Dairy Sci. 103 (5), 3980–3993.

Seth, K., Bajwa, U., 2015. Effect of acidulants on the recovery of milk constituents and quality of Mozzarella processed cheese. J. Food Sci. Technol. 52 (3), 1561–1569.

Shirashoji, N., Jaeggi, J.J, Lucey, J.A., 2006. Effect of trisodium citrate concentration and

cooking time on physicochemical properties of pasteurized process cheese. J. Dairy Sci. 89, 15–28 ISSN 0022-0302.
Shirashoji, N., Aoyagi, H., Jaeggi, J.J., Lucey, J.A., 2016. Effect of tetrasodium pyrophosphate concentration and cooking time on the physicochemical properties of process cheese. J. Dairy Sci. 99 (9), 6983–6994.
Shirashoji, N., Jaeggi, J.J., Lucey, J.A., 2010. Effect of sodium hexametaphosphate concentration and cooking time on the physicochemical properties of pasteurized process cheese. J. Dairy Sci. 93 (7), 2827–2837.
Singh, T.K., Drake, M.A., Cadwallader, K.R., 2003. Flavor of cheddar cheese: A chemical and sensory perspective. Comprehensive reviews Food Sci. Food Safe. 2 (4), 166–189. https://doi.org/10.1111/j.1541-4337.2003.tb00021.x.
Stangierski, J., Weiss, D., Kaczmarek, A., 2019. Multiple regression models and Artificial Neural Network (ANN) as prediction tools of changes in overall quality during the storage of spreadable processed Gouda cheese. Eur. Food Res. Technol. 245 (11), 2539–2547.
Swenson, B.J., Wendorff, W.L., Lindsay, R.C., 2000. Effects of ingredients on the functionality of fat-free process cheese spreads. J. Food Sci. 65, 822–825.
Tamine, A.Y., 2007. Structure of Dairy Products, 2007, pp. 1–288 ISBN: 978-1-4051-2975-6.
Tamine, A.Y., 2011. Processed cheese and analogues. Society of Dairy Technology. Ames: Wiley-Blackwell, USA. ISBN 978-1-4051-8642-1.
Tanaka, Y., Takei, T., Aiba, T., Masuda, K., Kiuchi, A., Fujiwara, T., 1986. Development of synthetic lung surfactants. J. Lipid Res. 27 (5), 475–485.
Topcu, A., Bulat, T., Özer, B., 2020. Process design for processed Kashar cheese (a pasta-filata cheese) by means of microbial transglutaminase: Effect on physical properties, yield and proteolysis. LWT 125, 109226.
Torres, F.R., Esmerino, E.A., Carr, B.T., Ferrão, L.L., Granato, D., Pimentel, T.C., Bolini, H.M.A., Freitas, M.Q., Cruz, A.G., 2017. Rapid consumer-based sensory characterization of requeijão cremoso, a spreadable processed cheese: Performance of new statistical approaches to evaluate check-all-that-apply data. J. Dairy Sci. 100 (8), 6100–6110.
Walstra, P., Walstra, P., Wouters, J.T.M., Geurts, T.J., 2005. Dairy Sci. Tech. CRC Press. https://doi.org/10.1201/9781420028010.
Weimer, B.C. (2007). Improving the Flavour of Cheese. ISBN 978-1-84569-007-6.
Weiserová, E., Doudová, L., Galiová, L., Žák, L., Michálek, J., Janiš, R., Buňka, F., 2011. The effect of combinations of sodium phosphates in binary mixtures on selected texture parameters of processed cheese spreads. Int. Dairy J. 21 (12), 979–986.
Weiss, D., Stangierski, J., Baranowska, H.M., Rezler, R., 2018. Kinetic models of quality parameters of spreadable processed Gouda cheese during storage. Food Sci. Biotechnol. *27* (5), 1387–1394.
Yılmaz, M.T., Karaman, S., Cankurt, H., Kayacier, A., Sagdic, O., 2011. Steady and dynamic oscillatory shear rheological properties of ketchup–processed cheese mixtures: Effect of temperature and concentration. J. Food Eng. 103 (2), 197–210. https://doi.org/10.1016/j.jfoodeng.2010.10.016.
Yun, J.J, Kiely, L.J, Kindstedt, P.S, Barbano, D.M., 1993. Mozzarella Cheese: Impact of coagulant type on functional properties. J. Dairy Sci. 76 (12), 3657–3663.
Yvon, M., Rijnen, L., 2001. Cheese Flavour Formation by Amino Acid Catabolism. Int. Dairy J. 11 (4), 185–201 ISSN:0958-6946.
Zehren, V.L., Nusbaum, D.D., 2000. Process cheese, Second ed. Cheese Reporter Publishing, Co, Inc., Madison, Wisconsin, USA.
Zerfiridis, G.K., 2001. *Τεχνολογία προιόντων γάλακτος*. *Θεσσαλονίκη*: Εκδόσεις *Γιαχούδη*, 2. *Εκδ*. 55, 58.

The use of vegetable ingredients in processed cheese

4

Terri D. Boylston
Department of Food Science & Human Nutrition, Iowa State University, Ames, Iowa, United States of America

4.1 Introduction

Processed cheese was developed in the early 20th century to increase the shelf-life of natural cheeses. Natural cheese, the primary ingredient in process cheese, is heated and blended with emulsifying agents and other ingredients to produce a homogeneous oil-in-water emulsion with unique functional properties and extended shelf life. In the processed cheese, the fat globules, ranging from <1 to 5 μm in diameter, are dispersed in the casein gel network. The key functional properties for unmelted processed cheese include firmness, fractureability, springiness, and adhesiveness. The key functional properties of melted processed cheese include meltability, viscosity, and stretching ability (Kapoor and Metzger, 2008).

In the United States, three categories of process cheese, pasteurized process cheese (PC), pasteurized process cheese food (PCF), and pasteurized process cheese spread (PCS), are defined by a standard of identity in the Code of Federal Regulations (CFR) (21CFR133.169 to 133.180; FDA, 2012). These products differ in their moisture and fat contents, pH, and permitted ingredients. The permitted ingredients in PC, PCF, and PCS are natural cheese, emulsifying agents, acidifying agents, milk fat, water, salt, spices or flavorings, mold inhibitors, and anti-sticking agents. Pasteurized process cheese spread may also contain added gums. Pasteurized process cheese products have a similar composition to process cheese, but do not meet the standards of identity defined in the CFR, because they contain additional ingredients that are not permitted in the standardized categories of pasteurized process cheese.

Cheese analogs are processed cheese products in which the dairy ingredients are partially or completely replaced by vegetable ingredients. Partial dairy cheese analogs use casein and caseinates to provide the structure of the cheese, but replace the dairy fat with vegetable oils. Non-dairy cheese analogs are produced using vegetable proteins and vegetable oils. Although the process for making the cheese analogs is similar to that of dairy cheeses, differences in the functionality of the dairy ingredients and vegetable ingredients has an impact on the quality attributes of the processed cheese products (Chavan and Jana, 2007).

Dairy-based cheese, fats, and protein powders are important contributors to the flavor and texture of processed cheese products. These ingredients contribute to important protein-protein and protein-fat interactions in the processed cheese. However, dairy ingredients are high in cost and have been linked to increased risks of cardiovascular disease due to their high saturated fat and cholesterol contents. These concerns have led

Processed Cheese Science and Technology: Ingredients, Manufacture, Functionality, Quality, and Regulations.
DOI: https://doi.org/10.1016/B978-0-12-821445-9.00005-4

to an interest in the replacement of these dairy-based ingredients with vegetable-based ingredients to reduce costs, reduce total and saturated fats, cholesterol, and calories, or produce a non-dairy processed cheese product. The successful replacement of these dairy-based ingredients in the processed cheese relies on identifying ingredients with comparable functional properties, which can be a challenge (Bennett et al., 2006; da Silva et al., 2016; Ferrão et al., 2016; Trivedi et al., 2008). With the substitution of ingredients, it is critical to maintain the stability of the emulsion and the important functional properties of the unmelted and melted processed cheese products.

An understanding of the definitions of the key textural properties of cheese will be helpful as we discuss the effect of the vegetable ingredients on the textural properties of processed cheeses. Hardness is a measure of the gel strength and force necessary to compress the sample. Cohesiveness is an indication of the strength of the internal bonds of the cheese structure and the ease with the cheese shreds or slices without compacting. Adhesiveness indicates the degree to which the cheese adheres to another surface. High adhesiveness indicates a resistance to separation from a contact material, which is not desirable in processed cheese products. The meltabililty of the cheese reflects changes in the casein network and liquefaction of the fat globules during heating to contribute to changes in viscosity.

In this review, the role of vegetable-based ingredients, including hydrocolloids and gums, starches and plant-based proteins, in the development and functionality of processed cheese products will be discussed.

4.2 Function of hydrocolloids and gums in processed cheese products

Food gums and hydrocolloids are allowed for use only in PCS and pasteurized process cheese products without a standard of identity. These products have higher moisture content than PC and PCF. A wide range of gums and hydrocolloids, singly or in combination, can be added at levels up to 0.8% of the finished product. The primary function of these ingredients is to bind water to increase viscosity of the product, to form a strong network, and to improve mouthfeel. The approved gums and hydrocolloids differ in their dispersibility, solubility, hydration properties, viscosity, and interactions with milk proteins, so the choice of the gums and hydrocolloids can affect the functional properties of the unmelted and melted processed cheese products (Kapoor and Metzger, 2008).

The addition of locust bean gum, xanthan gum, and κ-carrageenan (0%–0.21%) to processed cheese analogs with acid casein, whey protein isolate, and anhydrous milk fat increased the firmness and decreased the adhesiveness to improve the rheological properties. In addition to binding water, these gums can form complexes with the proteins to enhance the strength of the gel network. κ-carrageenan, a sulfated hydrocolloid, is recognized for its ability to interact with the κ-caseins on the surface of the casein micelle and whey proteins (Gustaw and Mleko, 2007).

The addition of hydrocolloids to fat-free processed cheese spread resulted in an increase in firmness and a decrease in meltability in comparison to a full-fat processed cheese spread. The thixotropic nature of guar gum contributed to a cheese spread that was softer and more spreadable than the other hydrocolloid treatments. The fat-free processed cheese spreads with carrageenan and locust-bean gum were firmer and less spreadable than the guar gum treatment, but had a low meltability, similar to the guar gum treatment. The unique ability of the sulfated groups of carrageenan to form crosslinks with casein contributes to the increased firmness and decreased meltability of the processed cheese spreads. The ability of the hydrocolloids to bind water contributed to an improved creaminess and consistency of the reduced fat processed cheese spreads (Swenson et al., 2000).

The addition of konjac glucomannan (0%–1.25%) to a processed cheese without added fat resulted in a product with hardness and adhesiveness comparable to the standard whole fat processed cheese. Hardness of the cheese increased linearly with konjac glucomannan content. There was no significant difference in the adhesiveness of cheese with 0%–0.5% konjac glucomannan. Higher concentrations of konjac glucomannan resulted in a linear increase in adhesiveness. The addition of 0.5% konjac glucomannan had a hardness and adhesiveness similar to the standard whole fat processed cheese. Cohesiveness of the reduced fat processed cheeses was not affected by concentration of konjac glucomannan and was significantly lower than the standard whole fat processed cheese. The viscosity of the reduced fat processed cheeses with konjac glucomannan was similar to the full-fat processed cheese, but did not exhibit the typical melting properties (da Silva, 2016).

Hydrocolloids function as stabilizers of emulsions through their water-binding abilities. In themselves, the polysaccharides are unable to interact with both oil and water phases to function as emulsifiers. However, formation of complexes between the hydrocolloids and proteins could provide the surface active properties needed to replace citrate- and phosphate-based emulsifying salts in processed cheeses. Cerníková et al. (2010) investigated numerous hydrocolloids, including modified starch, low methoxyl pectin, locust bean gum, κ-carrageenan, and ι-carrageenan, as alternatives to emulsifying salts in processed cheeses. With the exception of the carrageenans, the other hydrocolloids were unable to stabilize the emulsion of the processed cheeses. The interaction between the carrageenans with the milk proteins provided the hydrophobic characteristics needed for interaction with the fat phase for emulsification. Carrageenans are also able to form a stable gel structure. However, the quality of the processed cheese with carrageenans was adversely due to the formation of a firm gel that was difficult to spread.

4.3 Function of prebiotics in processed cheese products

There is increasing interest in the addition of probiotic bacteria to foods, especially dairy products, to improve gut health. The addition of prebiotics, such as inulin and resistant starch, increase the fiber content of foods and promote the viability of the probiotic bacteria through the digestive tract. Inulin, a polymer of fructose which can vary in the degree of polymerization and branching, is a soluble and fermentable fiber. The longer chain inulin compounds are effective in water binding and gel formation, which can provide desirable benefits, nutritionally and functionally, to processed cheeses (Ferrão et al., 2016). Resistant starch is a polymer of glucose that is not completely digested and provides the physiological benefits of soluble fiber (Bi et al., 2016).

Inulin (1.25%–3.44%) was added to an imitation to cheese (28% rennet casein, 24.5% vegetable fats) as a partial fat replacer (23%–63% reduction) and to provide a health benefit. Cheeses with a moisture content of 54% and added inulin had a significantly higher firmness, but not different melting characteristics in comparison to the control. The increased firmness was attributed to water binding and gel formation by the inulin and higher protein to moisture content in the reduced fat cheeses with added inulin (Hennelly et al., 2006).

The addition of inulin (0%–8%) to processed cheese spread improved the nutritional value and contributed to changes in textural and melting properties. Due to the ability of inulin to bind water, the addition of inulin decreased the water activity of the processed cheese spread, in addition to decreases in moisture, protein and fat content. The ability of inulin to bind water and form a gel contributed to the increase in firmness, shear, and solidification viscosity with level of added inulin. The stickiness of the processed cheese spread decreased with the addition of inulin, though the stickiness of the cheese spreads with 4% and 6% inulin were not different from the control. The melting properties of the control and cheese spread with 4% inulin were not significantly different, but higher inulin contents resulted in decreases in melting properties. The decrease in melting properties is attributed to the weakening of the gel structure by the interaction of inulin with the cheese proteins and the decrease in protein hydration and emulsification capacity due to water binding by inulin. Sensory attributes at inulin levels of up to 6% were not significantly affected. Addition of inulin at 8% resulted in decreases in body and texture, as well as spreadability due to water binding by inulin. The optimized processed cheese spread with 6% inulin had a fiber content of 4.9% (Giri et al., 2014, 2017).

Inulin and resistant starch were used as fat replacers in imitation Mozzarella cheese formulated with 22% rennet casein and 24% vegetable fat (control). Inulin or resistant starch directly replaced the fat at 2.4%, 4.8%, 7.2%, 9.6%, and 12%. The replacement of fat with inulin or resistant starch increased the moisture content and decreased the fat content of the imitation cheeses. Due to the increased water binding and decreased fat content, as the content of the inulin or resistant starch increased, hardness increased and springiness and meltability decreased with the greatest changes for the treatments with resistant starch. The decreased protein hydration with the addition of inulin or resistant starch, resulting in greater protein-protein interactions and reduced protein-fat interactions and emulsification, also contributed to the observed decreases in meltability (Bi et al., 2016).

A nonfat imitation cheese was developed through the addition of 43.2% resistant starch (based on dry matter content) formulated with rennet casein and a 60% moisture content. The decrease in fat content increased protein-protein interactions and the increase in resistant starch content contributed to gel formation to result in an increase in the firmness of the imitation cheeses. Through increasing the moisture content from 52% to 60%, the hardness of the imitation cheese decreased due to increased hydration of the protein matrix. The resulting imitation cheese had a firm texture, good meltability, and cohesive properties that facilitates slicing and shredding, with a reduced fat content and increased fiber content (Noronha et al., 2007).

4.4 Function of starches in processed cheese products

The addition of starches to processed cheese products, as a partial substitution for casein, provides economic benefits. Starches are polymers of glucose molecules, used as a form of energy storage in cereal and root crops. Amylose and amylopectin, which are linear and branched molecules, respectively, are the major starch molecules. Heating the starches in the presence of water results in starch gelatinization and thickening, the degree of which depends on the type of starch. Differences in the granule size and gelatinization properties of the starches affect the rheological properties of processed cheese products with added starches. Cereal starches, with a relatively higher content of the linear molecule, amylose, are able to form gels upon cooling. Root starches are unable to form gels due to their higher content of the branched molecule, amylopectin. Casein provides the dominant gel structure, with starch functioning as filler. The competition between the starch and proteins for water may also affect the gel structure of the processed cheese, emulsification capacity of the casein, and textural properties. The gelatinization temperature of the starch increases with an increase in the amylose content. The amylose interacts with the protein matrix to affect the texture of the processed cheese. The level of added starch should be optimized to provide the desired firmness without reducing cohesiveness due to dilution of the proteins (Bennett et al., 2006; Chavan and Jana, 2007; Ferrão et al., 2016; Mounsey, 2009; Mounsey and O'Riordan, 2001; Trivedi et al., 2008).

Wheat starch (0%–9%) has been shown to be an effective filler and replacement for rennet casein (0%–45% reduction) in imitation cheeses formulated with 24.5% rennet casein and 26% vegetable fat. Imitation cheeses with 3% added wheat starch and a 15% reduction in rennet casein were similar in textural properties to the control imitation cheese. Formulations of 5% starch and 25% reduction in rennet casein had an increased firmness, but reduced cohesiveness. Higher starch additions and reductions in rennet casein resulted in decreased firmness and cohesiveness due to a reduced protein content and weaker gel structure (Mounsey, 2009).

The addition of starch increased the firmness of a model processed cheese system in which the content of rennet casein was reduced to reduce ingredient costs. The rennet casein contributes to the structure of the processed cheese spread, with decreases in the rennet casein resulting in a decrease in elasticity and firmness and a loss of emulsifying power. The addition of starch resulted in an increase in firmness of the model processed cheese system without a change in the basic structure of the processed cheese. The cereal starches (rice and wheat), which have the ability to form a gel, contributed to slightly greater increases in firmness than the root starches (waxy corn and potato). The increase in viscosity was attributed to the absorption of water during starch gelatinization and gel formation. The addition of starch to processed cheese was suggested as a potential treatment to maintain the desired texture in reduced protein processed cheeses (Bennett et al., 2006).

The composition of native starches (3%; maize, waxy maize, wheat, potato, rice) as a partial replacement for rennet casein (15% reduction) does have an effect on the textural properties of imitation cheeses formulated with rennet casein (24.5%) and vegetable fat (26%). The hardness of the imitation cheeses was directly related to the amylose content. The higher firmness of the imitation cheeses made from the maize, wheat, and potato starch was attributed to the formation of a hydrogen-bonded network by the leached amylose molecules. The lower cohesiveness of the imitation cheeses with added starch was attributed to the lower protein content and the disruption of the casein network by the starch granules. The reduced meltability of the imitation cheeses with starch was attributed to the disruption of the casein network, decreased hydration and increased protein-protein interactions of the casein network, changes in the size distribution of the fat globules, and the presence of the network of leached amylose molecules. The imitation cheeses with the higher firmness had the lowest meltability. Of the starches evaluated, rice starch, with its low amylose content, limited swelling capacity and small granule size, was most effective as partial replacement for casein and had textural properties most similar to the control imitation cheese (Mounsey and O'Riordan, 2001).

The addition of crosslinked and stabilized waxy maize starches, acetylated distarch adipate and hydroxypropyl distarch phosphate, at levels of 1% to 3%, to processed cheese analogs made with acid casein and anhydrous milk fat increased the firmness, cohesiveness, and viscosity of the cheese analogs. The crosslinked and stabilized starches swell less than the native starches causing less disruption of the protein-fat matrix to increase firmness and cohesiveness. Crosslinked starches effectively bind water to increase viscosity of the cheese analogs, with the crosslinking providing increased stability to acid, heat, and agitation. Meltability of the cheese analogs is reduced with the addition of the crosslinked starches. Water binding by the starches increases protein-protein interactions and decreases protein-fat interactions and meltability. The texture and meltability of the processed cheese analogs with added crosslinked starches can be optimized with modifications in the pH (Solowiej et al., 2016).

4.5 Function of vegetable oils in processed cheese products

Dairy fats contribute to the creamy mouthfeel, flavor, and texture of processed cheese products. The incorporation of vegetable-based oils, with higher contents of unsaturated fatty acids, into processed cheese products has the potential to improve the nutritional value of processed cheeses through the reduction of saturated fat and cholesterol contents. However, the substitution of the saturated dairy fats with unsaturated vegetable oils adversely affects the texture due to the phase change of the fat source (Jeske et al., 2018).

Peanuts, with a content of approximately 50% monounsaturated fatty acids and 35% polyunsaturated fatty acids, are considered to have a heart-healthy fatty acid profile. The incorporation of peanut paste into processed cheese at levels of up to 15%, resulted in a decrease in saturated fatty acid content from 67.2% to 57.4% and an increase in polyunsaturated fatty acids from 2.7% to 10.7%, with no significant change in the content of monounsaturated fatty acids. However, processed cheese with 15% peanut paste resulted in decreases in firmness, meltability, and overall sensory properties. The increased fat and moisture content and insoluble peanut proteins alter the protein-protein and protein-fat interactions that contribute to the characteristic texture of processed cheese. The addition of 10% peanut paste resulted in a processed cheese product with a healthier fatty acid profile and acceptable quality attributes (Rafiq and Ghosh, 2017).

The fatty acid composition and melting points of fats used to formulate cheese analogs from sodium and calcium caseinates had a significant impact on the texture of the analogs. Butterfat (melting point 32–34°C), soybean fat (melting point 36–38°C) and soybean oil (liquid at room temperature) individually or in combination were evaluated. The analogs with soybean fat had higher hardness, but lower cohesiveness and springiness than the treatments with soybean oil and butterfat. In addition to differences in the melting points of these fats and oils, these fats and oils differed in their ability to adsorb to the protein. The butterfat, with its high saturated fat content adsorbs to the hydrophobic groups of the protein to stabilize the emulsion and contribute springiness. The soybean oil has a lower affinity for the protein, resulting in greater protein-protein and protein-water interactions and increased water holding capacity, which also contributes to springiness. Both protein-protein and protein-fat interactions were present in the analogs with soybean fat, which resulted in increased firmness, and decreased springiness and cohesiveness. A blend of 50% soybean fat and 50% soybean oil produced a cheese analog with textural properties most similar to butterfat, but with a lower cholesterol content (Lobato-Calleros, 1997).

The substitution of dairy fats with vegetable fat at 25% and 50% reduced the saturated fat and cholesterol contents of Requeijão cremoso, a Brazilian spreadable processed cheese. The vegetable fat contained 71% oleic acid, in which 27% of the total fat content was trans-oleic acid, in comparison to the milk fat, which contained a high proportion of short chain fatty acids and 63% saturated fatty acids. Although the total fat content was the same for all treatments, differences in the fatty acid profiles affected the textural properties of the spreadable processed cheeses. The substitution of the vegetable fat for the milk fat resulted in fewer, larger fat globules in the spreadable processed cheese. The milk fat containing a higher percentage of short chain fatty acids was less hydrophobic than the vegetable fat and better able to associate with the protein network, resulting in greater emulsification. The larger fat globules in the vegetable fat cheese contributed to a higher protein density and higher values for hardness and adhesiveness. All spreadable processed cheese treatments had acceptable sensory attributes, although the flavor of the milk fat spreadable processed cheese was more acceptable than the spreadable processed cheeses with vegetable fat. This is a significant challenge in cheese analogs in which the milk fat content is reduced due to the importance of short chain fatty acids in the milk fat in the development of the characteristics cheese flavor (Cunha et al., 2010).

One drawback of replacing milk fat with vegetable oils is the higher unsaturated fatty acid content of the vegetable oils results in larger fat globules and a softer texture in the processed cheese products. The use of oleogels made with unsaturated vegetable oils as a substitute for milk fat can be effective in developing a processed cheese product with reduced saturated fat and cholesterol contents. Oleogels are produced through the use of waxes that are able to gel vegetable oils to form semi-solids with characteristics similar to that of solid fats. The textural properties of processed cheese products made with milk fat, oleogels made with either rice bran wax (100% wax ester) or sunflower wax (77% wax ester) at 0.5% or 1.0%, or ungelled vegetable oil were compared. Although the oleogels don't completely replicate the properties of milk fat in processed cheese products, the size of the lipid particles and connectivity of the oil droplets is less in the processed cheese products made with oleogels than with ungelled vegetable oil. Oil-binding capacity and firmness of the processed cheese products increased with wax concentration with the oleogel made with 1.0 % rice bran wax having attributes similar to the milk fat processed cheese products. Meltability of all processed cheese products made with oleogels was lower than for the ungelled vegetable oils and comparable to the milk fat processed cheese products. The fatty acid composition of vegetable oils is different from that of milk fat, which impacts the interactions with the protein network of processed cheeses. The use of olegels does modify the properties of the fat to more closely replicate milk fat in processed cheeses (Huang et al., 2018).

The reduction of fat in processed cheeses results in an undesirable firm, rubbery texture due to the increased protein-protein interactions. Lecithin functions as an emulsifier to enhance protein-lipid interactions. The addition of soy lecithin at levels ranging from 0.025% to 0.2% decreased hardness and increased cohesiveness and smoothness of the reduced-fat processed cheeses. The textural properties of the reduced-fat processed cheese with lecithin was comparable to full fat processed cheese, based on sensory and instrumental analyses (Drake et al., 1999).

4.6 Function of plant-based proteins in processed cheese products

The interest in sustainability, improved nutritional value, and reduced production costs has contributed to growth in the interest in plant-based alternatives to dairy products. In dairy-based cheeses, the casein proteins provide the gel structure and emulsification to produce the characteristic structure of processed cheese products. Differences in the structures of the plant-based proteins and casein result in differences in emulsification and water absorption capacity. Often, additives, including stabilizers, emulsifiers, and acidulants are incorporated with the plant-based proteins and vegetable oils to provide the desired sensory characteristics comparable to the dairy processed cheese products. Off-flavors due to oxidation of the polyunsaturated fatty acids in plant-based protein powders may also adversely affect the quality of the processed cheese products (Bachmann, 2001; El-Sayed, 1997; Jeske et al., 2018).

Plant proteins are an economical partial replacement for dairy-based ingredients in processed cheese products formulated with cheese and butter oil. El-Sayed (1997) replaced skim milk powder with chickpea protein isolate, peanut protein isolate and sesame protein isolate. El-Neshawy et al. (1988) reduced the Cheddar cheese content of the processed cheese products by adding blends of whey protein concentrate and soy protein concentrate, and/or chickpea flour. In both studies, off-flavors were detected in the treatments with plant-based protein powders due to the formation of off-flavors through lipid oxidation reactions. Nevertheless, the treatments with plant-based protein powders were acceptable, indicating that plant-based proteins can be used to as a partial replacement for dairy proteins to increase the content of plant proteins in dairy-based processed cheese products (El-Neshawy et al., 1988; El-Sayed, 1997).

Soybeans and soy protein are often used in the processing of imitation cheeses because of the health benefits associated with soy products. The soy proteins are able to form a gel upon heating to compensate for the decreased gel formation due to the replacement or reduction of casein proteins. However, most soy-based processed cheese products have sensory characteristics that are not characteristic of dairy-based processed cheese products. The formation of hexanal and other carbonyl compounds formed through the oxidation of polyunsaturated fatty acids by lipoxygenase contribute to the beany, oxidized off-flavors associated with soy products. The soy-based processed cheeses products also have a more grainy texture than the dairy-based processed cheese products (Balkir and Metin, 2011; Li et al., 2013).

Imitation kashar cheeses processed with soy protein isolate were firmer, but had a lower meltability than a control processed cheese formulated with dairy products. Soy proteins have a lower emulsification capacity than the casein proteins, contributing to larger fat globules dispersed in the protein network, increased protein-protein interactions, and a lower meltability (Balkir and Metin, 2011).

Because the coagulation of milk proteins is much different from soy proteins, the processing method for soy-based processed cheese products must be modified to develop an imitation cheese product with characteristics similar to the dairy-based processed cheese products. These modifications include the use of bacterial cultures, acids, and divalent salts, rather than rennet to form coagulate the proteins and form the gel structure, use of lactic acid bacteria to remove undesirable beany flavors and hydrolyze flatulence-causing oligosaccharides, and the addition of hydrocolloids and emulsifiers to improve the textural properties (Li et al., 2013).

Li et al. (2013) evaluated several different processes including fermentation with lactic acid bacteria, coagulation with glucono-δ-lactone, and enzymatic hydrolysis to develop a soy cheese spread from soymilk, coconut oil, carrageenan. The soy cheese spreads fermented with the lactic acid bacteria had a firmer, more cohesive, and more elastic gel structure than the spreads coagulated with the addition of glucono-δ-lactone. The lactic acid fermentation resulted in degradation of proteins and fats, improving their distribution throughout the gel structure and reducing the grainy texture. Although all of the soy-based cheese treatments were different in overall sensory properties than the dairy-based cheese spread, the combination of glucono-δ-lactone with lactic acid bacteria, resulted in the most acceptable soy-based cheese spread (Li et al., 2013).

Enzymatic hydrolysis of the soy proteins improved the textural attributes of cheese analogs made by replacing 60% of the casein with soy protein isolates. The cheese analogs made with soy protein isolates were significantly harder and more fracturable than the dairy cheeses and lacked adhesiveness. Hydrolysis of the proteins with either alcalase or trypsin decreased hardness and fracturability and improved adhesiveness to comparable to dairy cheeses. Although melting spread in the soy protein isolate analogs was improved with the hydrolyzed soy protein isolates, it was significantly less than the melting spread of the dairy cheeses (Kim et al., 1992).

4.7 Future challenges

There is growing interest in the incorporation of vegetable-based ingredients in processed foods, including processed dairy products. The use of vegetable-based ingredients in processed cheese products to reduce or replace dairy-based ingredients has many potential benefits, including reduced ingredient costs, improved nutritional value through the reduction of saturated fat and cholesterol, and enhanced sustainability. However, the functional properties of the dairy based proteins and fats are unique and the substitution of vegetable-based ingredients for dairy-based ingredients does alter the quality of the processed cheese products. Through understanding the function of the dairy components and vegetable-based ingredients in processed cheese products, food scientists will be able to take advantage of the properties of the vegetable-based ingredients to develop processed cheeses products containing vegetable ingredients that are comparable to the dairy-based products.

References

Bachmann, H.-P., 2001. Cheese analogues: a review. Int. Dairy J. 11, 505–515.

Balkir, P., Metin, M., 2011. Physicochemical and textural properties of imitation fresh kashar cheeses prepared from casein, caseinates and soy protein. Gida 36, 17–24.

Bennett, R.J., Trivedi, D., Hemar, Y., Reid, D.C.W., Illingworth, D., Lee, S.K., 2006. The effect of starch addition on the rheological and microstructural properties of model processed cheese. Aust. J. Dairy Technol. 61, 157–159.

Bi, W., Zhao, W., Li, D., Li, X., Yao, C., Zhu, Y., Zhang, Y., 2016. Effect of resistant starch and inulin on the properties of imitation Mozzarella cheese. Int. J. Food Properties 19, 159–171.

Cerníková, M., Bunka, F., Pospiech, M., Tremlová, B., Hladká, K., Pavlínek, V., Brezina, P., 2010. Replacement of traditional emulsifying salts by selected hydrocolloids in processed cheese production. Int. Dairy J. 20, 336–343.

Chavan, R.S., Jana, A., 2007. Cheese substitutes: an alternative to natural cheese – a review. Int. J. Food Sci. Technol. Nutr. 2, 25–39.

Cunha, C.E., Dias, A.I., Viotto, W.H., 2010. Microstructure, texture, colour and sensory evaluation of a spreadable processed cheese analogue made with vegetable fat. Food Res. Int. 43, 723–729.

da Silva, D.F, da Souza Ferreira, S.B., Bruschi, M.L., Britten, M., Matumoto-Pintro, P.T., 2016. Effect of commercial konjac glucomannan and konjac flours on textural, rheological and microstructural properties of low fat processed cheese. Food Hydrocoll. 60, 308–316.

Drake, M.A., Truong, V.D., Daubert, C.R., 1999. Rheological and sensory properties of reduced-fat processed cheeses containing lecithin. J. Food Sci. 64, 744–747.

El-Neshawy, A.A., Farahat, S.M., Wahbah, H.A., 1988. Production of processed cheese food enriched with vegetable and whey proteins. Food Chem. 28, 245–255.

El-Sayed, M.M., 1997. Use of plant protein isolates in processed cheese. Nahrung 41, 91–95.

FDA, 2012. Code of Federal Regulations, 21CFR133.169-133.180. Office of the Federal Register. US Government Printing Office, Washington, DC.

Ferrão, L.L., Silva, E.B., Silva, H.L.A, Silva, R., Mollakhalili, N., Granato, D., Freitas, M.Q., Silva, M.C., Raices, R.S.L., Padilha, M.C., Zacarchenco, P.B., Barbosa, M.I.M.J, Mortazavian, A.M., Cruz, A.G., 2016. Strategies to develop healthier processed cheeses: Reduction of sodium and fat contents and use of prebiotics. Food Res. Int. 86, 93–102.

Giri, A., Kanawjia, S.K., Khetra, Y., 2014. Textural and melting properties of processed cheese spread as affected by incorporation of different inulin levels. Food Bioproc. Tech. 7, 1533–1540.

Giri, A., Kanawjia, S.K., Singh, M.P., 2017. Effect of inulin on physico-chemical, sensory, fatty acid profile and microstructure of processed cheese spread. J. Food Sci. Technol. 54, 2443–2451.

Gustaw, W., Mleko, S., 2007. The effect of polysaccharides and sodium chloride on physical properties of processed cheese analogs containing whey proteins. Milchwissenschaft 62, 59–62.

Hennelly, P.J., Dunne, P.G., O'Sullivan, M., O'Riordan, E.D., 2006. Textural rheological, and microstructural properties of imitation cheeses containing inulin. J. Food Eng. 75, 388–395.

Huang, H., Hallinan, R., Maleky, F., 2018. Comparison of different olegels in processed cheese products formulation. Int. J. Food Sci. Technol. 53, 2525–2534.

Jeske, S., Zannini, E., Arendt, E.K., 2018. Past, present and future: The strength of plant-based dairy substitutes based on gluten-free raw materials. Food Res. Int. 110, 42–51.

Kapoor, R., Metzger, L.E., 2008. Process cheese: Scientific and technological aspects – a review. Comp. Rev. Food Sci. Food Saf. 7, 194–214.

Kim, S.Y., Parat, P.S.W, Rhee, K.C., 1992. Textural properties of cheese analogs containing proteolytic enzyme-modified soy protein isolates. J. Am. Oil Chem. Soc. 69, 755–759.

Li, Q., Xia, Y., Zhou, L., Xie, J., et al., 2013. Evaluation of the rheological, textural, microstructural and sensory properties of soy cheese spreads. Food Bioprod. Process. 91, 429–439.

Lobato-Calleros, C., Vernon-Carter, E.J., Guerrero-Legarreta, I., Soriano-Santos, J., Escalona-Beundia, H., 1997. Use of fat blends in cheese analogs: influence on sensory and instrumental textural characteristics. J. Texture Stud. 28, 619–632.

Mounsey, J.S., 2009. Effect of wheat starch on imitation cheese texture. J. Food Technol. 7, 30–33.

Mounsey, J.S., O'Riordan, E.D., 2001. Characteristics of imitation cheese containing native starches. J. Food Sci. 66, 586–591.

Noronha, N., O'Riordan, E.D., O'Sullivan, M., 2007. Replacement of fat with functional fibre in imitation cheese. Int. Dairy J. 17, 1073–1082.

Rafiq, S.M., Ghosh, B.C., 2017. Effect of peanut addition on the fatty acid profile and rheological properties of processed cheese. J. Food Process Technol. 8, 690–695.

Solowiej, B., Dylewska, A., Kowalczyk, D., Sujka, M., Tomczynska-Mleko, M., Mleko, S., 2016. The effect of pH and modified maize starches on texture, rheological properties and

meltability of acid casein processed cheese analogues. Eur. Food Res. Technol. 242, 1577–1585.
Swenson, B.J., Wendorff, W.L., Lindsay, R.C., 2000. Effect of ingredients on the functionality of fat-free process cheese spreads. J. Food Sci. 65, 822–825.
Trivedi, D., Bennett, R.J., Hemar, Y., Reid, D.C.W., Lee, S.K., Illingworth, D., 2008. Effect of different starches on rheological and microstructural properties of (I) model processed cheese. Int. J. Food Sci. Technol. 43, 2191–2196.

Flavors, colors, and preservatives used in processed cheese

5

Miočinović Jelena and Miloradović Zorana
University of Belgrade, Faculty of Agriculture, Department of Animal Source Food Technology, Belgrade, Serbia

5.1 Introduction

Processed cheeses (PCs) are cheese-based food produced by the melting of one or more natural cheeses in the presence of suitable emulsifying salts (ES). The first PC was produced in the early 20th century and since then many different types have been manufactured around the world (Talbot-Walsh et al., 2018). PCs can be divided into groups depending on the type and amount of ingredients used and permitted. Classification varies from country to country but usually PC has to contain more than 51% of natural cheese. In the United States, the Code of Federal Regulations (Part 133) defines four types of PC: pasteurized PC, pasteurized processed cheese food (PCF), pasteurized processed cheese spread (PCS), and pasteurized blended cheese (PBC). In the UK there are two categories: PC and cheese spread (Guinee, 2017).

The use of heating during the production of PC serves to block the action of microorganisms and enzymes, but often causes the products to be poor in taste compared to natural cheeses. PC flavor mainly depends on raw material, especially the natural cheeses. More mature cheeses in PC formulation contribute to stronger taste, but they are often highly priced. This encourages a search for flavor alternatives. Flavorings substances, such as enzyme modified cheese, starter distillate (SDL), wood smoke extract, and spices are often added in order to improve the flavor of PC. The product range can be further broadened by the addition of numerous sterile preparations of meat, of fish, of vegetables, nuts and fruits. Also, in recent years, consumers' increased awareness of health and well-being has led to the development of PCs fortified with macronutrients and other ingredients (probiotics, prebiotics, vitamins, minerals, plant macromolecules) that offer additional health benefits (Talbot-Walsh et al., 2018).

Color is one of the most important characteristics of PC. It depends on the ingredients used in production, such as natural cheeses, etc. However, the color of PC is often adjusted by the addition of some natural ingredients, such as annatto, paprika, or artificial colors.

PC properties, such as flavor and texture, are mainly determined by raw material (mostly natural cheeses) and process parameters, but ingredients such as flavors, colors, and preservatives are also very important in order to achieve acceptable sensory properties and good product quality. The aim of this chapter is to provide an overview of recent literature regarding the use of these important ingredients in PC production.

For economic reasons, the shelf-life of a product is also very important. Preservatives are often used in PC production in order to prevent moulding and yeast growth as well as to maintain the product quality. The most common preservatives are bacteriocin,

Processed Cheese Science and Technology: Ingredients, Manufacture, Functionality, Quality, and Regulations.
DOI: https://doi.org/10.1016/B978-0-12-821445-9.00014-5

nisin, and potassium sorbate; in some cheese products, calcium or sodium propionates are also used.

5.2 Overview of processed cheese production

The technological for production of PC can be split into the following phases:

- Formulation, selection, and preparation of raw material (natural cheeses)
- Mixing of raw material with water and other ingredients
- Melting and shearing of the blend
- Final treatments (optional homogenization), packaging, and refrigeration

The production of natural cheeses involves the controlled destabilization of proteins, during which a gel is formed by paracasein network, containing milk fat. Conversely, PC production involves the decomposition of the thus formed calcium phosphate paracasein network of natural cheeses, followed by a partial hydrolysis of proteins, which enables the binding of free water and the emulsification of fats released during the heating and shearing process.

Raw materials for PC can be all types of rennet-coagulated cheeses with the addition of acid-coagulated cheeses and other dairy products (butter, skimmed milk, casein, various whey products); other ingredients (flavors, colors, spices, acidifying agents, preparations) and water. The main ingredients are ES, mainly phosphates and citrates that promote the desired physico-chemical changes in the blend and enable the production of stable product (Guinee, 2017). The processing (heating and shearing) can be carried out in batch cookers, or in continuous systems. The temperature-time treatments in batch processing are usually within 70–95°C for 4–15 min. and depend on the formulation and the desired product texture. In continuous systems, the temperature varies in the range of 70–90°C, while in UHT treatments (sterilization) the temperature can be as high as 145°C. Processing has two functions: it kills any potential pathogenic and spoilage microorganisms and it enables physicochemical changes that lead to a product with the desired properties and stability. Temperatures below 100°C are generally sufficient to kill vegetative cells, but they do not eliminate microbial spores (Smigic et al., 2018).

PC, in both heated and unheated forms, have applications that require different properties. Quality and characteristics of natural cheeses as the main ingredient, and the used processing parameters, are the most significant factors that determine PC properties such as texture, flavor, cooking attributes, and overall quality. However, the quality of other ingredients, such as flavors, colors, and preservatives, that are used in production, also influence PC properties and shelf life (Guinee, 2017).

5.3 Flavors and flavorings substances in processed cheese

Flavor plays an important role in PC, but, unfortunately, very little research has been done in that area. Most of the available literature is focused on the study of textural and functional properties of PC. However, in the text which follows, the overview of literature regarding PC flavor gives special attention to the flavors that are used in production. The preparations of meat, fish, vegetables, nuts, fruits, and spices, that enable product differentiation, will also be discussed.

Generally, the flavor profiles of cheeses are complex and variety and typemspecific. Cheese flavor is the result of the correct balance and concentration of a wide variety of volatile flavor compounds (Singh et al., 2003). Drake et al. (2010) conducted a study in order to identify and define the sensory properties of different PC products. They evaluated 55 commercially available PCs by descriptive analysis and described flavors as sweet aromatic, cooked/milky, caramelized/toasted cheese, diacetyl, whey, fatty, and brothy.

5.3.1 Influence of raw materials on processed cheese flavor

The flavor profile of PC is influenced by the type and amount of natural cheeses used in the manufacturing of the processed product, the heat treatment applied and the presence or absence of other ingredients (Drake et al., 2010). Almost all types of natural cheeses could be used as a raw material for PC production. The most common are Cheddar, Gouda, and Emmental cheeses, but other types, such as blue mould cheese, Gruyere, and Tilsit, could also be used (Kapoor and Metzger, 2008).

The type, maturity, and flavor profile of the natural cheeses used for PC production play an important role in setting both texture properties and flavor. Different proportions of mature and young cheeses in PC formulations have a significant influence on the final flavor (Kapoor et al., 2007; Burgos et al., 2020). Even within a single cheese type, differently aged cheeses could display many different flavor profiles, which indicate that there is huge potential in flavour variability for PC. Young cheeses serve as a source of intact casein to provide structure, while mature cheeses are responsible for the flavor of PC. The composition of PC, especially fat content, strongly influences the flavor. Muir et al. (1997) studied the sensory characteristics of 16 commercial PCS with a different fat content. When the fat content of PC was decreased, the creamy attribute was also decreased and there was an increase in the acidity and bitterness.

Other dairy ingredients, such as skim milk powder, milk protein concentrates, casein, caseinates, whey proteins, whey powders, and whey permeate powders, are used for standardization of composition and they contribute to the flavor, texture, and functional properties of PC (Guinee, 2017). However, these ingredients can impart both desirable and undesirable flavors to finished products. For that reason, the amount of addition and/or replacement of natural cheese with these ingredients in PC formulation should be carefully controlled (Kussy and Aylward, 2008).

Abd-El-Salam et al. (1996) studied the effect of addition (0%, 20%, and 40% of the final blend) of liquid WPC (28% total solids, 15% whey protein) on the composition and the rheological properties of PCS (57% moisture, 3% ES). They found that the flavor and other sensory properties of PCS improve when the amount of WPC is increased. Pinto et al. (2007) partially replaced cheese solids from Cheddar cheese with whey protein concentrate (WPC) at different levels (1.5%, 3%, and 4.5%) in the production of PCS. The addition of WPC significantly improved the texture and the body of product while the flavor became milder in comparison with the product made from Cheddar cheese. Doosh et al. (2014) reported that the use of concentrated buttermilk in PC improved product stability during storage and gave an acceptable flavour. Recently, Desouky et al. (2019) found that the addition of camel milk powder in different concentrations (5%–15%) improved the quality attributes of cheese sauces and could be considered as a new functional source that substitutes cheeses in the manufacture of PC.

The use of old or of extended-storage dairy-powder ingredients (nonfat dry milk, whey, WPC, milk protein concentrate) often causes stale, storage-like, cardboard-like, and/or oxidized off-flavors. Also, some dairy ingredients have their own associated flavors, such as cooked, scorched, and lacking freshness (stale). The strong off-flavors derived from these added ingredients might not be masked in PCs and hence it is desirable to know their flavor characteristics before they are used in production. Furthermore, some flavor defects might become even more pronounced in the finished products (Kussy and Aylward, 2008). High-fat ingredients (anhydrous milk fat, cream, dehydrated cream, butter) are used for composition standardization, but it is well known that fat plays an important role in flavor formation. Nevertheless, they may impart oxidized or definitive feed off-flavors. With these materials, the adverse storage conditions and handling practices can diminish their sensory properties in ways that may be detectable in PCs. Hence, flavor assessment of all these ingredients is desirable and should be conducted prior to their use in PC manufacture (Kussy and Aylward, 2008).

In recent decades, there has been a noticeable need and consumer demand for products, including PCs, with a reduced fat content. Fat reduction often results in flavor defects that require modification of processes and/or the addition of ingredients (Johnson et al., 2009). A low-fat PC could be produced by using a low-fat natural cheese in the formulation of cheese base. However, this approach results in unacceptable textural and functional properties and in poor flavor. Another approach is to use, at various levels, different fat replacers. Hassan et al. (2007) showed that the utilization of EPS-producing starter in the manufacture of a lower-fat PC base may be a promising avenue: the development of EPS in the cheese structure may lead to inherent fat mimetic properties and, therefore, may solve various textural issues associated with lower-fat PCs without significantly affecting the sensory properties of the final product.

The addition of ES affects the flavor of PC (Awad et al., 2004). Sometimes ES can impart undesired and off flavors. For example, some types of phosphate ES, or their overdose, may contribute to subtle ''emulsifier burn'' or pronounced ''soapy'' or bitter flavor defects (Mayer, 2001, Kussy and Aylward, 2008). The addition of salt raises the pH of PC and, if pH is above 6.0, some defects could appear, such as "alkaline" or "detergent" or "soapy" flavor (Hoffmann et al., 2012; Gupta et al., 1984).

PCs are often perceived as being high in sodium chloride (NaCl). NaCl, universally well-known as table salt, is an ancient food additive, broadly used to increase the storage stability and the palatability of foods. Nowadays, in accordance with modern diet and the recommendations of Dietary Guidelines, consumers prefer lower sodium products because excess of NaCl is linked to hypertension and cardiovascular diseases

(Bansal and Mishra, 2020). The sodium content of most types of natural cheeses varies from 40 to 800 mg 100 g^{-1}, whereas for PC, it can be as high as 1500 mg 100 g^{-1} (El-Bakry, 2012). The reduction of NaCl content is a difficult challenge for cheese makers since salt has many roles in cheese production and it influences flavor, texture, safety, and overall acceptability (Johnson et al., 2009; Bansal and Mishra, 2020). The three major ingredients that contribute to sodium in PC are sodium-based ES (approximately 44% to 48%), natural cheeses (approximately 28% to 37%), and added salt (approximately 15% to 24%). The reduction of sodium in PC, therefore, requires a modification of one or of all tree aspects during the formulation and production.

Only a few research projects have targeted the development of reduced- and/or low-sodium natural cheese as an ingredient for PC (Karahadian and Lindsay, 1984; Kapoor and Metzger, 2008). In order to achieve reduction of sodium some efforts have also been directed at the development of novel formulations with utilization of potassium-based ES (Karahadian and Lindsay, 1984; Hoffmann et al., 2012; Gupta et al., 1984; Mozuraityte et al., 2019), with various salt replacers and with other flavour enhancers (Karahadian and Lindsay, 1984; Kapoor and Metzger, 2008), all without causing a detrimental effect on the flavour and the textural and functional properties of the final PC. Karahadian and Lindsay (1984) found that appropriate selection of Cheddar cheese, along with potassium-based ES and some other minor ingredients that enhance different flavour notes (enzyme-modified cheese [EMC]), could be used for production of PC with up to 75% reduced-sodium content and still with good functional and sensory properties.

Another approach in an effort to reduce sodium content was to decrease the amount of NaCl by a partial or a full its replacement with other suitable salts like potassium chloride (KCl), calcium chloride ($CaCl_2$), or magnesium chloride ($MgCl_2$). El-Bakry et al. (2011) reported that lower flavor scores of PC were obtained when more than 50% of NaCl was substituted with KCl, whereas no effect was found on sensory properties with less than 50% replacement of NaCl. However, Patel et al. (2013) observed that complete replacement of NaCl by KCl had a significant negative impact on the sensory attributes of PC, while a combination of 50% NaCl and 50% KCl is recommended for use as a salt replacer.

In PC production sometimes small amounts of ingredients like acids, sweetening agents, yeast extracts, preservatives can be used and they can have some impact on product flavor. Acids are often used in PC production for flavour enhancement and/or pH adjustment. They can contribute either desirable or undesirable sensory properties. When acetic acid is used up to 0.25%, it may mimic sharp flavors of natural cheeses, but when used in excess it may contribute to vinegar-like flavors. Lactic acid is naturally present in cheeses used for the production of PCs, but if added in excess of 0.25%–0.30%, it may add metallic off-flavors. Lactic acid and acetic acid may be used in combination to adjust pH and to achieve desirable acidic flavors, but the amount should not exceed 0.40% of typical PC formula. Sweetening agents, such as sucrose, dextrose, corn syrup, and hydrolyzed lactose, are used in PC production in order to balance the sensory properties and to increase sweetness. However, these ingredients may sometimes impart atypical sweet taste notes. Certain yeast extracts may provide flavor notes that resemble aged natural cheese, but they may also impart an undesirable flavor or an aftertaste. Some preservatives, such as sorbic or propionic acids and their salts (potassium sorbate and sodium propionate), used in PC production because of their fungistatic effect, may contribute to undesirable acidic or mouthfeel "burning" sensory properties (Kussy and Aylward, 2008).

5.3.2 Influence of flavorings on processed cheese flavor

Flavorings are products added to food in order to impart or modify odor and/or taste. Various flavorings can be used in PC production, such as SDL, enzyme modified cheese, and wood smoke. Their role is to enhance the flavor, especially when a lot of young natural cheeses are used in the formulation. This could reduce the necessary amount of high priced mature cheeses and lower the costs of production.

SDL is generally recognized as safe (GRAS) food ingredient and is used to improve the buttery aroma of many foods, including PCs. The buttery aroma is formed by the presence of the diacetyl (butane-2,3-dione), which imparts a high level of buttery flavor notes and is the key component of SDL. It is a volatile product of citrate metabolism produced by certain bacteria, including *Lactococcus lactis* ssp. *diacetylactis* and *Leuconostoc citrovorum*. Commercial SDL is available as a water- or oil-based liquid or as an encapsulated powder. In PC production, a concentrated SDL that contains 14,700–15,300 $\mu g\ g^{-1}$ diacetyl is usually used (Rincon-Delgadillo et al., 2012). The addition of SDL to PC can impart fresh cheese or buttery flavor notes, but, if added in excess, the finished product may seem artificial and/or popcorn-like (Kussy and Aylward, 2008).

EMC in forms of pastes and powders are used as sources of cheese flavor in the production of processed of cheeses and cheese imitations (Wilkinson et al., 2011; Guinee and Kilcawley, 2017). They have approximately 15–30 times the flavor intensity of natural cheese. Lower-fat PCs often have issues with the flavor, in which case the addition of appropriately selected EMC and other permitted flavor enhancers can control and rectify the problem. However, there is still a lack of available literature on the success or otherwise of the application of EMC in lower-fat PCs (Johnson et al., 2009).

EMC is produced by adding proteolytic and lipolytic enzymes to young or downgraded cheeses in order to break down proteins and milk fat and to liberate flavor compounds. Natural cheeses are heated and polyphosphates are added in order to achieve the liquefying of cheese and to adjust the pH to 6.6–6.8. The liquefied cheese is homogenized and pasteurized to destroy cheese-starter bacteria and contaminant organisms and it is then cooled to the temperature optimal for the proteolytic and lipolytic enzymes. Incubation usually takes 10 and 50 h, at 40–50°C. The temperature/time regime is dependent on many factors, such as product type, type and activity of the enzyme and required degree of flavor development. When the required level of hydrolysis is reached, EMC is heat-treated at around 90°C in order to denature the enzymes and it is cooled before it is packaged as a paste. Alternatively, it can be spray-dried to produce a powdered ingredient that can be dusted over or incorporated into different food products (snack food, bakery products, soups, sauces) and the PC (Early, 2012; Kilcawley et al., 1998). The enzymes could have different origins (plant, animal, or microbial sources), but microbial ones are the most common. Enzymes derived from *Rhizopus oryzae, Aspergillus oryzae*, and *Aspergillus mellus* break down cheese curd and acid casein, while those derived from *Penicillium roqueforti, Penicillium camemberti, Aspergillus niger*, and *Candida rugosa* are active in the hydrolysis of triacylglycerols into fatty acids. Proteolysis products: peptides and amino acids are responsible for the savoury and the umami flavours, while the short chain (C2–C6) and the medium chain (C8–C14) free fatty acids produce the cheesy flavours and buttery notes, respectively. Long chain FFAs (>C16) can give a soapy mouthfeel (Noronha et al., 2008). EMC variants of most of the major natural cheeses varieties are commercially available, such as Cheddar, Blue, Romano, Parmesan, Colby, Gouda, Camembert, Mozzarella, Gruyere, Asiago, Gorgonzola, and Emmental. They contain NaCl in the range of 1%–2% that contributes to the saltiness of the product. The flavor profile of the PCs is affected by the EMC composition, especially the lipolysis level, and the pH of the cheese base (Noronha et al., 2008). Because of their high flavor intensity, EMCs are, typically, added as flavorings to foods at a low level (0.1%–1% w/w), but their concentration depends on the type of product and the required flavor (Guinee and Kilcawley, 2017).

PCS could be produced with the addition of rapidly ripened ultrafiltered retentates. In an initial study, Aly et al. (1995) have evaluated the effectiveness of some ripening agents in accelerating flavor development in the retentates. Good-quality flavor is obtained by using a combination of 0.05% of sodium dodecyl sulfate (SDS) and 0.005% of a commercial lipase preparation, after incubation at 32°C for 5 days. Then, the SDS lipase treated retentates were used in different concentrations (0%, 20%, 40%, 60%, 80%, or 100%) as a substitute for mature Ras cheese for PC. Replacing up to 80% of mature Ras cheese solids by SDS-lipase-treated retentate gave a product with flavor, consistency and color superior to the controls, both when fresh or after storage.

Smoked cheeses and PCs are very popular in some countries (Italy, Germany, Egypt, Serbia, India, Iran). Smoking of dairy products, like cheese, is one of the oldest methods used for preservation and in the production of flavored food. Traditionally, cheese is smoked by the application of natural smoke from wood; however, some manufacturers prefer liquid smoke that is derived from smoke condensate or prepared by mixing chemically-defined substances (Ismail, 2016). Smoke flavors, both natural and liquid, can be applied to cheese products by using various methods: (1) direct addition to milk, (2) addition to the brine, (3) addition to cheese, and (4) surface application to finished cheeses. However, in PC production, only the last two methods are applied. Smoked PC can also be produced by the addition of smoked natural cheeses. When smoke flavors are used the drying loss that occurs during the normal smoking process is avoided and the production time is reduced. The food does not lose weight, which affects both the quality and the price of the product. The use of liquid smoke flavor (LSF) has important advantages: the time necessary to reach the required sensory profile is considerably reduced; and it is easier to control contaminants, including polycyclic aromatic hydrocarbons. LSF is currently available in liquid form, or as water mixable emulsion, a powder, or a solution in vegetable oil.

Smoking has a preservative effect on food that is connected to antimicrobial properties of smoke components, such as organic acids (acetic, propionic, and benzoic acids). Also, carbonyls in smoke materials have a role as an antimicrobial, but in smoke condensate they could be inferred by other components, such as aldehydes and ketones. It is well known that commercial liquid smoke preparations are effective against various types of spoilage and pathogenic microorganisms. However, different forms of smoke extracts (dry and liquid) differ considerably in their ability to inhibit the growth of the pathogens (Suñen et al., 2001). Wendorff et al. (1993) reported that phenolic compounds found in smoke inhibit growth of moulds on smoked Cheddar cheese.

5.3.2.1 Influence spices and other ingredients on processed cheese flavor

PC belongs to the cheese products group and the expected flavor is primarily of cheese. However, additional flavors have been introduced over time in order to broaden the range of product.

Flavors and ingredients used in PC production depend highly on the market or on consumer habits and target groups. Products intended for children usually have milder flavor and they are usually enriched with sweeteners and kid-friendly flavors, such as fruit (strawberry, banana, and apple). On the other side, PCs for adults are often made with spicy and hot ingredients, such as peppers, jalapeno, wasabi (Osthoff et al., 2011). Today, there are many PC varieties that contain ingredients such as meat (ham, fish, and shrimps), vegetables and fruits preparations (Khider et al., 2017; El-Sayed, 2017). The added ingredients are usually up to 15% and have only a small impact on texture and functionality, but they greatly influence the flavor of PC (Osthoff et al., 2011).

Herbs and spices have been used throughout history to fortify foods, including PC, serving as preservatives and flavor and therapeutic agents (Shan et al., 2011). Spices that are used in PC production are numerous, such as green pepper (*Capsicum frutescens var. grosum*), Jalapeno Chipotle (smoke-dried jalapeno) (*Capsicum annuum*), Habanero (*Capsicum chinense*), Horseradish (*Armoraci arusticana*), and Dill (*Anethum graveolens*) (El-Sayed and Youssef, 2019). They are added in different amounts ranging from 0.01% to 1% and usually added at the beginning of the melting process in order to achieve uniform distribution through cheese mass. Spices must be presterilized to avoid a possible contamination of the final product (Osthoff et al., 2011). El-Sayed (2017) studied the effects of different concentrations (0.1%, 0.2%, 0.3%, 0.4%, and 0.5%) of thyme powder on the chemical composition, the meltability, the oil separation, and the sensory properties of block PC, immediately after production and after 3 months of storage. Thyme is usually used as ingredient for flavor, but it also serves as a preservative due to its antioxidant and antimicrobial effects. Authors found that the addition of thyme powder has a significant influence on PC properties: in blends of up to 0.1%–0.3% the ingredient improved the texture and enhanced the flavor. Shalaby et al. (2017) produced a PC sauce (75% WPC and 25% acid curd) with acceptable flavour with 2% of essential oils of turnip, shallots, *Capsicum,* and cardamom. The flavor of PC sauces with turnip and shallots essential oils scored better than the other two. Ye et al. (2009) produced PC fortified with fish oil as an excellent source of omega-3 long-chain polyunsaturated fatty acids (omega-3 LC PUFA). However, the use of straight fish oil has a limitation due to oxidation and the "fishy" flavor, while the fish oil emulsion made with a milk protein complex (encapsulated fish oil) gives an acceptable results.

In PC production, different kinds of mushrooms can be used for flavor preparations. Mushrooms are a rich source of many valued nutritive components, such as amino acids, vitamins, and minerals, but they also have antioxidant, antitumor, and antimicrobial properties. Today, they are often used as supplements in different kinds of food in order to improve their functional properties and to contribute to specific flavor. Fatma et al. (2005) used the edible mushroom *Agaricus campestris* in different concentrations (5%, 10%, and 15%) for PC production. The mushrooms were steeped in citric acid, boiled in emulsifying solution and then used for production. The addition of 15% of mushroom caused defects of the appearance, color, and texture of the PCS. The authors concluded that PCS with improved nutritional and functional properties and acceptable sensory quality could be obtained by incorporating 5% to 10% of mushrooms. Khider et al. (2017) produced PC supplemented with either fresh or dried edible mushroom (*Pleurotus ostreatus* Hk35) in order to improve its nutritional value and functional and sensory properties. Fresh mushroom, cut in pieces and in smashed form, was added to cheese blend at levels of 0%, 55, 10%, and 15%, while dried mushroom (mushroom powder) was added at levels of 0%, 1%, 1.5%, and 2%. The highest flavou scores were recorded for PC supplemented with 1% and 1.5% of mushroom powder, respectively.

Solhi et al. (2020b) studied the rheological, chemical and sensory characteristics of PC containing different quantities of tomato powder (0%, 1%, 2%, and 4% wt/wt), all with appropriate levels of bioactive components, and during 90 days of storage. The results showed that the PC samples containing tomato powder had higher levels of proteolysis rate, phenolic and lycopene contents and antioxidant activity. Sensory analysis showed that the PC with an average level of tomato powder (2%) got maximum scores for flavor, color, and total acceptance. Qualitative properties of PC fortified with different levels of asparagus powder (0.5%, 1%, and 1.5% wt/wt) were evaluated during storage. Asparagus is a rich source of flavonoids and other phenolic compounds, phytochemicals, hydroxycinnamic acids, and dietary fibers, which possess strong antioxidant and functional properties (Fuentes-Alventosa et al., 2013). The results obtained from the organoleptic tests indicated that the color and flavor of the PC samples did not change significantly with the increase of the amount of asparagus powder. Although the highest scores for flavor, color, and total acceptance belonged to the control sample, the fortified samples also got relatively high scores, and consumers were satisfied with the sensory properties of the fortified PC samples (Solhi et al., 2020a).

PCs are usually not rich in functional and bioactive ingredients, and they need to be enhanced by the addition of ingredients with a high content of bioactive components. The fortification of PC with bioactive compounds has become rather popular in recent years: it offers health benefits, it may contribute to the flavor and texture, and it increases consumer acceptance. The main cheese fortification methods are through the incorporation of probiotics and prebiotics/fibers, vitamin enhancement and fortification of the PC with other macronutrients. Unfortunately, there are few literature references about influence these ingredients on the PC flavor. Szafranska and Soloweij (2020) investigated the use of different types and concentrations (1%–4%) of fibers (bamboo, acacia, potato, or citrus) in acid casein PC sauces and their effect on texture and rheological and sensory properties. Dietary fibers are important health-promoting ingredients, low in calories and cholesterol, while PC sauces are produced on the basis of acid casein, whey protein, and rapeseed oil. The highest scores for overall sensory quality were received for cheese sauce with bamboo and citrus fibers, while with regard to flavor, PC sauces with acacia and potato fibers were given lower scores.

5.4 Colors used in processed cheese

Color is one of the most important food characteristics; it influences consumer preference, taste perception and purchasing choice. Color indicates food freshness and quality because discolouration is a consequence of cheese spoilage mostly due to microbial growth, but also to chemical reactions (Smigic et al., 2018; Oliveira et al., 2016).

The color of PC products depends on many factors, such as the type, composition, and properties of natural cheeses used for formulation, the ingredients and the process parameters (Sharma et al., 2019). Usually, the color of cheeses and of PC is yellow to orange and, in special cases such as moulds cheeses; it is blue, green, or grey (Osthoff et al., 2011). In general, yellow color is associated with foods that have a high fat content, but it also depends on the size of the fat globules dispersed in the solution or lodged in the protein matrix. Usually, whiter color of product is a consequence of a smaller diameter of fat globule that causes a greater dispersion of light. Despite its high fat content, the fat in PCs is usually highly emulsified and, therefore, the globules are small in size, resulting in a whitish color. The type and the amount of ES have a significant influence on the color parameters (L*, a*, and b*) of PC (Awad et al., 2004; Cunha et al., 2012). Cunha et al. (2012) found that Requeijao cremoso PCs, produced with sodium hexametaphosphate, had a whiter, less yellow color than cheeses with tetrasodium pyrophosphate, sodium tripolyphosphate, or sodium citrate. However, despite some difference in fat globule diameter, they concluded that the color was more influenced by the type of protein network. The type of ES influences the colour parameters of block PC. An increase of pyrophosphate content in the salt mixture contributed to a shinier and whiter color final product. Also, color was changing during storage and became darker, especially at room temperature (Awad et al., 2004).

Food color is a substance not normally consumed as a food, but as a food additive it can be used to add or restore color. They are artificial dyes or natural constituents of foods and other sources that are add in order to compensate for loss of color due to processing or storage, or to balance out variation in natural color. Also, it is often used to make food more attractive (Talbot-Walsh et al., 2018; Sharma et al., 2019). The choice of colors in the production of PC depends on product properties, consumer expectations, and market habits. Nowadays, natural colors are preferred because of their health benefits and their additional functional properties (antioxidativity, antimicrobial) as well as their effectiveness as chemical synthesized additives (Sharma et al., 2019). Sometimes, substances considered as food and spices, for example, fruit or vegetable concentrate and saffron, are used because of their coloring properties, but they do not fall within the scope of the food additives regulation (Lehto et al., 2017). In the EU, colors permitted for use in cheeses and PCs are authorized by the EFSA (European Food Safety Agency) and are listed in a set of regulations approved by the European Commission (No. 1333/2008; 1331/2008; 1129/2011). In the USA, these additives are regulated by the Code of Federal Regulations Title 21 CFR Part 70, 71, 73, 74, 80-82. The food colors allowed for use in PC production are carotenes (E160a), paprika extract (E 160c), annatto, bixin, and norbixin (E160b). The maximum level of annatto, bixin, and norbixin addition, according to the EU Regulation, is 15 mg/kg, while in the USA there is no set limit. Kang et al. (2010) reported that the estimated annatto content in flavored PC in the USA is in the range 0.2 to 21.4 mg/kg. Two other colors do not have a defined maximum level, but they should be used according to good-manufacturing practice, at a level not higher than necessary to achieve the intended purpose, and providing that they do not mislead the consumer. For flavored PC, it is also permitted to use other colors, such as curcumin, tartrazine, lutein, carmine, lycopene, and β carotene. According to the EU Regulation, their maximum allowed level of addition is 100 mg/kg.

The main color used in the manufacture of cheese and PC is the natural color annatto. It is derived from seeds of *Bixa orellana* plant and is available in both lipid soluble (bixin) and water soluble (norbixin) forms. The three commercial processes that are most often used to extract color pigments from dried annatto seeds are: direct extraction into oil; direct extraction into aqueous alkali; and indirect extraction using solvents. The direct extractions, either into oil or into alkali, are natural and environmentally friendly. Lipid soluble annatto, obtained by direct extraction into oil, contains mostly bixin, while extraction into alkali, usually sodium or potassium hydroxide, produces water-soluble annatto with a dissociated form of norbixin in alkaline solution. Emulsified annatto contains both bixin and norbixin with an emulsifier. Depending on the method employed, the products obtained have different amounts of bixin and/or norbixin and can be further processed into solution, suspension, or powder, intended for different purposes. Also, different commercial additives can give the final product different colors, ranging from red to yellow (Sharma et al., 2019).

The oil-soluble annatto and the oil-based suspension usually contain 0.05%–1% and 1%–8% of bixin, respectively. These formulations are often used in high fat products including some types of PCs. Water-soluble annatto used in cheese making contains norbixin, and it imparts the desired yellow/orange colour. Water-soluble extracts contain 0.1%–0.4% norbixin, while powders contain much more (1%–15%). Emulsified annatto is suitable for food containing both water and an oil phase. They are normally used in emulsified dairy and food products, such as PC.

Another color that is often used in PC is paprika. Paprika is a well-known spice with a deep red color and pungent flavor that is obtained from red pepper *Capsicum annum*. Paprika oleoresin contains mainly capsanthin and capsorubin carotenoids that occur primarily as esters of lauric acid. It is available in different forms, such as in edible vegetable oil and in water soluble extract. The fat/oil-soluble version of paprika extract (oleoresin) is obtained by using hexane as solvent. Paprika is a common additive, used in many normally savory products, and should be compatible with the product flavor. Its use in natural cheese is limited because of its strong influence on flavor. However, the use of paprika in PC is quite common, serving to adjust color or flavor (Nogueira de Oliveira et al., 2011). Also, a mixture of oil and paprika is spread on the surface of some special cheese varieties. As mentioned above, there is no limit on the amount of paprika used in PC.

The pigment β-carotene is naturally present in the milk drawn from grass-fed lactating cows. It can be either chemically synthesized or extracted from natural sources. In the past, this color had been used in cheese making because of its natural presence in milk. Carotene is fat soluble and is not suitable for water based complex foods. Nowadays, some carotenoid based colors have been developed in order to prevent the loss of colour and to make additives suitable for special applications. Additionally, carotene plays a significant role in human health because it is a precursor to vitamin A, known to contribute to the prevention of cardiovascular diseases and cancer.

Lutein is another of the colors used in flavored PCs. It is a carotenoid, found in green leafy vegetables and yellow carrots, that contain xanthophylls, the natural yellow–red-colored fat-soluble pigment. There is no literature data about the use of this food color in PC, but it has been successfully added to Cheddar cheese (Jones et al., 2005) and Brazilian Prato cheese (Kubo et al., 2013; Sobral et al., 2016). Lutein could be used as a replacement for annatto because of its antioxidant properties. Lycopene is a naturally red pigment found in many vegetables and fruits. As a food color it can be used in flavored PC. Lycopene is one of the most potent antioxidants. Hassan et al. (2019) used lycopene extracted from tomato waste for fortification of PCS. Besides giving PCs an acceptable red color and flavor, the addition of lycopene contributes to their high content of iron, antioxidant and phenolic compounds.

5.5 Discoloration of processed cheese

Discoloration of PC is an ever-present problem in the dairy industry, always causing a significant economic loss to producers (Daly et al., 2012). A loss and/or a change of color could be the consequence of both the inferior quality of the raw materials, especially natural cheeses and other ingredients such as ES, and of bad processing parameters (Osthoff et al., 2011). The most important is the pink discoloration of PC made with colors, usually annatto. It is not located only on the surface, it penetrates throughout the cheese block (Shumaker and Wendorff, 1998). There are numerous reasons for this phenomenon. The discoloration could be caused by the alteration of annatto due to the varying of pH within the cheese, by the oxidation of bixin under light, by the presence of oxygen, by interactions between nitrates and annatto. The high heat treatment applied during cheese production could also be a reason for this defect.

The type of annatto preparation may have an influence on the pink discoloration. Alkaline extracts of annatto used in PC production have shown a higher propensity to causing pink discoloration compared to other extracts. It has also been reported that the emulsion-based annatto colorants in the PC formulation were more susceptible to pink discoloration compared to suspension-based annatto colorants

(Shumaker and Wendorff, 1998). PC has a greater tendency to develop pink discoloration if it contains an increased amount of colored natural cheese (cheese containing annatto) or if it has a higher ratio of aged cheese (that has undergone an extensive proteolysis) in the process blend. A possible alteration of cheese color could be due to Mailllard browning as a consequence of relatively high temperatures applied in the production as well as the composition of natural cheeses that were used (sugar and salt content). The alteration could be accelerated by high pH levels in the final product (>5.9) and by storage at relatively high temperatures (35°C).

5.6 Preservatives and their role in processed cheese

The safety of dairy products is a global public-health issue and their spoilage is important both for economic reasons as well as in order to reduce food waste. Preservatives are substances which prolong shelf-life by protecting foods from deterioration caused by pathogenic microorganisms. The most common preservatives permitted in the production of PC are nisin, potassium sorbate, and calcium and sodium propionate.

Spoilage of PC caused by microbial growth is usually detected by the occurrence of sensory defects such as gas production, abnormal odors, discoloration and unusual texture, or pH variations. Generally, PCs are considered to be a low-risk food in terms of safety. During the long history of their production only few outbreaks have ever occurred, usually related to PCS with high pH and high water activity (Oliveira et al., 2016; Smigic et al., 2018; Glass and Doyle, 2005). The microbiological quality and safety of PCs is dependent on the safety and quality of ingredients and also on other factors, such as pH, moisture, fat, salts, and/or the presence of food additives (Glass and Johnson, 2004).

Heat treatments that are usually applied during PC production are appropriate for destroying vegetative forms of present microbiota, but they are not sufficient to kill bacterial spores. This type of contamination originates from gram-positive spore-forming bacteria of the genus *Bacillus* and *Clostridium* (Oliveira et al., 2016; Glass and Johnson, 2004). In spoiled PCS (42 samples), *Clostridium sporogenes* were identified in 33% of products, followed by *Clostridium cochlearium* and *C. tyrobutyricum* (Lycken and Borch, 2006). Spores of anaerobic species could have been present in some ingredients used in production, particularly in the cheese. Contamination of PCs after pasteurization could include the presence of *Listeria monocytogenes, Salmonella spp., Staphylococcus aureus*, and *E. coli* O157:H7. However, Glass et al. (1998) found that PC slices allowed the survival but not the rapid growth of *S. aureus*, whereas populations of *L. monocytogenes, E. coli* O157:H7, and *Salmonella* serotypes had decreased during the 96 h storage at 30°C.

5.6.1 *Nisin*

One of the most extensively studied natural antimicrobial agents is nisin, and it has gained widespread application in the dairy industry (de Arauz et al., 2009; Silva et al., 2018). Nisin (nisin A) is a polypeptide composed of 34 amino acids and it belongs to class I bacteriocins. The use of bacteriocins as natural food preservatives fulfills consumer demand for high quality and safe foods without the use of chemical preservatives. However, the application of bacteriocins as food additives can be limited for various reasons, such as effectiveness of pathogen elimination and their high cost (And and Hoover, 2003).

Nisin is produced by certain strains of *Lactococcus lactis* subsp. *lactis* (A, Z, F, and Q variants) and *Streptococcus* strains (U, U2, P, and H variants) (O'Connor et al., 2015). The commercial form of nisin as food preservative is Nisaplin, with nisin A (2.5%) as an active ingredient and other ingredients such as NaCl and nonfat dry milk (And and Hoover, 2003). Nisin has an inhibitory effect on the growth of numerous gram-positive bacteria, including LAB (*Lactococcus, Lactobacillus, Micrococcus),* on pathogens, such as *Listeria* and *Staphylococcus*, and on the spore forming bacteria, *Bacillus* and *Clostridium* (And and Hoover, 2003).

Nisin (E 234) is an authorized food additive and recognized to be safe by the Joint Food and Agriculture Organization/World Health Organization (FAO/WHO) Expert Committee on Food Additives (Favaro et al., 2015; Silva et al., 2018). In the European Union (EU), its use in several food categories is regulated by Annex II of Regulation (EC) 1333/2008. In PC production, the use of 12.5 ppm of nisin is allowed (EU, 2008). In the USA, it is permissible to use nisin preparation in an amount which results in not more than 250 ppm in the food (FDA 133.179). Nisin can be used alone, in combination with other preservatives, or also combined with several physical treatments. When applied in combination with other bacteriocins, using thermal processing, high pressure, low-temperature storage, reduced pH and/or aw, nisin can prolong shelf-life. Despite a small amount of data, the following text revises the literature on the use of nisin and its effect on the production of PCs.

Nisin has been widely applied in cheese and pasteurized cheese spreads as a replacement for nitrate, in order to prevent the outgrowth of clostridia spores (And and Hoover, 2003). The potential for growth and toxin production by *C. botulinum* in PC products, and particularly in spreads, is significant. Trials in the USA have indicated that, in PCS, nisin is effective in delaying or preventing the growth and subsequent toxin production by inoculated spores of *C. botulinum* types A and B. The study of Somers and Taylor (1987) also showed that addition of nisin in the preparation of the pasteurized cheese spread prevented the botulinum growth and toxin production. Their results indicated that the cheese spreads formulated with 52% of moisture, pH 5.9 and with 2.5% disodium phosphate, require only 12.5 ppm nisin to inhibit toxin production.

Trials with PC products have been carried out in the UK using a cocktail of spores of the before mentioned Clostridium spp. at inoculation levels of approximately 200 spores per gram. Control samples without nisin readily became spoiled, while partial control was achieved with 2.5 mg kg^{-1} and spoilage was prevented during storage at 37°C by 6.25 mg kg^{-1} of nisin. Zottola et al. (1994) reported that cheese spreads, inoculated with *Clostridium sporogenes* at 22°C and containing 300–390 IU nisin g^{-1}, had a significantly longer shelf life, and compared to control samples without nisin. Results indicated that nisin delayed spoilage to 87 days, compared to 14 days for samples without nisin. Nisin can also be effective in the control of different pathogens in dairy products, such as *L. monocytogenes* and *Staphylococcus aureus* (Sobrino-López and Martín-Belloso, 2008). In addition, Zottola et al. (1994) reported that control of food-borne pathogens *L. monocytogenes* and *S. aureus* can be achieved even by using a nisin-containing cheese (Cheddar) in pasteurized PC or in cold pack cheese spread formulation. Facultative aerobic Bacillus spp. (*Bacillus cereus* and *Bacillus subtilis*) can also cause spoilage problems in PC products, and these organisms could be also inhibited by the addition of 5–20 mg kg^{-1} of nisin (Sobrino-López and Martín-Belloso, 2008).

Smigic et al. (2018) studied the effect of nisin (12.5 ppm) and storage temperature (4, 8, and 20°C) on the microbiological, physicochemical and sensorial quality of processed cheese. Aerobic colony count (ACC) was significantly increased in processed cheese samples prepared without nisin, while no significant increase was detected in nisin supplemented samples. Along with the ACC, psychrotrophic colony count, moulds and yeast were also checked, but they were not significantly different in the samples with and the samples without nisin. Also, the addition of nisin had no effect on the sensory properties and the texture of PC over the 12 weeks experimental period. The authors concluded that PC samples were a rather stable product, but the most effective combination in controlling ACC was the addition of nisin (12.5 ppm) and low storage temperature (4°C). After 30 weeks storage, the nisin retention in the cheese spread decreased with the increase in storage temperature, falling from 80% for storage at 20°C to 60% at 25°C and 40% at 30°C (Delves-Broughton and Weber, 2011).

The antimicrobial efficacy of nisin, when directly added into the food matrix, depends on several factors such as pH, temperature, composition, structure, and the natural microbiota of food. Nisin shows lack of stability at neutral pH and greatly loses activity at higher pH (de Arauz et al., 2009). Low storage temperature is more suitable for nisin activity. A limitation on the use of nisin in dairy products could be due to its interaction with fat and other components in the food matrix (Favaro et al., 2015). Also, some processes used in food production play an important role in the efficacy of bacteriocins (Bhatti et al., 2004). Applying temperatures of 80–105°C for 5–10 min. during the production of PCS with pH 5.6–5.8 resulted in a loss of about 20%–30% of nisin. A loss of nisin activity was found also in other cheeses such as Ricotta (10%–32% after 10 weeks of storage). In order to overcome the above-mentioned limitations Ibarra-Sanchez et al. (2020) suggested new and advanced strategies, including nisin encapsulation technology, the addition of active packaging, bioengineering, and combination with other antimicrobials, all with the aim of expanding and improving the use of nisin as a dairy preservative.

5.6.2 Potassium sorbate

Mould spoilage is one of the major problems causing deterioration of cheese. The development of fungal growth may cause undesirable defects, such as off odors and flavors, abnormal textures, discolorations and accumulation of mycotoxins (Sengun et al., 2008), which are all economically important because they lead to consumer rejection of products. There are quite a few strategies to prevent mould growth, such as a good cleaning and sanitizing program, brushing of cheese wheels and the use of vacuum packaging. However, the addition of food preservatives with fungistatic effects is still the best approach to improving cheese quality. However, fungal growth also depends on environmental factors such as temperature, pH, and availability of nutrients. Marín et al. (2014) reported that water activity (aw) is a key factor that modulates the growth of fungi associated with cheese.

Sorbic acid or its salt potassium sorbate are well known preservatives that are often used in cheese and PC production. Sorbic acid is present in nature and was first isolated in 1859 by German chemist A.W. von Hofmann from the fruit of *Sorbus aucuparia.* Later, in 1939, its antimicrobial activity was discovered and sorbates have been used as food preservatives since the early 1950s. They are effective in inhibiting the growth of yeast and moulds in food, but less effective against bacteria. Potassium sorbate is more soluble in water and has obvious advantages over sorbic acid in foods. The antimicrobial activity of sorbate is pH dependent and increases as the pH approaches its dissociation constant.

According to USA regulation, pasteurized process cheese products in the form of slices or cuts in consumer-sized packages may contain an optional mould-inhibiting ingredient consisting of not more than 0.2% by weight of sorbic acid, potassium sorbate, sodium sorbate, or any combination of these. The EU Regulation allows the use of 2000 mg/kg of sorbate in PC. Sorbates could be either incorporated in cheese mass or impregnated in the rind. The usage of doses of preservatives lower than allowed is of interest because modern consumer demands call for a reduction of their concentration in food.

The exact mechanism by which sorbic acid and its potassium salt inhibit microbial growth is not entirely understood. No single mechanism can explain the range of toxicity to various spoilage organisms. In literature, there is a lot of data about the use of potassium sorbate and about its effect on different kinds of cheeses (Ozdemir and Demirci, 2006; Yilmaz and Kurdal, 2008; Babacan and Özdemir, 2018), but a very small part of it relates to PCs (Alrabadi, 2017; Han and Floros, 1998), despite the fact that potassium sorbate is widely used for those products. The effect of potassium sorbate (0.05 kg/100 kg) during the ripening of Kashar cheese at 12°C was studied by Ozdemir and Demirci (2006). The results of the study showed that the potassium sorbate did not affect total counts of aerobic mesophilic bacteria, lactic acid bacteria, proteolytic microorganism, and lipolytic bacteria, but decreased the counts of coliform, *S. aureus* and yeast and mould. In the study, cheeses were treated with potassium sorbate using different methods which are dry salting, dipped plunge, and spraying. Similar to this, Poullet et al. (2009) found that addition of potassium sorbate influence the growth of colifom bacteria and *S. aureus* during the ripening of Casar de Caceres cheese. The addition of 0.1% sorbic acid to West African soft cheese delays spoilage during storage at 7–9°C by causing a decrease of mesophilic bacteria, yeasts, and moulds (Aworh and Egounlety, 2009).

The influence of different sorbic and potassium sorbate concentrations (300, 500, and 700 ppm) added to brine was investigated during the ripening of Turkish white cheese (Yilmaz and Kurdal, 2008). During the ripening, preservative diffusion to cheese significantly affected total counts of aerobic mesophilic bacteria, coliform bacteria, yeast, and mould of cheese ($p < 0.01$). Therefore, authors concluded that they can be successfully used as preservative agents in production of white cheese. Deeb et al. (2010) investigated the effect of potassium sorbate (0.02%, 0.05%, 0.1%, and 0.2%) and probiotic bacteria on spoilage bacteria (*Pseudomonas fluorescens)* during cold storage of soft cheese. Sorbate concentration above 0.1% is highly effective against bacteria, but can cause unusually sweet flavor indicating that excessive use of preservative is not acceptable. Hence, they concluded that addition of potassium sorbate at a concentration of 0.1% and *B. longum* had a great inhibitory effect on existing microorganisms and also improved the sensory quality of cheese. The study of Alrabadi (2017) showed the effect of activated polyethylene films with potassium sorbate as an antifungal agent at different concentrations (0.0%, 1.0%, 2.0%, 3.0%, 4.0%, 5.0%, and 6.0%) on *Aspergillius niger* in Gouda PC. The results showed the effectiveness of potassium sorbate as a packaging material for preserving Gouda PC, and the minimum efficient concentration was found to be 2%.

The diffusivity of potassium sorbate is important in order to reach or to keep the residual concentration above the critical fungistatic level. Han and Floros (1998) found that American PC had two times faster diffusion of potassium sorbate than Mozzarella. Solubility and diffusivity of potassium sorbate can enable the prediction of the overall concentration profile after potassium sorbate is applied on the surface of cheese, before packaging.

5.6.2.1 Propionate

Propionate such as calcium and sodium propionates are often used as preservatives for surface treatments of cheese products. According to USA regulation, it is permitted to use sodium propionates at levels of $\leq$0.3% (w/w) of the finished product if it is in a form of slices or cuts and in consumer packs. Sodium propionate (E281) is currently an authorized food additive in the European Union, under Annex II of Regulation (EC) 1333/2008, for use in several food categories, including cheese products, up to 3,000 mg/kg (Additives and Food, 2016).

The use of preservative combinations, that is, in mixtures, has a high preservation effect. The effect of different preservative mixtures, such as nisin, nisin + natamycin, nisin + potassium sorbate, or nisin + natamycin + potassium sorbate, in the production of PC sauce was investigated in the study (Saad et al., 2015). The nisin + natamycin + potassium sorbate mixture was the most effective for the product shelf life. However, mixtures with natamycin have a slight negative effect on the sensory properties, while the nisin + potassium sorbate mixture has no undesirable effect on the chemical, physical and sensory properties even after 3 months of storage, and so it could be used as a preservative system in cheese sauce products.

Recently, consumers are increasingly demanding healthy food, free from synthetic preservatives. As a result, natural preservative are attracting more and more attention. Many natural ingredients have shown antimicrobial activity which could inhibit or delay growth of spoilage and pathogenic bacteria. Different spices act against various pathogenic microorganisms in different ways (Josipović et al., 2015). The additions of extracts from cinnamon, garlic, lemon grass, cress, rosemary, sage, and oregano have all inhibited the population of *L. monocytogenes* in dairy based media, including PC

(Tayel et al., 2015). Cayenne and green pepper were able to reduce *S. aureus* population in Egyptian Kareish cheese (Wahba et al., 2010).

5.7 Future remarks

Flavors, colors, and preservatives are important ingredients with a significant role in the production of PC. Their addition enables the production of products with different properties in terms of taste and color, while preservatives are important from the perspective of shelf life. In the last few years, producers have challenges in the formulation and production of PC with natural ingredients in order to meet the requirements of the modern market that are looking for products with increased health benefits. Also, further work should be directed in a way to look for new ingredients from natural sources as well as possibilities to fortify products in order to make them healthier and more acceptable for consumers.

References

Abd-El-Salam, M.H., Al-Khamy, A.F., El-Garawany, G.A., Hamed, A., A., K., 1996. Composition and rheological properties of processed cheese spread as effected by the level of added whey protein concentrates and emulsifying salt. Egypt. J. Dairy Sci. 24, 309–322.

Additives, E., Panel O, F.Food, N. S. A. T., 2016. Safety of the extension of use of sodium propionate (E 281) as a food additive. EFSA J. 14, e04546.

Alrabadi, N.I., 2017. The effect of potassium sorbate on *Aspergillius Niger* in Gouda processed cheese. J. Appl. Environ. Biol. Sci. 7, 1148 –123.

Aly, M.E., Abdel-Baky, A.A., Farahat, S.M., Hana, U.B.B, 1995. Quality of processed cheese spread made using ultrafiltered retentates treated with some ripening agents. Int. Dairy J. 5, 191–209.

And, H.C., Hoover, D.G, 2003. Bacteriocins and their food applications. Compr. Rev. Food Sci. Food Saf. 2, 82–100.

Awad, R.A., Abdel-Hamid, L.B., El-Shabrawy, S.A., Singh, R.K, 2004. Physical and sensory properties of block processed cheese with formulated emulsifying salt mixtures. Int. J. Food Prop. 7, 429–448.

Aworh, O.C., Egounlety, M., 2009. Preservation of West African soft cheese by chemical treatment. J. Dairy Res. 52, 189–195.

Babacan, A., Özdemir, S., 2018. The effect of sorbate on microbiological, sensory properties and ripening parameters of Kashar cheese. Food Health 4, 147–158.

Bansal, V., Mishra, S.K, 2020. Reduced-sodium cheeses: Implications of reducing sodium chloride on cheese quality and safety. Compr. Rev. Food Sci. Food Saf. 19, 733–758.

Bhatti, M., Veeramachaneni, A., Shelef, L.A, 2004. Factors affecting the antilisterial effects of nisin in milk. Int. J. Food Microbiol. 97, 215–219.

Burgos, L., Pece, N., Maldonado, S., 2020. Textural, rheological and sensory properties of spreadable processed goat cheese. J. Food Stud. 9, SI62–SI74.

Cunha, C.R.D., Alcântara, M.R., Viotto, W.H, 2012. Effect of the type of emulsifying salt on microstructure and rheological properties of "requeijão cremoso" processed cheese spreads. J. Food Sci. 77 (8), E176–E181.

Daly, D.F.M., Mcsweeney, P.L.H., Sheehan, J.J, 2012. Pink discolouration defect in commercial cheese: a review. Dairy Sci Technol. 92, 439–453.

De Arauz, L.J., Jozala, A.F., Mazzola, P.G., Vessoni, P., T., C., 2009. Nisin biotechnological production and application: a review. Trends Food Sci. Technol. 20, 146–154.

Deeb, A.M.M. & Ahmed, H.F. Effect of potassium sorbate and/or probiotic bacteria on spoilage bacteria during cold storage of soft cheese. 2010.

Delves-Broughton, J., Weber, G., 2011. 3 - Nisin, natamycin and other commercial fermentates used in food biopreservation. In: Lacroix, C. (Ed.), Protective Cultures, Antimicrobial Metabolites and Bacteriophages for Food and Beverage Biopreservation. Woodhead Publishing, Sawston, United Kingdom.

Desouky, M.M., Salama, H.H., El-Sayed, S.M., 2019. The effects of camel milk powder on the stability and quality properties of processed cheese sauce. Acta Sci. Pol. Technol. Aliment. 18, 349–359.

Doosh, K.S., Alhusyne, L.A., Almosawi, B.N, 2014. Utilization of concentrated buttermilk in functional processed cheese manufacturing and studying some of its physicochemical properties. Pak. J. Nutr. 13, 33–37.

Drake, S.L., Yates, M.D., Drake, M.A, 2010. Development of falvor lexicon for processed and imitation cheeses. J. Sens. Stud. 25, 720–739.

Early, R., 2012. 17 - Dairy products and milk-based food ingredients. In: Baines, D., Seal, R. (Eds.), Natural Food Additives, Ingredients and Flavourings. Woodhead Publishing, Sawston, United Kingdom.

El-Bakry, M., 2012. Salt in cheese: a review. Curr. Res. Dairy Sci. 4, 1–5.

El-Bakry, M., Beninati, F., Duggan, E., O'riordan, E.D., O'sullivan, M., 2011. Reducing salt in imitation cheese: Effects on manufacture and functional properties. Food Res. Int. 44, 589–596.

El-Sayed, H.A., 2017. Effect of thyme on the quality and shelf life of block processed cheese. J. Food Dairy Sci. 8, 335–340.

El-Sayed, S.M., Youssef, A.M, 2019. Potential application of herbs and spices and their effects in functional dairy products. Heliyon 5, e01989.

EU, 2008. Regulation (EC) No 1333/2008 of the European Parliament and of the Council of 16 December 2008 on food additives.

Hassan, FA.M., Zaky, WM., Farahat, ES.A., Zayan, AF., Mohamed, A.G., Abdulalim, T.S, 2019. Processed cheese spreads fortified with lycopene extracted from tomato wastes. World J. Dairy Food Sci. 14, 87–93.

Fatma, A.F., Gehan, A.M.H., A.G., M., 2005. Fortification of processed cheese spread with accustomed edible mushroom. Arab Univ. J. Agric. Sci. 13, 825–839.

Favaro, L., Barretto, P., A., L., Todorov, S.D, 2015. Bacteriocinogenic LAB from cheeses – application in biopreservation? Trends Food Sci. Technol. 41, 37–48.

Fuentes-Alventosa, J.M., Jaramillo-Carmona, S., Rodríguez-Gutiérrez, G., Guillén-Bejarano, R., Jiménez-Araujo, A., Fernández-Bolaños, J., Rodríguez-Arcos, R., 2013. Preparation of bioactive extracts from asparagus by-product. Food Bioprod. Process. 91, 74–82.

Glass, K., Doyle, M.E, 2005. Safety of processed cheese. FRI Briefings.

Glass, K.A., Johnson, E.A., 2004. Factors that contribute to the botulinal safety of reduced-fat and fat-free process chesse products. J. Food Prot., 67, 8, 1687–1693.

Glass, K.A., Kaufman, K.M., Johnson, E.A, 1998. Survival of bacterial pathogens in pasteurized process cheese slices stored at 30 degrees C. J. Food Prot. 61, 290–294.

Guinee, T.P., 2017. Pasteurized processed and imitation cheese products. Cheese, *Fourth ed* Academic Press, San Diego.

Guinee, T.P., Kilcawley, K.N, 2017. Chapter 29 - Ingredient Cheese and Cheese-Based Ingredients. In: Mcsweeney, P.L.H., Fox, P.F., Cotter, P.D., Everett, D.W. (Eds.), Cheese, Fourth ed. Academic Press, San Diego.

Gupta, S.K., Karahadian, C., Lindsay, R.C, 1984. Effect of emulsifier salts on textural and flavor properties of processed cheeses1. J. Dairy Sci. 67, 764–778.

Han, J.H., Floros, J.D, 1998. Potassium sorbate diffusivity in American processed and mozzarella cheeses. J. Food Sci. 63, 435–437.

Hassan, A.N., Awad, S., Mistry, V.V, 2007. Reduced fat process cheese made from young reduced fat cheddar cheese manufactured with exopolysaccharide-producing cultures. J. Dairy Sci. 90, 3604–3612.

Hoffmann, W., Gärtner, J., Lück, K., Johannsen, N., Maurer, A., 2012. Effect of emulsifying salts containing potassium on the quality of block-type processed cheese. Int. Dairy J. 25, 66–72.

Ibarra-Sánchez, L.A., El-Haddad, N., Mahmoud, D., Miller, M.J., Karam, L., 2020. Invited review: advances in nisin use for preservation of dairy products. J. Dairy Sci. 103 (3), 2041–2052.

Ismail, M.M., 2016. Smoking of dairy products. J. Nutr. Health Food Eng. 4.

Johnson, M.E., Kapoor, R., Mcmahon, D.J., Mccoy, D.R., Narasimmon, R.G, 2009. Reduction of sodium and fat levels in natural and processed cheeses: scientific and technological aspects. Compr. Rev. Food Sci. Food Saf. 8, 252–268.

Jones, S.T., Aryana, K.J., Losso, J.N, 2005. Storage stability of lutein during ripening of cheddar cheese. J. Dairy Sci. 88, 1661–1670.

Josipović, R., Knežević, Z.M., Frece, J., Markov, K., Kazazić, S., Mrvčić, J., 2015. Improved properties and microbiological safety of novel cottage cheese containing spices. Food Technol. Biotechnol. 53, 454–462.

Kang, E.J., Campbell, R.E., Bastian, E., Drake, M.A, 2010. Invited review: annatto usage and bleaching in dairy foods. J. Dairy Sci. 93, 3891–3901.

Kapoor, R., Metzger, L.E, 2008. Process cheese: scientific and technological aspects—a review. Compr. Rev. Food Sci. Food Saf. 7, 194–214.

Kapoor, R., Metzger, L.E., Biswas, A.C., Muthukummarappan, K., 2007. Effect of natural cheese characteristics on process cheese properties. J. Dairy Sci. 90, 1625–1634.

Karahadian, C., Lindsay, R.C, 1984. Flavor and textural properties of reduced-sodium process American cheeses1. J. Dairy Sci. 67, 1892–1904.

Khider, M., Seoudi, O., Fathy, A., Y., F., 2017. Functional processed cheese spreads with high nutritional value as supplemented with fresh and dried mushrooms. Int. J. Nutr. Food Sci. 6, 45–52.

Kilcawley, K.N., Wilkinson, M.G., Fox, P.F, 1998. Enzyme-modified cheese. Int. Dairy J. 8, 1–10.

Kubo, M.T.K., Maus, D., Xavier, A.A.O., Mercadante, A.Z., Viotto, W.H, 2013. Transference of lutein during cheese making, color stability, and sensory acceptance of prato cheese. Food Sci. Technol. 33, 82–88.

Kussy, D., Aylward, E., 2008. Pasteurized process cheese. In: Clark, S., C., M., Drake, M., Bodyfelt, F. (Eds.), The Sensory Evaluation of Dairy Products. Springer, New York, NY.

Lehto, S., Buchweitz, M., Klimm, A., STRAßBURGER, R., Bechtold, C., Ulberth, F., 2017. Comparison of food colour regulations in the EU and the US: a review of current provisions. Food Addit Contam. Part A 34, 335–355.

Lycken, L., Borch, E., 2006. Characterization of Clostridium spp. isolated from spoiled processed cheese products. J. Food Prot. 69, 1887–1891.

Marín, P., Palmero, D., Jurado, M., 2014. Effect of solute and matric potential on growth rate of fungal species isolated from cheese. Int. Dairy J. 36, 89–94.

Mayer, H.K., 2001. Bitterness in processed cheese caused by an overdose of a specific emulsifying agent? Int. Dairy J. 11, 533–542.

Mozuraityte, R., Berget, I., Mahdalova, M., Grønsberg, A., Øye, E.R., Greiff, K., 2019. Sodium reduction in processed cheese spreads and the effect on physicochemical properties. Int. Dairy J. 90, 45–55.

Muir, D.D., Williams, S.A.R., Tamime, A.Y., Shenana, M.E, 1997. Comparison of the sensory profiles of regular and reduced-fat commercial processed cheese spreads. Int. J. Food Sci. Technol. 32, 279–287.

Nogueira De Oliveira, M., Ustunol, Z., Tamime, A., 2011. Manufacturing practices of processed cheese. In: Tamime, A. (Ed.), Processed Cheese and Analogues, Blackwell Publishing Ltd., New Jersey, United States.

Noronha, N., Cronin, D.A., O'riordan, E.D., O'sullivan, M., 2008. Flavouring of imitation cheese with enzyme-modified cheeses (EMCs): Sensory impact and measurement of aroma active short chain fatty acids (SCFAs). Food Chem. 106, 905–913.

O'connor, P.M., Ross, R.P., Hill, C., Cotter, P.D, 2015. Antimicrobial antagonists against food pathogens: a bacteriocin perspective. Curr. Opin. Food Sci. 2, 51–57.

Oliveira, R.B.A., Margalho, L.P., Nascimento, J.S., Costa, L.E.O., Portela, J.B., Cruz, A.G., Sant'ana, A.S., 2016. Processed cheese contamination by spore-forming bacteria: a review of sources, routes, fate during processing and control. Trends Food Sci. Technol. 57, 11–19.

Osthoff, G.E., Slabber, W., Dürrschmid, K., 2011. Flavours and flavourants, colours and pigment. In: Tamime, A.Y. (Ed.), Processed Cheese and Analogues. Wiley Publishing, New Jersey, United States.

Ozdemir, C., Demirci, M., 2006. Selected microbiological properties of Kashar cheese samples preserved with potassium sorbate. Int. J. Food Prop. 9, 515–521.

Patel, J.B., Patel, P.V., Lee, E.J, 2013. The effect of potassium chloride as a salt replacer on the qualities of processed cheese. J. Stud. Res. 13, 313–329.

Poullet, B., Huertas, M., Sánchez, A., Cáceres, P., Larriba, G., 2009. Microbial study of Casar de Cáceres cheese throughout ripening. J. Dairy Res. 58, 231–238.

Rincon-Delgadillo, M.I., Lopez-Hernandez, A., Wijaya, I., Rankin, S.A, 2012. Diacetyl levels and volatile profiles of commercial starter distillates and selected dairy foods. J. Dairy Sci. 95, 1128–1139.

Saad S.A., EL-Mahdi L.D., Awad R.A., Hassan Z.M.R., 2015. Nutrition processed cheese sauces with different preservative systems. Integr. Food Nutr. Metab. 2, DOI:10.15761/IFNM.1000116.

Sengun, I.Y., Yaman, D., Gonul, S., 2008. Mycotoxins and mould contamination in cheese: a review. World Mycotoxin J. 1 (3), 291–298.

Shalaby, S.M., Mohamed, A.G., Bayoumi, H.M, 2017. Preparation of a novel processed cheese sauce flavored with essential oils. Int. J. Dairy Sci. 12, 161–169.

Shan, B., Cai, Y.-Z., Brooks, J.D., Corke, A.H, 2011. Potential application of spice and herb extracts as natural preservatives in cheese. J. Med. Food 14, 284–290.

Sharma, P., Segat, A., Kelly, A.L., Sheehan, J.J, 2019. Colorants in cheese manufacture: production, chemistry, interactions, and regulation. Compr. Rev. Food Sci. Food Saf 1–23 n/a.

Shumaker, E.K., Wendorff, W.L, 1998. Factors affecting pink discoloration in annatto-colored pasteurized process cheese. J. Food Sci. 63, 828–831.

Silva, C.C.G., Silva, S.P.M., Ribeiro, S.C, 2018. Application of bacteriocins and protective cultures in dairy food preservation. Front. Microbiol. 9, 594.

Singh, T.K., Drake, M.A., Cadwallader, K.R, 2003. Flavor of cheddar cheese: a chemical and sensory perspective. Compr. Rev. Food Sci. Food Saf. 2, 166–189.

Smigic, N., Miocinovic, J., Tomic, J., Tomasevic, I., Rajkovic, A., Djekic, I., 2018. The effect of nisin and storage temperature on the quality parameters of processed cheese. Mljekarstvo 68, 182–191.

Sobral, D., Bueno Costa, R.G., Machado, G.M., Jacinto De Paula, J.C., Martins Teodoro, V.A., Nunes, N.M., Dos Santos Pires, A.C. & Pinto, M.S. 2016. Can lutein replace annatto in the manufacture of Prato cheese? LWT - Food Sci. Technol.,68,349-355.

Sobrino-López, A., Martín-Belloso, O., 2008. Use of nisin and other bacteriocins for preservation of dairy products. Int. Dairy J. 18, 329–343.

Solhi, P., Azadmard-Damirchi, S., Hesari, J., Hamishehkar, H., 2020a. Effect of fortification with asparagus powder on the qualitative properties of processed cheese. Int. J. Dairy Technol. 73, 226–233.

Solhi, P., Azadmard-Damirchi, S., Hesari, J., Hamishehkar, H., 2020b. Production of the processed cheese containing tomato powder and evaluation of its rheological, chemical and sensory characteristics. J. Food Sci. Technol. 57, 2198–2205.

Somers, E.B., Taylor, S.L, 1987. Antibotulinal effectiveness of nisin in pasteurized process cheese spreads. J. Food Prot. 50, 842–848.

Suneeta Pinto, S., Rathour, A.K., Prajapati, J.P., Jana, A.H., Solanky, M.J, 2007. Utilization of whey protein concentrate in processed cheese spread. Indian J. Nat. Prod. Resour. 6, 398–401.

Suñen, E., Fernandez-Galian, B., Aristimuño, C., 2001. Antibacterial activity of smoke wood condensates against Aeromonas hydrophila, Yersinia enterocolitica, and Listeria monocytogenes at low temperature. Food Microbiol. 18, 387–393.

Szafrańska, J.O., Sołowiej, B.G, 2020. Effect of different fibres on texture, rheological and sensory properties of acid casein processed cheese sauces. Int. J. Food Sci. Technol. 55, 1971–1979.

Talbot-Walsh, G., Kannar, D., Selomulya, C., 2018. A review on technological parameters and recent advances in the fortification of processed cheese. Trends Food Sci. Technol. 81, 193–202.

Tayel, A.A., Hussein, H., Sorour, N.M., El-Tras, W.F., 2015. Foodborne pathogens prevention and sensory attributes enhancement in processed cheese via flavoring with plant extracts. J. Food Sci. 80, M2886–M2891.

Wahba, N.M., Ahmed, A.S., Ebraheim, Z.Z, 2010. Antimicrobial effects of pepper, parsley, and dill and their roles in the microbiological quality enhancement of traditional Egyptian Kareish cheese. Foodborne Pathog. Dis. 7, 411–418.

Wendorff, W.L., Riha, W.E., Muehlenkamp, E., 1993. Growth of Molds on cheese treated with heat or liquid smoke. J. Food Prot. 56, 963–966.

Wilkinson, M.G., Doolan, I.A., Kilcawley, K.N, 2011. Cheese | enzyme-modified cheese. In: Fuquay, J.W. (Ed.), Encyclopedia of Dairy Sciences, Second Ed. Academic Press, San Diego.

Ye, A., Cui, J., Taneja, A., Zhu, X., Singh, H., 2009. Evaluation of processed cheese fortified with fish oil emulsion. Food Res. Int. 42, 1093–1098.

Yilmaz, L., Kurdal, E., 2008. Effect of sorbic acid and potassium sorbate addition to the brine on microbiological and chemical properties of Turkish white cheese during ripening. Food Sci. Technol. Res. 14, 437–444.

Zottola, E.A., Yezzi, T.L., Ajao, D.B., Roberts, R.F, 1994. Utilization of cheddar cheese containing nisin as an antimicrobial agent in other foods. Int. J. Food Microbiol. 24, 227–238.

Functionality of salts used in processed cheese manufacture

6

František Buňka[a,b], *Michaela Černíková*[a] *and Richardos N. Salek*[a]
[a] Department of Food Technology, Faculty of Technology, Tomas Bata University in Zlín, Zlín, Czech Republic, [b] Food Research Laboratory, Department of Logistics, Faculty of Military Leadership, University of Defense, Brno, Czech Republic

6.1 Introduction

Emulsifying salts (ES) are crucial ingredients in processed cheese manufacture (sometimes the term "emulsifying agents" is also used). In the literature, there are many definitions of emulsifying salts. The relevant definition, which can be used, is published in EU Regulation No. 1333/2008 on food additives – "emulsifying salts are substances which convert proteins contained in cheese into a dispersed form and thereby bring about homogenous distribution of fat and other components".

When a mixture of raw materials containing common ingredients, for example, cheese (as fundamental component – see Chapters 2 and 3), dairy fat (butter, cream, and/or anhydrous milk fat), water and other dairy and/or nondairy substances, is processed (heat treated up to normally used temperature, recently around 90–100°C – see Chapters 8 and 9) in the absence of ES, the following chain of interactions and processes are expected. Firstly, dairy fat droplets membranes (mainly originating from the cheese applied, but also some other dairy ingredients, such as milk powder, etc.) are destructed and the fat is melted. Secondarily, the melted and uncoated fat starts to coalesce forming larger units unprotected by lipoprotein membranes. Furthermore, under the temperature used during processing and actual pH of the mixture (approximately 5.0–5.5), aggregation and contraction of caseins is anticipated. The above-mentioned processes result in the separation of hydrophilic (especially water and proteins) and hydrophobic (especially fat) phases and an undesirable heterogenic mixture which is unaccepted by the consumer would be developed (Carić et al., 1985; Carić and Kaláb, 1997; Guinee, 2003). Practically, the same or very similar interactions are expected also when milk powders (commonly skim milk powder) or reconstituted milk (especially milk powder dispersed in water) are used as the predominate raw materials.

For the development of a homogenous and smooth final product, effectively acting emulsifiers should be present in the formulation. In the matrix of cheese, there are available some emulsifiers, especially α_{S1}-casein, α_{S2}-casein, β-casein and their high molecular parts (originated by hydrolysis of caseins during cheese ripening by proteolytic enzymes – see Chapters 2 and 3). In addition, in the matrix of cheese, the above-mentioned biopolymers are trapped in a network, in which are bonded (besides other generally minor interactions) by calcium ions. Generally, the latter mentioned bivalent

Processed Cheese Science and Technology: Ingredients, Manufacture, Functionality, Quality, and Regulations.
DOI: https://doi.org/10.1016/B978-0-12-821445-9.00011-X

bonds (so called "calcium bridges") make their action as effective emulsifiers however impossible. The emulsifying capacity of these proteins is practically eliminated. Therefore, in traditionally produced processed cheese, the application of substances that are able to "disrupt" calcium bridges and liberate caseins and their high molecular hydrolysates (for simplifying, only the term "caseins" will be used in the text below) are necessary. The substances that can possess this "disrupt" ability are the ES (Dalgleish, 1992; Ginger and Grigor 1999; Fox, 2001; Kapoor and Metzger, 2008; Kwak et al., 2002; Farrell et al., 2004).

6.2 Principles of processed cheese production

In a simplified way, the principles of processed cheese production could be divided into two phases. In the first phase, caseins are liberated and dispersed, respectively, by the action of ES and then could serve as emulsifiers (so called "ion-exchange phase"). In the second phase, the new protein network is established, in which the proteins are hydrated and the fat is emulsified (so called "creaming phase").

The release of caseins is caused by action of ES, during which the ion-exchange phenomenon occurs. Polyvalent anions (especially phosphates, polyphosphates, and citrates; tartrates are also rarely presented) and monovalent cations (especially sodium, but in some situations and in some countries also potassium and/or aluminum) are used as ES. Multivalent anions have more intensive affinity to bivalent ions (in this situation calcium ions) in comparison to monovalent ions. Therefore, when, for example, sodium salt of phosphate is added into the mixture of the raw materials, in presence of sufficient amount of water, under heat treatment and agitation, ion-exchange of calcium and sodium ions is realized. Moreover, the insoluble calcium paracaseins (in the cheese protein network) are transformed into more soluble sodium paracaseins and the latter proteins could be dispersed in the melt. Then, the dispersed caseins could effectively act as emulsifiers. The process of the ion-exchange is schematically depicted in Fig. 6.1. Consequently, the first role of ES is to modify the environment in the heated mixture of raw materials in a manner that the present caseins can exhibit their emulsifying capability. In most of the available literature, this role is defined as the "main role" of ES (Carić et al., 1985; Guinee et al., 2004; Guinee, 2003; Kapoor a Metzger, 2008; Kawasaki, 2008; McIntyre et al., 2017; Mizuno a Lucey, 2005a,b; Shirashoji et al., 2006; Templeton and Sommer, 1936).

In some literature sources, ES (salts of phosphates, polyphosphates, and citrates) are called as emulsifiers (as surface active substances); however, this is not the correct definition because these substances their structure does not possess amphiphilic properties. Therefore, the term emulsifying salts/agents, which help caseins to emulsify fat and bond water, is the most appropriate (Carić et al., 1985; Guinee et al., 2004).

Moreover, during processing (so called "melting") under heat treatment (recently usually in the range of 90–100°C) and mechanical agitation in presence of water, polyvalent anions (especially phosphates) start to bind onto caseins via calcium ions and increase the hydrophilic character of the present proteins. Caseins possess the ability to bind water and the caseins modified in the above-mentioned way are even

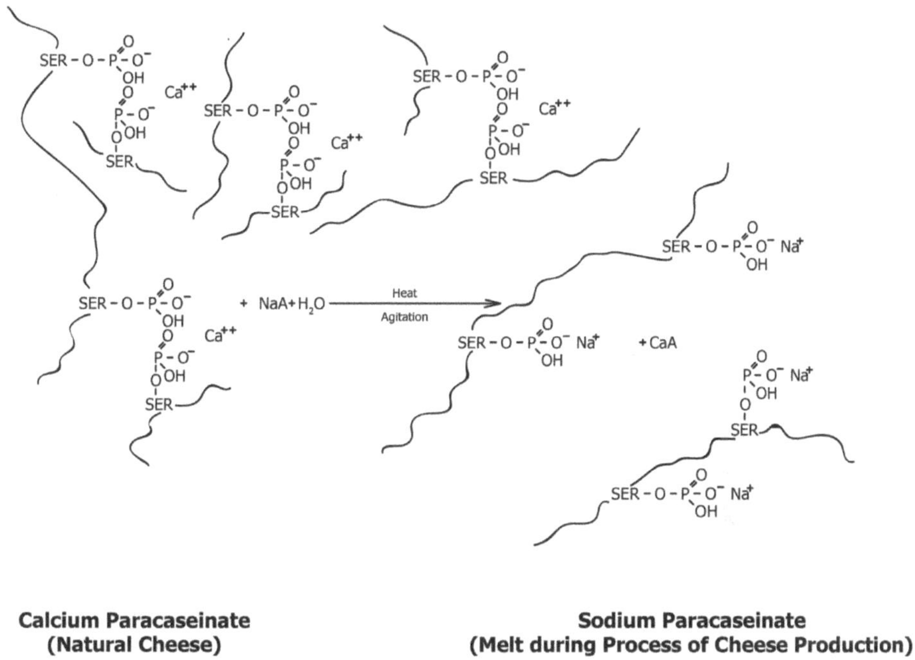

Figure 6.1 Schema of the first role of emulsifying salts (ion exchange) during melting of raw material contains cheese (Na – monovalent cation; Ca – bivalent cation; A – polyvalent anion – especially phosphates, polyphosphates and/or citrates, SER – serine residues). Modified according to Carić and Kaláb (1997).

more effective in terms of present water binding. Simultaneously, fat is emulsified by dispersed caseins and partially also by the residues of original lipoprotein membranes (covered fat droplets before processing). Additionally, caseins with bonded phosphates obtain more intensive negative charge which help to disperse proteins. Generally, the ES addition increases the pH value from approximately 5.0–5.5 (the mixture of raw materials) to the target value in the range of 5.6–6.0 (for spreadable processed cheese), which is also supported by the raising of caseins negative charge. Both the above-mentioned processes (proteins hydration and fat emulsification) cause the increase of viscosity of the melt. Additionally, during holding time (generally several minutes – see Chapter 9) under a target melting temperature and during cooling of the hot melt, the new structure is established – the new hydrated protein network is formed, in which fat is emulsified. The above-mentioned processes are called "creaming." There are many interactions, which participate in the process of the final network structure formation and thus supporting creaming, for example, new calcium bridges, phosphates-calcium complexes, hydrophobic interactions, hydrogen bonds, and/or disulfide bonds. The establishment of the new protein network can be observed as the intensive increase of melt viscosity as is described below. Some authors limited the term "creaming" only to the phase of holding time of the melt under the target melting temperature, and then

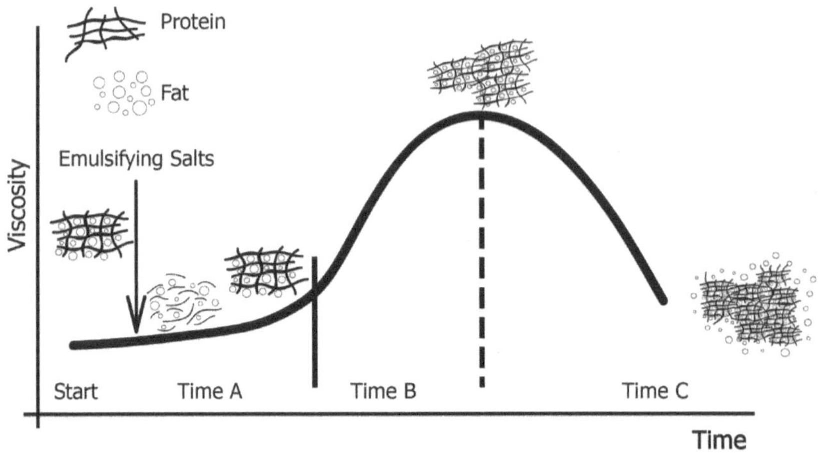

Figure 6.2 Viscosity and the structure changes during the melting and subsequent creaming processes. Time A – ion exchange and caseins dispersion; Time B – caseins hydration, fat emulsification, start and follow-up arrangement of the new protein network; Time C – overcreaming – fat and water releasing from the casein matrix. Modified according to Lee et al. (2003).

they spoke about "aggregation and gel formation" (Berger et al., 2002; Bowland and Foegeding, 2001; Buňka et al., 2012; Culler et al., 2017; Hoffmann, 2016; Hougaard et al., 2015; Fu et al., 2018; Kalliapan and Lucey, 2011; Kapoor et al., 2007; Kapoor and Metzger, 2008; Kawasaki, 2008; Lucey et al., 2003; Lucey et al., 2011; Marchesseau and Cuq, 1995; Mizuna and Lucey, 2005b, 2007).

The successful melt creaming for the establishment of the network (a final product) with desirable properties for the consumers requires appropriate time, in which a target melting temperature is maintained under continuous agitation. The melting and creaming processes could be schematically depicted in Fig. 6.2 by the melt viscosity development. In particular, during the first phase (Time A), the ion exchanging and caseins dispersion occur by the activities of ES under heat treatment (see above and also Chapters 8 and 9) and constant agitation. Subsequently (Time B), the caseins are hydrated, the fat is emulsified and the new protein network starts to establish. During prolonging time of processing, the viscosity of the melt is increasing due to multiplying of interactions connected with caseins chains in the newly formed matrix and decreasing of fat droplets up to the maximum available viscosity under actual conditions (for details see Chapter 9). In praxis, producers operate in Time B and the maximum point of viscosity is not reached. The heat treatment and holding time under the target melting temperature is stopped when the melt possesses the appropriate properties for the manufacture of a final product with desirable properties for the customers. When the processing is still extended, the aggregations of proteins (increasing number of interactions between caseins) will continue and from a certain point fat and generally later also water will be released from the matrix and the whole system will become unstable (Time C). Generally, the decreasing of the viscosity could be also observed. Nevertheless, this undesirable phase is commonly named as "overcreaming" and is evaluated as a quality defect (Kawasaki, 2008; Klostermeyer and Buchheim, 1988; Lee et al., 2003; Shirashoji et al., 2006).

The cheese ripening (especially proteolysis) plays also role in the new protein network establishment. When high amount of short-chained caseins and low concentration of intact casein are available (the proteolysis is very intensive – see Chapters 2, 3, and 9), the risk of the unstable protein matrix forming would arise and the undesirable final product releasing fat or even also water would be manufactured (Buňka et al., 2013; Carić and Kaláb, 1997; Salek et al., 2016).

According to Lee et al. (2003), the creaming process (as the formation of a new network) is mainly the interactions of caseins. The authors supported their statement in an experiment with full-fat processed cheese and also with a sample with no fat content. In both products, the creaming took place. On the other hand, the presence of fat significantly influences the protein matrix and therefore the consistency of the final product. Generally, fat disturbs the continuity and the compactness in such way that decreases the number of interactions between caseins and a processed cheese with lower viscosity (more spreadable) is obtained. The proteins interact with fat, emulsify it and a compact network is achieved. A similar effect on the continuity and the compactness of the protein matrix possesses water and some other ingredients usually or rarely used during the manufacture processed cheese (see Chapters 3 and 9). Also the cooling rate of the processed cheese has a significant effect on the network formation (see Chapter 10). When the cooling rate is too fast, the viscosity of the products decreases especially because the interactions between proteins take place, especially when the melt is hot (Awad et al., 2002; Bowland and Foegeding, 2001; Černíková et al., 2017; Guinee, 2003; Heertje et al., 1981; Kapoor and Metzger, 2008; Lee et al., 2003; Lucey et al., 2011; Piska and Štětina, 2004; Raval and Mistry, 1999).

Generally, this section could be concluded by the statement that the final structure of processed cheese is influenced especially by:

1. the effectiveness of ion-exchange and proteins dispersion (including the actual pH-value and its impact on the negative charge of caseins);
2. the intensity of protein hydration;
3. the intensity of fat emulsification; and
4. the number of interactions between caseins.

Of course, an addition of other ingredients can contribute to the increase of the latter mentioned interactions between proteins, but on the other hand, can also disrupt the continuity and the compactness of the casein network.

It should be added that despite many studies dealing with processed cheeses and similar products have been done and published for the last 20 years, however, there are not answered all the questions and clarified all the tasks about the final structure formation and all factors influencing it.

6.3 Phosphate emulsifying salts

Phosphates are salts derived from phosphoric acid (H_3PO_4). Food grade phosphoric acid is obtained by industrial (raw) phosphoric acid by purification, mostly by elimination of impurities. Phosphates constitute large groups of compounds, in which the anion consists of PO_4 tetrahedra, which may be linked together by the sharing of corners. The phosphate series begins with single PO_4 group, which may exist as the triply charged monophosphate (previously named orthophosphate) anion PO_4^{-3}, or may participate in molecules where one, two or three of the four oxygens are covalently bonded to other atoms. The PO_4 group is a tetrahedron formed by four oxygen atoms surrounding a phosphorus atom. The present oxygen may constitute a bridge between the phosphorus atom and other atoms, including also other phosphorus atoms. Therefore, a long series of two- or three-dimensional phosphates can originate through P-O-P linkage. A term often used in relation to polyphosphates (formally, including diphosphates) is that of "condensed phosphates." The origin of the term lies in some of the industrial processes used in the manufacture of polyphosphates, which involve elimination of water at high temperatures. Under very specific conditions, monophosphates or longer chain phosphates having terminal OH groups lose water between hydroxyl groups. The dehydrated tetrahedra are thereby brought together through a sharing of corners (are condensed) in order to form polymers. Similar processes may result in two monophosphate molecules coming together to form a diphosphate (previously named pyrophosphates) or a monophosphate and a diphosphate may link into a triphosphate, and so on. The term ultraphosphate includes any phosphate having a tridimensional structure. On the other hand, in processed cheese industry, linear (two-dimensional) polyphosphates are utilized (Molins, 1991; Montchamp, 2015).

In addition, sodium, potassium, ammonium, calcium, magnesium, and sodium aluminum salts of phosphates can be used worldwide in food industry. In processed cheese industry, sodium salts of phosphates are predominantly applied. Potassium salts of phosphates are generally not utilized due to possible bitter taste of the final products (Guinee et al., 2004; Kapoor and Metzger, 2008; Mayer, 2001). In some countries, the sodium aluminum salt of phosphate is permitted (see Section 6.3.4) and some recent papers showed that bitter taste could not be problem and pointed out lower sodium content (Chavhan et al., 2015; Khetra et al., 2015; Mozurairyte et al., 2019; Nogueira et al., 2018; Schatz et al., 2014).

6.3.1 Properties of individual emulsifying salts

Phosphate and polyphosphate salts with varying number of linearly bonded phosphate units in the molecule possess different properties. The first property that is significantly influenced by the latter mentioned factor is the ion exchange ability. The capability of sodium phosphate-based ES to exchange sodium and calcium between ES and a natural cheese casein structure is crucially influenced by two factors: (i) the length of phosphate-based salt and (ii) target temperature during manufacturing. When longer phosphate-based salt is used and higher target melting temperature applied, the ion-exchange ability and subsequently casein dispersion is enhanced (Buňka et al., 2012; Carić and Kaláb, 1997; Gupta et al., 1984; Lee et al., 1996; Meyer, 1973; Molins, 1991; Nagyová et al., 2014; Sádlíková et al., 2010; Salek et al., 2015; Templeton and Sommer, 1936; Weiserová et al., 2011).

Table 6.1 The pH-values[a] of aqueous solutions (1% w/v) and final processed cheeses (3% w/w) produced using selected sodium phosphate based emulsifying salts regularly used in industry (Nagyová et al., 2014; Sádlíková et al., 2010; Salek et al., 2015).

Chemical formula	Name	P_2O_5 content	E-code	pH-value of aqueous solution (1% w/v)	pH-value of processed cheese (3% w/w)[b]
NaH_2PO_4	Monosodium phosphate	59.15	339 (i)	4.5	4.9
Na_2HPO_4	Disodium phosphate	50.00	339 (ii)	9.1	6.5
Na_3PO_4	Trisodium phosphate	49.94	339 (iii)	11.9	6.9
$Na_2H_2P_2O_7$	Disodium dihydrogen phosphate	63.95	450 (i)	4.1	4.7
$Na_4P_2O_7$	Tetrasodium diphosphate	53.38	450 (iii)	10.2	6.7
$Na_5P_3O_{10}$	Pentasodium triphosphate	57.88	451 (i)	9.7	6.6
$(NaPO_3)_n$[C]	Sodium polyphosphate	69.61	452 (i)	5.9	5.4

[a] Expressed as mean; standard deviations for aqueous solutions and processed cheeses were in range of 0.01–0.03 and 0.02–0.06, respectively.
[b] Model spreadable processed cheese with 40% (w/w) and 50% (w/w) fat in dry matter manufactured using Dutch-type cheese.
[c] The mean length of sodium polyphosphate was at the level of n ≈ 20.

The ion-exchange activity of sodium phosphate-based ES is also affected by the actual pH of the mixture of the raw materials, which is influenced by selection of (i) cheese type, (ii) other ingredients (see Chapters 2–5), and (iii) also by the used ES. When the pH value of the mixture of the raw materials is slightly lower, the ion-exchange ability of sodium phosphate-based ES is enhanced. The latter mentioned relationship acts, when the pH-values of the mixture raw material and the melt under processing, respectively, are in the range of approximately 5.0–6.5. The pH-values of aqueous solution (1% w/v) prepared using different sodium phosphate-based ES (regularly used in the industry) are presented in Table 6.1 and can be ordered as follows:

$$Na_2H_2P_2O_7 < NaH_2PO_4 < (NaPO_3)_n < Na_2HPO_4 < Na_5P_3O_{10} < Na_4P_2O_7 < Na_3PO_4$$

In Table 6.1, the pH-values of processed cheese manufactured with the individual sodium phosphate-based ES (in concentration of 3% w/w) are also shown. The order of pH-values of processed cheeses produced is the same as in the case of the aqueous solutions (1% w/v). It should be added that the actual pH-value of the raw material mixture influences also the aggregation processes of caseins. The caseins aggregation is more intensive, when the pH-value of the mixture is near to its isoelectric point (pH ≈ 4.6; for more details, see Chapter 9). Excessive aggregation of caseins and poor ion-exchange ability of the used ES and therefore insufficient dispersion of proteins could lead to unsatisfactory fat emulsification and protein hydration. It is obvious that the selection of proper ES is very important for reaching the target pH-value of the final product (Culler et al., 2017; Guinee et al., 2004; Marchesseau et al., 1997; Lee and Klostermeyer, 2001; Meyer, 1973; Molins, 1991; Muslow et al., 2007; Nagyová et al., 2014; Shirashoji, 2016; Sádlíková et al., 2010; Salek et al., 2015).

With respect to the above-mentioned relations and all references presented above, the ion-exchange ability of sodium phosphate-based ES could be ordered as follows:

$$NaH_2PO_4 < Na_2HPO_4 < Na_3PO_4 < Na_2H_2P_2O_7 < Na_4P_2O_7 < Na_5P_3O_{10} < \text{sodium polyphosphates}(n \leq 9) < \text{sodium polyphosphate}(n > 9)$$

The usage of sodium phosphate-based ES does not limit only on the pH-value shift of the final products but some of these salts possess also buffer capacity and are able to stabilize the pH of the developed processed cheese. The buffer capacity of sodium phosphate-base ES is higher when the length of phosphates and polyphosphates is decreasing. Therefore, monophosphates and partially diphosphates are evaluated as substances with effective buffer ability (Carić and Kaláb, 2004; Guinee et al., 2004; Lucey et al., 2011; Muslow et al., 2007).

The length of phosphate-based ES influences also the ability to support the casein network (gel). Diphosphates are evaluated as very effective substances for dairy gel forming followed by triphosphates. In some articles, it is expected that also units consisting of calcium with diphosphates or triphosphates help to establish firm casein networks, due to the support to hydrophobic interactions between proteins. On the other hand, polyphosphates are considered that do not support gel forming especially due to bonding with caseins and increasing their negative charge that could act more repulsively for the proteins (Buňka et al., 2012; Carić and Kaláb, 2004; El-Bakry et al., 2011; Guinee et al., 2004; Lu et al., 2008; Lucey et al., 2011; Mizuno and Lucey, 2007; Muslow et al., 2007; Shirashoji et al., 2010, 2016).

The effect of individual (sodium) phosphate-based ES on the properties of processed cheese group has been detailly studied for last years. On the other hand, the most of the referenced studies is not directly comparable because different manufacturing conditions were used (e.g., raw materials, the target dry matter and fat in dry matter contents, process parameters, such as the target temperature and holding time, agitation speed, storage, etc.). On the other hand, indirectly, we could use the observed trends, which are of course also very important.

Fig. 6.3 presented hardness (N; unpublished data) of processed cheese samples with singly used nine ES (target parameters: 40 % w/w dry matter content; 50 % w/w fat in dry matter content), which were produced under the same conditions. Each ES was added in amount of 2 % w/w. Besides of trisodium citrate (TSC) (Fig. 6.3 will be used also for Section 6.4), hardness of model processed cheese is increased with the rising number of phosphate units in the sodium phosphate and polyphosphate molecule. Fig. 6.3 also showed that the hardness increased between samples with sodium polyphosphates in comparison with products in which was used pentasodium triphosphate ($P < 0.001$). Hardness of processed cheeses also increased when the number of phosphates units in sodium polyphosphates raised up to $n \approx 13$ (the mean value of the polyphosphate length; $P < 0.001$). Then, hardness of the samples, in which sodium polyphosphate with $n \approx 13$ (POLY13), $n \approx 20$ (POLY20), and $n \approx 28$

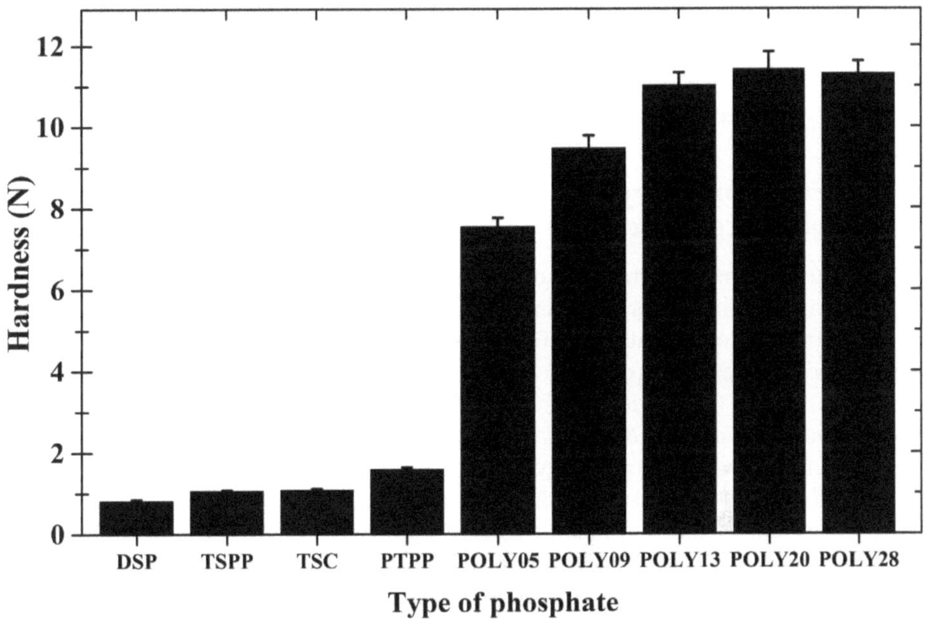

Figure 6.3 Hardness (N) of model processed cheese samples (40% w/w dry matter content; 50% w/w fat in dry matter content) manufactured using nine individually applied emulsifying salts (DSP, disodium phosphate; TSC, trisodium citrate; TSPP, tetrasodium diphosphate; PTPP, pentasodium triphosphate; sodium salt of polyphosphate with the mean number of phosphate units $n \approx 5$ – POLY05, $n \approx 9$ – POLY09, $n \approx 13$ – POLY13, $n \approx 20$ – POLY20, $n \approx 28$ – POLY28).

(POLY28), remained practically similar ($P \geq 0.05$). When ES were used individually, the effect on the consistency (expressed as hardness of the samples) is controlled and can be explained by the capability of the certain phosphate and polyphosphate to ion exchange as was in detail described above.

With respect to the Section 6.2, Section 6.3.1 could be summarized by the following statements. Monophosphates are low-molecular substances with the capability to bind onto caseins and increase their hydration. Diphosphates cause gel formation of caseins and act as cross-linking agents while bonding with calcium ions. The latter mentioned phenomenon leads to reduce of charge repulsion, which can facilitate hydrophobic interactions between hydrophobic segments of the proteins. On the other hand, excessive amounts of diphosphate could bond too much amount of calcium, making it unavailable for cross-linking interactions. Long-chain polyphosphates can strongly bind calcium and intensively support to caseins to disperse. On the other hand, they do not effectively support gel formation due to their multiple negative charges making it difficult to have reassociation of caseins through their hydrophobic segments.

At the end of this subsection, it should be mentioned that ES are not used singly in practice but as mixtures of several phosphate salts sometimes with citrate salts (see below). On the other hand, the appropriate understanding of the effect of the single ES addition on the casein network and/or the whole system (with other components) is helpful for understanding and explanation of the role of individual ES in the binary and ternary mixtures.

6.3.2 Effect of binary and ternary mixture of emulsifying salts on processed cheese

In the industry, especially mixtures of individual phosphates and also citrates are used. Commonly, the mixture for commercial manufacturing of processed cheese contains two or even more individual ES. Before year 2010, very few articles had systematically studied the effect of mixtures of ES on processed cheese properties.

El-Bakry et al. (2011) dealt with the influence of the binary mixture composition of disodium phosphate (DSP) and TSC. They found out that the firmness of processed cheese increased when the relatively amount of TSC raised.

Weiserová et al. (2011) studied the effect of the six binary mixtures contained DSP, tetrasodium diphosphate (TSPP), pentasodium triphosphate (PSTP) and sodium salt of polyphosphate with the mean amount of phosphates units was $n \approx 20$ (POLY20). Fig. 6.4 presented hardness (N; unpublished data) of processed cheese samples with the latter mentioned six binary mixtures (target parameters: 40% w/w dry matter content; 50% w/w fat in dry matter content). Two percent (w/w) of ES was used (calculated on the total weight of the raw material). When the binary mixture of DSP:POLY, TSPP:POLY20, PSTP:POLY20 and TSPP:PSTP (Fig. 6.4, parts A–D) was used, the hardness of the samples increased, when the relative amount of phosphate with higher ion-exchange capability raised. Furthermore, a different trend of hardness development was observed when the binary mixtures of DSP:TSPP and DSP:PSTP were studied (Fig. 6.4, parts E and F). In samples produced using DSP:TSPP or DSP:PSTP, the maximum hardness was obtained, when the ration between DSP and TSPP or DSP and PSTP ranged in the interval 1:1 – 2:3. These findings leaded to the opinion that ion-exchange capability (and subsequently the intensity of caseins dispersion) is not the sole effect of ES on processed cheese properties. Weiserová et al. (2011) suggested the hypothesis that ES (especially in some ratios) intensively influenced the creaming phase and generally, the new protein network forming.

Buňka et al. (2012) continued with studying of the ternary mixture of DSP, TSPP and POLY20 and its influence on processed cheese properties. Fig. 6.5 presented hardness (N; unpublished data) of processed cheese samples with the latter mentioned ternary mixture (target parameters: 40% w/w dry matter content; 50% w/w fat in dry matter content). The concentration of the whole mixture of ES was 2% (w/w; calculated on the total weight of the raw material).

In samples with the relative amount of POLY20 ≤ 60 %, processed cheese hardness rapidly increased (in comparison with the products with the same relative amount of POLY20), when the mutual ratio of DSP and TSPP was in the interval of 1:1–2:3 (see Fig. 6.5). When the ratio of DSP and TSPP was different ($>$1:1 or $<$2:3) the hardness of processed cheese decreased. The effectiveness of the specific combination DSP:TSPP was increasing, when the relative amount of POLY 20 was decreasing. The hardest

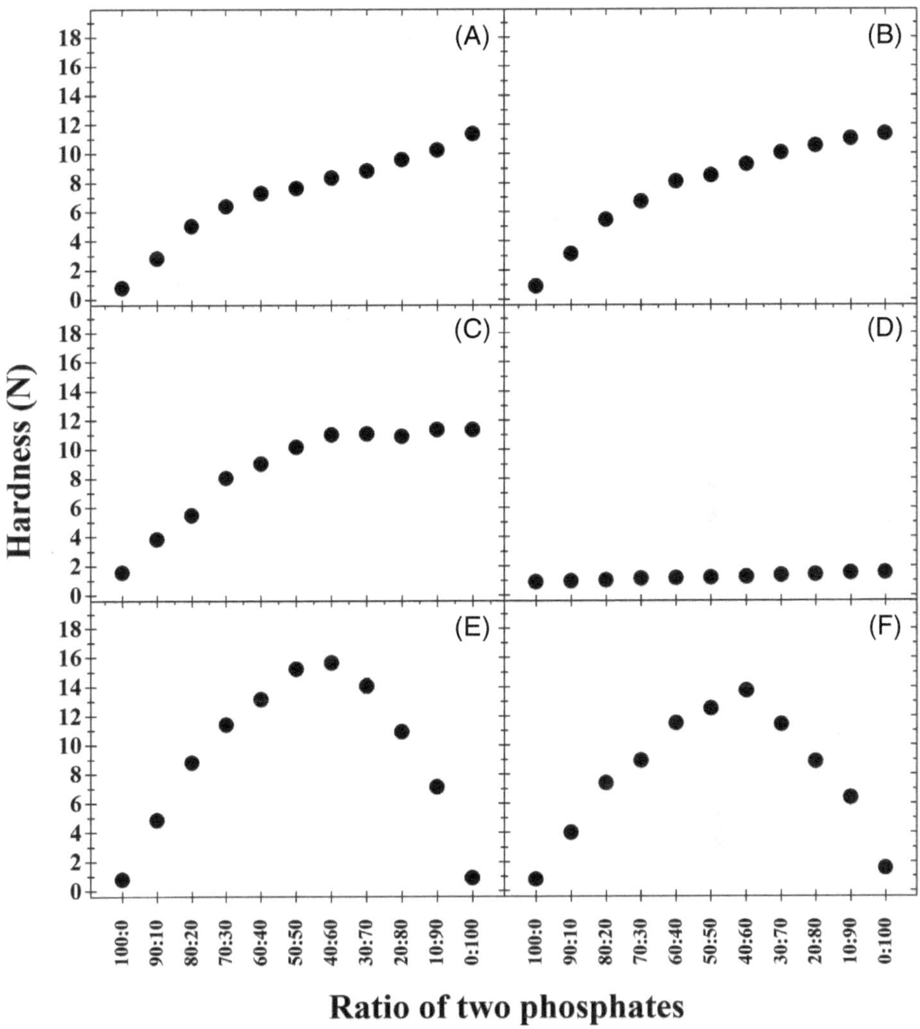

Figure 6.4 Hardness (N) of model processed cheese samples (40% w/w dry matter content; 50% w/w fat in dry matter content) manufactured using the six binary mixtures contained disodium phosphate (DSP), tetrasodium diphosphate (TSPP), pentasodium triphosphate (PSTP), and sodium salt of polyphosphate with the mean amount of phosphates units was n ≈ 20 (POLY20). (A) DSP:POLY20; (B) TSPP:POLY20; (C) PSTP:POLY20; (D) TSPP:PSTP; (E) DSP:TSPP; and (F) DSP:PSTP.

processed cheese were obtained when the ratio of DSP:TSPP was 2:3 and POLY20 was not included. However, in samples with the relative amount of POLY20 > 60 %, the significant influence of the specific ratio of DSP:TSPP was not practically observed (Buňka et al., 2012).

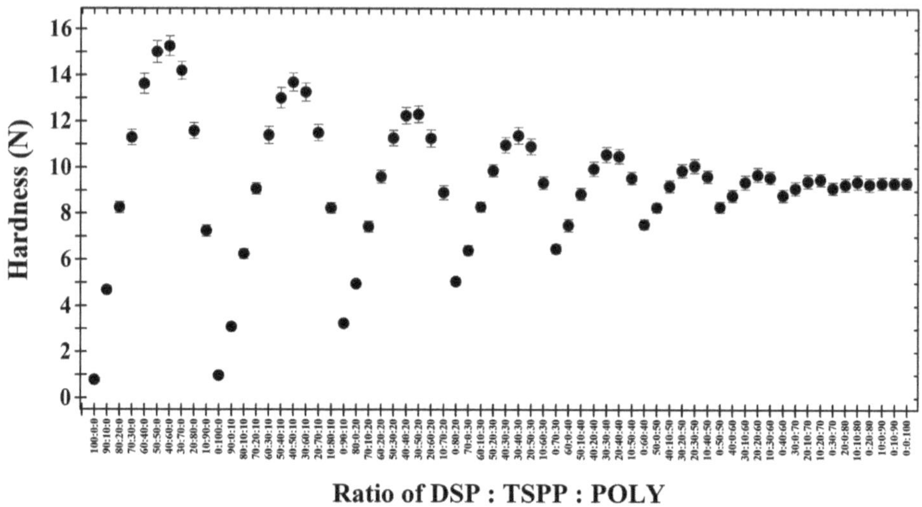

Figure 6.5 Hardness (N) of model processed cheese samples (40 % w/w dry matter content; 50 % w/w fat in dry matter content) manufactured using the ternary mixtures contained disodium phosphate (DSP), tetrasodium diphosphate (TSPP) and sodium salt of polyphosphate with the mean amount of phosphates units was n $\approx$ 20 (POLY20).

Nagyová et al. (2014) followed up with a study, where the authors used sodium salts of polyphosphates with the different mean number of phosphate units in the molecules (n $\approx$ 5 – POLY05, n $\approx$ 9 – POLY09, n $\approx$ 13 – POLY13, n $\approx$ 20 – POLY20 and n $\approx$ 28 – POLY28) in ternary mixtures with TSPP and DSP. Generally, the trend of the influence of the ternary mixtures composition, described by Buňka et al. (2012) only for POLY20 (with DSP and TSPP), on the firmness of processed cheese remained unchanged also for the other polyphoshate with 5–28 units of phosphates linearly bonded in the molecules. When the relative amount of the above mentioned phosphates were $\leq$ 60 %, the effect of the DSP:TSPP combinations 1:1–2:3 still existed regardless of the polyphoshate used. However, the absolute values of hardness were affected. When POLY05 or P0LY09 were added, the absolute values of the model processed cheeses hardness were slightly higher with the application of ternary mixtures with the relative amount of polyphosphate $<$ 50% compared to those samples with POLY13, POLY20 and POLY28 (excluding for the amount of polyphosphates was set at zero). The latter mentioned phenomenon was observed not only in products, where the combination of DSP and TSPP was at 1:1–2:3, but also with other ration of DSP and PTPP. With the application of the relative amount of polyphosphate $\geq$ 60%, the hardness of the processed cheeses decreased slowly as the number of phosphate units (linearly bonded in the polyphosphates) declined.

The explanation of the phenomena in samples manufactured using the binary and ternary mixtures of monophosphates, diphosphates, and polyphosphates could be based on the specific properties of the individual phosphates (described in subsection 6.3.1) or their mutual synergic relationship (Buňka et al., 2012, 2013;

El-Bakry et al., 2011; Hoffmann and Schrader, 2015; Lu et al., 2008; Mizuno and Lucey, 2005b, 2007; Molins, 1991; Nagyová et al., 2014; Sadat et al., 2017; Shirashoji et al., 2010, 2016; Weiserová et al., 2011; Salek et al., 2015) as follows. When zero or the low relative amount of polyphosphates (<10 %) are in the mixtures of ES with monophosphates and the appropriate amount of diphosphates is available (in the ratio of approximately 1:1–2:3), strong gel formation of caseins caused by the presence of diphosphates occurred. Simultaneously, small molecules of monophosphates have the ability to permeate among cross-linked proteins and strongly bind water. A sufficiently short distance between caseins supports their hydrophobic associations resulting in the increase of processed cheeses hardness. Zero or very low proportions of polyphosphates do not give caseins such a strong negative charge leading to the repulsion of proteins. Within a low relative amount of diphosphates, there is not a sufficient concentration of crosslinking agents and thus strong gel cannot be formed. On the other hand, an excessive relative amount of diphosphates leads to very strong binding of a large amount of calcium, which makes the formation of diphosphate-calcium crosslinks (bridges) between caseins more difficult (Figs. 6.4 E and 6.5). Simultaneously, a relative low content of monophosphate permeating among crosslinked caseins is not sufficient to bond of an adequate amount of water.

When the relative amount of polyphosphates is higher (practically $\leq$ 60%), the effect of polyphosphates on the processed cheese network becomes stronger. Their capability to give caseins multiple negative charges reduces the effects of the specific ratio of monophosphates to diphosphates. With the rising relative amount of polyphosphates, the hardness of processed cheese with the specific ratio of monophosphates and diphosphates (approximately 1:1–2:3) is decreasing (Fig. 6.5). When the relative amount of polyphosphates is $>$60%, the effect of polyphosphate on the caseins network formation is predominant while the influence of monophosphates and diphosphates (regardless of the ratio) is only minor. In addition, polyphosphates strongly disperse caseins, to which polyphosphates bind during the creaming and therefore, increase their hydration. The intensively dispersed caseins emulsify the fat present within the matrix. The low relative amount of monophosphates and/or diphosphates cannot show their specific properties. The above-mentioned phenomena lead to the formation of products which have similar hardness (under the constant relative amount of polyphosphates) regardless of the remaining amount and mutual ratio of monophosphates and diphosphates (Fig. 6.5).

When the mean number of phosphate units linearly bonded in the polyphosphates decreases (under the condition of the relative amount of polyphosphates $<$50 %), the hardness of the processed cheese slightly increases. The explanation could be based on the different capability of polyphosphates (with a varying mean number of phosphate units) to give caseins multiple negative charges, which increase with the length of the linear polyphosphate. Shorter polyphosphates (especially, the mean number of phosphate units $n \leq 9$) cause a lower intensity of the negative charge on caseins, which can lead to a higher intensity of hydrophobic interactions of the dispersed caseins and thus increase of hardness of the processed cheese. When the relative amount of polyphosphates is $>$60 % (regardless of the length), the influence on the samples hardness is predominantly driven by the mean number of phosphate units linearly bonded in the polyphosphates. Therefore, product firmness slightly increases with the raising phosphate units in the molecule, practically up to n $\approx$ 13. Then (n $\geq$ 13) the hardness of processed cheese is very similar regardless of the length of the polyphosphates.

Many factors are able to influence the consistency of processed cheese, including the type and the maturity (a concentration of intact casein) of the applied natural cheese, the total concentration of emulsifying salts used and also the storage period (in more details, it is described in Chapters 3 and 9). Buňka et al. (2013) and Salek et al. (2016, 2017, 2020) used different types of natural cheese in varying maturity level (Dutch-type cheese, matured 2–8 weeks; Swiss-type cheese, matured 4–16 weeks; Mozzarella-type cheese, matured up to 4 weeks; Cheddar-type cheese, matured 4–16 weeks; and white brined type cheese, matured 2–24 weeks). The absolute values of the processed cheese hardness varied due to different cheese type used. On the other hand, the general trend (described above) of the influence of the phosphate-based ternary mixture composition on the processed cheese hardness remained uninfluenced. When the degree of maturity raised (and the intact casein concentration declined), the hardness of the processed cheese decreased, but the general trend of functionality of the ternary mixtures of phosphates emulsifying salts was not changed. The processed cheese hardness increased during the storage period (up to 60 days) but without significant effect on the general trend of functionality of ternary mixtures of phosphates emulsifying salts. Buňka et al. (2014) also found out that also concentration of phosphate-based ternary mixture (in the interval 2%–3% w/w) did not influence the above described general trend. The absolute values of the processed cheese hardness were affected only. When the concentration of the ternary mixture raised, the hardness of the processed cheese increased (in the internal 2%– 3% w/w).

6.3.3 Other properties of emulsifying salts

Phosphates possess an inhibitory effect on the growth of some microorganisms. As far as the growth inhibition of gram-positive bacteria is concerned, the inhibitory effect of phosphates is dependent on the length of their chain. The effect of long-chain phosphates is more significant than the inhibitory effect of short-chain phosphates. Moreover, the antimicrobial effect is also influenced especially by temperature and pH value of the environment. As it was described in the subsection 6.2.1, long-chain polyphosphates have a high affinity for bivalent metal ions. These ions are essential for maintaining the integrity of the cell wall of gram-positive bacteria because they form transverse bridges between the molecules of teichoic acids in the cell wall. Binding of bivalent ions can also be manifested by the unavailability of these ions for some essential physiological processes of the growth. Cleavage of the above-mentioned ions will result in bactericidal or bacteriolytic effects. When spore-forming bacteria are exposed to the effects of polyphosphates, the spore germination process is suppressed. The bactericidal or bacteriolytic effect of polyphosphates was studied in model laboratory conditions but also in real processed cheese with similar results. On the other hand, the inhibitory effect of phosphates on gram-negative bacteria are rarely described in the literature (Buňková and Buňka, 2017;

Ehsannia and Sanjabi, 2016; Jen and Shelef, 1986; Lee et al., 1994; Loessner et al., 1997; Lorencová et al., 2012; Maier et al., 1999; Velazquez et al., 2001; Zaika et al., 1997).

The hydrolysis of phosphates in food systems is gaining attention because of its implication in the possible loss of functional phosphate properties. Phosphate hydrolysis in foods could occur due to causes as varied as enzyme activity, high temperature, pH-value and also a result of bacteria metabolisms. Generally, the linkages between phosphate units in linear polyphosphates are more stable at 25°C in comparison to temperature above 70°C. On the other hand, when the temperature of the environment is $<25°C$, the hydrolysis could slowly run. Linear polyphosphates are relatively stable in alkaline or neutral solution, their hydrolysis could be accelerated by decreasing pH to acid area. The latter mentioned relationship is important, because the optimal pH-value of spreadable processed cheeses generally range in the interval of 5.6–6.0 (see Chapter 9) and for processed cheese block and slices slightly lower pH-values (in many cases 5.0–5.5) are required. Polyphosphate hydrolysis proceeds usually by split terminal phosphate groups rather than by middle group link rupture. On the other hand, the studies described also "random cleavage in the middle" of linear polyphosphates rarely exist. As a result of hydrolysis, breakdown of polyphosphate is accompanied by a continuous built up of monophosphate, which is the end product of the hydrolytic process. Generally, the hydrolysis rate raises with the increase of the number of phosphate units in the linear polyphosphate molecule. Increasing water content, concentration of bivalent ions (especially calcium cation), storage temperature, and actual pH-value could also accelerate the hydrolysis rates (Barth et al., 2017; Carić and Kaláb, 1997; Guinee et al., 2004; Lucey et al., 2011; Molins, 1991).

6.3.4 Legislation of emulsifying salts usage

There are many regulations that limit the use of phosphate-based ES in processed cheese and/or its analogues. Firstly, the Codex Alimentarius and the General Standard for Food Additives should be mentioned. Phosphoric acid and phosphates (sodium dihydrogen phosphate, disodium hydrogen phosphate, trisodium phosphate, potassium dihydrogen phosphate, dipotassium hydrogen phosphate tripotassium phosphate, calcium dihydrogen phosphate, calcium hydrogen phosphate, tricalcium phosphate, ammonium dihydrogen phosphate; diammonium hydrogen phosphate, magnesium dihydrogen phosphate, magnesium hydrogen phosphate, trimagnesium phosphate, disodium diphosphate, trisodium diphosphate, TSPP, tetrapotassium diphosphate, dicalcium diphosphate, calcium dihydrogen diphosphate, magnesium dihydrogen diphosphate, pentasodium triphosphate, pentapotassium triphosphate, sodium polyphosphate, potassium polyphosphate, sodium calcium polyphosphate, calcium polyphosphate, ammonium polyphosphate) can be used into processed cheese and cheese analogues up to the maximal level of 9,000 mg/kg calculated as phosphorus. When sodium aluminum phosphates are used, the maximal level is lower at 1,600 mg/kg calculated as aluminum. The latter mentioned phosphate is dedicated for using into American cheese only.

For the USA, CFR-regulations (Code of Federal Regulations) are issued for processed cheese. For example for "pasteurized processed cheese," the allowed amount of ES is 3% calculated on the total weight of the "pasteurized process cheese." Eight phosphates can be used for "pasteurized processed cheese" manufacturing (monosodium phosphate, DSP, dipotassium phosphate, trisodium phosphate, sodium metaphosphate [sodium hexametaphosphate], disodium diphosphate, TSPP, and sodium aluminum phosphate) (CFR 133.169). The same conditions are valid also for similar products from a group defined by regulations 21CFR133.170 – 21CFR133.180, where ES application is allowed.

In the European Union, Regulation No. 1333/2008 on food additives is issued. The maximum level that can be used is 20,000 mg/kg calculated as P_2O_5. Phosphoric acid, monophosphates, diphosphates, triphosphates, and polyphosphates could be used (Brickley et al., 2008). The latter mentioned regulation is supported by Regulation No. 231/2012 laying down specifications for food additives listed in Annexes II and III to Regulation (EC) No. 1333/2008 of the European Parliament and of the Council.

6.4 Citrate emulsifying salts

The salts obtained from citric acid are named as citrates. To obtain citrate salts the acidic hydrogen atoms must be replaced by cations from the tribasic citric acid. For instance, H^+ ions of citric acid neutralization with Na^+ have as a result three types of salts: mono-, di-, and TSC. The most important emulsifying salt used over processed cheese producing is TSC. One percent aqueous solution of TSP (the code according to Regulation No. 1333/2008 is E331) had the pH-value 8.4 ± 0.1 and the pH value of processed cheese (target parameters: 40% w/w dry matter content; 50% w/w fat in dry matter content; solely applied TSC in amount of 2% w/w.) was 6.4 ± 0.1. The monosodium and the disodium citrates are not usually applied solely because the final product is acidic and with poor texture, including the oiling-off defect. Usually, they are applied in blends as processed cheese pH correctors (see subsection 6.6). TSC is used for the manufacture of slices or sliceable blocks of processed cheese. Potassium or ammonium citrates are used for the production of processed cheese with reduced sodium content (allowed in some countries). However, at high levels of the latter mentioned substances, the processed cheese could have the tendency to become bitter. According to Fig. 6.3, TSC presents better calcium ion-exchange ability and casein dispersion properties in comparison to sodium mono and similar compared to diphosphates. TSC is not bind to caseins and therefore is not directly part of the novel protein network. The role is especially in the ion-exchange capability (Dimitreli and Thomareis, 2009; El-Bakry et al., 2011; Lu et al., 2008; Mizuno and Lucey, 2005, Shirashoji et al., 2006; Molins, 1991; Salek et al., 2015).

Fig. 6.6 presented hardness (N; unpublished data) of processed cheese samples with different ternary mixtures of DSP, TSPP, P20, and TSC (target parameters: 40% w/w dry matter content; 50% w/w fat in dry matter content), which were produced under the same conditions. Each ES was added in amount of 2% w/w.

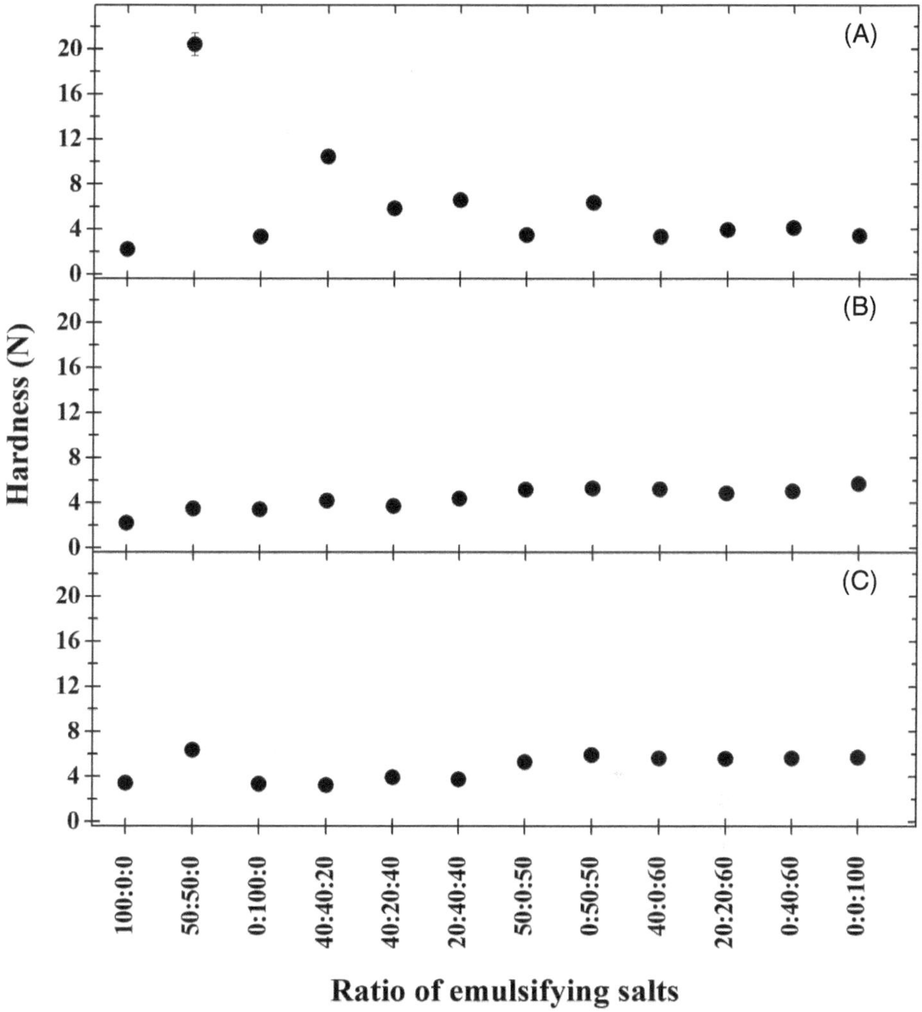

Figure 6.6 Hardness (N) of model processed cheese samples (40% w/w dry matter content; 50% w/w fat in dry matter content) manufactured using ternary mixture emulsifying salts (DSP, disodium phosphate; TSC, trisodium citrate; TSPP, tetrasodium diphosphate; POLY20 – sodium salt of polyphosphate with the mean number of phosphate units n ≈ 20 – POLY20). (A), (B), and (C) presented the ternary mixture of DSP:TSPP:TSC (A), DSP:TSC:P20 (B), and TSC:TSPP:P20 (C), respectively.

Fig. 6.6 A depicts the development of hardness of the processed cheese samples, depending on the composition of the emulsifying salts composed DSP, TSPP, and TSC. The specific ratio of DSP:TSPP (approximately 1:1–2:3) was also identified and was discussed in the subsection 6.3.2, leading to a significant increase in hardness of the processed cheese samples. However, the influence of the latter specific ratio rapidly decreased, with the increasing proportion of TSC in the ternary mixture. The influence of this specific ratio of DSP:TSPP on hardness of the processed cheese samples even at 40% levels of TSC was insignificant. On the other hand, if the proportion of TSC was higher than 40% in the ternary mixture of emulsifying salts, the hardness of the samples increased with the increasing proportion of TSPP and TSC and with the decreasing levels of DSP (Salek et al., 2015).

Fig. 6.6 B shows the development of hardness of the processed cheese samples, depending on the composition of the emulsifying salts in ternary mixtures composed of DSP, TSC, and POLY20. From the results it can be concluded that with the increasing proportion of TSC and POLY20 (and with reducing concentrations of DSP), the hardness of the processed cheese samples also increased. More intensive hardness increase was particularly observed when the proportion of POLY20 increased in the ternary mixtures of emulsifying salts (Salek et al., 2015).

Fig. 6.6 C presents the values of hardness of the processed cheese samples of the ternary mixtures of ES of TSC:TSPP: POLY20. When the proportion of POLY20 was at zero levels in the ternary mixtures of ES, the firmer observed processed cheese samples were that composed of TSC and TSPP at a ratio approximately 1:1. Any deviation from this ratio resulted in significant hardness decrease of the processed cheese samples. However, this phenomenon was only observed in the absence of POLY20 in the ternary mixtures of the ES. With the increasing proportion of POLY20, the hardness of the model samples also increased. On the other hand, at constant levels of POLY20 the hardness of the model samples slightly reduced parallel to TSC proportion decrease (Salek et al., 2015).

The explanation of the development in samples produced using the ternary mixtures of monophosphates, diphosphates, polyphosphates, and TSC could be based on the specific properties of the individual phosphates and/or citrates (described in Section 6.3.1 and in this chapter) or their mutual synergic relationship (Beykont and Kilic Akyilmaz, 2014; El-Bakry et al., 2011; Lu et al., 2008; Mizuno and Lucey, 2005b, 2007; Shirashoji et al., 2010; Salek et al., 2015) as follows.

The development of the processed cheese hardness with the ternary mixtures containing TSC could be explain using the ability of the mixture of emulsifying salts to disperse casein. When the mixture of emulsifying salts with more intensive ability to disperse casein was used, the harder processed cheese was manufactured (Buňka et al., 2013; Lu et al., 2008; Mizuno and Lucey, 2007; Shirashoji et al., 2010).

In case of TSC and TSPP binary mixtures, a specific ratio was observed (approximately 1:1), which increased the hardness of the samples monitored. Probably there is no clear explanation including the interactions between TSC and TSPP and subsequently their involvement in the casein network. The TSC does not participate in creating new networks, and therefore the effect of diphosphates on casein crosslinking would not be influenced by TSC (Kalliapan and Lucey, 2011; Lu et al., 2008; Mizuno and Lucey, 2005). A possible explanation may be that diphosphates are effective on enhancing casein proteins gel formation only when their concentration is at optimum levels relatively to the protein content (Mizuno and Lucey, 2007).

TSC is used especially in the mixture of trade emulsifying salts and is applied in block-and slice-type processed cheese and/or in products where enhanced meltability is required (Gupta et al., 1984; Carić and Kaláb, 1997; Guinee et al., 2003; Mizuno and Lucey, 2005a, 2007; Shirashoji et al., 2006).

Salek et al. (2015, 2016, 2017, 2020) used different types of natural cheese with varying maturity degrees (Swiss-type cheese, matured 4–16 weeks; Mozzarella-type cheese, matured up to 4 weeks; Cheddar-type cheese, matured 4–16 weeks; and white brined type cheese, matured 2–24 weeks). The absolute values of the processed cheese hardness varied due to the different cheese type used. On the other hand, the general trend (described in the subsection 6.3.2 and this chapter) of the effect of the ternary mixture composition (DSP, TSPP, POLY20, and TSC) on the processed cheese hardness remained unchanged. When the degree of the maturity raised (and the intact casein concentration declined), the hardness of processed cheese decreased, but the general trend of functionality of ternary mixtures (including TSC) of phosphates emulsifying salts was not influenced. The processed cheese hardness increased during the storage period (up to 60 days) but without significant effect on the general trend of functionality of ternary mixtures of phosphates emulsifying salts (Dimitreli and Thomareis, 2009; El-Bakry et al., 2011; Lu et al., 2008; Salek et al., 2015, 2016, 2017; Shirashoji et al., 2006).

Apart from phosphate emulsifying salts, citrate-based emulsifying salts are also used during the production of processed cheese and their influence on processed cheese microbiota was observed. The inhibitory effects of citrates, mainly on microorganisms involved in the spoilage of processed cheese due to gas production was studied. The antimicrobial effect of the addition of polyphosphates and orthophosphates monophosphates is better in comparison with citrates (Buňková and Buňka, 2017; Glass and Doyle, 2013; Karahadian et al., 1985; ter Steeg et al., 1995).

According to Regulation No. 1333/2008, sodium citrates could be used as *quantum satis* ("the amount which is enough").

6.5 Crystallization of emulsifying salts

Crystallization of emulsifying salts in processed cheese is a well-known defect of this product group. The defect could be observed on the surface and/or in the interior of the processed cheese. Fig. 6.7 shows a crystal of phosphate emulsifying salt in the interior of the processed cheese with 45% (w/w) dry matter content and 50% (w/w) fat in dry matter in different magnification (part A – 5000x; part B – 10,000x; part C – 20,000x; part D – 25,000x). The undissolved phosphate crystals could interact with proteins (especially caseins) and form different structures in the crystal core and on the crystal edge. In the part A and B, there are visible also circular place after fat globules which were removed by preparation of the sample.

The crystals could be also visible to the naked eye and possess usually a white powdery covering. Crystals on the processed cheese surface are created more intensively by citrate-based salts in comparison with phosphate-based salts. A coating of crystals impart a grainy and unacceptable appearance to the surface and also in the "body" of the processed cheese might occur. Not only phosphate- and citrate-based emulsifying salts could crystalize. Crystals of tyrosine, lactose and salts of proteins and fatty acids

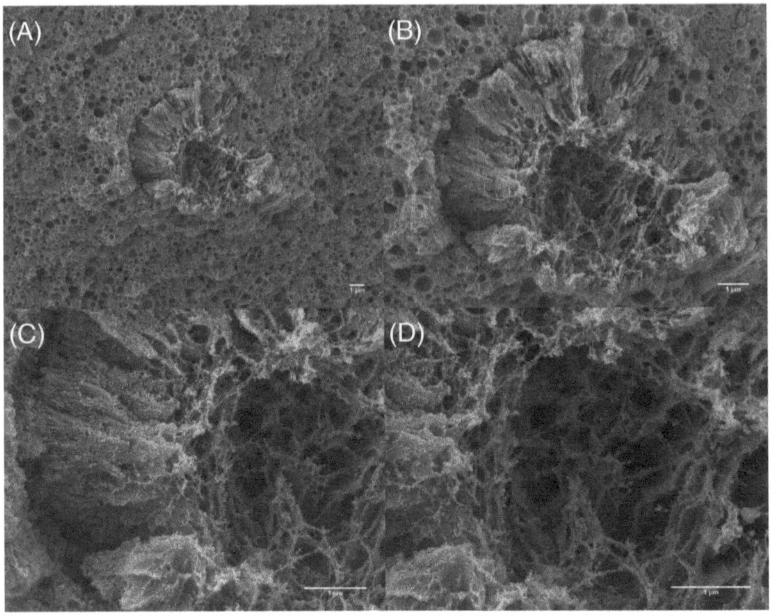

Figure 6.7 Microstructure of the model processed cheese with 45% (w/w) dry matter content and 50% (w/w) fat in dry matter with a crystal of phosphate emulsifying salt in different magnification (A) 5000x; (B) 10,000x; (C) 20,000x; (D) – 25,000x). Bars in the figure parts represent the scales 1 μm. Photo by Michaela Černíková (2015).

can be also observed on the surface and/or in the body of processed cheese samples (Carić and Kaláb, 1997; Guinee et al., 2004; Guinee, 2017; Lucey et al., 2011; Molins, 1991). According to Guinee (2017), there are three main causes of crystallization in processed cheese:

1. The formation of insoluble calcium phosphate crystals (as a result of the interaction between the anion of the emulsifying salts and the calcium paracaseinate);
2. The incomplete dissolution of the emulsifying salts (especially when excess of emulsifying salts is added); and
3. Use of excess lactose (e.g., in added skim milk powder of whey), which could lead to supersaturation and formation of lactose crystals.

Generally, the risk of crystal formation in the resultant processed cheese decreases as the levels of calcium and phosphate in the natural cheese used are low. The phosphate-based (especially monophosphates) emulsifying salts (under pH appropriate for spreadable processed cheese pH $\approx$ 5.5–6.0) crystallize, especially in the body of processed cheese in comparison to citrate-based emulsifying salts, whose crystals are usually observed on its surface. Refrigerated storage of processed cheese increase the susceptibility of the formation of calcium citrate units (so called "bloom" of "haze"). The increase in the water content of processed cheese would reduce the risk that the cheese serum would become supersaturated with phosphates and/or citrates (Carić and Kaláb, 1997; Guinee et al., 2004; Guinee, 2017; Lucey et al., 2011; Molins, 1991).

6.6 Usage of emulsifying salts in industry

In the dairy industry, there are not used the sole emulsifying salts. On the other hand, "trade emulsifying salts," which usually contain 2–5 individual chemical substances (generally, phosphate-based and/or citrate based emulsifying salts) are applied. The exact composition of trade emulsifying salts is under "trade secrete" and the processed cheese producers know only chemical substances without the accurate ratios. Suppliers of trade emulsifying salts specify only the capability to (i) ion-exchange of bivalent and monovalent cations, (ii) pH-shift of processed cheese, and (iii) creaming action. The ability to the latter mentioned properties is usually explained by using the number of pluses ("+" low, "++" medium, "+++" strong) and for the pH-shift, the expected interval is presented. Processed cheese manufacturers usually use two or more trade emulsifying salts for final product in order to reach the desired properties. In fact, "mixtures of mixtures" of emulsifying salts are applied. Type and maturation-degree of natural cheese, the rework addition (see Chapter 9), type and target composition of processed cheese, process parameters (the speed of agitation, the melting temperature and holding time, the cooling rate), the price of emulsifying salts and some other factors are taken into account by processed cheese producers. Also the concentration of the emulsifying salts used is very important. Generally, the optimal amount lies within the interval of 1.5%–3.5% w/w and the factors influencing the selected concentrations are similar as was described in this sub-section. In the most products, the range of emulsifying used is within the range of 2%–3% w/w. Nevertheless, if the emulsifying salts concentration applied is not enough, the ion-exchange ability is not appropriate and pieces (smaller or larger) of unmelted raw materials (especially, cheese, acid cheese, quark etc.) could occur or fat could not be emulsified well. On the other hand, overdosing of emulsifying salts could lead to organoleptic defects, such as bitterness, other untypical taste, crystals, forming, etc.

There is a specific group of trade emulsifying salts, in which is presented only an emulsifying salt – so called "correction salts." Their role is to increase or decrease the pH-value of the melt and also the final product with the minimal addition of emulsifying salts. Trisodium phosphate is especially used for increasing of the product pH-value. On the other hand, disodium dihydrogen diphosphate, monosodium dihydrogen phosphate, and/or disodium citrate and monosodium citrate are used for decreasing of the pH-value of processed cheese. Among the frequently emulsifying salts worldwidely used in the dairy industry belong, for example, (alphabetically ordered): BK Giulini Corp. (e.g., JOHA-lines), Fosfa (e.g., SYROFOS-lines), Prayon (KASOMEL-lines).

6.7 Processed cheese-type products with reduced or without emulsifying salts

Many people like processed cheese and they would also like it with minimum additives. The trend of reducing the amount of phosphorus and sodium in food also exists. In the literature, we can find works and articles describing partial or complete replacement of emulsifying salts in the processed cheeses production. The using approaches are different from author to author. In principle, it could be mention four type of methods of partial or fully complete substitution of traditional emulsifying salts (phosphate-and/or citrate-based). The first approach is only reducing the amount of emulsifying salts and replacing part of them with other substances. The second approach created the complete replacement of emulsifying salts by another food additive or mixture of food additives, which is not based on phosphates or citrates. The third option concerns the partial removal of calcium ions from the processed mixture, respectively from the raw material by physical or physical-chemical processes, whereby it is possible to subsequently reduce the amount of emulsifying salts or eliminate them completely. The fourth approach exists especially in the European Union. It could be called the "legislation approach." There are used different raw materials in the processed cheeses producing e.g. mineral concentrates, which contain phosphates and/or citrates with similar function as that of emulsifying salts. This raw material is not a food additive entered in the List of Food Additives (Regulation EU No. 1333/2008) and the producers do not have to mention this raw material as a food additive (emulsifying salt) on the processed cheese packaging. Producers label this raw material as a part of the used "ingrediencies."

6.7.1 Partial replacement

Carić and Kaláb (1997) described the first partial replacement. They replaced the half amount of emulsifying salts by 1.0% (w/w) of monoacylglycerol. Similarly, Kwak et al. (2002) achieved 50% replacement of emulsifying salts. In their case, partially hydrolyzed casein was used. The use of vegetable hydrocolloids (low methylated pectin, modified starch, locust been gum, xanthan gum, guar gum) was described by Pluta et al. (2000) in their work. The samples containing 0.8% (w/w) locust been gum and samples with 2.0% (w/w) of modified starch were the best evaluated in this article. However, in the raw material mixture was used also rework in an amount 2.0% (w/w). Rework contained phosphate emulsifying salts, so this experiment dealt only with partial replacement of emulsifying salts. Another positive aspect reported in this work was the reduction of fat content and thus the production of processed cheese with lower content of fat and emulsifying salt.

6.7.2 Complete replacement

Another possibility is to completely replace the traditional phosphate or citrate type of the emulsifying salts. There are used other food additives or raw material without phosphates or citrates. Shäffer et al. (1999,2001) examined two groups of products, one of them was with detailed unspecified plant hydrocolloids and the second one was a control group with emulsifying salts including a mixture of monophosphates and polyphosphates. They observed insufficient peptization of the protein in products without emulsifying salt. Therefore, these products did not form a homogenous mass in comparison with the second group with emulsifying salts. Černíková et al. (2010) applied hydrocolloids (low methoxy pectin, locust bean gum, kappa-carrageenan, iota-carrageenan, modified starch with bound sodium octenyl succinate, and combination of low esterified pectin with lecithin) as possible replacements of emulsifying salts based on phosphates or citrates. The concentration of each hydrocolloid was ≤1.0 % (w/w) in the final product. The most homogenous products were identified cheeses with kappa- and iota-carrageenan, which were found homogenous with fracture able (not spreadable) structure. Hladká et al. (2014) described selected properties of processed chesses with 1.0% (w/w) kappa-carrageenan without traditional emulsifying salts in comparison to traditional processed cheeses and they also observed properties in dependence on the raw material maturity (1–16 weeks). They observed that samples without emulsifying salts were almost five times harder than processed chesses with emulsifying salts and their meltability was extremely low and did not change (compared with traditional processed cheeses, where meltability was increasing) with the growing level maturity of the raw material (Edam cheese block). In both cases, harness decreased and adhesiveness raised with the increasing level of maturity of the raw material.

Other studies in this field are the works of Neudek (2013, 2014), who used 1.0%–8.0% (w/w) amaranth flour and corn starch in concentration 1.0%–4.0% (w/w). On the other hand, amaranth flour contains phosphorus in amount approximately 557.00 ± 46.16 mg 100 g^{-1} and a part of it is in the phytic acid form (Arentd and Zannini, 2013). Advanced calcium bioavailability is not exactly established, because phytic acid bound calcium to the strongly fixed compounds.

6.7.3 Partial removal of calcium

Another possibility for production of processed cheeses with a minimum amount or zero content of phosphate or citrate emulsifying salts is based on reducing or completely eliminate calcium ions from the raw material, which thus becomes more meltable. Smith and Rivera (2016, 2017) worked on this hypothesis. They used raw material with reduced amount of calcium content, which can be obtained by ultrafiltration, diafiltration, ion-exchange, acidification, or other techniques or a combination thereof. In addition, their experiment included different type of raw material mixture (various natural cheese, casein with reduced calcium content, whey proteins, water) depended on the required functional properties of the final products, whether they were determined for slicing, or a spreadable product. These products can contain in addition other milk ingredients (e.g., skimmed milk powder, milk protein concentrate, anhydrous milk fat), sodium chloride for preservation of salty taste, food additives, such as colors, flavorings, preservatives (e.g., sorbic acid), and also hydrocolloids (plant gums, starch, gelatine, and others). Smith and Rivera (2016, 2017) manufactured these products at a melting temperature in the range of 70–90°C for 5 min and some of them were homogenized by one-stage or two-stage homogenization. Authors declared that the product were homogenous.

6.7.4 Legislation approach

The last approach to the production of processed cheeses without traditional emulsifying agent is based on the law definition of the processed cheese and the emulsifying salts as a food additive in the European Union. There exists products placed on the market which content for example milk concentrates, which include, inter alia, phosphates, and/or citrates. These substances function as emulsifying salts (see above), eliminate calcium form the calcium caseinate and allowing hydration and subsequent processes as traditional emulsifying salts. The sense of using these materials is in legal significance. European Union Food Law does not include these materials (e.g. milk concentrates) as food additives (Regulation No. 1333/2008). So, they do not assign an E-code within the international nomenclature. In this case, producers label the product as "processed cheese E-free" (that could be debatable). Professional and scientific discussion about this topic is going on.

6.8 Conclusion

The emulsifying salts are the crucial components during processed cheese manufacture. In this chapter, the roles of phosphate- and citrate-based emulsifying salts were described. Ion-exchange, pH-shift, pH-stabilization, and creaming processes were explained for individually used emulsifying salts and also their binary and ternary mixtures. Legislation of phosphate- and citrate-based emulsifying salts was also taken into account. Last but not least, the possibility of traditional emulsifying salts substitutions were described and discussed.

References

Arendt, E.K., Zannini, E., 2013. Amaranth. In Cereal Grains for the Food and Beverage Industries. Woodhead Publishing Series in Food Science, Technology and Nutrition. No. 248. Woodhead Publishing, Cambridge Ltd, UK, pp. 439–473.

Awad, R.A., Abdel-Hamid, L.B., El-Shabrawy, S.A., Singh, R.K., 2002. Texture and microstructure of block type processed cheese with formulated emulsifying salt mixtures. Lebensm. Wiss. Technol. 35, 54–61.

Barth, A.P., Tormena, C.F., Wiotto, W.H, 2017. pH influences hydrolysis of sodium polyphosphate in dairy matrices and the structure of processed cheese. J. Dairy Sci. 100, 8735–8743.

Beykont, E., Kilic-Akyilmaz, M., 2014. Physical properties of an imitation cheese as affected by emulsifying salts an citric acid. J. Food Process. Preserv. 38, 1918–1925.

Berger, W., Klostermayer, H., Merkenich, K., Uhlmann, G., 2002. Process Cheese Manufacture – A JOHA Guide. BK Giulini Chemie GmbH & Co. OHC, Ladenburg.

Bowland, E.L., Foegeding, E.A., 2001. Small Strain Oscillatory Shear and Microstructural Analyses of a Model Processed Cheese. J. Dairy Sci. 84, 2372–2380.

Brickley, C.A., Govindasamy-Lucey, S., Jaeggi, J.J., Johnson, M.E., McSweeney, P.L.H., Lucey, J.A., 2008. Influence of emulsifying salts on the textural properties of nonfat process cheese made from direct acid cheese bases. J. Dairy Sci. 91, 39–48.

Buňka, F., Doudová, L., Weiserová, E., Kuchař, D., Michálek, J., Slavíková, S., Kráčmar, S., 2012. The effect of different ternary mixtures of sodium phosphates on hardness of processed cheese spreads. Int. J. Food Sci. Technol. 47, 2063–2071.

Buňka, F., Doudová, L., Weiserová, E., Kuchař, D., Ponížil, P., Začalová, D., Nagyová, G., Pachlová, V., Michálek, J., 2013. The effect of ternary emulsifying salt composition and cheese maturity on the textural properties of processed cheese. Int. Dairy J. 29, 1–7.
Buňka, F., Doudová, L., Weiserová, E., Černíková, M., Kuchař, D., Slavíková, Š., Nagyová, G., Ponížil, P., Grůber, T., Michálek, J., 2014. The effect of concentration and composition of ternary emulsifying salts on the textural properties of processed cheese spreads. LWT – Food Sci. Technol. 58, 247–255.
Buňková, L., Buňka, F., 2017. Microflora of processed cheese and the factors affecting it. Crit. Rev. Food Sci. Nutr. 57, 2392–2403.
Carić, M., Gantar, M., Kaláb, M., 1985. Effects of emulsifying agents on the microstructure and other characteristics of process cheese – a review. Food Microstruct. 4, 297–312.
Carić, M., Kaláb, M., 1997. Processed cheese products. In: Fox, P.F. (Ed.), Cheese: Chemistry, Physics and Microbiology. Elsevier, Amsterdam, pp. 467–505.
Chavhan, G.B., Kanawjia, S.K., Khetra, Y., Puri, R., 2015. Effect of potassium-based emulsifying salts on sensory, textural, and functional attributes of low-sodium processed Mozzarella cheese. Dairy Sci. Technol. 95, 265–278.
Černíková, M., Buňka, F., Pospiech, M., Tremlová, B., Hladká, K., Pavlínek, V., Březina, P., 2010. Replacement of traditional emulsifying salts by selected hydrocolloids in processed cheese production. Int. Dairy J. 20, 336–343.
Černíková, M., Salek, R.N., Kozáčková, D., Běhalová, H., Luňáková, L., Buňka, F., 2017. The effect of selected processing parameters on viscoelastic properties of model processed cheese spreads. Int. Dairy J. 66, 84–90.
Culler, M.D., Saricay, Y., Harte, F.M., 2017. The effect of emulsifying salts on the turbidity of a diluted milk system with varying pH and protein concentration. J. Dairy Sci. 100, 4241–4252.
Dalgleish, D.G., 1992. Bovine milk protein properties and the manufacturing quality of milk. Livest. Prod. Sci. 35, 75–93.
El-Bakry, M., Duggan, E., O´Riordan, E.D., O´Sullivan, M., 2011. Effect of chelating salt type on casein hydration and fat emulsification during manufacture and post-manufacture functionality of imitation cheese. J. Food Eng. 102, 145–153.
Dimitreli, G., Thomareis, A.S., 2009. Instrumental textural and viscoelastic properties of processed cheese as affected by emulsifying salts and in relation to its apparent viscosity. Int. J. Food Prop. 12, 261–275.
Eshannia, S., Sanjabi, M.R, 2016. Physicochemical, microbiological and spoilage analysis of probiotic processed cheese analogues with reduced emulsifying salts during refrigerated storage. J. Food Sci. Technol. 53, 996–1003.
Ginger, M.R., Grigor, M.R., 1999. Comparative aspects of milk caseins. Comp. Biochem. Physiol. B 124, 133–145.
Farrell, H.M., Jimenez-Flores, R., Bleck, G.T., Brown, E.M., Butler, J.E., Creamer, L.K., Hicks, C.L., Hollar, C.M., Ng-Kwai-Hang, K.F., Swaisgood, H.E., 2004. Nomenclature of the proteins of cow´s milk – sixth revision. J. Dairy Sci. 87, 1641–1674.
Fox, P.F., 2001. Milk proteins as food ingredients. Int. J. Dairy Technol. 54, 41–55.
Fu, W., Watanabe, Y., Satoh, H., Inoue, K., Moriguchi, N., Fusa, K., Yanagisawa, Y., Mutoh, T., Nakamura, T., 2018. Effects of emulsifying conditions on creaming effect, potasmechanical properties and microstructure of processed cheese using a rapid visco-analyzer. Biosci. Biotechnol. Biochem. 82, 476–483.
Glass, K., Doyle, M.E., 2013. Safety of Processed Cheese. FRI Food Safety Review, Food Research Institute, University of Wisconsin, Madison.

Guinee, T.P., 2003. Pasteurized processed cheese products. In: Roginski, H., Fuquay, J.W., Fox, P.F. (Eds.), Encyklopedia of Dairy Science, 1. Elsevier, Amsterdam, pp. 411–418.
Guinee, T.P., Carić, M., Kaláb, M., 2004. Pasteureized processed cheese and substitute/imitation cheese products. In: Fox, P.F. (Ed.), Cheese: Chemistry, Physics and Microbiology. Elsevier, Amsterdam, pp. 349–394.
Guinee, T.P., 2017. Pasteurized processed and imitation cheese products. In: McSweeney, P.L.H., Fox, P.F., Cotter, P.D., Everett, D.W. (Eds.), Cheese: Chemistry, Physics and Microbiology. 4. ed. Elsevier, Amsterdam, pp. 1133–1184.
Gupta, S.K., Karahadian, C., Lindsay, R.C., 1984. Effect of emulsifier salts on textural and flavour properties of processed cheese. J. Dairy Sci. 67, 764–778.
Heertje, I., Boskamp, M.J., Van Kleef, F., Gortemaker, F.H., 1981. The microstructure of processed cheese. Neth. Milk Dairy J. 35, 177–179.
Hladká, K., Randulová, Z., Tremlová, B., Pomížil, P., Mančík, P., Černíková, M., Buňka, F., 2014. The effect of cheese maturity on selected properties of processed cheese without traditional emulsifying agents. LWT – Food Sci. Technol. 55, 650–656.
Hoffmann, W., 2016. Potential restrictions of photocentrifugation for determining the emulsion stability of melted spreadable processed cheese. Dairy Science Technol. 96, 251–259.
Hougaaard, A.B., Sijbrandij, A.G., Varming, C., Ardö, Y., Ipsen, R., 2015. Emulsifying salt increase stability of cheese during holding. LWT – Food Sci. Technol. 62, 362–365.
Hoffmann, W., Schrader, K., 2015. Dispersion analysis of spreadable processed cheese with low content of emulsifying salts by photocentrifugation. Int. J. Food Sci. Technol. 50, 950–957.
Jen, C.M.C., Shelef, L.A., 1986. Factors affecting sensitivity of *Staphylococcus aureus* 196E to polyphosphates. Appl. Environ. Microbiol. 52, 842–846.
Kaliappan, S., Lucey, J.A., 2011. Influence of mixtures of calcium-chelating salts on the physicochemical properties of casein micelles. J. Dairy Sci. 94, 4255–4263.
Kapoor, R., Metzger, L.E., Biswas, A.C., Muthukummarappan, K., 2007. Effect of natural cheese characteristics on process cheese properties. J. Dairy Sci. 90, 1625–1634.
Kapoor, R., Metzger, L.E., 2008. Process Cheese: Scientific and Technological Aspects – A Review. Compr. Rev. Food Sci. Food Saf. 7, 194–214.
Karahadian, C., Lindsay, R.C., Dillman, L.L., Deibel, R.H., 1985. Evaluation of the potential for botulinal toxigenesis in reduced-sodium processed American Cheese foods. J. Food Prot. 48, 63–69.
Kawasaki, Y., 2008. Influence of "creaming'' on the properties of processed cheese and changes in the structure of casein during cheese making. Milchwissenschaft 63, 149–153.
Klostermeyer, H., Buchheim, W., 1988. Die Mikrostruktur von Schmelzkäse-erzeugnissen. Kiel. Milchwirtsch. Forschungsber. 40, 219–231.
Kwak, H.S., Choi, S.S., Ahn, J., Lee, S.W., 2002. Casein hydrolysate fraction act as emulsifiers in process cheese. J. Food Sci. 67, 821–825.
Khetra, Y., Chavhan, G.B., Kanawjia, S.K., Puri, R., 2015. Storage changes in low sodium-processed Mozzarella cheese prepared using potassium-based emulsifying salts. Dairy Sci. Technol. 95, 639–646.
Lee, R.M., Hartman, P.A., Olson, D.G., Williams, F.D., 1994. Bactericidal and bacteriolytic effects of selected food-grade phosphates, using *Staphylococcus aureus* as a model system. J. Food Prot. 57, 276–283.
Lee, S.K., Klostermeyer, H., Schrader, K., Büchheim, W., 1996. Rheological properties and microstructure of model processed cheese containing low molecular weight emulsifiers. Nahrung - Food 40, 189–194.
Lee, S.K., Klostermeyer, H., 2001. The effect of pH on the rheological properties of reduced-fat model processed cheese spreads. Lebensm. Wiss. Technol. 34, 288–292.

Lee, S.K., Buwalda, R.J., Euston, S.R., Foegeding, E.A., Mckenna, A.B., 2003. Changes in the rheology and microstructure of processed cheese during cooking. Lebensm. Wiss. Technol. 36, 339–345.

Loessner, M.J., Maier, S.K., Schiwek, P., Scherer, S., 1997. Long chain polyphosphates inhibit growth of Clostridium tyrobutyricum in processed cheese spreads. J. Food Prot. 60, 493–498.

Lorencová, E., Vltavská, P., Budinsky, P., Koutný, M., 2012. Antibacterial effect of phosphates and polyposphates with different chain length. J. Environ. Sci. Health A 47, 2241–2245.

Lu, Y., Shirashoji, N., Lucey, J.A., 2008. Effects of pH on the textural properties and meltability of pasteurized process cheese made with different types of emulsifying salts. J. Food Sci. 73, 363–369.

Lucey, J.A., Johnson, M.E., Horne, D.S., 2003. Perspectives on the basis of the rheology and texture properties of cheese. J. Dairy Sci. 86, 2725–2743.

Lucey, J.A., Maurer-Rothmann, A., Kaliappan, S., 2011. Functionality of ingredients: emulsifying salts. In: Tamine, A.Y. (Ed.), Processed Cheese and Analogues. John Wiley & Sons, Ltd, Chichester, pp. 110–132.

Maier, S.K., Scherer, S., Loessner, M.J., 1999. Long-chain polyphosphate causes cell lysis and inhibits Bacillus cereus septum formation, which is dependent on divalent cations. Appl. Environ. Microbiol. 65, 3942–3949.

Marchesseau, S., Cuq, J.L., 1995. Water-holding capacity and characterization of protein interactions in processed cheese. J. Dairy Res. 62, 479–489.

Marchesseau, S., Gastaldi, E., Lagaude, A., Cuq, J.L., 1997. Influence of pH on protein interactions and microstructure of process cheese. J. Dairy Sci. 80, 1483–1489.

Mayer, K.H., 2001. Bitterness in processed cheese caused by an overdose of a specific emulsifying agent? Int. Dairy J. 11, 533–542.

McIntyre, I., O´Sullivan, M., O´Riordan, D., 2017. Monitoring the progression of calcium and protein solubilisation as affected by calcium chelators during small-scale manufacture of casein-based food matrices. Food Chem. 237, 597–604.

Meyer, A., 1973. Processed Cheese Manufacture. Food Trade press Ltd, London, UK.

Mizuno, R., Lucey, J.A., 2005a. Effects of emulsifying salts on the turbidity and calcium-phosphate-protein interaction in casein micelles. J. Dairy Sci. 88, 3070–3078.

Mizuno, R., Lucey, J.A., 2005b. Effects of two types of emulsifying salts on the functionality of nonfat pasta filata cheese. J. Dairy Sci. 88, 3411–3425.

Mizuno, R., Lucey, J.A., 2007. Properties of milk protein gels formed by phosphates. J. Dairy Sci. 90, 4524–4531.

Molins, R.A., 1991. Phosphates in Food. CRC Press, Boca Raton.

Montchamp, J.L., 2015. Phosphorus Chemistry I. Springer International Publishing, London.

Mozuraityre, R., Berget, I., Mahdalova, M., Goonsberg., O., E., R., Greiff, K., 2019. Sodium reduction on processed cheese spreads and the effect on phycicochemical properties. Int. Dairy J. 90, 45–55.

Muslow, B.B., Jaros, D., Rohm, H., 2007. Processed cheese and cheese analogues. In: Tamine, A.Y. (Ed.), Structure of Dairy Products. Blackwell Publishing Ltd., Oxford, pp. 210–235.

Nagyová, G., Buňka, F., Salek, R.N., Černíková, M., Mančík, P., Grůber, T., Kuchař, D., 2014. Usage of sodium polyphosphates with different linear length in the production of spreadable processed cheese. J. Dairy Sci. 97, 111–122.

Neudek, P. 2013. Tavený sýr s amarantovou moukou. Užitný vzor., č. dokumentu 25124. p. 4. http://spisy.upv.cz/UtilityModels/FullDocuments/%20FDUM0026/uv025124.pdf

http://spisy.upv.cz/UtilityModels/FullDocuments/FDUM0025/uv025124.pdf. (Accessed 2 August 2017) (in Czech).

Neudek, P., 2014. Tavený sýr s amarantovou moukou. Užitný vzor. 2014, č. dokumentu 26324. p. 4. http://spisy.upv.cz/UtilityModels/FullDocuments/FDUM0026/uv026324.pdf. (Accessed 2 August 2017) (in Czech).

Nogueira, E.B., Costa-Lima, B.R.C., Torres, F., Regazone, A.V., Melo, L., Franco, R.M., Cortez, M.A.S, 2018. Effect of potassium-based emulsifying salts on the sensory and physicochemical parameters of low-sodium spreadable processed cheese. Int. J. Dairy Technol. 71, 717–800.

Piska, I., Štětina, J., 2004. Influence of cheese ripening and rate of cooling of the processed cheese mixture on rheological properties of processed cheese. J. Food Eng. 61, 551–555.

Pluta, A., Ziarno, M., Smolińska, A., 2000. Możliwości zastosowania hydrokoloidów w produkcji serów topionych. Przemysl Spożywczy 5, 42–44.

Raval, D.M., Mistry, V.V., 1999. Application of ultrafiltred sweet buttermilk in the manufacture of reduced fat process cheese. J. Dairy Sci. 82, 2334–2343.

Sádlíková, I., Buňka, F., Budinský, P., Voldánová, B., Pavlínek, V., Hoza., I., 2010. The effect of selected phosphate emusifying salts on viscoelastic properties of processed cheese. LWT – Food Sci. Technol. 43, 1220–1225.

Sadat., A., Ezzatphanah, H., Bakhoda, H., 2017. Solubility and structure of milk powders manufactured with the addition of disodium phosphate and tetrasodium pyrophosphate mixtures. Int. J. Food Prop. 20, 2645–2657.

Salek, R.N., Černíková, M., Nagyová, G., Kuchař, D., Bačová, H., Minarčíková, L., Buňka, F., 2015. The effect of ternary mixtures composition containing phosphate and citrate emulsifying salts on selected textural properties of spreadable processed cheese. Int. Dairy J. 44, 37–43.

Salek, R.N., Černíková, M., Maděrová, S., Lapčík, L., Buňka, F., 2016. The effect of different composition of ternary mixtures of emulsifying salts on the consistency of processed cheese spreads manufactured from Swiss-type cheese with different degrees of maturity. J. Dairy Sci. 99, 3274–3287.

Salek, R.N., Černíková, M., Pachlová, V., Bubelová, Z., Konečná, V., Buňka, F., 2017. Properties of spreadable processed Mozzarella cheese with divergent compositions of emulsifying salts in relation to the applied cheese storage period. LWT – Food Sci. Technol. 77, 30–38.

Salek, R.N., Černíková, M., Lorencová, E., Pachlová, V., Kůrová, V., Šenkýřová, J., Buňka, F., 2020. The impact of Cheddar or White brined cheese with various maturity degrees on the processed cheese consistency: a comparative study. Int. Dairy J. 111, 104816.

Schatz, K., Hoffmann, W., Schrader, K., Maurer, A., 2014. Effect of emulsifying salts containing potassium on the melting properties of block-type dairy cheese analogues. Int. J. Dairy Technol. 67, 202–210.

Schäffer, B., Lörinzy, D., Belágyi, J., 1999. DSC and electronmicroscopic investigation of dispersion-type processed cheeses made without peptization. J. Therm. Anal. Calorim. 56, 1211–1216.

Schäffer, B., Szakály, S., Lörinzy, D., Belágyi, JD., Schäffer, B., 2001. Processed cheeses made with and without peptization. J. Therm. Anal. Calorim. 64, 671–679.

Shirashoji, N., Jaeggi, J.J., Lucey, J.A., 2006. Effect of trisodium citrate concentration and cooking time on physicochemical properties of pasteurized process cheese. J. Dairy Sci. 89, 15–28.

Shirashoji, N., Jaeggi, J.J., Lucey, J.A., 2010. Effect of sodium hexametaphosphate concentration and cooking time on the physicochemical properties of pasteurized process cheese. J. Dairy Sci. 93, 2827–2837.

Shirashoji, N., Aoyagi, H., Jaeggi, J.J., Lucey, J.A, 2016. Effect of tetrasodium pyrophosphate concentration and cooking time on the physicochemical properties of process cheese. J. Dairy Sci. 99, 6983–6994.

Smith, G.F., Rivera, E., 2016. Processed cheese without emulsifying salts. United States Patent. US 20090092730 A1. http://patft.uspto.gov/netacgi/nph-Parser?Sect2=PTO1&Sect2=HITOFF&p=1&u=/netahtml/PTO/search-bool.html&r=1&f=G&l=50&d=PALL&RefSrch=yes&Query=PN/9232808. (Accessed 2 August 2017).

Smith, G.F., Rivera, E., 2017. Processed cheese without emulsifying salts. United States Patent. US 20160081362 A1. http://patft.uspto.gov/netacgi/nph-Parser?Sect2=PTO1&Sect2=HITOFF&p=1&u=/netahtml/PTO/search-bool.html&r=1&f=G&l=50&d=PALL&RefSrch=yes&Query=PN/9532584. (Accessed 02 August 2017).

Templeton, H.L, Sommer, H.H., 1936. Studies on the emulsifying salts used in processed cheese. J. Dairy Sci. 19, 561–572.

Steeg, T., F., P., Cuppers, H.G.A.M., Hellemons, J.C., Rijke, G., 1995. Growth of proteolytic *Clostridium botulinum* in process cheese products. 1. Data acquisition for modeling the influence of pH, sodiumchloride, emulsifying salts, fat dry basis and temperature. J. Food Prot. 58, 1091–1099.

Velazquez, L.D., Escudero, M.E., De Guzman, A.M., 2001. Antibacterial effects of different food-related phosphates using Aeromonas hydrophila. J. Food Prot. 64, 195–200.

Weiserová, E., Doudová, L., Galiová, L., Žák, L., Michálek, J., Janiš, R., Buňka, F., 2011. The effect of combinations of sodium phosphates in binary mixtures on selected texture parameters of processed cheese spreads. Int. Dairy J. 21, 979–986.

Zaika, L.L., Scullen, J., Fanelli, J.S., 1997. Growth inhibition of *Listeria monocytogenes* by sodium polyphosphate as affected by polyvalent metal ions. J. Food Sci. 62, 867–872.

Low salt and low sodium processed cheeses

7

Gustavo Luis de Paiva Anciens Ramos[a,b], *Adriana Torres Silva e Alves*[c], *Leila Maria Spadoti*[c], *Patrícia Blumer Zacarchenco*[c] *and Adriano Gomes da Cruz*[a]
[a] Universidade Federal Fluminense (UFF), Faculdade de Veterinária, Niterói, Rio de Janeiro, Brazil, [b] Instituto Federal de Educação, Ciência e Tecnologia do Rio de Janeiro (IFRJ), Departamento de Alimentos, Rio de Janeiro, Brazil, [c] Instituto de Tecnologia de Alimentos (ITAL), Centro de Tecnologia de Laticínios (Tecnolat), Campinas, São Paulo, Brazil

7.1 Introduction

The history of processed cheeses goes back to homemade Swiss cheese fondue, *Kochkase* (German melted cheese), *Cancoillotte* or French *Canquillotte* preparations, among others. The fondue is made with Swiss cheese, which is an enzymatically coagulated cheese. In the preparation of *Kochkase*, sodium bicarbonate is used, while in *Cancoillotte,* eggs are used. For fondue, wine and/or beer is used. Thus, it appears that, since its origins, processed cheeses, in addition to the sodium commonly present in the base cheese that will undergo the melting process, other sodium-based salts, such as sodium bicarbonate, were added. Later in the history of the advance of processed cheese technology, with the definition of the importance of using melting salts containing sodium in its formulation, the need to reduce this mineral in the composition of these cheeses has become a greater challenge (Van Dender, Zacarchenco, 2014).

The World Health Organization has been working to reduce sodium consumption, one of the measures adopted being the reduction of the content of this mineral in the formulation of processed foods, as is the case with processed cheeses. In this context, several countries, including Brazil, have also established recommendations for sodium reductions in various food categories. In order to understand this measure, a brief summary of the importance and problems of sodium consumption is necessary.

Sodium (Na) forms 2.8% of the earth's crust, being the main cation in the hydrosphere. However, it is a mineral that does not appear free in nature, but linked to Cl and P, with sodium chloride or common salt (NaCl), the most rich in sodium (40%).

Sodium has numerous functions in the body, being considered an essential mineral for the human species. Together with potassium (K) and chlorine (Cl), Na is involved in maintaining osmotic pressure and buffer systems in intra and extracellular fluids, in the transport of nutrients and in the transmission of nerve impulses. Na makes up 90% of the total cations and Cl, ⅔ of anions in the blood. Both minerals are crucial in maintaining the relationship between circulating volume and osmotic pressure in blood plasma. The presence of Na is indispensable for the adequate intestinal absorption of glycids and amino acids. That is why, in Na deficiency, the energy-protein utilization of

Processed Cheese Science and Technology: Ingredients, Manufacture, Functionality, Quality, and Regulations.
DOI: https://doi.org/10.1016/B978-0-12-821445-9.00001-7

the ingested nutrients decreases. The intestinal absorption of some vitamins (riboflavin, thiamine and ascorbic acid) is coupled to Na. Particular properties of Na include its effect on the expansion capacity of colloidal proteins, the maintenance, together with K, of normal cardiac muscle activity, and participation in the process of nervous and muscular excitation.

However, despite its importance for health, its frequent and excessive consumption is harmful. According to Pan American Health Organization (OPAS, 2015), adults who consume more than 2000 milligrams of sodium daily (5000 milligrams of NaCl) are at a higher chance of developing high blood pressure, the main risk factor for cardiovascular disease, as well as renal insufficiency. The official guidelines of the World Health Organization recommend that these limits be adjusted downwards so that, when considering the consumption of children and adolescents, they generally consume less daily calories than adults. High salt consumption, even during childhood, has an effect on blood pressure and can predispose children to diseases such as hypertension, osteoporosis, asthma and other respiratory diseases, obesity and stomach cancer.

In 2013, in Brazil, an agreement was signed between the Minister of Health and ABIQ (Brazilian Cheese Industry Association) that established for the "*requeijão*" (a typical spreadable processed cheese produced on large scale in Brazil) cheese category a maximum limit of 597 mg of sodium per 100 g of the cheese until 2014 and 541 mg/100g until 2016, actions included in the so-called "National Plan for the Reduction of Salt Consumption" (Van Dender et al., 2014b).

When addressing cheeses in general, salt has effects in several aspects, such as enhancement of whey separation, control of the humidity of the final product, control of the metabolism and viability of the starter culture, control of secondary microorganisms that can deteriorate or produce volatile compounds, control of enzymatic activity and enhancement of cheese flavor. It can be said that salt is one of the biggest obstacles to microbial development, helping to maintain the safety of cheese consumption (Bae et al., 2017).

Based on the above, measures to reduce the consumption of this mineral in processed foods are necessary, and its reduction in processed cheeses is the subject of this chapter.

7.2 Functions of NaCl and melting salts in processed cheeses

Sodium in cheeses comes from two sources: sodium present in milk and which increases in cheese (cheeses are a concentrated form of milk), and sodium chloride which is added (Cruz et al., 2011). Processed cheeses are added with melting salts that also contain sodium.

Studies pointed out that these products could be shelf-stable and safe from *Clostridium botulinum* toxin formation when formulated with appropriated values of pH, water activity, and antimicrobial ingredients. More specifically, processed cheese depends on salt and emulsifying salts, such as sodium phosphate, for safety, flavor, and functionality, but these also add to the sodium content. The processed cheese products used in the US school lunch program contain approximately 350–400 mg of sodium per 1-ounce serving, which provides microbial safety but is greater than the school lunch program target, from 2012, of 200–300 mg of sodium per 1-ounce serving. If sodium is reduced to achieve this determination, the product falls into the unsafe zone unless validated alternative preservatives are added (Taylor et al., 2018; USDA, 2012).

In cheddar cheese, a high salt content is necessary to avoid bitter taste. A salt content below 1.12% results in more acidity, less flavor intensity and unpleasant aftertaste. NaCl inhibits the hydrolysis of β-casein, which is associated with bitterness in cheddar cheese. Also, this type of cheese with a lower salt level can present calcium lactate crystals, due to the lower pH (Bansal and Mishra, 2020).

Starter cultures and nonstarter cultures, such as lactic acid producing bacteria, serve as competitive bacteria against pathogenic and spoilage microorganisms (Taylor et al., 2018). Because processed cheeses must undergo a heating step, the levels of the starter culture present that may be protective against pathogen growth are greatly reduced or eliminated. This characteristic reinforces the need for preservatives.

Nisin is an effective antibotulinum agent but it can be cost prohibitive if added in the amounts necessary for antimicrobial activity. If sodium chloride in processed cheese is replaced with potassium chloride, there are limitations due to the metallic/bitter aftertaste, a possibly reduced water-holding capacity, and, if added on a percentage basis, reduced microbiological activity. However, sorbic acid/potassium sorbate is another potential alternative that is being studied for its ability to reduce sodium in processed cheese. Based on several studies, sodium reduction by substitution with potassium salts on a molar basis, rather than a percentage basis, could achieve a 50% reduction in sodium (Glass et al., 2016). Fat reduction is also being examined; the lower the fat content, the more stable the product, as antimicrobial activity could be sequestered in the fat phase if the antimicrobial is fat soluble (Taylor et al., 2018).

The functions of NaCl and melting salts in processed cheeses are detailed below.

7.2.1 Fundamentals of the melting process

To obtain the fusion of the enzymatic clot (or basic mass obtained by enzymatic coagulation), it is necessary to remove the divalent calcium (mainly), which is linked to the casein molecules, which occurs through an ion exchange between the paracaseinate calcium and sodium. This sodium comes from the added NaCl and the addition of flux salts based on sodium citrates or phosphates. These salts have the ability to sequester bivalent calcium making paracaseinate more soluble by the entry of monovalent sodium in its molecule. It is important to highlight that if the process of obtaining the basic mass is not enzymatic, but by acidification, the resulting mass may be more demineralized, which may allow lower contents of melting salts due to the easier melting of the mass. The objective of a study realized in 2019 (Hammam and Metzger, 2019) was to manufacture process cheese products using a combination of acid curd cheese and micellar casein that would provide the required emulsion capacity without the use of emulsifying salts.

Besides that, the use of high-protein formulations such as WPI (whey protein isolate) and WPC80 (whey protein concentrate containing 80% of protein) allowed for the partial reduction of the emulsifying salts (i.e., a reduction in disodium phosphate content from 2.0% to 0.8%) (Królczyk et al., 2016)

7.2.2 NaCl

NaCl, since ancient times, has been a target of desire by humans and animals. At some time it was a bargaining chip, to the point that the term "salary" derives from the Latin expression salarium argentum, which means "payment in salt". Sodium chloride is added to impart a salty taste to the processed cheese, aid in its preservation and the sodium ion participates in the ion exchange with the calcium ion mainly helping the fusion of the protein matrix and its emulsification with the water and fat of the formulation (Gonzalez and Silva, 2019). In processed cheeses, NaCl is used in a lower amount compared to "*requeijão*" (a typical spreadable processed cheese), since the cheeses used in the fusion usually already have salt. In a processed cheese for culinary use called culinary "*requeijão*", the salt (NaCl) content varies from 0.6% to 1.0%, considering a product with a pH range of 5.4–5.7, total dry extract (TDE) from 39% to 41%, fat from 26% to 28%, and fat in the dry extract from 64% to 71% (Sobral et al., 2018).

NaCl in processed cheeses can be used dry or diluted in water and can be added in the cold homogenization stage (at the beginning of the process), at the time of melting or at the end of the process. In general, the percentage of NaCl added is calculated in relation to the basic mass. To illustrate, in the formulation of "*requeijão cremoso*" made from mass (or curd) obtained by hot acid precipitation, the NaCl content (in relation to the basic mass) used varies from 1.3% to 2% (Sobral et al., 2017).

In the case of processed cheeses, one way to reduce the consumption of sodium by the consumer may be through the production of this food with the addition of a lower sodium chloride content. However, this decrease is not a simple task, due to the contribution of NaCl to the taste and preservation of food. This reduction must be balanced with the prevention of the development of pathogenic and deteriorating microorganisms and the maintenance of the proper flavor and texture (Van Dender et al., 2014b).

In addition to adding smaller amounts of NaCl, processed cheese makers can use other resources to adjust the sodium content in these cheeses, such as: (1) applying common salt substitutes (KCl, for example), (2) using melting salts with lower or sodium-free contents, and (3) using combinations of these practices with the addition of flavor modifying or enhancing ingredients (Van Dender et al., 2014b).

Research has shown that whey and milk permeate can contribute to giving salty taste without significantly contributing to the sodium content. A study was carried out to explore the sensory characteristics and the composition of different permeates and their application in a product with low sodium content. The results demonstrate the potential of using whey permeate and delactosed whey permeate as a salt substitute in product applications (Smith et al., 2016).

7.2.3 Melting salts

The process of melting in processed cheeses is based on the removal of calcium from the protein system; protein solubilization or peptization and dispersion; hydration and swelling; pH stabilization and formation of a new protein structure during cooling.

For the processed cheese to be a stable emulsion without separating the protein and fat phases, the presence of an emulsifier is required. Casein molecules are known to be "natural emulsifiers," whose emulsifying properties are lost in cheese due to the binding effect of divalent cations (calcium, magnesium). Consequently, the divalent cations must be extracted during the manufacture of processed cheese and replaced by monovalent cations with strong bases, so that the caseins can return to act as emulsifiers. During the fusion process, ion exchange agents (melting salts) induce this replacement, which must be rapid (Berger et al., 1997; Mauer and Scheurer, 2005).

Thus, the use of melting salts is fundamental in the process of manufacturing processed cheeses, and in order to be functional during the process of melting cheese and processed cheeses, these salts must have, in addition to rapid ion exchange capacity, some properties, such as being soluble in water so that it can reach the micelles and submicelles; not have high molecular weight and, most importantly, be neutral in terms of physiology and taste (Sobral et al., 2017). The sodium and potassium salts of polybasic acids meet these requirements, but with different comparative efficiency, with an increasing order of efficiency as follows: tartrate < citrate < orthophosphate < polyphosphate (Mauer and Scheurer, 2005). For reasons of cost and taste, sodium is the preferred monovalent cation (Berger et al., 1997).

The melting salts must also have the following functions in the melting process:

- Remove calcium (Ca^{2+}) from the protein system
- Solubilize and hydrate the protein
- Promote fat emulsification and stabilize the emulsion
- Control and stabilize the pH
- Keep free Ca^{2+} concentrations low
- Assist in the formation of a characteristic structure after cooling

A melting salt with adequate characteristics must have the ability to retain the released Ca^{2+} ion, preventing it from returning to the protein and undoing the emulsion. The negative charges of the released caseinates are neutralized by sodium ions. The mechanism of action of emulsifying salts can be summarized as follows: (1) dissolution of the melting salt in the aqueous phase, (2) acting on the breaking of the calcium phosphate bridges that exist between the micelles and the sub-layers of casein, and (3) promotion ion exchange and replacement of calcium from the micelle by sodium from the melting salt (Van Dender et al., 2014a).

The new protein structure (sodium paracaseinate) formed by ion exchange is more open and disaggregated, more susceptible to binding to a greater volume of water.

More hydrated, casein is able to remain dispersed in a colloidal suspension, exposing its polar and nonpolar groups, which allows it to act as an emulsifier at the water-oil interface, binding to water and fat and promoting a stable structure, without separation of phases. Therefore, melting salts do not have an emulsifying capacity, but they do have properties to modify caseins so that they are actually the emulsifiers of the emulsion that is processed cheese (Sobral et al., 2017).

Thus, there is a technological challenge in replacing sodium with other ions in the formulation of processed cheeses, since this mineral has relevant technological functions in relation to the sensory, microbiological and physicochemical characteristics of the foods in which it is used (Sobral et al., 2017; Van Dender et al., 2014a).

7.2.4 Microstructure of processed cheese

The microstructure of processed cheese is described as a stable oil-in-water emulsion maintained by a gel network of hydrated and emulsified casein proteins. During the manufacturing of processed cheese, the addition of emulsifying salts, such as sodium salts of phosphates and citrates, improves the emulsification properties of caseins in natural cheese by disrupting the calcium–phosphate protein network. The disruption of the natural cheese casein-protein network results in the exposure of hydrophilic and hydrophobic sections of the individual caseins. Subsequently, when shear and heat are applied, the disrupted caseins become hydrated due to hydrophilic interactions with the aqueous phase, which also allow them to interact with the fat phase via hydrophobic interactions. Emulsifying salts and other ingredients used while formulating a processed cheese, in combination with the processing conditions (cook temperature, cook time, rate of agitation during cooking, and the rate of cooling after manufacture), are the key factors that provide the final processed cheese with its unique microstructure. The microstructure of processed cheeses gives them many functional properties (unmelted and melted texture) (Johnson et al., 2009).

In the study of Ferrão et al. (2018), the researchers produced seven formulations of a spreadable processed cheese ("*requeijão cremoso*") made with cow milk, as follows: (I) full-fat + regular salt content, (II) full-fat + reduced salt content, (III) low-fat + regular salt content, (IV) low-fat + reduced salt content, (V) low-fat + reduced salt content + 3.3% xylo-oligosaccharides (XOS), (VI) low-fat + reduced salt content + 3.3% XOS + 1% arginine, and (VII) low-fat + reduced salt content + 3.3% XOS + 1% yeast extract. They analyzed the microstructure, rheology, physico-chemical characteristics and the sensory profile of the samples. The images of scanning electron microscopy, at Fig. 7.1, present the characteristics of samples with regular and reduced salt content. The sensorial results showed that the formulations I and VI received the higher scores for the attributes homogeneity. In Formulation IV, the fat reduction originated products with lower typical "*requeijão cremoso*" processed cheese aroma, acid taste, and oil film when compared to the Formulation II. The salt reduction in full-fat formulation (Formulation II) resulted in products with higher scores for sweet aroma, bitter taste, and adhesiveness, and lower scores for salty taste, acid taste and homogeneity when compared to the traditional Formulation I.

Figure 7.1 Scanning electron microscopy (SEM) of "*requeijão cremoso*" processed cheese. Increase of 5000 ×, 10 kV. G = void space of fat globule, MP = protein matrix. Formulations: (I) full-fat + regular salt content, (II) full-fat + reduced salt content, (III) low-fat + regular salt content, (IV) low-fat + reduced salt content (Adapated from Ferrão et al., 2018).

7.3 Sensory aspects

Although reducing sodium in the diet can lower blood pressure in adults, studies have shown that many people do not understand all the health risks associated with a high salt diet, many of whom are resistant to changing their salt intake, as this negatively affects the taste of the food (Smith et al., 2016). Since the main driving forces behind the purchase intention and taste of food are taste and texture, which are strongly associated with the salt content, reducing sodium intake has proved to be quite challenging.

7.3.1 NaCl reduction

NaCl contributes positively to the flavor of the cheese, as most consumers appreciate the salty taste in foods, and cheeses without salt tend to be considered tasteless and watery. The concentration and distribution of salt in cheeses has a great influence on various aspects of cheese quality. Among its main effects, Fox et al. (2000) highlighted the following:

1. Salt inhibits or slows down the growth and activity of microorganisms;
2. Inhibits the activity of several enzymes;
3. Affects the flavor of cheeses directly and indirectly due to its influence on microorganisms and enzymes present in it;
4. Causes changes in proteins that influence the texture of cheeses. Salt plays an essential role in hydrating protein and modifying the casein's ability to bind water within the cheese matrix, with an effect on texture and stability properties (El-Bakry, 2012).

Considering that salt has a great effect on microbiology, enzymology, pH and that it causes changes in proteins, it is not surprising that its concentration in cheese affects its sensory quality.

7.3.2 Replacement of NaCl by KCl

The use of salt substitutes is necessary, in general, when the amount of salt to be reduced is greater than 33%. The most popular salt substitute is potassium chloride (KCl), mixed with NaCl (Van Dender et al., 2014b). Cations such as potassium, magnesium, and calcium give sourness and less saltiness compared to sodium, which offers what is really recognized as pure salty taste. Sodium and lithium are the only cations that have a primarily salty taste. However, KCl is usually adopted as a substitute for sodium, as other compounds, such as ammonium chloride, calcium chloride, and magnesium sulfate have unacceptable undesirable flavors when added to cheeses. KCl is used to replace up to 30% NaCl, because above this content the metallic flavor becomes more accentuated and sensorially perceptible (Buňková and Buňka, 2015).

In sensory panels, some tasters perceive the presence of KCl if the mixture contains more than 50% of this salt, although the sensory perception of individuals varies. NaCl can be replaced by KCl, which has a diuretic effect and helps to lower blood pressure. However, people with kidney disease should be alerted to the presence of KCl in the products. Chemically, KCl does not cause bitterness. It imparts a metallic taste different from the classic bitter taste of caffeine, which is used in the stages of selection of tasters in sensory analysis panels. The response of these panels to the descriptor "bitterness" in cheeses with a low sodium content with or without the addition of KCl is due in part to the metallic note of the KCl and, in part, to the hydrophobic peptides, which classically confer bitterness (Sobral et al., 2017).

A modeling study was conducted using Dutch National Food Survey data to evaluate the dietary impact of reformulation ($n = 2106$). Product-specific sodium criteria, to enable a maximum daily sodium chloride intake of 5 grams/day, were applied to all foods consumed in the survey. The impact of replacing 20%, 50%, and 100% of sodium chloride from each product with potassium chloride was modeled. At baseline median, potassium intake was 3334 mg/day. An increase in the median intake of potassium of 453 mg/day was seen when a 20% replacement was applied, 674 mg/day with a 50% replacement scenario, and 733 mg/day with a 100% replacement scenario. Replacement of sodium chloride by potassium chloride, particularly in key contributing product groups, would result in better compliance to potassium intake guidelines (3510 mg/day). Moreover, it could be considered safe for the general adult population, as intake remains compliant with European Food Safety Authority (EFSA) guidelines. Based on current modeling, potassium chloride presents as a valuable and safe replacer for sodium chloride in food products (Van Buren et al., 2016).

In a study carried out in Brazil, it was found that "*requeijão*" cheese produced without added fat, with common salt (NaCl) and melting salt Joha S9 (mixture of sodium phosphates, with 30.5% sodium) presented 593 mg of sodium/100 g of the cheese. However, when the product was made without the addition of fat, replacing 40% of sodium chloride with potassium chloride and the same melting salt Joha S9, the product showed 476.86 mg sodium/100 g, which met the target 541 mg/100 g of so-called "National Plan for the Reduction of Salt Consumption" (Van Dender et al., 2014b).

Another study conclude the reduction of the amount of sodium phosphate emulsifying salt reduced the sodium content in the final product, but also resulted in lower pH and viscosity. Replacing 15% of sodium phosphate with potassium phosphate or citrate gave no changes in pH, viscosity and rheological properties in the final processed cheese. Addition of citrate (both sodium and potassium) improved the textural properties. This indicated that 15% of sodium can be reduced by replacing sodium salts by potassium salts. However, 30% replacement of sodium by potassium phosphate and citrate led to slightly lower pH and viscosity of the processed cheese. Prolonged creaming time (from 1.5 min to 4.5 min) slightly improved the viscosity of the processed cheese (Mozuraityte et al., 2020).

Cheeses such as Gouda and Cheddar with reduced NaCl levels may present a more malleable texture than desired, leading to operational and technological problems with regard to packaging and shape maintenance. In addition, a more hydrated protein matrix is more susceptible to enzymatic hydrolysis. The lower viscosity is due to the increased mobility of casein in the cheese matrix, caused by the reduced salt content. To mitigate this fact, processing conditions and raw materials are adjusted (such as reducing the residual lactose content in cheddar cheeses), and adjusting the composition (reducing the fat and moisture content). Magnesium chloride and calcium chloride are not indicated for texture, as they cause even less viscous and greasy textures when compared to the effects of adding KCl (Buňková and Buňka, 2015; Henneberry et al., 2015).

7.3.3 Substitution of melting salts without sodium or with reduced content

In processed cheeses, most of the sodium in the formulation is contained in the melting salt used as an ingredient, making a total of 44%–48% of the sodium contained in the final product. The manufacture of products with reduced sodium meets public policies to reduce consumption by the population and also aims to serve the consumer market composed mainly of hypertensive patients.

To achieve such a reduction in the sodium content of processed cheeses, melt salts can be used in which sodium is replaced by potassium (Maurer-Rothmann and Scheurer, 2005). These potassium salts are shown in Table 7.1.

Table 7.1 Properties of potassium-based melting salts.

Type of melting salt	Chemical formula	P_2O_5 (%)	pH	K (%)	Solubility (g/100g water) 20°C/80°C
		Orthophosphates			
Monopotassium phosphate	KH_2PO_4	52.1	4.5	28.7	22.7/68.0
Dipotassium phosphate	K_2HPO_4	40.8	8.9	44.9	159.7/266.0
Tripotassium orthophosphate	K_3PO_4	33.4	11.8	55.2	98.8
		Diphosphates			
Tetrapotassium diphosphate	$K_4P_2O_7$	42.9	10.3	47.9	184.8/>240
		Triphosphates			
Pentapotassium triphosphate	$K_5P_3O_{10}$	47.5	9.7	43.6	200
		Polyphosphates			
Potassium polyphosphate	$(KPO_3)_n$	60.1	3–5	33.9	<0.1
		Citrates			
Tripotassium citrate Hydrated	$K_3C_6H_5O_7$ x H_2O	—	7.2	33.5	178.5/199.5

The substitution of sodium emulsifying salts for potassium salts is, in general, partially done. The amount used is reduced and is not made frequently, since the potassium-based emulsifying salts can promote a bitter taste in the processed cheeses and this characteristic can be accentuated with the storage time, leading to product rejection. However, the effect of sodium hydrogen phosphate against *C. botulinum* is better than the effect of potassium hydrogen phosphate, which has 75% of the inhibitory activity compared to sodium salt (Buňková and Buňka, 2015).

Studies have also been carried out in Brazil aiming to reduce the sodium content in processed cheese without adding fat, replacing part of the NaCl with KCl (40%) and using melting salts with reduced levels of sodium and without sodium. In this study, the cheese obtained had sodium contents that varied from 374 to 447 mg/100 g, all below the value defined in the "National Plan for the Reduction of Salt consumption" (Van Dender et al., 2010).

A study realized at India successfully reduced sodium in processed Mozzarella cheese using a combination of KCl and potassium-based emulsifying salts. They produced low-sodium processed Mozzarella cheese by partial replacement of sodium chloride with potassium chloride (KCl) and also by using potassium-based emulsifying salts, i.e., potassium citrate and dipotassium phosphate, in place of sodium-containing salts. Full-sodium and low-sodium cheeses had 905 mg/100 g and 260 mg/100 g sodium, respectively. Thus, the study resulted in 71.27% reduction in overall sodium content of processed Mozzarella cheese. The authors pointed out that with increasing levels of KCl, flavor score of processed Mozzarella cheese was shown to decrease with enhanced bitterness perception. Utilization of potassium-based emulsifying salts in preparation of processed cheese was also encountered to affect the flavor such as a decrease in the overall saltiness and development of bitter and chemical-metallic off flavors. Textural aspects perceived in the mouth play a primitive role in determining overall quality of products. In this study, increase in the level of emulsifying salts in the formulation also increase the sensorial scores for body and texture of processed cheese. Increasing KCl content had the inverse role and the cheese was softer with pasty body. Replacement of NaCl with KCl must be carefully studied as at high concentrations (>1%), KCl tends to decrease cheese firmness (Chavhan et al., 2015).

7.3.4 Use of flavor enhancers

One way to adjust the sodium content in cheeses is to use combinations of reducing the amount of NaCl, using substitutes such as KCl, and adding modifying ingredients or flavor enhancers. Studies have shown that flavor enhancers that increase the perception of salty taste include nucleotides, such as monocalcic and monosodium glutamate, nonprotein nitrogen (NPN) compounds and lactic, citric, and orotic acids (Smith et al., 2016).

Among the flavor enhancers are yeast extracts that contain considerable levels of nucleotides. Umami taste is found in various foods, such as fish, meat, milk, tomatoes and some vegetables and is produced by glutamate ions and by ribonucleotides (inosine and guanine) present in these foods. In addition to their effects as unique flavors, umami substances also have synergism with each other. The characteristic synergism of umami occurs, for example, when monosodium glutamate is combined with the ribonucleotide inosinate 5-monophosphate (IMP) (or its guanosine equivalent). Another possibility to obtain good sensory results in products with sodium reduction is the use of whey permeate (Sobral et al., 2017).

Due to the increased use of whey proteins in the food industry, whey permeate, also known as deproteinized whey, has been generated in abundance. This permeate contains approximately 65% to 85% lactose, 8% to 20% minerals, 3% to 8% protein (mainly NPN), and not more than 1.5% fat. Currently, whey permeate is used mainly in fermentations to produce lactic acid or ethanol, sprayed in the fields as fertilizers or used in animal feed. However, permeate can also be used in many food applications (Smith et al., 2016).

As both the permeate resulting from the serum ultrafiltration process and that resulting from the ultrafiltration of milk have been indicated as a salt substitute, a study was carried out with the objective of determining the sensory properties and the composition of permeate in order to verify whether others elements (in addition to sodium) would contribute to the salty taste of the permeate. In this study, eighteen permeates of whey and reduced lactose were produced industrially, in duplicate. Analysis of composition, specific mineral content, non-protein nitrogen, organic acids, nucleotides, aromatic volatiles and descriptive analysis of permeates and model solutions were performed using a trained sensory panel. The permeates were characterized in this panel by baked, milky, sweet, slightly salty and "brothy flavor" flavors. The organic acids in higher concentrations in the permeate were lactic and citric acids. The volatile compounds included aldehydes, compounds containing sulfur and diacetyl. Comparative sensory tests with sodium chloride solutions confirmed that the salty taste of permeates with reduced lactose content was not due only to the presence of sodium.

Permeate models were created with NaCl, KCl, urea and lactic, citric, hippuric, uric, and orotic acids. The researchers concluded that, in addition to NaCl, KCl, lactic and orotic acid contributed to the salty taste. In the case of delactated permeates, the salty taste was more pronounced than in the standard permeate due to concentrated minerals and organic acids during the lactose removal process. The hypothesis is that permeates obtained from acid whey would be more salty than those produced from sweet whey, due to the higher concentration of organic acids (lactic acid) in acid whey (Frankowski et al., 2014; Smith et al., 2016)

Other research has demonstrated the possibility of using Cheddar whey permeate as a salt substitute, due to its high content of minerals and lactic acid.

7.4 Aspects of legislation in the world and in Brazil

7.4.1 World

The impressive reductions in cardiovascular disease (CVD) seen in high-income countries over recent decades have not been seen in low and middle-income countries, where both rates of CVD and, given population growth and ageing, the total disease burden are continuing to increase. The success in high-income countries has been attributed to risk factor modification among most, although not all of the population, especially reductions in smoking, treatment of hypertension and hypercholesterolemia. Yet there are many barriers to extending this success to low and middle-income countries reflecting both limited capacity to develop and implement healthy public policies in areas such as diet and tobacco control and slow progress in extending coverage with effective health care (Mancia et al., 2017).

WHO has published guidelines on approaches to develop, implement, and monitor sodium intake, and integrate them with approaches to eliminate iodine deficiency. However, such programs have been adopted or implemented mainly in higher income counties. Interventions can be divided into those seeking to change individual choices, such as consumer education and front of pack labelling, and those limiting the choices that are available, such as reformulation or changes to cooking methods in institutional settings (e.g., schools, workplaces). The latter can be further divided into those that are voluntary and those that are imposed through legislation (Johnson et al., 2014).

Some countries have implemented legislative intervention to reduce sodium intake, including mandatory salt targets (e.g. Paraguay), taxation on high salt foods (e.g., Portugal), regulations to ensure front of pack labeling (e.g., Ecuador, Indonesia) and standards for salt in procurement policies (e.g., Argentina, Latvia). Other low and middle-income countries are developing approaches to inform strategies to reduce sodium intake (Mancia et al., 2017).

National salt campaigns can increase consumer awareness and make a noticeable impact on salt intake at the population level. For example, following a national campaign in the United Kingdom, self-report data indicated that the number of people adding salt to food at the table fell by almost 10% in the subsequent five years. A South African national campaign promoting salt reduction in diets, Salt Watch was launched in 2014. It has been shown that an education program improves the adherence to a low-sodium diet in heart failure patients. A study published in 2017 highlighted that mass media campaigns may be an effective tool to use as part of a strategy to reduce discretionary consumption of salt among the population along with other methods. The South African government has adopted a legislative approach (implemented June 2016) that requires the food industry to comply with maximum targets for salt levels in a wide range of food categories, with a further reduction required by 2019 (Wentzel-Viljoen et al., 2017).

According the European Regulation (EC) N° 1924/2006 (nutrition claims and conditions applying to them) the nutrition claim "Low sodium/salt" may only be applied where the product contains no more than 0.12 g of sodium, or the equivalent value for salt, per 100 g or per 100 mL. This regulation also establish that the claim "Reduced (name of the nutrient)" may only be made where the reduction in content is at least for sodium, or the equivalent value for salt, 25% compared to a similar product (Sweetman, 2016).

7.4.2 Brazil

In Brazil, according to RDC 54/2012, the food to be classified as having a reduced sodium content must present a 25% reduction in relation to the sodium content present in the traditional version of the same food. In order to use the low sodium attribute, the food must contain a maximum of 80mg of this mineral per portion when these are greater than 30 g or 30 ml and for portions less than or equal to 30 g or 30 ml the condition must be met in 50 g or 50 mL. Considering the portions of processed cheeses as being 30 g and those of processed skimmed cheeses as being 50 g, these products could contain a maximum of 160 mg of sodium to be considered low in sodium. However, the reduced sodium or light sodium condition can be applied at higher levels, considering the salt content of the traditional version of the product (Brasil, 2012).

A technical group composed of representatives from the Health Care Secretariat, the Health Surveillance Secretariat, the National Health Surveillance Agency (ANVISA) and the Brazilian Association of Food Industries (ABIA) was established in 2007, in Brazil, by the Ordinance no 3092/07, from the Ministry of Health, to propose joint actions and establish a gradual strategy to reduce the levels of free sugars, sodium, saturated fats and trans fatty acids in processed foods. The Minister of State for Health, ABIA, ABIQ (Brazilian Association of Cheese Industries) and three associations of meat products industries signed, in 2013, the "Term of Commitment Extract" which established, among other points, which for the "*requeijão*" cheese category should meet the maximum sodium content of 587 mg/100 g until 2014 and 541 mg/100 g until 2016. These actions were part of the Ministry of Health's so-called "National Salt Reduction Plan" (Van Dender et al., 2014b).

7.5 Examples of research on processed low-sodium cheese and commercial products

Several studies on processed cheeses have already been developed or are underway with the aim of studying the physical-chemical, microbiological, biochemical, sensory

Table 7.2 Specification of the emulsifying salts used in the formulations of the different "*requeijã*" cheese types.

	Joha S9	Joha SK75	Joha B9	Joha B50	Joha S9K
Composition	Blend of Na phosphates	Blend of Na and K phosphates	Blend of Na, Ca, and K phosphates	Blend of Na and K phosphates	Blend of K phosphates
(P_2O_5) (%)	59.7 ± 1.0	50.00 ± 1.0	47.0 ± 1.0	43.5 ± 1.0	42.8 ± 1.0
Theoretical sodium (%)	30.5	8.2	8.81	5.4	–
Theoretical calcium (%)	–	–	1.94	–	–
Theoretical potassium (%)	–	31.5	31.44	38.7	45.6

Reference: BKG Adicon at Alves et al., 2011

and rheological changes that occur when manufacturing this product with reduced sodium content.

In a study evaluating the composition of cheddar cheese against different levels of NaCl addition, it was found that in cheeses salted at 1.8%–3.0% w/w of NaCl there was no statistically significant difference in composition. In cheeses salted at 0.5% and 1.25%, there was an increase in water content, and a consequent reduction in fat, protein and ash content, including a specific drop in calcium content (Rulikowska et al., 2013). Regarding the replacement of NaCl by KCl, most studies report that there is no significant difference in composition (Bansal and Mishra, 2020).

When the concentration of free amino acids in reduced-sodium full-fat cheddar cheeses was evaluated, a reduction in the concentration of cystic acid, proline, glycine, lysine, glutamate, aspartate, and threonine was observed, with a NaCl reduction content of 0.9%–1.9% (McCarthy et al., 2017). The impact of reducing NaCl on pH depends on the type of cheese. It is observed that the pH value drops 0.2-0.3 in low sodium processed cheeses (Møller et al., 2013; Ganesan et al., 2014; Murtaza et al., 2014).

In the evaluation of meltability in processed cheeses based on the replacement of traditional salt with KCl, no significant effects were observed through the Schreiber test, obtaining values of 5.6 for cheeses salted with KCl only, 6.0 for cheeses salted with NaCl only, and 5.9 when a mixture of NaCl:KCl is used (Patel et al., 2013).

In Brazil, a study was carried out aiming to reduce the sodium content in processed cheese without the addition of fat. Therefore, the sodium content was reduced by replacing 40% of sodium chloride with potassium chloride and also by exchanging part of the melting salt traditionally used in the manufacture of a Brazilian spreadable processed cheese (called "*requeijão cremoso*") (Joha S9) for different percentages of melting salts with less sodium content (Joha B9, Joha B50, Joha SK75) or without sodium (Joha S9K) (Alves et al., 2011). The melting salts used in this study are shown in Table 7.2.

In the optimization step of the different "*requeijão*" cheese formulations using Response Surface Methodology, 11 processes were elaborated for each melting salt combination, as described below: 11 cheese processes (9 different formulations and 3 central points) using different percentages of the melting salts Joha S9 and B9, called RAn (n = 1 to 11); 11 cheese processing using different contents of the Joha S9 and S9K melting salts, called RBn (n = 1 to 11); 11 cheese processing using different levels of the melting salts Joha S9 and B50, called RCn (n = 1 to 11) and 11 cheese processing using different percentages of the melting salts Joha S9 and SK75, called RDn (n = 1 to 11), in a total of 44 processes. These formulations were evaluated, after 1 to 3 days of storage, according to their physical-chemical, microbiological, instrumental and sensory characteristics and based on these evaluations, the best formulations were determined using, respectively, the melting salts Joha S9 + B9 (optimized AR), Joha S9 + S9K (optimized RB), Joha S9 + B50 (optimized RC) and Joha S9 + Joha SK75 (optimized RD). After this stage, the optimized formulations RAotim., RBotim., RCotim. and RDotim. were again elaborated together with 2 control treatments, being the control treatment 1 (RP1) a cheese without the addition of fat, where the replacement of 40% of sodium chloride by potassium chloride was not performed and only the Joha S9 melting salt was used, and the control treatment 2 (RP2) a cheese without added fat, in which the replacement of 40% of sodium chloride by potassium chloride and only traditional Joha S9 melting salt was carried out. Finally, these 6 treatments (RAotim., RBotim., RCotim., RDotim., RP1 and RP2) were compared among themselves in terms of physical-chemical, microbiological, sensory composition and texture profile, during 90 days of refrigerated storage, in order to determine which is the best formulation of low-fat "*requeijão*" cheese with low sodium content (RSGTRS) that could be obtained using the RP1 curd as a quality standard. The study of the economic viability of this research project was also carried out (Van Dender et al., 2014b).

From the results obtained in this study it was concluded that the best formulation of RSGTRS elaborated using the mixture of Joha S9 and Joha B9 melting salts was RA6 (1.2% Joha S9 and 0.8% Joha B9), with the use of the mixture Joha S9 and Joha S9K was the RB4 (1.0% Joha S9 and 0.8% Joha S9K), with the use of the mixture Joha S9 and Joha B50 was the RC5 (1.0% Joha S9 and 1, 2% Joha B50) and using the mixture Joha S9 and Joha SK75 was RD5 (1.0% Joha S9 and 1.2% Joha SK75). Still, among the 4 selected formulations, the one that presented similar quality to the standard (curd without added fat with normal sodium content) was RA6, which is the formulation that could be offered as an alternative for curd consumers looking for foods with smaller fat and sodium contents. Regarding the economic viability of this project, the results of the economic profitability indicators point to a viable enterprise under the conditions of the premises considered in this study (Vieira et al., 2014).

The evaluation of the use of KCl and its emulsifying salts in place of sodium salts in the spreadable processed cheese industry has also been studied by researchers in Iraq. Three treatments were carried out: a control in which NaCl was added as a salt and sodium phosphate was used as an emulsifying salt; a treatment in which potassium salts were used to replace sodium salts and a third treatment containing sodium and potassium salts in equal proportion. All treatments were maintained at 6 ± 2°C and at 20°C (room temperature) for 60 days. The results obtained showed that the use of KCl in place of NaCl was successful and that there was no significant difference between the treatments regarding the values of humidity, fat and protein contents, pH and acidity. Regarding microbiological characteristics, there was no significant difference between the control and the samples of treatment with potassium salts stored at 6°C. However, for those stored at 20°C/60 days, there was an increase in the total bacterial count in all treatments. The sensory evaluation showed that during the storage of the product at 6 ± 2°C/60 days there was no significant effect, while in storage at 20°C/60 days there was a significant difference in the evaluations compared to those stored at 6 ± 2°C (Al-Mosway et al., 2017).

An study from India researchers investigate the effect of partial substitution of sodium chloride (NaCl) with potassium chloride (KCl) and total replacement of sodium-based emulsifying salts with potassium-based emulsifying salts in the production of experimental processed Mozzarella cheese (PMC). A blend of potassium citrate and dipotassium phosphate (1:1) was used as an emulsifying salt. Changes in sensory attributes, physico-chemical characteristics, and functional properties of PMC were analyzed and compared with the corresponding changes in a control PMC made with NaCl and trisodium citrate. Products were evaluated after regular intervals of 7 days up to 56 days. Significant differences in flavor (28–56 days), sensory body and texture, acid degree value (0–42 days), soluble protein, titratable acidity (7–56 days), pH (7–56 days), meltability (0–56 days), and stretchability (4–49) were evident between the control and experimental PMC over storage. It appears that potassium replacement of sodium resulted in a low decrease in casein hydration and an increase in pH and bitterness. This subsequently resulted in a lower rate of proteolysis, lower flavor scores, lower meltability, lower titratable acidity, lower acid degree values, and greater stretchablity and sensory body and texture scores (Khetra et al., 2015).

In cheddar cheese formulated with 60% replacement of NaCl by KCl with the addition of flavor enhancer and bitter blocker, it appears that the flavor enhancer has a great impact on the sensory attributes of the cheese. The use of 5′-insoniate disodium generated lower sensory scores, due to brothy and umami flavors (Grummer et al., 2013). In formulation tests also on cheddar cheese, it was found that a formulation with 2 g/L of hydrolyzed vegetable protein and yeast extract as flavor enhancers, lysine, glycine and adenosine-5′-monophospate as bitter blocker and replacement of 75% of NaCl by KCl generated the best sensory acceptance, similar to the control, indicating effective alternatives to increase the content of substitution of traditional salt with KCl (Khetra et al., 2016).

In other study, process cheese products (PCP) were produced using a combination of acid curd cheese and micellar casein that would provide the required emulsion capacity without the use of emulsifying salts. The acid curd utilized was produced from micellar casein concentrate that was standardized to 3%, 6%, and 9% of protein. In the formulation, the acid curd was blended with micellar casein so that the formula contained a 2:1 ratio of protein from acid curd relative to micellar casein. Additional dairy and non-dairy ingredients (Cheddar cheese, butter, and salt) were also used in the formulation. The PCP was prepared by mixing all ingredients in a kitchen aid to produce a homogeneous paste. Approximately 25 g of the mixture was cooked in a RVA for 3 min at 95°C with a 1000 rpm stirring speed during the first 2 min and 160 rpm for the final min. The cooked PCP was then transferred into molds and refrigerated until further analysis. This trial was repeated three times using acid curd (produced from 3%, 6%, and 9% protein MCC). No significant differences ($P > 0.05$) were detected in the cooked viscosity (402–483 cP), hardness (354–384 g), melting area (30.0–31.4 mm) and melting temperature (48–51 °C) between PCP made from different acid curd and was similar to typical process cheese produced with conventional ingredients and emulsifying salt. The authors concluded that PCP could be made with no emulsifying salts when the formulation utilizes a 2:1 ratio of protein from acid curd relative to micellar casein (Hammam and Metzger, 2019).

7.6 Conclusion

It is a consensus that the levels of sodium intake must be reduced. The education programs are important, but legal restrictions also work. There are technological solutions to reduce sodium levels in processed cheeses. The simple reduction of NaCl amount could cause reduction in the period of shelf life and risks of pathogen growth. One of these solutions, the substitution of NaCl by KCl must be at partial levels because of the alterations in the taste of the products. This substitution and the alterations in the taste is true also to other foods. In the case of processed cheeses, the typical solutions are the substitution of the part of NaCl by KCl and addition of mixtures of traditional emulsifying salts and emulsifying salts reduced or without sodium. However, the addition of emulsifying salts reduced or without sodium can alter the traditional characteristics of processed cheeses like taste, texture and melting properties. The studies evolving the application of protein concentrates and the tests with acid curd cheese and micellar casein could represent others alternatives to reduce or eliminate the use of emulsifying salts.

References

Al-Mosway, A.J., Al- Kzaz, A.A., Assam, R., 2017. Replacement of sodium salt with potassium salt in the processed spread cheese and study its effect on the chemical, microbial and sensory properties. Iraqi J. Agric. Sci 48 (6), 1485–1492.

Alves, A.L.V.T., Spadoti, L.M., Van Dender, A.G.F., Alves, A.T.S., Zacarchenco, P.B., Trento, F.K.H.S., Ormenese, R.C.S.C., Yotsuyanagi, K., Morgano, M., 2011. Efeito do uso de diferentes combinações de sais fundentes nas principais características de requeijão sem adição de gordura e com teor reduzido de sódio. In: Congresso Interinstitucional De Iniciação Científica.

Bae, I., Park, J.H., Choi, H.Y., Jung, H.K., 2017. Emerging innovations to reduce the salt content in cheese; effects of salt on flavor, texture, and shelf life of cheese; and current salt usage: a review. Korean J. Food Sci. An. 37 (6), 793.

Bansal, V., Mishra, S.K., 2020. Reduced-sodium cheeses: Implications of reducing sodium chloride on cheese quality and safety. Compr. Rev. Food Sci. Food Saf. 19 (2), 733–758.

Berger, W., Klostermeyer, H., Merkenich, K., Uhlmann, G., 1997. Elaboración de queso fundido – Uma guia JOHA. BK Giulini Chemie, Ladenburg, p. 237.

Brasil, 2012. Resolução - RDC N° 54, de 12 de novembro de 2012. Regulamento Técnico sobre Informação Nutricional Complementar. http://portal.anvisa.gov.br/documents/%2033880/2568070/rdc0054_12_11_2012.pdf/c5ac23fd-974e-4f2c-9fbc-48f7e0a31864. Accessed on 12 December 2020.

Buňková, L., Buňka, F., 2015. Microflora of processed cheese and the factors affecting it. Crit. Rev. Food Sci. Nutr. 57 (11), 2392–2403.
Chavhan, G.B., Kanawjia, S.K., Khetra, Y., Puri, R., 2015. Effect of potassium-based emulsifying salts on sensory, textural, and functional attributes of low-sodium processed Mozzarella cheese. Dairy Sci. Technol. 95, 265–278.
Cruz, A.G., Faria, J.A., Pollonio, M.A., Bolini, H., Celeghini, R., Granato, D., Shah, N.P., 2011. Cheeses with reduced sodium content: effects on functionality, public health benefits and sensory properties. Trends Food Sci. Technol. 22, 276–291.
El Bakry, M., 2012. Salt in cheese: a review. Curr. Res. Dairy Sci. 4 (1), 1–5.
Ferrão, L.L., Ferreira, M.V.S., Cavalcanti, R.N., Carvalho, A.F.A., Pimentel, T.C., Silva, H.L.A., Silva, R., Esmerino, E.A., Neto, R.P.C., Tavares, M.I.B., Freitas, M.Q., Menezes, J.C.V., Cabral, L.M., Moraes, J., Silva, M.C., Mathias, S.P., Raices, R.S.L., Pastore, G.M., Cruz, A.G., 2018. The xylooligosaccharide addition and sodium reduction in *requeijão cremoso* processed cheese. Food Res. Int. 107, 137–147.
Frankowski, K.M., Miracle, M.E., Drake, M.A., 2014. The role of sodium in the salty taste of permeate. J. Dairy Sci. 97 (9), 5356–5370.
Fox., P.F., Guinee, T.P., Cogan, T.M., Mcsweeney, P.L.H., 2000. Fundamentals of Cheese Science. Aspen Publication, Gaithersburg, Maryland, p. 557.
Ganesan, B., Brown, K., Irish, D.A., Brothersen, C., McMahon, D.J., 2014. Manufacture and sensory analysis of reduced-and lowsodium Cheddar and Mozzarella cheeses. J. Dairy Sci. 97 (4), 1970–1982.
Glass, K.A., Mu, M., Rossi, F., LeVine, B., McCoy, D., 2016. Modeling the inhibition of Clostridium botulinum in reduced-sodium pasterurized cheese products. J. Food Prot. 79 (Sup A), 79.
González, F.H.D., Silva, S.C., 2019. Minerais e vitaminas no metabolismo animal. Laboratório de Análises Clínicas, Faculdade de Veterinária. Universidade federal do Rio Grande do Sul, Porto Alegre, Rio Grande do Sul, p. 135.
Grummer, J., Bobowski, N., Karalus, M., Vickers, Z., Schoenfuss, T., 2013. Use of potassium chloride and flavor enhancers in low sodium Cheddar cheese. J. Dairy Sci. 96 (3), 1401–1418.
Hammam, A.R.A., Metzger, L.E., 2019. Manufacturing of process cheese without emulsifying salt using acid curd. J. Dairy Sci. 102 (1), 81–82.
Henneberry, S., Wilkinson, M.G., Kilcawley, K.N., Kelly, P.M., Guinee, T.P., 2015. Interactive effects of salt and fat reduction on composition, rheology and functional properties of mozzarella-style cheese. Dairy Sci. Technol. 95 (5), 613–638.
Johnson, C., Mohan, S., Praveen, D., Woodward, M., Maulik, P.K., Shivashankar, R., Amarchand, R., Webster, J., Dunford, E., Thout, S.R., MacGregor, G., He, F., Reddy, K.S., Krishnan, A., Prabhakaran, D., Neal, B., 2014. Protocol for developing the evidence base for a national salt reduction programme for India. BMJ Open 4, e006629.
Johnson, M.E., Kapoor, R., McMahon, D.J., McCoy, D.R., Narasimmon, R.G., 2009. Reduction of sodium and fat levels in natural and processed cheeses: scientific and technological aspects. Compr. Rev. Food Sci. F. 8, 252–268.
Khetra, Y., Chavhan, G.B., Kanawjia, S.K., Puri, R., 2015. Storage changes in low sodium-processed Mozzarella cheese prepared using potassium-based emulsifying salts. Dairy Sci. Technol. 95, 639–649.
Khetra, Y., Kanawjia, S.K., Puri, R., 2016. Selection and optimization of salt replacer, flavor enhancer and bitter blocker for manufacturing low sodium Cheddar cheese using response surface methodology. LWT—Food Sci. Technol. 72, 99–106.
Królczyk, J.B., Dawidziuk, T., Janiszewska-Turak, E., Sołowiej, B., 2016. Use of whey and whey preparations in the food industry – a review. Pol. J. Food Nutr. Sci. 66 (3), 157–165.

Mancia, G., Oparil, S., Whelton, P.K., McKee, M., Dominiczak, A., Luft, F.C., AlHabib, K., Lanas, F., Damasceno, A., Prabhakaran, D., La Torre, G., Weber, M., O'Donnell, M., Smith, S.C., Narula, J., 2017. The technical report on sodium intake and cardiovascular disease in low- and middleincome countries by the joint working group of the World Heart Federation, the European Society of Hypertension and the European Public Health Association. Eur. Heart J. 38, 712–719.

Maurer-Rothmann, A., Scheurer, G., 2005. Estabilização dos sistemas proteicos do leite. Landerburg, BK Giulini, 51 pp.

McCarthy, C.M., Kelly, P.M., Wilkinson, M.G., Guinee, T.P., 2017. Effect of fat and salt reduction on the changes in the concentrations of free amino acids and free fatty acids in Cheddar-style cheeses during maturation. J. Food Compos. Anal. 59, 132–140.

Møller, K.K., Rattray, F.P., Bredie, W.L.P., Høier, E., Ardö, Y., 2013. Physicochemical and sensory characterization of Cheddar cheese with variable NaCl levels and equal moisture content. J. Dairy Sci. 96 (4), 1953–1971.

Mozuraityte, R., Berget, I., Mahdalova, M., Grønsberg, A., Ruud, E., Greiff, K., 2020. Sodium reduction in processed cheese spreads and the effect on physicochemical properties. Int. Dairy J. 2020 in press.

Murtaza, M.A., Huma, N., Sameen, A., Murtaza, M.S., Mahmood, S., Mueen-ud-Din, G., Meraj, A., 2014. Texture, flavor, and sensory quality of buffalo milk Cheddar cheese as influenced by reducing sodium salt content. J. Dairy Sci. 97 (11), 6700–6707.

OPAS, 2015. Semana Mundial pela Conscientização do Consumo de Sódio. https://www.paho.org/bra/index.php?option=com_content&view=article&id=4797:semana-mundial-pela-conscientizacao-do-consumo-de-sodio&Itemid=820.

Patel, J.B., Patel, P.V., Lee, E.J., 2013. The effect of potassium chloride as a salt replacer on the qualities of processed cheese. J. Students Res. 13, 313–329.

Rulikowska, A., Kilcawley, K.N., Doolan, I.A., Alonso-Gomez, M., Nongonierma, A.B., Hannon, J.A., Wilkinson, M.G., 2013. The impact of reduced sodium chloride content on Cheddar cheese quality. Int. Dairy J. 28 (2), 45–55.

Smith, S.T., Metzger, L., Drake, M.A., 2016. Evaluation of whey, milk, and delactosed permeates as salt substitutes. J. Dairy Sci. 99 (11), 8687–8698.

Sobral, D., Costa, R.G.B., Oliveira, R.B.A., Paula, J.C.J., Silva, H.L.A., Esmerino, E.A., Guimarães, J.T., Cappato, L.P, Zacarchenco, P.B., Van Dender, A.G.F., Chaves, A.C.S.D., Cruz, A.G., 2017. Queijos Processados chapter. 3, p. 71-113. In: Cruz, A.G., Zacarchenco, P.B., Oliveira, C.A.F., Corassin, C.H. (Eds.), Processamento de produtos lácteos: Queijos, Leites Fermentados, Bebidas Lácteas, Sorvetes, Manteiga, Creme de Leite, Doce de Leite, Soro em Pó e Lácteos Funcionais. Elsevier, Rio de Janeiro, p. 330.

Sobral, D., Costa, R.B.G., Oliveira, R.B.A., Silva, H.L.A., Esmerino, E.A., Guimarães, J.T., Cappato, L.P., Zacarchenco, P.B., Dender, A.G.F.V., 2018. Queijos Processados. In: A.G. Cruz, P. Zacarchenco, C.A.F. Oliveira, C. Corassin (Eds.). Processamento de produtos lácteos. 1st ed. Rio de Janeiro: Elsevier. 320 pp.

Sweetman, J., 2016. Commercialization of foods for customers with specific dietary needs. In: Osborn, S., Morley, W. (Eds.), Developing Food Products for Consumers with Specific Dietary Needs. Woodhead Publishing Series in Food Science, Technology and Nutrition. London: Woodhead Publishing. 298 pp.

Taylor, C., Doyle, M., Webb, D., 2018. The safety of sodium reduction in the food supply: a cross-discipline balancing act"—Workshop proceedings. Crit. Rev. Food Sci. Nutr. 58, 1650–1659.

USDA, 2012. Nutrition standards in the National School Lunch and School Breakfast Programs. Final rule. Fed. Regist. 77, 4088–4167.

Van Buren, L., Dötsch-Klerk, M., Seewi, G., Newson, R.S., 2016. Dietary impact of adding potassium chloride to foods as a sodium reduction technique. Nutrients 8, 235.

Van Dender, A.G.F., Spadoti, L.M., Zacarchenco, P.B., Trento, F.K.H.S., Alves, A.T.S., Mendes, T.Q., Ormenese, R.C.S.C., Morgano, M.A., Yotsuyanagi, K., 2010. Optimisation of the manufacturing of processed cheese without added fat reduced sodium. Aust. J. Dairy Technol. 65 (3), 217–221.

Van Dender, A.G.F., Gallina, D.A., Zacarchenco, P.B., 2014a. Características dos Sais Fundentes e seu Papel na Obtenção de Queijos Processados. In: VanDender, A.G.F. (Ed.), Requeijão Cremoso e Outros Queijos Fundidos - Tecnologia de Fabricação, Controle de Processo e Aspectos de Mercado. 2nd revised and expanded. Setembro Editora Ltda, São Paulo, pp. 29–56.

Van Dender, A.G.F., Spadoti, L.M., Bosi, M.G., Moreno, I., Zacarchenco, P.B., 2014b. *Requeijão* Cremoso e Outros Queijos Fundidos: Inovações e Alternativas Tecnológicas. In: VanDender, A.G.F. (Ed.), Requeijão Cremoso e Outros Queijos Fundidos - Tecnologia de Fabricação, Controle de Processo e Aspectos de Mercado. 2ªnd revised and expanded. Setembro Editora Ltda, São Paulo, pp. 261–326.

Van Dender, A.G.F., Zacarchenco, P.B., 2014. Requeijão e outros queijos fundidos: história e aspectos gerais. In: VanDender, A.G.F. (Ed.), Requeijão Cremoso e Outros Queijos Fundidos - Tecnologia de Fabricação, Controle de Processo e Aspectos de Mercado. 2and revised and expanded. Setembro Editora Ltda, São Paulo, pp. 21–28.

Vieira, M.C., Cavichiolo, J.R., Van Dender, A.G.F., Spadoti, L.M., Zacarchenco, P.B., Gomes, R.A.R., Van Dender, A.G.F., 2014. Análise de Viabilidade Econômica da Produção de Formulações de Requeijão Cremoso sem Adição de Gordura e com Teor Reduzido de Sódio. Informações econômicas 44 (3), 36–50.

Wentzel-Viljoen, E., Steyn, K., Lombard, C., De Villiers, A., Charlton, K., Frielinghaus, S., Crickmore, C., Vash Mungal-Singh, V., 2017. Evaluation of a mass-media campaign to increase the awareness of the need to reduce discretionary salt use in the South African population. Nutrients 9, 1238.

Manufacture of processed cheese: Equipments used

8

Mustafa Ozturk[a] *and Meral Kilic-Akyilmaz*[b]
[a] Sakarya University, Department of Food Engineering Sakarya, Turkey,
[b] Istanbul Technical University, Department of Food Engineering Istanbul, Turkey

8.1 Introduction

Processed cheese manufacturing lines can be as simple as a single cooker or a collective group of many specific processing equipments harmonized by an automated control system. Employing correct equipments is essential to achieve target output, meet legal standards and produce high quality and long shelf life product meeting consumer demands. The major equipments used in the manufacture vary in design and capacity but they carry out the same purpose, heating and mixing the ingredients to create a homogeneous stable emulsion.

Processed cheese manufacturing lines consist of size reduction, blending, cooking, and packaging equipments (Fig. 8.1). Cooker is the heart of process cheese manufacture, batch or continuous cookers can be employed to heat and agitate the processed cheese blend. Today, almost all cookers use direct heating for its advantages on quality associated with rapid heating (e.g., less Maillard reaction) and low energy requirements (Dixon, 2011). Batch cookers can be divided into two categories based on mechanical action capabilities; low shear cooker and high shear cooker. Steam injection nozzles of continuous cookers also apply high shear on the process cheese blend creating small size fat droplets similar to homogenized products (Dixon, 2011).

Portioning and packaging of processed cheese products show great variation worldwide. Bulk products are portioned and packaged by direct filling of the hot processed cheese into the container, while processed cheese slices, shreds, or cubes are manufactured by forming, cooling, shaping, and packaging (Buys and Mostert, 2011).

8.2 Equipments used in preparation

The main ingredient in processed cheese manufacture is natural cheese. Selection and proportion of various natural cheeses are conducted by trained employees to create products with desired sensory and physical properties. Natural cheeses with hard rind, wax layer or yeast/mold contamination need to be cleaned or removed before manufacture manually or mechanically (Gouda and Abou El-Nour, 2003). Rapid motor driven scrappers are employed to remove the rind or contaminants from the outer surface of natural cheeses. Counter rotating stainless steel rollers create fine particles from the hard rind, thus the rind can be incorporated into the blend (Guinee et al., 2004).

Processed Cheese Science and Technology: Ingredients, Manufacture, Functionality, Quality, and Regulations.
DOI: https://doi.org/10.1016/B978-0-12-821445-9.00015-7

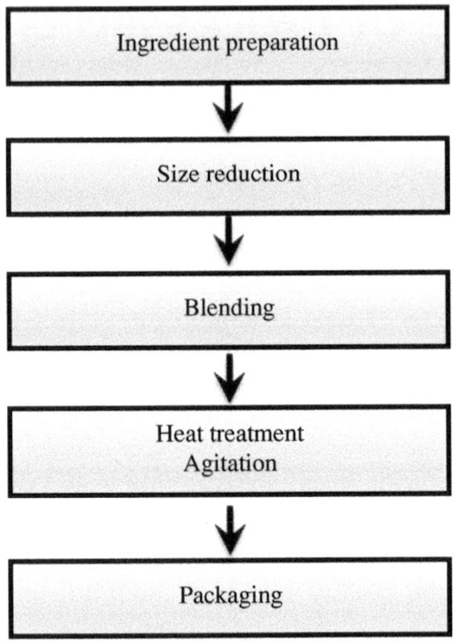

Figure 8.1 Flow chart for manufacture of processed cheese.

Most natural cheese used in processed cheese making are packaged with plastic films that reduces the need for cleaning or scrapping, which is easier for processing and no cheese waste is produced. Natural cheeses are carried and stored on wooden pallets or cardboard boxes in some plants. These bulk (or secondary) packages are removed after the cheeses are transferred to manufacturing area. Primary package of natural cheeses needs to be removed closer to manufacture line to prevent contamination risk (Dixon, 2011).

Formulation of the ingredients is a critical step in the processed cheese manufacture to achieve target physical, chemical, sensory and regulatory standards. Dry matter and fat contents are taken into consideration in determination of formulation. Weighing platforms or table top scales are the fundamental elements in every processed cheese plant. The weight of bulk products (e.g. cheese, butter) are generally weighed on motorized conveyor belts with load cells. Weighed ingredients are transferred to a dump hopper by motorized conveyor belts for initial size reduction (Dixon, 2011).

8.3 Size reduction

Large cheese blocks and butter need to be converted into small pieces to be able to uniformly mix them along with other ingredients and create a homogeneous blend. Earlier equipments involved special knives that cut the cheese into small pieces which are then fed into a mincer or shredder for further size reduction. Particulated cheese is then ground by roller mills into flakes that can be mixed with other ingredients effectively (Gouda and Abou El-Nour, 2003). In recent years, many process cheese manufacturing lines employ curd breakers to carry out initial size reduction of bulk ingredients. Curd breaker with claws breaks the cheese and butter blocks into smaller chunks. Employment of a curd breaker ensures less downtime in production line, flexibility on grinder selection, less strain on the grinder, and less heat production during grinding (Dixon, 2011). The throughput of the curd breaker is transferred to grinders generally via motorized conveyor belts. The feed level from the conveyor is automatically controlled by photoelectric cells to prevent product built up.

Grinding cheese is an important part of the manufacturing process as it helps the producers to better control the final composition of the product and efficacy of the heating process. Various cheese grinders have found their place in market varying in the strength, number of augers (single screw, double screw), capacity and extrusion hole plate options. The size of extrusion hole plates vary from 2–20 mm and selection of proper extrusion hole plate size for the grinder is important to achieve desired textural properties in final product (Dixon, 2011). In the manufacture of sterilized high moisture cheese spreads, small extrusion hole plates are preferred to ensure sufficient heat transfer in cheese particles, no cold center remains thereby reaching desired microbial inactivation and long shelf life. Large extrusion hole plates are generally used in the manufacture of process cheeses where more integrated and natural looking texture is desired in the final product (Dixon, 2011). The extruded paste from grinders is transferred into blenders via bins and lifting devices or by inclined conveyors in automated production lines.

8.4 Blenders

Ingredients used in processed cheese manufacture are large in quantity and varies in composition and chemistry. Properties of natural cheese, such as composition, pH, mineral content, and level of intact casein affect the processing conditions. Homogeneous and even distribution of ingredients before cooking is essential to create a standard final product. Insufficient blending may cause great variations in chemical composition and texture of processed cheese produced. Blending or pre-mixing the ingredients in presence of water increases protein hydration thereby leading to better cooking process and processed cheese characteristics (Guinee et al., 2004).

Ground and extruded cheese and butter are mixed with other dry ingredients (i.e., dairy powders, salt, emulsifying salts) and water in blenders. Dairy powders or other powdered ingredients can be carried into blender with water or sucked in by vacuum. The water used in the pre-mix must be carefully calculated by considering the condensed water incorporated into the cheese during steam injection cooking. Feed systems are programmable logic controlled (PLC) and product inlet is set up depending on the final product characteristics. Twin shaft ribbon blenders are the most commonly used equipment in blending (Fig. 8.2A and B).

The size of the blender should be selected carefully to minimize the differences between sub-batches. To ensure homogeneous distribution of ingredients and reduce downtime mostly more than one blender is employed in factories and they work in alternates. Properly blended and standardized mix from of one blender fed into cooker generally with an auger dump hopper, while a second blender would be mixing and

(A)

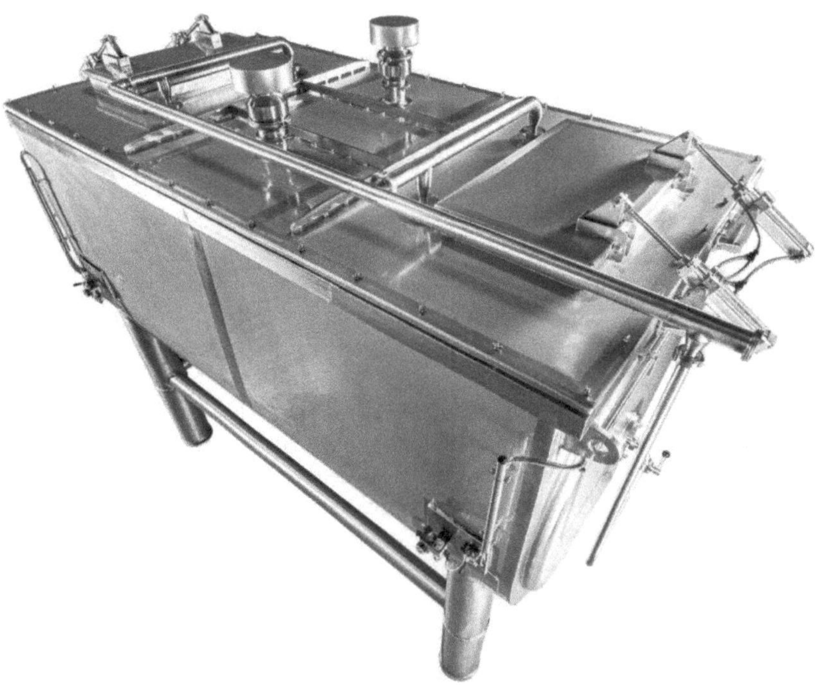

(B)

Figure 8.2 One shaft ribbon blender. (A) Ribbon assembly, (b) whole equipment (Reproduced by courtesy of GEMAK Inc., Sincan, Ankara, Turkey).

standardizing ingredients. Dump hopper needs to have at least the capacity to hold the standardized product from one blender. Generally two parallel screw augers of dump hopper push the mix into the positive displacement pump while a third screw ribbon positioned above the other augers controls feed to the pump and prevent the product build-up. Cookers with interchangeable hoppers can also be used to provide continuous cooking process. Blended pre-mix should be processed without long hold times as time dependent changes may occur in physico-chemical properties depending on degree of mixing (e.g. protein-protein, protein-fat and protein-water interactions) and composition of the pre-mix. Exposure of premix to air also needs to be kept minimal to maintain good manufacturing practice (Meyer, 1973; Dixon, 2011).

8.5 Cookers

Heating and creating a homogeneous mass is an essential stage in processed cheese manufacture. Physico-chemical changes in the protein network are created and also the blend is heat-treated to yield a microbiologically safe and long shelf life product in a cooker. Cookers used in process cheese industry vary in capacity, mixing system, heating mechanism and operating conditions. Cookers can work independently (batch cookers) or as a part of automated process (continuous cookers). Heating in cookers can be done with indirect or direct heating with steam injection; however, currently most cookers in processed cheese industry use direct steam for heating.

The parameters, temperature, agitation speed and time, applied in cooker determine physical properties of processed cheese. Large fat globules increase flow and meltability of the cheese and there is an inverse relationship between degree of homogenization and meltability (Guinee et al., 2004). As the agitation speed is increased, size of fat particles is reduced and a fine emulsion is obtained, this in turn causes whitening of the cheese with a reduction in meltability and an increase in cohesiveness and hardness (Fox et al., 2017). Excessive heat treatment or sterilization of the blend results in extended shelf life at the expense of degradation of texture, color and flavor, therefore minimal heating for obtaining a safe and stable product is recommended.

8.5.1 Batch cookers

Batch cookers are generally used in the manufacture of pasteurized processed cheese products (Fig. 8.3A–D). The processing temperature and time for batch cookers vary in the range of 70–95°C for 4–15 min depending on the desired final physical, chemical, and microbial characteristics of the product (Guinee, 2011). Pressurized batch cookers with long holding time can be used to eliminate microbial spores along with vegetative cells to manufacture sterilized products. Batch cookers generally are heated with direct steam injection (Dixon, 2011). Steam injection and cooking duration in modern cooking equipment are mostly controlled with software to avoid excessive churning, heating, and energy losses. Low shear batch cookers can have single or twin screw augers creating the mechanical action on the blend generating mixing speeds from 50 to 150 rpm. High shear cookers are designed with high speed cutting blades enabling

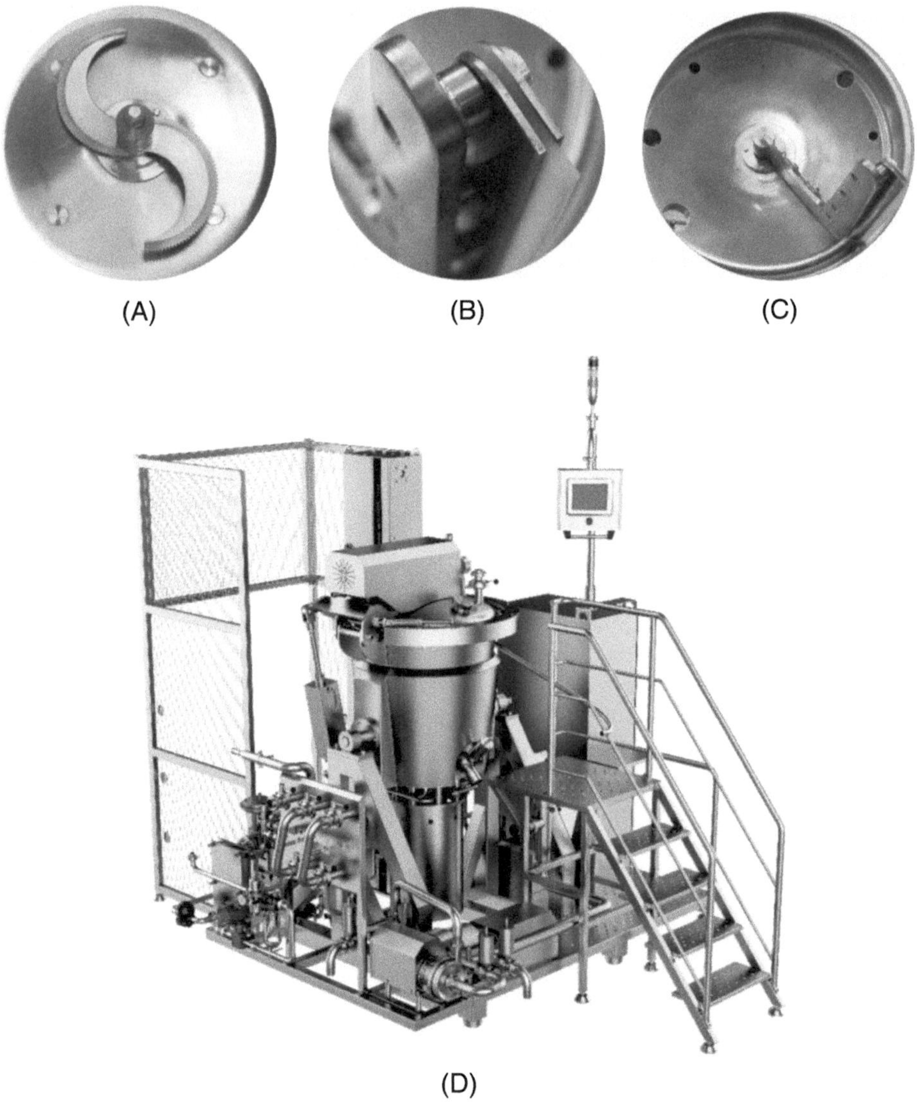

(A) (B) (C) (D)

Figure 8.3 Batch cooker. (A) High shear blade, (B) scraper, (C) inner view of scraper and lama scraper, (D) batch cooker unit with PLC control (Reproduced by courtesy of Kromel Inc., Arifiye, Sakarya, Turkey).

them to reach mixing rates as high as 3000 rpm (Kapoor and Metzger, 2008). Some high shear batch cookers generally referred to as combi cookers that can carry out complete process line of manufacturing such as size reduction, blending ingredients, heating, cooling and deaerating (Fig. 8.4A and B). In such a process, emulsifying salts

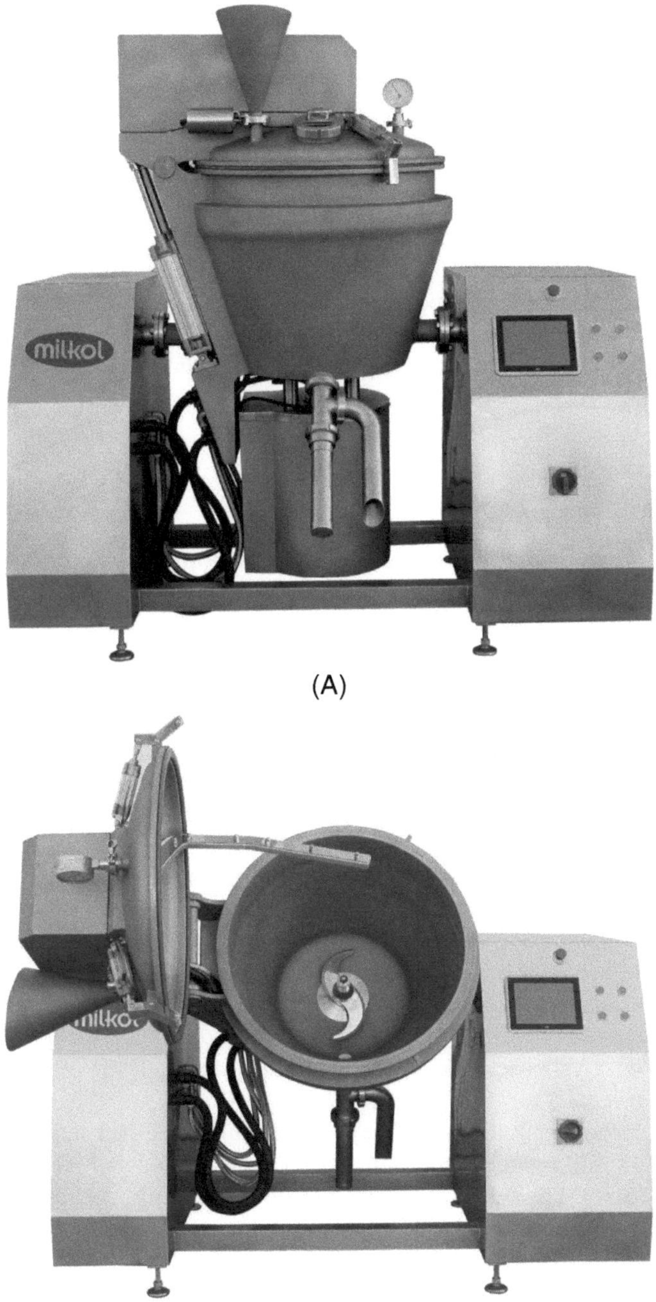

Figure 8.4 Batch cooker. (A) Side view, (B) tilted (Reproduced by courtesy of Intermak Inc., Karatay, Konya, Turkey).

may be dissolved in water before addition to the mix to provide better distribution and reduce processing time (Guinee et al., 2004).

The shear created during mechanical action reduces fat droplet size and creates the desired emulsion from the blend. High shear cookers create small fat droplet size similar to homogenization and they are quite useful in the manufacture of process cheese spreads. In the manufacture of block type processed cheeses, low shear extrusion cookers are preferred to create fibrous texture and large fat droplets yielding a texture similar to that of pasta-filata cheeses (Dixon, 2011).

Emulsifying salts disperse and solubilize casein in cheese by chelating calcium that allows emulsification of fat/oil by them. They increase and stabilize the pH of the cheese from 5.7 to 6.0 by the end of the cooking cycle. Final pH of the mix is generally adjusted by using organic acids and this can be done manually or automatically at the end of cooking.

8.5.2 Continuous cookers

Continuous cookers can be used to manufacture uniform processed cheese products with continuous flow in the process. A continuous cooker with the acid and caustic tanks of CIP system assembled on the same platform is shown in Fig. 8.5. The standardized mix from the blender is pumped into the continuous cooker where heat is applied via steam injection nozzles. Steam injection nozzles distribute the steam into the mix and the temperature of the mix is raised to pasteurization or sterilization temperature in a very short time compared to that in indirect cookers. Various designs of steam injection nozzles also provide the necessary agitation and yield small fat droplet size in the product similar to homogenization.

Direct steam injection continuous cookers are generally preferred in the manufacture of sterilized processed cheeses and high moisture cheese spreads (Fox et al., 2017). The blend is heated to 135–145°C and held for less than 10 s and vacuum-flash cooled to ~90°C. The blend can then be transferred into a buffer tank to collect and agitate the blend to obtain a thick creamy structure ready for hot-filling (Fox et al., 2017). The purpose of sterilization in process cheese manufacture is not to manufacture a sterile cheese product as aseptic filling and packaging are also required for sterilized product. The aim is to inactivate Clostridia spores and extend the shelf life of processed cheese (Dixon, 2011).

Horizontal or vertical tubes can be used for heating by direct steam injection. The cooking process and product conditions are PLC-controlled and recorded in modern continuous cookers. Pressure and temperature in the holding tube are controlled by a back pressure valve and two thermometers at the beginning and the end. The length of holding tube and flow rate of the product are important for applying desired pasteurization or sterilization conditions. Pulsing movement created by the pumps results in only 50% product in the holding tube, which should be taken into account when equipment software is set up for calculating the holding time for the product (Dixon, 2011). Extension of holding time means that more shear is applied to the blend forming a fine emulsion, this translates to a firmer product with lower meltability (Kapoor and Metzger, 2008; Shirashoji et al., 2010).

Figure 8.5 Continuous cooker unit with CIP system (Reproduced by courtesy of Kromel Inc., Arifiye, Sakarya, Turkey).

Extrusion technology has been used for manufacture of process and analogue cheeses. The technology was reported to allow mixing, melting, and emulsifying for production of various forms of processed cheese depending on moisture and lipid contents (Zuber et al., 1987). Adhikari et al. (2009) evaluated effects of two barrel temperatures (80 and 90°C), two moisture levels (44%–48%) and two levels of emulsifying salts (1.0%–1.5%) on textural and sensory properties of processed cheese product with Cheddar cheese by using a corotating and intermeshing extruder. Authors recommended use of extrusion technology for production of processed cheese with lower moisture content at lower temperature which led to higher sensory scores due to firmer/chewier structure and more Cheddar flavor compared to other samples.

Canning (110–125°C/3.2–100 min) was applied in processed cheese manufacture to inactivate microorganisms and spores (Lazárková et al., 2011). However, high temperature caused degradation of amino acids, flavor, and color in the product. Authors recommended application of a regime of 120°C/10 min or 125°C/3.2 min for maintaining quality.

8.6 Cooling and homogenization

Cooling system and rate have direct influence on the physical properties of processed cheese products. In general, slow cooling is preferred for block type processed cheeses while processed cheese spreads are cooled rapidly. Vacuum cooling is preferred in processed cheese manufacture for its effectiveness to reduce temperature rapidly and prevent Maillard browning (Dixon, 2011). Air entrapped in the network is also removed from the matrix to prevent holes in the final product. The feed to the vacuum flash tank is controlled by product diverting valve, no product is directed to vacuum flash tank until the target temperature is achieved. After completion of heating process, the product is directed to vacuum flash tank for cooling and deaeration of the product. The temperature of the pasteurized and sterilized product can be rapidly decreased to ~20° C and ~50° C by vacuum cooling, respectively. Two-stage homogenization (15 and 5 MPa) may be applied to the processed cheese mass after vacuum cooling to create better fat dispersion, smoother texture and initiate creaming especially in processed cheese spreads (Guinee et al., 2004). Homogenization is not common in processed cheese manufacture as homogenizers require capital investment, downstream homogenization has contamination risk and high shear cooking systems provide sufficient reduction in fat droplet size (Fox et al., 2017). The viscosity of vacuum-cooled sterilized product is generally adjusted in creaming tank by holding, agitating or blending the product with re-work. Creaming is an important step in processed cheese manufacture as correct consistency is critical for hot filling and development of final texture of the product. Rework is prepared by mixing previous products with some product from vacuum flash cooling tank. Prior to use of rework, high shear cookers are used and rework is heated to 85° C before addition to the creaming tank. The proportion of fresh product and the re-work in the creaming tank must be controlled to ensure standard production (Dixon, 2011; Fox et al., 2017; Guinee et al., 2004).

Creaming of processed cheese is prevented by quick cooling in some cases. Cooling of block processed cheeses is slower which increases risk of browning, excessive hardness and growth of spore-forming microorganisms (Gouda and Abou El-Nour, 2003). Slow cooling causes a firm structure with low meltability which is associated with protein-protein interactions and fat crystallization former being primarily responsible (Zhong et al., 2004).

Before filling or packaging, the molten processed cheese is filtered to remove contaminants (e.g. piece of plastic bags from natural cheese or butter) and any unwanted physical deformities such as cheese lumps or calcium lactate crystals. The mash size of the filters lies within 200–800 μm depending on the product viscosity (Dixon, 2011).

8.7 Portioning and packaging

Processed cheese products are produced in many size and shapes including portions (spreadable and sliceable), blocks, sausage-shaped, shreds, cubes, in metal cans, tubes and jars, collapsible tubes and slices. The portioning of processed cheese products

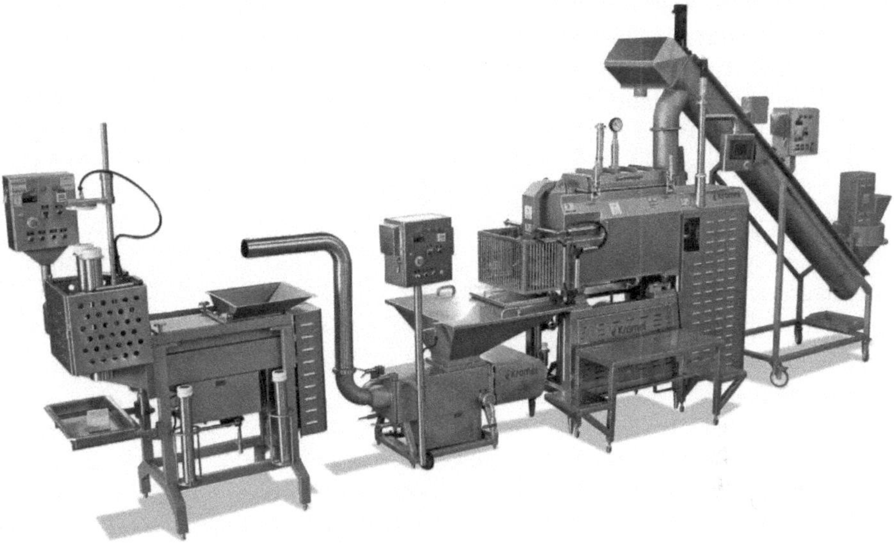

Figure 8.6 Cooker with portioning unit for block type processed cheese (Reproduced by courtesy of Kromel Inc., Arifiye, Sakarya, Turkey).

follows similar processes; hot molten cheese in a stirred hopper is pumped into the filling machine and then transferred into the packaging material at desired volumetric dose. Filling machines can be automatic or semi-automatic. Hot molten cheese is filled into packages in the absence of oxygen to inhibit mold growth. Hermetically sealed metal cans and glass tubes can be sterilized post production (Buys and Mostert, 2011). Portions, blocks or triangle, is packaged in foil, while bulk products are filled into drums or buckets. A process line for block type processed cheese from cooker to portioning unit is shown in Fig. 8.6.

Manufacture of slices requires further processing of hot molten cheese mass. In the past, processed cheese slices were manufactured by slicing the cooled bulk product. Producing slices from processed cheese block had several disadvantages such as post process contamination, inconsistent texture, poor melting and difficulty in weight control (Dixon, 2011). Currently slicing process is conducted by four main steps; creating a sheet of cheese, cooling, cutting and portioning. Chill roll, horizontal stainless steel belts and vertical stainless steel belts can be used to create processed cheese slices (Buys and Mostert, 2011). Hot molten processed cheese mass is distributed evenly on steel rolls to squeeze cheeses into desired thickness. In chill roll technology, one large and two small steel rolls carry out shaping and cooling. In horizontal and vertical steel belt technology processed cheese sheet formation and cooling occur on an infinite motorized belt and redirecting rolls. Cooled (below 15°C) processed cheese sheet is scraped from the roll or belt and directed to slitter blades. Ribbons created by slitter blades are twisted and overlapped onto each other to create stacks. Processed cheese stacks created in all three technologies move towards a motorized conveyor and portioned with a vertical moving wire. The number of stacks can be increased by a sequentially redirecting the conveyor (Buys and Mostert, 2011; Dixon, 2011).

Individually wrapped processed cheese slices are popular for retail sale worldwide. In the manufacture of individually wrapped process cheese, processed cheese mass is filled into a plastic film, formed into the desired thickness, cooled, heat sealed and stacked. In some processes, cooling is performed in water bath at the end of the process. Unlike slice on slice, formulation of individually wrapped processed cheese slices do not require high amount of natural cheese as the plastic film provides the necessary body/support for the molten cheese mass. Thus, high lactose containing ingredients such as whey powder and skimmed milk powder can be incorporated into product formulation (Buys and Mostert, 2011). Processed cheese needs to be stored at nonfreezing temperatures below 20°C to inhibit growth of Clostridia (Gouda and Abou El-Nour, 2003).

8.8 Conclusion

The basic processed cheese manufacturing lines and equipments remained fairly the same in the past twenty years. Primary focus of the modifications in equipment design and changes in manufacturing materials was to decrease energy requirement, reduce contamination risks and improve CIP efficacy. Utilization of computer software in automated processing lines decreased problems such as product overheating and overworking and allowed better control of the thermal process efficiency. Newly developed portioning and packaging equipments provide a laminar and positive air flow in the environment to decrease microbial contamination. Implementation of sensors in processed cheese equipment could allow better process control, especially in blenders and cookers in the future. Today, there are some efforts to control mixing speed and duration by artificial intelligence (by the response logs obtained from the motor) to prevent over- or under- mixing. In addition, sensors that can allow process control based on viscosity of the blend in the cooker could be a solution to over-heating problems. In the manufacture of high quality process cheese, the control systems are as much important as the equipments used. Units in processing line and the control systems must be working in harmony to manufacture process cheese with desired quality and minimize energy losses.

References

Adhikari, K., Cole, A., Grün, I., Heymann, H., Hsieh, F., Huff, H., 2009. Physical and sensory characteristics of processed cheeses manufactured by extrusion technology. J. Sci. Food Agric. 89, 1428–1433.

Buys, E.M., Mostert, J.F., 2011. Packaging materials and equipment. In: Tamime, A.Y. (Ed.), Processed Cheese and Analogues. Wiley-Blackwell Publishing Ltd., West Sussex, UK, pp. 199–218.

Dixon, S., 2011. Processed cheese plants and equipment: a practical overview. In: Tamime, A.Y. (Ed.), Processed Cheese and Analogues. Wiley-Blackwell Publishing Ltd., West Sussex, UK, pp. 179–198.

Fox, P.F., Guinee, T.P., Cogan, T.M., McSweeney, P.L., 2017. Fundamentals of Cheese Science, second ed., Springer, New York, NY, pp. 589–627.

Gouda, A., Abou El-Nour, A., 2003. Processed cheese. In: Caballero, B., Finglas, P., Toldra, F. (Eds.), Encyclopedia of Food Sciences and Nutrition. Academic Press, San Diego, CA, pp. 1108–1115.

Guinee, T., 2011. Pasteurized Processed Cheese Products. In: Faquay, J.W. (Ed.), Encyclopedia of Dairy Sciences. Academic Press, San Diego, CA, pp. 805–813.

Guinee, T.P., Carić, M., Kalab, M., 2004. Pasteurized processed cheese and substitute/imitation cheese products. In: Fox, P.F., McSweeney, P.L.H., Cogan, T.M., Guinee, T.P. (Eds.), Cheese: Chemistry, Physics and Microbiology, 2. Academic Press, San Diego, CA, pp. 349–394.

Kapoor, R., Metzger, L.E., 2008. Process cheese: scientific and technological aspects-a review. Comp. Rev. Food Sci. Food Safe. 7 (2), 194–214.

Lazárková, Z., Buňka, F., Buňková, L., Holáň, F., Kráčmar, S., Hrabě, J., 2011. The effect of different heat sterilization regimes on the quality of canned processed cheese. J. Food Proc. Eng. 34, 1860–1878.

Meyer, A., 1973. Processed Cheese Manufacture. Food Trade Press Ltd., London, UK.

Shirashoji, N., Jaeggi, J.J., Lucey, J.A, 2010. Effect of sodium hexametaphosphate concentration and cooking time on the physicochemical properties of pasteurized process cheese. J. Dairy Sci. 93 (7), 2827–2837.

Zhong, Q., Daubert, C.R., Farkas, B.E, 2004. Cooling effects on processed cheese functionality. J. Food Proc. Eng. 27, 392–412.

Zuber, F., Mégard, D., Cheftel, J.C., 1987. Continuous emulsification and gelation of dairy ingredients by HTST extrusion cooking: production of processed cheese. Int. J. Food Sci. Technol. 22, 607–626.

Technological aspects of processed cheese: Properties and structure

9

Michaela Černíková[a], *František Buňka*[a,b] *and Richardos N. Salek*[a]
[a] Department of Food Technology, Faculty of Technology, Tomas Bata University in Zlín, Zlín, Czech Republic, [b] Food Research Laboratory, Department of Logistics, Faculty of Military Leadership, University of Defense, Brno, Czech Republic

9.1 Introduction

The processed cheese matrix is created by a continuous protein network in which the fat and the serum phase together with other optional ingredients are dispersed (Udyarajan et al., 2007).

Exists a huge range of processed cheeses with different consistency (block type, spreadable, sauce), flavor, packaging type, and possibility of application. The choice of the end product parameters and the derived raw material composition is the first step in determining the final consistency of the processed cheese. Each consumer prefers different processed cheese and therefore it is important to know the parameters that can affect processed cheese consistency.

Moreover, exist four basic parameters affecting the properties and the structure of processed cheeses. The first of them could be the raw material composition, which is described in detail in Chapters 2–7. In this chapter will be described only the influence of selected raw materials on the processed cheese consistency. The second group of factors is the so-called target parameters. In particular, target parameters are based on the raw materials and they include dry matter content, fat in dry matter content, protein content, calcium content, etc. The third group of factors with influence on the hardness or softness of processed cheeses is the processing parameters applied during processed cheese production. Particularly, the agitation rate, heating rate, melting temperature, holding time at the melting temperature and cooling rate belong to this third group. Processed cheese storage (for more details see Chapter 10) is also one of the most important parameters influencing the consistency and the structure of processed cheeses. Hence, this could be reported as the fourth group of parameters which affected processed cheese consistency.

Furthermore, consistency as a crucial property of processed cheese can be evaluated by rheological, textural, or other methods. Inner structure can be observed by different microscopy techniques. Some of the terms used in the chapter are explained in following text.

Storage (G′) and loss (G″) moduli are determined as a function of frequency during the measurement of rheological parameters, which can provide us with information

Processed Cheese Science and Technology: Ingredients, Manufacture, Functionality, Quality, and Regulations.
DOI: https://doi.org/10.1016/B978-0-12-821445-9.00002-9

about consistency. The complex modulus (G*) can be calculated using Eq. 9.1.

$$G^* = \sqrt{(G')^2 + (G'')^2} \tag{9.1}$$

The changes in the viscoelastic properties of the processed cheeses can be also explained using the Winter's critical gel theory. According to the following relation (9.2) the complex modulus can be expressed:

$$G^*(\omega) = A_F \cdot \omega^{\frac{1}{z}} \tag{9.2}$$

where A_F is the strength of the gel ($Pa \cdot s^{1/z}$) and z (dimensionless) is the interaction factor defined as the number of structure units interacting with one another in a three-dimensional network (Gabriele et al., 2001; Sádlíková et al., 2010; Winter and Chambon, 1986). The increasing values of G′, G″, G*, and A_F correlate to the rising firmness of processed cheeses (generally semi-solid materials, including various food matrices). Moreover, the rising values of interaction factor indicate more intensive internal bonds; thus, more intensive interactions can occur in the tested sample matrix (Gabriele et al., 2001; Sádlíková et al., 2010). Additionally, texture profile analysis usually includes parameters such as hardness, adhesiveness and cohesiveness. Hardness can be defined as the force needed to achieve a given deformation (maximum force during the penetration event (N) with defined depth, probe speed, trigger force) (Piska and Štětina, 2004; Jooyandeh, 2009; Genovese et al. 2010; Buňka et al., 2012).

Microstructure is usually used as a supported technique to other instrumental methods (rheology, texture). The microstructure of processed cheeses could provide information about the detailed structure especially about the size and distribution of fat globules, the structure of the protein matrix and other ingredients, which could be observed. Each of the microscopy techniques has some specific utilization and their application depends on laboratory equipment and also on the aim of the study. Optical microscopy was used by Da Silva et al. (2016), Černíková et al. (2008), Hladká et al. (2014), scanning electron microscopy (SEM) was applied by Kaláb and Modler (1985), Noronha et al. (2008b), Cunha et al. (2010), Černíková et al. (2017a). Lee et al. (2003), Zhang et al. (2011), and Hoffmann and Schrader (2015) used in their works transmission electron microscopy (TEM). Other authors have utilized other microscopy techniques including confocal laser scanning microscopy (Hosseini-Parvar et al., 2015; Lee et al., 2015; Talbot-Walsh et al., 2018).

9.2 The effect of raw materials composition on the processed cheeses consistency

There are many raw ingredients to use for the manufacture of processed cheeses, which are described in Chapters 2–7. Moreover, an important parameter is not only the amount of individual ingredients, but also, for example the type and the maturity degree of natural cheese, type of fat, the concentration and composition of the emulsifying salts (see

Chapter 6), the rework content, the concentration, and the structure of stabilizers. Many of these ingredients can affect the resulting consistency and structure of processed cheese, as they can affect the dry matter, proteins, fat content as well as the pH value.

9.2.1 The effect of natural cheeses and dairy fat

Many types of nature cheese could be used in processed cheese manufacture (see Chapter 3). Different countries use various types of natural chesses, for example, Dutch-type, Swiss-type, Cheddar-type, Mozzarella-type chesses are applied in Europe. White-brined-type cheeses are frequently used in the Near and the Middle East. Furthermore, for processed cheese structure and consistency is not important only the type of natural cheese, but also the "age" of the natural cheese (so-called degree of ripening) (Guinee et al., 2004; Carić et al., 1985; Kapoor et al., 2007; Kapoor and Metzger, 2008; Salek et al., 2015, 2016, 2017, 2020). Many biochemical and microbiological changes are taking place during the natural cheeses ripening. These biochemical and microbiological changes are responsible for the organoleptic and physico-chemical properties of natural cheeses, which are also crucial for processed cheeses properties. The studies of Brickley et al. (2007, 2008), Buňka et al. (2013, 2014), Salek et al. (2016, 2017, 2020) and many others confirmed that, when the ripeness of the applied natural cheeses increases, the firmness of the resultant processed cheese decreases (the most obvious changes were observed during the first two up to three months). In addition, more mature natural cheese with more hydrolyzed protein chains entail better meltability and aroma fullness of the processed cheeses. However, there is a risk of a sharp to pungent taste and above all, instability of the final product, because short protein chains are not able to bind water sufficiently. In addition, there is a risk of fat release, because short protein chains do not possess the appropriate emulsifying ability (Berger et al., 2002; Guinee et al., 2004; Piska and Štětina, 2004). In practice, this problem is usually solved by processing raw material with a higher proportion of the so-called intact (nonhydrolyzed or very slightly hydrolyzed) casein (for example, curd or "younger" cheeses). Young natural cheeses are cheaper than middle or high mature cheeses. These young raw materials can cause poorer meltability in processed cheese, but the resulting gel-network is stable without a tendency to release water. Consistency of processed cheeses manufactured from young raw material is from stiff to very firm (solid). Therefore, a larger proportion of younger natural cheese is used to produce block-type processed cheese or processed cheese intended for slicing (Guinee et al., 2004; Brickley et al., 2007; Lu et al., 2007; Salek et al., 2016). For spreadable processed cheeses could be used more mature natural cheeses. In practice is necessary to utilize also young natural chesses. They are generally in the form of less mature cheeses, cuttings made in blocks or slices of natural cheeses. A certain amount of ripened raw material is usually added to prevent defects associated with the lack of sensory active substances and fullness of taste in order to create the desired consistency (Berger et al., 2002).

Producers use in the processed cheese manufacture different sources of milk fat, containing various amounts of surfactants, such as phospholipids, lipoproteins, glycoproteins, etc. In the case of butter and anhydrous milk fat, it is possible (due to the technology of their production) to except extremely small amounts of surfactants.

Conversely, in the case of cream and natural cheeses, the fat is in the form of fat globules with lipoproteins membranes containing surfactants. These surface-active substances can cooperate in the fat re-emulsification over processed cheese production and their concentration can affect the consistency of processed cheese. The influence of dairy fat source with various surface-active compounds on the viscoelastic properties of processed cheeses was studied by Černíková et al. (2018b). Moreover, the various fat content of natural cheese (ingredient containing fat globules coated by surface-active compounds) and butter (fat without surface-active compounds) were used in different ratios to reach identical fat content in the final processed cheese (50% [w/w]). Significantly effect of the various fat content originated from natural cheese (whole fat content was balanced by butter without fat globule membranes) was not found. This is particularly an important fact for processed cheese producers. They can utilize natural cheeses with different fat in dry matter content without concern about final consistency of their product. The basic "rule" is to keep percentage of the amount of emulsifying agent to the protein unit constant.

9.2.2 The effect of emulsifying salts and pH value

The total amount and ratio of individual emulsifying salts is crucial for processed cheeses consistency. However, the functionality of emulsifying salts is affected by the processing parameters, in particular by the melting temperature, the holding time and the cooling rate. Emulsifying salts are used in the processed cheese production to achieve a homogenous mass. They play many roles in the manufacture of processed cheeses. The most important properties include the ion exchange ability, pH value adjustment, buffering capacity (stabilization of pH against environmental influences), and also have an effect on the creaming (for more details see Chapter 6) process. The choice of emulsifying salts also affects the so-called remeltability (the ability of processed cheese to melt when is reheated).

The influence of emulsifying salts on the processed cheese consistency is given due to the ability of emulsifying salts to modify the pH value of processed cheeses. Caseins have isoelectric point at pH $\approx$ 4.6. The balanced positive and negative charges on the casein molecules cause maximum approach of chain and subsequent interaction and aggregation of the casein fractions. Furthermore, proteins are not able to adequately emulsify the fat, and therefore the fat component may separate (Gupta et al., 1984). Spreadable processed cheeses have a pH optimum in the range of 5.6–6.0, processed cheese blocks are characterized by pH in the range of 5.0–5.5 (Berger et al., 2002) and processed cheese sauces have pH in the interval of 6.5–7.0. This means that a lower pH of processed cheese than the optimum (specific for each type of product) might result in a solid or even crumbly consistency of processed cheeses (Marchesseau et al., 1997; Lee and Klostermeyer, 2001; Muslow et al., 2007; Barth et al., 2017), which exhibited a harder and more brittle appearance than processed cheeses with higher pH (6.0–6.8). Processed cheeses with higher pH showed a more liquid texture, resembling a very viscous liquid (Barth et al., 2017). Higher pH than the optimal can cause not only a slushy consistency, but there is also a risk of developing undesirable microflora. Therefore, it is necessary to carefully consider the choice of emulsifying salts for each type of product and choose specific combination in order to achieve the optimal pH, with consideration of final product consistency and with regards to the raw material composition. If the raw materials have low pH (e.g., quark cheese), it is absolutely essential to choose emulsifying salts which will increase the pH of the system and vice versa. Moreover, if more mature natural cheeses are used for the production, it is advisable to apply emulsifying salts adjusting the pH to a more acidic range. Emulsifying salts, with are able to modify pH value and to achieve the optimal pH value for the desired consistency, are named correction salts. Furthermore, it is important to take in advantage the hydrolysis of polyphosphate emulsifying salts. Barth et al. (2017) reported that the pH influences the hydrolysis of sodium polyphosphate in dairy matrices and the structure of processed cheese. In addition, the same authors stated that hydrolysis of sodium polyphosphates increased with the decreasing pH value. In cheeses with an initial pH of 5.2 and 5.6, the final pH remained practically constant, because the higher rate of hydrolysis at this pH increased the percentage of monophosphates and potential buffering agents, leading to constant pH values. Decreasing pH value can cause the rising in the values of hardness, adhesiveness, and gumminess of the processed cheese. On the other hand, decreased electrostatic repulsion between caseins and increasing the protein-protein interactions can result in a more compact structure with large agglomeration. Higher pH (above 6.0) was provided a concentrate protein emulsion with long texture and less adhesiveness. The negative charges of the caseins allow the formation of hydrogen bonds and electrostatic interactions, which enhances the absorption and binding of water by the protein matrix. The long polyphosphate chains can add multiple negative charges to the caseins and thus decreased hardness of the matrix can be expected.

However, emulsifying salts do not influence the consistency only due to pH value. A lot of studies established specific ratios of emulsifying salts. Each emulsifying salt has specific properties and in unique ratios can lead to different consistency of the manufactured processed cheeses (Mizuno and Lucey, 2005, 2007; Shirashoji et al., 2006, 2010; Sádlíková et al., 2010; El-Bakry et al., 2011; Weiserová et al., 2011; Buňka et al., 2012, 2013; Nagyová et al., 2014; Salek et al. 2015, 2016, 2017, 2020). In these works, the authors described the use of individual emulsifying salts, and/or binary and ternary mixtures and their effect on processed cheese consistency (for more details see Chapter 6). In addition, the total concentration of emulsifying salts has an influence on the processed cheese consistency. Buňka et al. (2014) stated that the consistency of processed cheese increased with the rising concentration of emulsifying salts mixture, when applied in the range of 2.0%–3.0% (w/w). The same conclusion was also presented by Mozuraityte et al. (2019), who stated that aless amount of added of emulsifying salts can result in processed cheeses with lower viscosity values, due to less calcium bounds and less casein solubilization. Mozuraityte et al. (2019) studied not only the effect of the total amount of emulsifying salts, but also the effect of sodium reduction in processed cheese spreads and they found out, that when 15% of sodium in phosphate salt was replaced by the same molar amount of potassium in phosphate form or citrate salts, the viscosity of the model processed cheese was similar. A slightly lower viscosity of the model processed cheese was detected, when 30% of sodium in the phosphate salt was replaced by potassium in phosphate form or citrate salts. The reason of change in viscosity could be subtly higher pH value. Higher pH was also obtained by El-Bakry et al. (2011) when sodium was replaced by potassium equivalent salts. Based on their opinion, higher solubility of potassium phosphate (than sodium phosphate) could be the reason of the slightly higher pH of the final processed cheese products.

9.2.3 The effect of rework

Rework (also called precooked cheese) is the processed cheese that has already been manufactured and the creaming (see Chapter 6) process occurred. Producers use rework as a raw material for processed cheeses production. There are a few types of rework. Rework is created in the industry either intentionally (expected) or unintentionally. Hence, intentionally means production of processed cheese for rework or residue of processed cheese in the production equipment. Unintentionally involves processed cheese originally intended for the market network but finally this product was not released for sale, for example due to incorrect dry matter content and/or false fat in dry matter content or because of unsuitable packaging, or excessive stickiness on the packaging material. Rework is usually applied as fresh (processed cheese residue in the production equipment), 3 or 14 days after manufacture (it could be intentionally or unintentionally produced as processed cheese) in an amount up to 20% (w/w), depending on the type of rework and also target parameters of the final processed cheese (Meyer, 1973; Kaláb et al., 1987; Guinee et al., 2004). The reasons for rework use are on the one hand economic and on the other hand due to the increase of viscosity (with the rising age of the rework and with its increasing concentration), increase of firmness, meltability improvement, and possibility of emulsifying salts reduction (because rework already contents emulsifying salts) (Lauck, 1972; Meyer, 1973; Kaláb et al., 1987; Pluta et al., 2000; Kapoor and Metzger, 2008; Černíková et al., 2018a). Kaláb et al. (1987) described in their work three types of rework: fresh rework (produced and rapidly frozen immediately after production), regular rework from previous production batches and the so-called hot melt, simulating processed cheese subjected to extreme stress conditions (cooled from 82°C to 4°C in 5 h). Moreover, the last rework type was referred by Meyer (1973) as over creamed. Each rework type can be used for various reasons and in different quantities. The first type of rework (fresh) can be applied in an amount of 1%–2% (w/w) to increase the creaminess of processed cheese spreads, which contain much ripened natural cheese or its high proportion with intact casein content less than 65%. The second type of rework is used for processed cheese blocks or block-type processed cheese in concentrations of 2%–30% (w/w) to increase product firmness and elasticity. The third type of rework (over creamed) is utilized only in a quantity less than 1% because it has an extraordinarily strong creaming effect and could easily lead to over creaming of the whole production. On the other hand, Lauck (1972) recommended that the rework addition should be only in a quantity of 2%–15% (w/w). Guinee et al. (2004) explained the application of individual rework types. In particular, additional heating of rework can cause a higher degree of temperature-induced dehydration and aggregation of the para-casein (particularly in third type of rework) and thus increase the elasticity of the final product. Emulsifying salts in rework can create a more effective dispersion which leads to faster hydration, higher

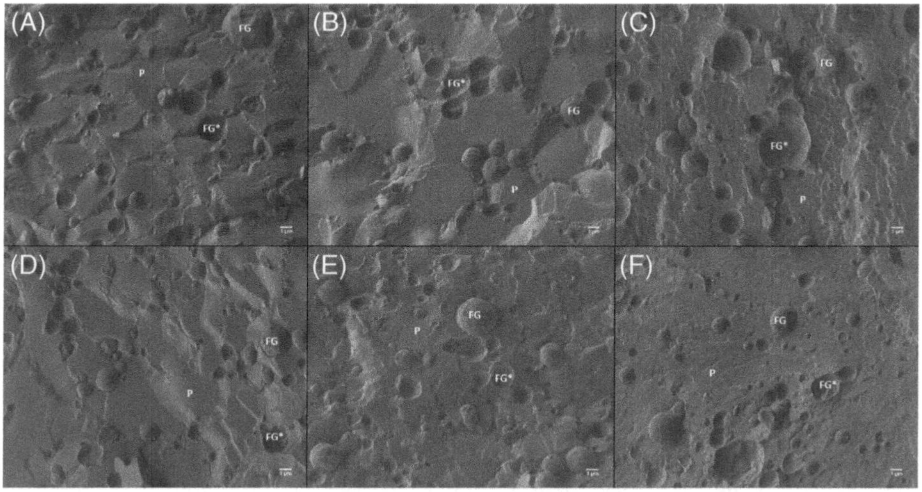

Fig. 9.1 Microstructure of the model processed cheese with 36% (w/w) dry matter content and 45% (w/w) fat in dry matter with different rework content after 60 days of storage at 6 ± 2°C. Part A – containing 0.0% w/w rework, part B – containing 2.5% w/w rework, part C – containing 5.0% w/w rework, part D – containing 10.0% w/w rework, part E – containing 15.0% w/w rework, part F – containing 20.0% w/w rework. Magnification 5000x; bars in the figure parts represent the scales 1 μm; FG – fat globules, FG* – places after remove fat globules, P – protein matrix.

degree of emulsification (effective fat dispersion) and finally to viscosity increase. The effect of different rework amounts on the viscoelastic properties and microstructure of spreadable processed cheeses was studied by Černíková et al. (2018a). Moreover, the above-mentioned authors used 0%–20% (w/w) of rework under real industrial conditions. The rework was prepared from the same raw material as the examined processed cheeses and was added to the raw material 3 days after their manufacture and stored at 6 ± 2°C. The consistency of processed cheeses without rework addition was soft (the lowest value of elastic and loss moduli and the biggest size of fat droplets) in comparison to samples with the addition of rework. The processed cheese firmness increased with the quantity of rework 2.5, 5.0 and 10.0% (w/w) and simultaneously, the size of the fat droplets gradually decreased (Fig. 9.1 part B, C, D), which means that increasing emulsification properties with further rework addition occurred. Higher concentration of rework from 10.0% to 20.0% (w/w) had not a significant effect on the processed cheese consistency, evaluated by the viscoelastic properties (complex modulus; G*) and also the median of the fat droplets size was smaller than in case of lower level of rework addition (Fig. 9.1). The addition of rework affected the intensity of the emulsification of the present fat, which was demonstrated by the reduction of the fat droplets size. In addition, as the degree of fat emulsification grew, processed cheese firmness increased, which corresponds to the findings presented in the work of Kapoor and Metzger (2008), Lee et al. (2015), and Černíková et al. (2017a). However, these results are not in agreement to that reported by Kaláb et al. (1987), who used block-type processed cheese with higher dry matter content. In particular, the above-mentioned authors observed changes in the structure and consistency also in products with higher than 10.0% (w/w) of rework content.

Fu et al. (2018) used four types of rework (precooked cheese) in an amount of 1.5% (w/w) with different agitation speeds (400 or 1500 rpm [rounds per minute]) and stirring times (10 or 30 minutes), PCLS – precooked cheese of low stirring speed and short stirring time (400 rpm, 10 min), PCLL – precooked cheese of low stirring speed and long stirring time (400 rpm, 30 min), PCHS – precooked cheese of high stirring speed and short stirring time (1500 rpm, 10 min), PCHL – precooked cheese of high stirring speed and long stirring time (1500 rpm, 30 min), final melting temperature was 90°C and rework was used 24 h after production and was stored at 5°C in comparison to processed cheese without rework. The viscosities of the products with short time mixing were smooth and without any increasing of viscosity (the creaming effect did not occurred). The addition of long time (30 min) rework (PCLL, PCHL) affected the viscosity of the processed cheese, which increased and promoted the creaming effect during cooking. The using PCHL showed higher viscosity increase probably due to advanced agitation speed. This study showed that rework in an amount less than 1.5% (w/w), which was produced with longer mixing time had a capacity to increase the viscosity (to induce the creaming effect) of the processed cheese manufactured only for ordinary 15 min, due to fine-stranded network. Lenze et al. (2019) described that the addition of rework (they used 5% and 10% (w/w)) dramatically accelerated the creation of the new structure. During the creaming process (see Chapter 6) was observed a very short initial phase (to 10 min), when the exponential phase started earlier, and the apparent viscosity of the matrix increased very sharply. Plateau phase was shorter and in higher level of apparent viscosity compared to samples without rework addition. Moreover, rework addition in an amount of 10% (w/w) caused faster structure formation than 5% (w/w) supplement (Lenze et al., 2019).

9.2.4 The effect of hydrocolloids

Hydrocolloids are classified also as biopolymers. They include two categories of molecules – proteins and polysaccharides that are nowadays widely used for their physicochemical characteristics, primarily water binding, increasing of system viscosity (thickening), ability to create gel structures, stabilizing foams and emulsions or dispersions and decreasing water release (control of syneresis), secure temperature stability, control release of flavors etc. Hydrocolloids can be used also to produce cheaper final product. Even though they are generally present in extremely low concentrations, less than 1.0% (w/w), they can have a significant influence on the textural and organoleptic properties of the final food products (Lu et al. 2007; Williams and Phillips, 2009). Hydrocolloids can be used in dairy systems (not only in processed cheeses) separately or in mixtures (Syrbe et al., 1998; Swenson et al., 2000; Spanguolo et al., 2005; Vega et al., 2005; Černíková et al., 2008; Macků et al., 2008; Agoda-Tandjawa et al., 2017; Chatziantoniou et al., 2019).

Proteins in processed cheeses are largely presented by caseins or caseins fragments. Whey proteins, or all milk protein could be components of processed cheeses, when someone add for example liquid or dried whey, or skim milk powder to the raw material. Dairy ingredients are described in Chapter 2. From time to time, it may be encountered in the application of other animal proteins (gelatin) or plant origin proteins (soybean, pea protein) to processed cheeses or their analogues (vegetable ingredients are described in Chapter 4). Furthermore, gelatin in an amount of 2% (w/w) was applied by Swenson et al. (2000) for the production fat-free processed cheeses. Their consistency was firmer than the full-fat reference sample. Native whey protein isolate and polymerized whey protein isolate in amounts of 2%, 45, 6% (w/w) were also investigated as partial fat replacements in processed cheese. However, the final structure of processed cheese was different. The greater addition of native whey protein isolate promoted the hydrophilic cheese surface and less rigid and less dense structure in comparison to samples with polymerized whey protein isolate with more hydrophobic surface (Nastaj et al., 2020) were observed.

The second group of hydrocolloids is polysaccharides. Producers of processed cheeses can used polysaccharides from plants (native or its subsequently modification starch and pectin), from seeds (guar or locust been gum), tree exudates (gum arabic, gum karaya, gum tragacanth), or polysaccharides from algae (agar, carrageenan, alginate). The last group of polysaccharides, which are used in processed cheeses originate from microorganisms (xanthan gum) (Williams and Phillips, 2009). Hydrocolloids have to comply with food law, which content different requirements in EU (e.g., Regulation (EC) No 1333/2008 of the European Parliament and of the Council on food additives, as amended) and in USA (requirements Code of Federal regulation). In both cases, all of the ingredients must be safe for consumer's health and used in permitted amounts (Europa, 2008; USA, 2020). Locust bean gum, guar gum, xanthan gum, gum karaya, tragacanth, carrageenan, carboxy-methylcellulose, sodium alginate, propylene-glyco-alginate, gelatin are allowed as additives to spreadable processed cheeses in concentration up to 0.8% (w/w) due to specific interaction with caseins proteins (USA, 2020). Processed cheeses containing xanthan gum, guar gum, locust bean gum and mixture of xanthan and guar gums showed less dense and solid structure, while their spreadability and smoothness increased. The largest fat globules and the most spreadability were established in the samples with xanthan gum addition. On the other hand, the smallest fat globules were observed in the sample with a mixture of locust bean gum with κ-carrageenan, leading to samples with low spreadability and higher values of firmness. Similar results were reported in the samples containing a mixture of locust bean and xanthan gums, respectively. In both cases were noted synergistic effects, resulting in increased firmness and decreased spreadability. In case of partial substitution of κ-carrageenan with guar or xanthan gums were produced samples with decreased firmness and increased smoothness and spreadability (Chatziantoniou et al., 2019). Furthermore, carrageenan provides a consistency suitable for slicing or grating of processed cheese. Exist many studies focused on the application (0.1–1.0% [w/w]) of carrageenan into the processed cheeses. According to these authors the increasing carrageenan concentration up to 1.0% (w/w) an increase in the firmness of processed cheese was observed (Syrbe et al., 1998; Ribeiro et al., 2004; Gustaw and Mleko, 2007;

Černíková et al., 2008). Swenson et al. (2000) used higher carrageenan concentration (2.0% [w/w]) for fat-free processed cheeses. The increase of firmness was similar, as in case of other applied hydrocolloids (locust-bean gum, guar gum). In addition, approximately 0.2% (w/w) carrageenan causes a reduction in stickiness to the packaging material (Gustaw and Mleko, 2007). Černíková et al. (2008) stated that a concentration of 0.05% (w/w) was too low for the effective increase of processed cheeses rigidity. However, higher concentrations (0.15% and 0.25% [w/w]) caused significant increase of firmness of the model processed cheeses in comparison to the control processed cheese without carrageenan addition. This fact was also supported by Bourriot et al. (1999). There exists a minimum concentration of carrageenan for effective formation of a carrageenan network. Hence, this amount is depending on many factors, such as strength of the protein matrix gel, degree of protein hydrolysis, pH, ionic environment, etc. On the other hand, Syrbe et al. (1998) published information about possible destabilization and disintegration of the system with further increase of the carrageenan concentration. Carrageenan in an amount ion the range of 0.5%–3.0% (w/w) can be also used to reduce the natural cheese content during the processed cheese manufacture with excellent mouthfeel and good grating and slicing properties (Williams and Phillips, 2009).

Other group of polysaccharides is used in low-fat products. As a stabilizer for low fat processed cheeses can be applied also hydrocolloid konjac glucomannan (gum), locust bean gum, guar gum, etc. (Swenson et al., 2000; Da Silva et al., 2016). Da Silva et al. (2016) used 0.5% (w/w) konjac glucomannan or konjac flavor in processed cheese samples with fat reduction (25%, 50%, 75%, 100%). The highest value of hardness account samples with 50% fat reduction. The effect of fat replacement by inulin on the physico-chemical properties of processed cheese were studied by Sołowiej et al. (2015), who observed decreased hardness and adhesiveness of low-fat processed cheese in comparison to control full fat processed cheeses. On the other hand, Giri et al. (2017) described decreasing spreadability with the rising concentration of inulin (in amount 0%–8% [w/w]).

Other hydrocolloid applied into the processed cheeses could be pectin. A gel created by low esterified pectin is dependent on many factors such as, solubility and concentration of pectin, pH of system, content and availability bivalent cations (especially Ca^{2+} ions) and is realize noncovalent ion interaction between residues of galacturonic-acid in main chain and Ca^{2+} ions. The affinity of pectin chain to calcium grows with the decreasing degree of esterification and increasing concentration of polymer (Smewing, 1999; Garnier et al., 1994; May, 2000; Löfgren and Hermansson, 2007). Addition of carbohydrates facilitates gel formation although the concentration of Ca^{2+} ions is low (Grosso and Rao, 1998; Fu and Rao., 1999). The gel strength increases with the rising calcium content (El-Nawawi and Heikal, 1995; Lootens et al., 2003). Moreover, the growing pectin content (0.0%–0.8% [w/w]) resulted in increased firmness of processed cheese, but this dependence (processed cheese rigidity on pectin concentration) is not linear. This range of pectin concentration did not affect the appearance and the flavor of the model processed cheeses (Macků et al., 2008). The grow of gel rigidity due to increasing pectin amount was also observed by El-Nawawi and Heikal (1995).

There exist many types of starch which can be used as ingredients to the processed cheeses for their effect on the consistency. They are usually used for improvement the texture, respectively influencing the development of the three-dimensional network in processed cheese matrix. Some producers take advantage of starch to replace part of the natural cheese used and to manufacture processed cheese analogues. Mounsey and O´Riordan (2001) described lower meltability of the final processed cheese after application of various types of starch with low amount of amylose (wheat, potato, rice starch) in a total concentration of 3% (w/w). They explained the meltability decrease by partial protein dehydration and more intensive protein interaction to each other. According Mounsey and O´Riordan (2001) and Bennett et al. (2006) is important not only the botanical source, but also the starch structure. A starch with low content of amylose (rise approx. 13% (w/w), maize approx. 3,8% (w/w)) cause the decreasing of hardness, when is used in an amount of 3% (w/w). On the other hand, when this type of starch will be applied in higher quantity, more than 5% (w/w), increased rigidity of processed cheese could be expected, analogously as it will be utilized starch with higher amylose concentration (waxy maize, wheat, potato). Trivedi et al. (2008a) reported increasing values of complex modulus and viscosity in dependence to the starch type and level. In addition, the same authors demonstrated the possibility of partial replacement of natural cheese from the raw materials by cheaper ingredients like starch. Commercial processed cheese with partial protein were replaced by starch assign in high starch concentration a pasty texture with tendency to stick to the wrapper (Trivedi et al., 2008b). The effect of tapioca and potato starch on the mechanical properties and microstructure of processed cheese with different types (random or fine structure) of casein network were studied by Fu and Nakamura (2018). The tapioca starch addition to the processed cheese with random type of casein structure led to the dispersion of the casein network structure, so the firmness increased. The addition of potato starch become to create fine-stranded structure (from initial random structure) the casein network in processed cheese samples and the most firmness final product was generated. When the authors used as an original casein network fine structure, the influence of both starches type on the processed cheese firmness was not proved.

Hosseini-Parvar et al. (2015) focused their experiment on the effect of basil seed gum on textural, rheological, and microstructural properties of processed cheese. Basil seed gum is a novel surface-active hydrocolloid and showed promising stabilizing and emulsifying properties and can be used as a functional ingredient in the food industry. Increasing level of basil seed gum led to more elastic behavior in the processed cheese structure (value of elastic and loss moduli increased), the ratio protein/solid content was retained. So, basil seed gum contributed to the fat emulsification.

9.2.5 The effect of low-molecular saccharides

Lactose and sucrose are among substances commonly used in the processed cheese production, for example in the form of whey, or other ingredients during the manufacture flavored processed cheese (e.g., chocolate, when producers usually use common sugar). Macků et al. (2009) described the addition of glucose, galactose, lactose, and sucrose into processed cheeses also containing pectin. In conclusion, the latter authors stated that the addition of 1.0% (w/w) of saccharide caused a reduction in the gel stiffness, regardless of its reducing or nonreducing properties, structure (aldose versus ketose; monosaccharide versus disaccharide), or steric arrangement (epimers). Moreover, with the increasing lactose content, the firmness of the processed cheese decreased (Macků et al.,2009). According to the Code of Federal regulation is in processed cheese food permitted an amount of 7.48% (w/w) lactose content whereas in processed cheese spread, 10.2% (w/w) (USA, 2020). Higher lactose content can cause crystallization of lactose with a negative result on the consistency of the final product (Kapoor and Metzger, 2008).

9.2.6 The effect of emulsifiers

Addition of emulsifiers supports the emulsification of fat. Hence, the fat droplets will become smaller which can cause increasing of firmness and decreasing of spreadability of the manufactured processed cheese. Buňka et al. (2007) reported a negative influence 0.25% (w/w) monoacid-glycerol's with low chain fatty acid on the organoleptic parameters of processed cheese, but no significant effect on the processed cheese consistency. On the other hand, monoacid-glycerol's with C16 and C18 did not have any effect on taste and aroma of the final processed cheese, whereas increasing gel strength of processed cheese was observed. Moreover, Awad et al. (2002) reported rising firmness of processed cheeses with the application of compounds, which increase the emulsification of fat. In these cases, it was about block-type processed cheeses (not spreadable as in case Buňka et al. [2007]). Lee et al. (1996) described that the influence of emulsifiers is not dependent only on their emulsifying ability, but also on other properties, e.g., the character of the emulsifiers (ionogenic or non-ionogenic emulsifiers). Ionogenic emulsifiers can change dominantly the charge of the protein fractions in processed cheese, which can affect the final consistency of the processed cheese.

9.2.7 The effect of bioactive substances

In the recent years, many authors fortified processed cheeses or processed cheese analogues with the so-called biological active (bioactive) substances, which could improve some nutritional benefits of processed cheeses. Bioactive ingredients could be applied in a dried material form, extracts or essential oils of herbs and include polyphenols, or other ingredients with antioxidant capacity. Flavonoids (rutin, quercetin, asparagus powder) were used for example by Přikryl et al. (2018) and Solhi et al. (2020), carrot paste by Mohamed et al. (2016), apricot pulp by Mohamed and Shalaby (2016), tomato juice by Mehanna et al. (2017). Moreover, the alga *Chlorella vulgaris* is a source of antioxidants, proteins, fatty acids, fiber, vitamins, and minerals and it was applied by Tohamy et al. (2018). In addition, Sadek et al. (2017) applied probiotics into processed cheese. It is possible also to apply vitamins, but there is problem with the stability of water-soluble vitamins during heating (similar problem as by other bioactive stuffs). All these ingredients could affect processed cheese consistency in different ways. In particular, tomato juice caused higher phase separation, especially of oil, leading to a less stable product (Mehanna et al., 2017). Přikryl et al. (2018) applied rutin and quercetin in amount 0.5% (w/w), which significantly decreased the gel strength of the processed cheese, when the holding time was one minute at the temperature of

80°C. Higher temperatures (90°C) caused significantly lower values of the complex modulus at all examined times in comparison with samples without rutin or quercetin addition. Moreover, Solhi et al. (2020) fortified model processed cheeses with different levels of asparagus powder, which is a rich source of flavonoids and other phenolic compounds with strong antioxidant and functional properties. Asparagus powder in amounts of 0.5%, 1.0%, 1.5% (w/w) led to a processed cheese structure with a more elastic character, the rigidity increased and the spreadability of the processed cheese decreased.

A special group of ingredients are prebiotics. Belsito et al. (2017) and Ferrão et al. (2018) used galactooligosaccharides, respectively xyloologosaccharides into a unique type of processed cheese, the *requeijão crema*. The addition of galactooligosaccharides in an amount of 1.5%, 3.0%, and 4.0% (w/w) led to lower values of complex modulus and the values of A_F also decreased, whereas the lowest G* was observed in samples with 1.5% (w/w) galactooligosaccharides addition. Processed cheeses with 3.0% (w/w) showed the most compact structure (evaluated by scanning electron microscopy) with reduced size of fat globules (Belsito et al., 2017). Furthermore, the addition of xyloologosaccharides in an amount of 3.3% (w/w) into low fat and low salt content processed cheeses, created denser, with more compact structure end products. Moreover, application of xylooligosaccharides resulted in increasing values of apparent viscosity, elasticity (G′) and firmness (G*), but in compare to low fat and normal salt content was G* lower (Ferrão et al., 2018).

9.2.8 The effect of calcium content

Calcium is one of the most noteworthy mineral elements in processed cheese. Experts discuss calcium deficiency in human population for long time (Kasper and Burghardt, 2015; Vannucci et al., 2017) and also discuss about calcium bioavailability from processed cheeses (Schäffer et al., 1999). However, in this chapter will be described only the influence of calcium content on processed cheese consistency, because calcium is a divalent cation which plays important role in processed cheese production, especially via its reaction with emulsifying salts (Chapter 6). In addition, higher calcium content could result in stiffer processed cheese due to more effective binding between proteins molecules, creating in turn more intensive gel network in processed cheese (Olson, et al., 1958; Marchesseau, et al., 1997). This fact was supported by Lucey and Fox (1993), Acharya and Mistry (2007), Kapoor et al. (2007), and Brickley et al. (2008), who reported that with an increasing amount of calcium a decreased meltability of natural cheese might occur. However, increasing calcium content in order to reach harder product has itself some limits, because excessive calcium addition can cause the development of an unstable system with many proteins aggregates (Mizuno and Lucey, 2007). From another point of view decreasing or elimination of calcium ions from natural cheeses (by the use of ultrafiltration, diafiltration, chelation, dialysis, ion exchange, acidification or other techniques) is one of the possibilities to produce processed cheese with lower emulsifying salts content or even without emulsifying salts addition (Smith and Rivera, 2016, 2017; Chapter 6). Recently, some works appeared delineating with the so called dry mixes, which are based on casein, for the production of processed cheeses. Also, in these studied it was reported that calcium and calcium binding salts affected the final consistency of the manufactured product (McIntyre et al., 2017).

9.2.9 The effect of flavorings and other raw materials

The consistency of processed cheeses can be affected, except of the above-mentioned raw material (such as natural cheese, emulsifying salts, rework, hydrocolloids, saccharides, emulsifiers, bioactive stuffs, calcium content, and also other substances, which belong to different chemical groups), by the possible addition of flavorings. In general, various substances can be used for fat replacement, respectively to obtain products with lower fat in dry matter content, or to increase the percentage of mono-and/or polyunsaturated fatty acids, to achieve a heathier product. These different substances can have various influences on the processed cheese consistency. However, some authors dealt more about the organoleptic and chemical changes than can alter the structure of the final product (Saadi, 2018).

Furthermore, raw materials such as flavoring agents (smoked meat, sweet capsicum, champignons, or mushrooms) could also increase the nutrient value of processed cheeses and can be also labeled as bioactive substances (USA, 2020). They usually disturbed the integrity of processed cheese matrix and have an effect on the processed cheese consistency similar as other ingredients. Their higher concentration can lead to softening and a more spreadable processed cheese might be developed. Rafiq and Ghosh (2017) studied potato incorporation into processed cheeses for carbohydrate enrichment and decreasing the fat content, leading to cost effective production of processed cheese. In particular, the use of 40% (w/w) led to significantly ($P < 0.05$) lower values of hardness and gumminess.

Decrease of hardness was indicated with increasing xylitol addition up to 4% (w/w) as a possible replacement of fat (Kommineni et al., 2012). Moreover, different oleogels in processed cheese production were investigated by Huang et al. (2018). The incorporation of oleogels reduced oil agglomeration in the un-gelled processed cheeses. Hardness, storage modulus, and meltability values of the control samples were comparable to samples with oleogels. In addition, the replacement of milk fat with canola oil was previously described by Ramel and Marangoni (2018). According to the latter authors, it could be possible to replace 49% milk fat by canola oil and 5% oat fiber, retaining similar values of hardness, springiness, and meltability. Moreover, the new product was evaluated as cheaper with improved nutritional value; however, it cannot be called process cheese.

9.3 The effect of target parameters on the processed cheeses consistency

The second big group of factors, which significantly can affect the consistency of processed cheeses are the target parameters, especially the dry matter content, fat in dry matter content, nonfat dry matter, percentage of protein content in dry matter content, and pH (see section 9.2.2). As a source of protein could be applied natural cheeses which are also a source of fat, quark cheese, caseins, and other ingredients described in previous chapters. Similarly, it can be about the fat sources. It can be used milk fat in different forms such as butter, cream, natural cheeses, anhydrous milk fat and then could be used nondairy fat sources, such as different types of oils. About the target parameters are not available many comparable studies performed in the last 20 years. The existing articlesDimitreli and Thomareis (2004, 2007, 2008), Bayarri et al. (2012), Guinee and O'Callaghan (2013), Chatziantoniou et al. (2015), Lee et al. (2015), and Černíková et al. (2017a) are relatively difficult to compare to each other and to search for clear trends in influencing the consistency of processed cheese, because these individual experiments were usually performed under different conditions (in terms of final processed cheese parameters (especially dry matter and fat in dry matter content) and the composition of the raw material mixture, as well as in terms of the processing parameters used in over the production of processed cheese model samples.

The dry matter content and the fat in dry matter content are one of the most important parameters. The limits or ranges of these parameters are usually incorporated to the food law. These are also the basic parameters according to which processed cheeses are often divided in terms of consistency into processed cheese blocks, processed cheeses intended for slicing, spreadable processed cheeses, or processed cheese sauce.

Lee et al. (2015) studied the influence of protein content (10%–20% [w/w]) and fat content ([0%–40% {w/w}) on the viscoelastic properties of model samples of processed cheeses manufactured from sweet (rennet) precipitated casein, melting temperature 85°C, and were stored for 24 h. In particular, with the increasing protein content and decreasing fat content (constant ratio protein: water, changeable dry matter content) growing processed cheese firmness was reported. The influence of protein content was found out to be more significant in comparison to the fat content. Guinee and O´Callaghan (2013) prepared for their work processed cheeses from Cheddar and fat-free cheese at the melting temperature of 80°C. The final processed cheeses had different contents of fat (14%–33% [w/w]) and different contents of protein (25%–12% [w/w]), whereas the dry matter content (46%–47% [w/w]) was kept at a constant level. The cheeses were stored for a maximum of 4 days. Moreover, the firmness decreased with the increasing fat content (reducing ratio protein: fat). Dimitreli and Thomareis, (2004, 2007, 2008) examined the viscosity, the textural and the viscoelastic properties of processed cheeses (manufactured from Gouda at a melting temperature 80°C) with different dry matter content (38%–62% [w/w]), protein content (11%–30% [w/w]), and fat content (12%–23% [w/w]). All cheeses were storied only for 24 h. The firmness of the processed cheeses decreased with the decreasing dry matter content and the protein content and with the increasing fat content. Furthermore, Bayarri et al. (2012) and Chatziantoniou et al. (2015) solved specific processed cheeses made from whey cheese and at the same time examined the influence of the content components on the consistency of commercially available processed cheeses. The above-mentioned authors reported increasing firmness with the decreasing fat content. Černíková et al. (2017a) studied the consistency and microstructure model processed cheese with two divergent dry matter content (35% and 45% [w/w]), two different fats in dry matter content (40% and 50% [w/w]), melting temperature of 86°C, storage time was 14 days. When assessing the effect of one parameter (e.g., dry matter content), the other parameter (in this case the fat content) was constant and vice versa. It was found that with the increasing dry matter content (with constant fat in dry matter content) and the decreasing fat in dry matter content (while maintaining dry matter content) the gel strength increased. At the same time, the interaction factor (z) increased, which means that the number of interacting structural units in the protein network was higher. Furthermore, it has been shown that with increasing fat in the dry matter content (at a constant dry matter content), the diameter of the fat globules increased and at the same time the firmness of the processed cheeses decreased. The protein to fat ratio was ~ 1: 0.85 for model processed cheeses with 40% (w/w) fat in dry matter content and ~1:1.30 for products with 50% (w/w) fat in dry matter content. The relatively lower amount of proteins, especially caseins, which act as "true" emulsifiers in the processed cheese matrix, caused a lower degree of fat emulsification, which was reflected in increasing of the fat globule's diameter. Hence, larger fat globules can disrupt the continuity of the protein matrix much more than smaller fat globules and produce a softer, more spreadable processed cheese. On the other hand, processed cheeses containing a larger number of smaller fat globules could give rise to firmer and less spreadable processed cheese (Kapoor and Metzger, 2008; Lee et al., 2003; Noronha et al., 2008a). Additionally, increasing the dry matter content (with a constant fat content in the dry matter) and thus increasing the relative concentration of proteins, which in turn increase the emulsifying ability of the system, was reflected in a decrease in the diameter of the fat globules and at the same time an increase in the stiffness of the processed cheese. Dalgleish (1997) and Lee et al. (2015) also stated that with the increasing protein content the fat globules presented lower sizes. Increasing of apparent viscosity with rising to protein concentration in the model samples was described also by Lenze et al. (2019), who proved the necessity minimally 15% (w/w) of protein to reach an apparent viscosity over then 1.0 $Pa{\bullet}s^{-1}$ at a time approximately of 200 min.

The effect of fat reduction on the viscoelastic properties was examined by Subramanian et al. (2006), who described a synergistic effect of the fat reduction and moisture increase on the dynamic and transient rheological properties during heating and following cooling. Moreover, these authors reported that the samples with 80% fat reduction (4.76% [w/w] fat, 55.3% [w/w] moisture) showed higher dynamic and transient linear viscoelastic properties in comparison to the regular-fat processed cheeses (23.81% [w/w] fat, 51.1% [w/w] moisture) and that the reduction of fat content influenced the rheological properties more than the increase in the moisture content (Subramanian et al., 2006).

Increasing nitrogenous substances (under the same condition) can lead to growing firmness values and decrease of spreadability. This fact is explained by the rising intensity of the developed protein network, because nitrogenous substances are primarily "network building or supporting compounds" and they need to "meet" other protein molecules to increasing the number of interactions with following protein aggregation and gel (network) formation. Lenze et al. (2019) also assumed that the increasing protein concentration including para-caseinate can better and quickly adsorb on the substances surface at the oil-water interface, thereby decreased interfacial tension and enhanced emulsion stabilization might be expected. On the other hand, under the same conditions, increasing water amount and level of water soluble saccharides, or fat content, or other ingredients containing water can lead to a protein matrix with a damaged three-dimensional network (processed cheese) with a more spreadable character (see 9.2.5 and 9.2.7).

9.4 The effect of processing parameters on the processed cheeses consistency

The large group of processing parameters in processed cheese production involves many factors having effect on the processed cheese consistency. These factors include the agitation rate, heating rate, holding temperature, time at holding temperature and cooling rate. Several research papers have been published in this area for the last 20 years, however, are difficultly comparable to each other, because in these works were prepared/developed different types of final processed cheese products (including blocks, spreads, or imitation of processed cheeses), different processing parameters and divergent raw materials.

9.4.1 The effect of holding time

Sutheerawattananonda et al. (1997) studied the effect of the holding time in the range of 0–15 min at the melting temperature of 65.5°C on the distribution of fat globules in processed cheese blocks and observed that the fat globules size decreased (emulsification improved) in the first five minutes of holding time at the melting temperature. The reduction of fat globules size had a consequence harder processed cheese and was stopped after 5 min of holding time. Further extension of holding time did not have any significant effect on the size of the fat globules. Bowland and Foegeding (1999) studied the effect of processing time on the consistency of processed cheese blocks prepared from rennet casein. The authors chose 10, 20, and 30 min of holding time; 80°C as melting temperature and constant agitation speed (first 3 min 300 rpm [only rennet casein] and, than other ingredients were added and samples were mixed at 2000 rpm). The strength of the processed cheese gel increased with the prolonged melting time. On the other hand, Swenson et al. (2000) studied the effect of holding time (0–20 min) on the final consistency of fat-free processed cheese spread. They used a lower melting temperature, specifically 75°C, than Bowland and Foegeding (1999), the agitation speed was constant but not exactly mentioned. In this case, the opposite result was observed, the extending of the holding time caused a decreasing in the firmness of the processed cheese product. Lee et al. (2003) described viscosity changes of the holt melt. In this work, the viscosity riced for the first 25 min and then, after reaching the local maximum, the viscosity of the hot melt slowly fell to the end of the measurement (50 min). The final consistency of the cooled products was not evaluated. Noronha et al. (2008a) investigated the effect of different agitation speeds in the range of 100–1500 rpm at the point of 80°C with a constant endurance of 2 minutes on the consistency of melted cheese blocks. The final consistency of the processed cheese block was stiffer with increasing agitation speed. Similar results were published also by Přikryl et al. (2018), who stated rising value of complex modulus with prolonged holding time (0, 5, 10 min) at melting temperatures of 80 and 90°C and constant agitation speed 1500 rpm. In the articles mentioned above was studied separately one of the most important processing parameters, the holding time, or the melting temperature (see section 9.4.4), or the influence of special ingredients (see sections 9.2.4 and 9.2.7).

A comprehensive work, which would consider both the effect of the mixing speed and the holding time at the melting temperature was published by Černíková et al. (2017b, 2018c). Differences in these articles were in the usage of processed cheese samples with various fat in dry matter content (40% and 50% [w/w]).

Furthermore, Černíková et al. (2017b) dealt with the combined effect of the agitation speed of the knives of the production equipment (1000, 1500, and 3000 rpm) and the holding time (0–20 min) at a melting temperature of 90°C on the consistency of spreadable processed cheeses with 35% (w/w) dry matter content and 40% (w/w) fat in dry matter during 60 days of storage. The holding time of the melting temperature was divided into shorter intervals than in earlier publications. First, a holding time of 0 minutes was selected, when the melting process was completed after reaching the melting temperature. Furthermore, the melting temperature endurance was extended to 1, 3, 5, 7, 9, 11, 13, 15, 17, and 20 min. After reaching the melting temperature, the melt was constantly stirred at the set speed during the mixing. The monitored residence time intervals were chosen to evaluate the effect of blending time more precisely on the consistency of processed cheeses. It was found that both parameters significantly affected the consistency of processed cheeses. As the melting temperature continued, the firmness of processed cheeses generally increased, similarly like previously reported by Bowland and Foegeding (1999). The exception was the lowest endurance times (one day after manufacturing), when model samples produced with an endurance of 1 min showed lower stiffness compared to samples produced without any endurance, which were not reported by previous authors. A downward trend in stiffness (weakening of the protein matrix) for 1 min was observed for all three mixing speeds used. The decreasing complex modulus, which determinates the hardness of processed cheese, was supported by the detected decrease in the gel strength (A_F) and the interaction factor (z). With a further increase in the holding time of the melting temperature above 3 min, the stiffness of the model samples increased with the increasing holding time for all observed knife speed intensities (1000, 1500, and 3000 rpm). There was an exception in 3000 rpm and 1 day after storage. In this case was noticed any further decrease of the complex modulus value. However, there was no significant difference between processed cheese production at the agitation speeds of 1000 and 1500 rpm (melting temperature 90°C 1 and 3 min). A similar trend in the rising of the strength and coherence of processed cheese was observed by Sutheerawattananonda et al. (1997) and Bowland and Foegeding (1999). However, Černíková et al. (2017b, 2018c) used higher melting temperature than the abovementioned authors; Sutheerawattananonda et al. (1997) applied 65.5°C and Bowland and Foegeding (1999) used 80°C, which is closer to the typical melting temperature applied in the industry. The increase in firmness of the processed cheeses was explained by Sutheerawattananonda et al. (1997) by reducing the size of fat globules, where a larger number of small fat globules disrupted the continuity of the protein matrix less intensively compared to the presence of a smaller number of order of magnitude larger fat globules. Simultaneously with the decrease in the size of fat globules, the authors found that the stiffness of the monitored samples also increases with increasing holding time. However, the above-mentioned authors also stated that the reduction in the size of the fat globules stops after 5 min. Therefore, in this matrix, where the increase in stiffness occurred even with longer melting temperature durations (in addition to the effect of decreasing the size of fat globules), other interactions can be expected (especially more intensive hydration of caseins and increase in the number of their bonds), which contribute to increasing stiffness and are supported by more extensive melt mechanical stress (Bowland and Foegeding, 1999; Lee et al., 2003). The biggest and rapid increase of z-factor was observed at the highest used agitation speed (3000 rpm) from the 10th min and further. Similar trend of initial firmness decreases up to 3rd minutes and following increase of firmness (determined G*), as in the work Černíková et al. (2017b) who research semi-fat processed cheeses, was noted by processed cheese samples with higher fat in dry matter content (full fat processed cheeses) in the study of Černíková et al. (2018c). The results are in detail presented in Figs. 9.2 and 9.3.

9.4.2 The effect of agitation speed

As mentioned in sections above, not only the holding time, but also the agitation speed fundamentally can affect the consistency of the model processed cheeses. However, in the available literature exists two different conclusions. In particular, Noronha et al. (2008a) reported more rigid processed cheese block with an increasing agitation speed. They used a melting temperature of 80°C, agitation speed in the range of 100–1500 rpm (including 750, 1125, and 1500 rpm in the Stehpan equipment and 100 or 200 rpm in the Blentech cooker) and constant holding time of 2 min. On the other hand, Černíková et al. (2017b, 2018c) reported that the softest samples were produced at 1500 rpm, followed by those at 1000 rpm of agitation speed. With the use of the highest agitation speed (3000 rpm) were prepared the most rigid spreadable processed cheese samples (for more details see section 9.4.1). Nevertheless, the differences in the obtained results may be due to the type of the developed processed cheeses (blocks versus spreadable processed cheeses, respectively), divergent melting temperature, lower mixing speeds, and shorter holding time.

9.4.3 The effect of homogenization and 3D print

Homogenization is usually used for processed cheese sauces and not for spreadable or block-type processed cheeses, because most of the equipment (especially high-shear cookers) applied can simulate the homogenization effect. Mohammadi and Fadaei (2018) studied three different dual-stage homogenization ((1) 70, 30 bar, (2) 110, 40 bar, (3) 150, 50 bar)) regimes and their effect on the textural properties of processed cheese with dry matter content reduction (lower dry matter content is expected to lessen

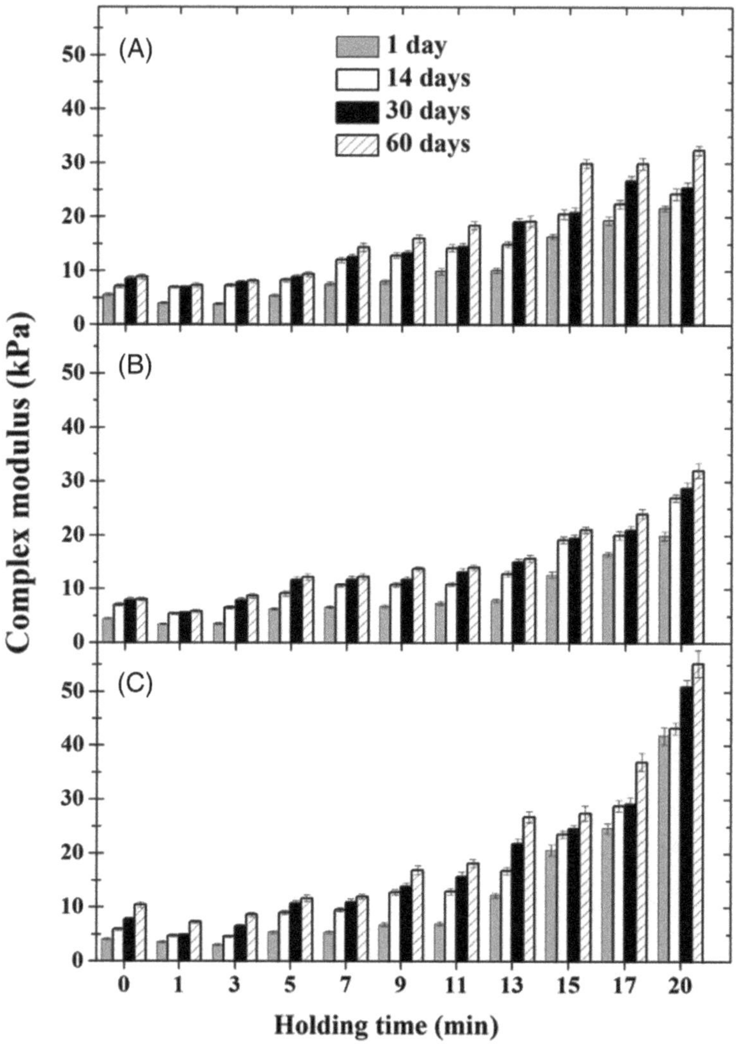

Fig. 9.2 Dependence of the complex modulus (G*, kPa) of the processed cheese with 35% (w/w) dry matter content and 40% (w/w) fat in dry matter content on the holding time (maintaining in Stephan kettle after reaching the melting temperature) at frequency 1 Hz, for 1 month storage at 6 ± 2°C. Part A – using 1000 rpm, part B – using 1500 rpm, part C – using 3000 rpm.

the cost of the final product). From their work it was stated that the homogenization did not affect the acceptability of the consumers, whereas all used homogenization regimes improved the processed cheese texture. Furthermore, higher pressure resulted in more intensive influence and gave the most firm processed cheese. Cream homogenization was applied by Lenze et al. (2019) as a pre-treatment during preparation of model

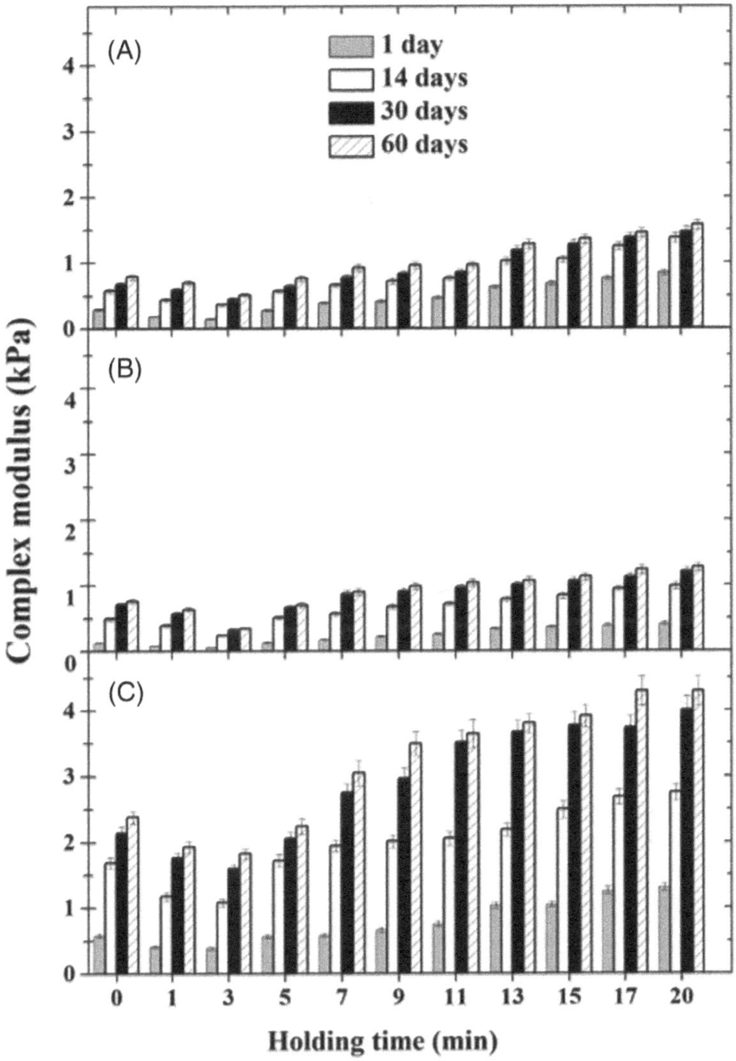

Fig. 9.3 Dependence of the complex modulus (G*, kPa) of the processed cheese with 35% (w/w) dry matter content and 50% (w/w) fat in dry matter content on the holding time (maintaining in Stephan kettle after reaching the melting temperature) at frequency 1 Hz, for 1 month storage at 6 ± 2°C. Part A – using 1000 rpm, part B – using 1500 rpm, part C – using 3000 rpm.

processed cheese samples. The above-mentioned authors used six pressure types ((1) 0 bar, (2) 50 bar single-stage; and dual-stage, (3) 110, 20 bar, (4) 150, 30 bar, (5) 200, 40 bar, (6) 250, 50 bar). Final fat globules sizes were from 1.1 to 3.7 μm. In addition, the higher-pressure treatment intensively affected the kinetics of the creaming curve.

Hence, smaller fat globules led to shorter and higher level of plateau phase. That can be explained due to greater surface of fat globules which can mediate more interactions with the proteins and better water immobilization (see section 9.4). Recently, a new technology, the 3D printing, was applied on processed cheeses (Le Tohic et. al, 2018).

9.4.4 The effect of melting temperature

Processed cheeses can be produced at various melting temperatures in the ranges of 85–110°C (Kapoor and Metzger, 2008) or 72–145°C (Tamine, 2011). Some authors described also lower temperature (65°C; Sutheerawattananonda et al., 1997). Higher temperatures (135–145°C) are used especially during continuous production in comparison to lower melting temperatures which are applied when discontinuous equipment is used. Generally, for processed cheese blocks can be used lower temperatures than for spreadable processed cheeses. But it is also possible to produce block-type processed cheeses using higher melting temperature with the adaptation of other production conditions, such as the type of the processing equipment and agitation speed (Guinee, 2003; Guinee et al., 2004; Tamine, 2011).

The influence of the melting temperature on the consistency of processed cheese was studied by Kaláb et al. (1987) and Dimitreli and Thomareis (2004), who reported that when the melting temperature increases, the viscosity of the melt decreases and the size of the fat globules decreases too, which is also accompanied by an increase in the stiffness of the final product. On the other hand, Swenson et al. (2000) studied the effect of temperature in the range of 60–90°C on the consistency of fat-free processed cheese spreads. Their work showed decreasing in the firmness of the fat-free processed cheeses, with the increasing melting temperature in the range from 60–80°C. The increasing firmness of these fat-free processed cheese spreads was observed at a melting temperature of 90°C. An increase of processed cheese spread firmness produced at 90°C was also observed by Přikryl et al. (2018) in all batches produced (control samples without biologically active substances as well as samples containing rutin and quercetin) compared to samples produced at 80°C. This conclusion was valid for the fat-free processed cheeses and also for full fat processed cheeses with 50% (w/w) fat in dry matter content. Similarly, Mozuraityte et al. (2019) studied different creaming times (1.5, 3.0, 4.5 min) at various creaming temperatures (45, 55, 65°C) and at a final melting temperature of 95°C with a holding of 6 min. The current authors stated that the creaming temperature was not crucial for the final consistency of the processed cheese product. In this case is important to mention that in chapter 6 is the creaming process described as something different.

9.4.5 The effect of cooling rate

The final consistency of processed cheese is also affected by the cooling rate. Piska and Štětina (2004) stated that a high cooling rate leads to less firm, more spreadable, and more sticky processed cheese. In addition, fast cooling of processed cheeses provides more extensively presence of β' polymorph in compared to β polymorph which is usually observed in bulk milk fat (Ramel and Marangoni, 2017). The authors also observed that the presence of protein and/or other ingredients led to higher changes from the β' polymorph to the β polymorph (Ramel and Marangoni, 2017). On the other hand, when the conditions and other parameters (such as dry matter content, fat in dry matter content, melting temperature, holding time, and agitation speed) are the same, slow cooling gives rise to stiffer products. It means that faster cooling is required for processed cheese spreads, while relatively slow cooling rate is important for processed cheese blocks (Muslow et al., 2007). In the case of cheese spreads and especially sauces, slow cooling causes the product to remain at temperatures suitable for the growth of microorganisms for a longer period of time, and this may lead to undesired multiplication of any contaminating microflora (Görner and Valík, 2004)

9.5 The effect of storage on the processed cheese consistency

Processed cheese storage is, together with other processing parameters, such as agitation speed, holding time, melting temperature, also one of the most important factors which can influence the consistency, properties, and structure of processed cheese. The possible sources of processed cheese changes (like appearance, structure, color, and flavor) during storage are, for example, loss of water, hydrolysis of polyphosphates, changes in ionic equilibria, fat polymorphism, formation of crystals, oxidation, nonenzymatic browning, enzyme activity, interactions with packaging materials (for other changes than consistency see Chapter 10). The changes related to the aging of processed cheeses can be induced by ingredients, processing parameters, packaging type and storage conditions.

In the study of Černíková et al. (2017b, 2018c) the basic trends (described above in section 9.4.1 and in Figs. 9.2 and 9.3), the effects of agitation speed and holding time at melting temperature, did not change during the two months of storage. However, there was an increase in the absolute values of the elastic and loss moduli (firmness) of the model processed cheese samples, which was also described in many other studies (Awad et al., 2002; Weiserová et al., 2011; Nagyová et al., 2014; and Salek et al., 2015). In particular, the explanations of this phenomenon could be the possible hydrolysis of emulsifying salts – diphosphates and polyphosphate, the possible change in the dissociation of the salts or other compounds present, the decrease in the pH values of processed cheeses and also the polymorphism of milk fat and ongoing changes in its crystalline form, where β' polymorph change to β polymorph (Awad et al., 2002; Muslow et al., 2007; Dimitreli and Thomareis, 2009; Shirashoji et al., 2010; Weiserová et al., 2011; Nagyová et al., 2014). Increasing dry matter content during processed cheese storage period (due to water evaporation in dependence on the permeability of the packaging material) could be another reason explaning the increase in the firmness of the processed cheese (Guinee at al., 2004; Mohammadi and Fadaei, 2018). Weiss et al. (2018) wanted to prepare mathematical models as a predictive tool for the assessment of changes in the quality parameters during the storage of spreadable processed cheese at 8, 20 and 30°C. Thus, during 4 months at

8°C were found only minor and not significant changes. At the storage temperature of 20°C were identified dynamic changes which significantly decreased the quality of the tested processed cheese product, but the highest rate of changes was registered at the storage temperature of 30°C. Reactions in the processed cheeses led to undesirable flavor and changes in the consistency. Authors recommended the approximate shelf-life limit for spreadable cheeses stored at 8, 20 and 30°C without significant changes in its quality on the 49, 28, and 4 days (Weiss et al., 2018). However, the above-mentioned proposal does not fully correspond to the commonly applied storage conditions for processed cheeses, because processed cheeses under certain circumstances could be stored also at a room temperature for 4 months or stored for 6 months in a refrigerator. Also, Schär and Bosset (2002) described processed cheese as a "semi-preserved food" with a limited shelf-life. Furthermore, they reported that products without bacterial spoilage can retain their quality for 4–12 months at room temperature and at lower storage temperatures for longer time because lower temperature causes more slowly changes in quality processed cheeses.

9.6 Sterilized processed cheese

As it was described in previous chapters, processed cheese ranks among typical technological nonacid foods (pH > 4) providing favorable conditions to many microorganisms, including spore-forming bacteria (see following Chapter 14). Common processed cheeses have a shelf life of a few months, depending on several factors, for example, on the used ingredients, melting temperature and holding time, packaging technology and material, storage temperature, etc. To ensure longer shelf life of processed cheeses heat sterilization of hermetically sealed products can be applied. Sterilized processed cheese could be used as common canned food for usual using, or in combat rations (for boarding of army/soldiers and/or the parts/members of rescue systems logistic). In the case of the combat rations, special requirements exist, for example, the shelf-life shall be minimally 24 months at ambient temperature (Mafart et al. 2001; Schär and Bosset 2002; Buňka et al., 2004).

For a numerical expression of the lethal effect of a given sterilization mode (a combination of temperature and holding time), the so called "F value" (expressed in minutes) can be used. The F value can be defined as the time of heat treatment at a constant reference temperature (generally T = 121.1°C) or as any equivalent heat treatment which would cause the same extent of destruction:

$$F = \int_0^t 10^{\frac{T-121.1}{z}} dt \tag{9.3}$$

Where, t is a sterilization time of the sterilization mode (min), T is a real temperature of the sterilization mode at time t (°C), and z is a value expressing the increase of temperature, which is necessary to obtain the same lethal effect at 1/10 of time. The z-value is usually set at ≈ 10°C. Heat sterilization does not only affect microorganisms, but it also can influence the chemical and/or biochemical reactions of food components, resulting in their partially deterioration. Therefore, the nutrition value and organoleptic properties (especially color and consistency) are also affected. The nutritional and organoleptic quality of sterilized products can be characterized by the so called "C value," which is practically equivalent to the relation of the F value. On the other hand, the kinetic parameters of chemical and/or biochemical reactions differ in comparison to the lethal effect of heat treatment. The difference is expressed by another z value (for the C value is usually signed as z_c value), which is often set at $\approx$ 31.4°C. The latter mentioned value is the z_c value for thiamine deterioration. It means that in the Eq. 9.3 the z-value is substituted by the z_c value; the rest of the equation (relationship) is the same (Bylund, 1995; Contarini et al., 1997; Efigênia et al., 1997; Gaillard et al., 2005; Lazárková et al., 2011; Özilgen, 1998; Rufián-Henares et al., 2004; Zanoni et al., 2003).

9.6.1 The effect of heat sterilization on processed cheese quality

In years 2003–2008, some experiments dealing with the effect of heat sterilization on processed cheese quality were performed. The sterilization mode of 117°C with the holding time of 20 min was chosen for the model samples (Buňka et al., 2004, 2008; Tremlová et al., 2006). The authors concluded that the basic chemical properties such as the dry matter, the fat and the crude protein contents were not significantly influenced by heat sterilization. Furthermore, no aerobic and/or anaerobic microorganisms were detected, resulting in appropriate sterilized processed cheese products. On the other hand, the applied sterilization mode affected the samples pH-values, which decreased by 0.1–0.2 pH units. The possible reason for this lowering could be found in the hydrolysis of the applied polyphosphate emulsifying salts, which can cause a change in their buffering capacity (see Chapter 6), and in protein-protein interactions resulting in proton release. Another reason could lie in a well-known reaction also releasing protons:

$$3\ Ca^{2+} + 2\ HPO^{-} \rightarrow Ca_3(PO_4)_2 + 2\ H^{+}$$

The sterilization caused a slightly decrease of some amino acids' contents, especially cysteine, methionine, serine, threonine, and tyrosine. The reduction of the bioavailable lysine was also recorded. The decline was at a maximum of 10% (w/w), but in the most of the tested cases it was around 5% (w/w) in comparison with the original content in nonsterilized products. The main reason could lie in the Maillard´s reactions as interactions of carbonyl compounds (especially reducing saccharides) with amino components (especially amino acids). In addition, other reactions also could contribute to amino acids reduction, for example, (1) interactions of bound asparagine and glutamine with lysine (free and/or bound) during which ammonia is released and an isopeptide bond (which cannot be cleaved by enzymes in the human digestive tract) is formed, (2) elimination of hydrogen sulphide from bound cystine, (3) Strecker's degradation of amino acids (oxidative decarboxylation) results in the formation of ammonia, carbon dioxide and aldehyde, etc. The course of the above-mentioned reactions can be supported also by the significantly increased content of ammonia, which was observed in sterilized processed cheese in comparison with nonsterilized products. A slight hydrolysis of caseins was also observed due to sterilization treatment (Adamiec et al., 2001; Buňka et al., 2004, 2008; Friedman, 1996; Gaucheron et al., 1999; Guinee et al., 2004; Lazárková et al., 2010, 2011; Molins, 1991; Rufián-Henares et al., 2006; Tremlová et al., 2006).

The heat sterilization affected also the organoleptic properties of the tested processed cheese. The brightness (L*) of the sterilized samples was decreased in comparison with non-sterilized products. From a point of view of chromaticity, a significant shift from the green to red (a*) and from the blue to yellow areas (b*) were recorded. The latter mentioned results leaded to a significant increase of the yellowness index of the sterilized products (compared with nonsterilized samples), which is considered as a marker of product degradation. The reason could lie in the above-mentioned reactions, especially in the complex of Maillard´s reactions and their products, influencing the colour of the samples. Furthermore, the sterilization also caused a significant increase of the firmness of the processed cheese. The reason could be related to the hydrolysis of polyphosphate emulsifying salts and the decrease in their affinity to calcium ions. Potentially liberated calcium ions could be incorporated in a more intensive networking of the protein matrix (see Chapter 6). Some of the above-mentioned interactions of proteins could also affect the change of the casein network. Furthermore, the decline of the pH-value of the sterilized processed cheese (in comparison with nonsterilized samples) could contribute to the increase of products firmness, too. The instrumentally measured results (for both colour and firmness) were also confirmed by the results obtained from sensory analysis between the sterilized and nonsterilized processed cheese (Adamiec et al., 2001; Buňka et al., 2008; Friedman, 1996; Kristensen et al., 2001; Molins, 1991).

9.6.2 The effect of storage time and temperature on sterilized processed cheese quality

Bubelová et al. (2015) and Buňka et al. (2008) studied the quality changes of sterilized processed cheese stored at the temperature of 6°C, 23°C, and 40°C, respectively for a period of 2 years. Due to hermetically closed doses, no changes in the dry matter, the fat and the crude protein contents were observed. On the other hand, the pH-values decreased gradually over storage (by 0.2–0.3 pH units). Moreover, the highest decline was reported at the storage temperature of 40°C. The decrease of the pH-value was less intensive when the samples were stored at 6 or 23°C, respectively. The reason for the lowering of the pH values could be very similar as it was described above due to heat sterilization. Only the intensity and/or rate of this processes was slower.

The amino acid profile was also influenced during the 2 years storage period and due to the applied storage temperature. The most intensive decline of some amino acid contents was recorded at the storage temperature of 40°C. Moreover, from the point of view of the total amino acid content, the decrease was caused by both storage temperature and length of storage time. In particular, it reached 2% (w/w) after

1-year and 4.5% (w/w) after 2-year storage at 6°C; and 7.5% (w/w) after 1-year and 12% (w/w) after 2-year storage at 40°C (calculated on the initial amino acid content at the beginning of the storage). The amino acid losses in samples stored at 23°C were similar in comparison to sterilized processed cheese stored at 6°C. Tyrosine, serine, threonine, histidine, cysteine, methionine, and lysine were undergone the most intensive deterioration processes. On the other hand, alanine, leucine, isoleucine, and valine were the most stable amino acids, regardless of the storage time and/or temperature. These observations are in accordance to those presented in the work of Buňka et al. (2009), who dealt with individual amino acid stability. The decreasing of essential amino acid contents could be expressed also by so called the essential amino acid index of samples. For the 2 years storage period, the decline of the essential amino acid index at a level of approximately 4%, 6%, and 12% were recorded when sterilized processed cheese were stored at 6, 23, and 40°C, respectively. The main reason could lie also in the complex of Maillard´s reactions and/or the Strecker's degradation. The latter mentioned statement could be supported by the increase of the ammonia content in the sterilized processed cheese during the 2 years storage period. The most intensive growth of ammonia content was recorded at the storage temperature of 40°C.

The chemical and/or biochemical changes influenced also the organoleptic quality of the sterilized processed cheese during storage. The firmness of samples increased, and the colour and flavor were deteriorated during the 2-year storage. The changes were more intensive at elevated temperature. Additionally, the best ratings were obtained when the samples were stored at 6°C. The organoleptic properties of the sterilized processed cheese were also evaluated by the acceptability for the sensory assessors when the storage temperature of 23°C was used (during the whole 2-year period). On the other hand, the samples stored at 40°C were evaluated as unacceptable (firmness, colour and/or flavor) by the sensory assessors even after 6 or 12 months of storage (Bubelová et al., 2015; Buňka et al., 2008).

9.6.3 The strategy for minimization of deterioration effects of heat sterilization

Lazárková et al. (2010, 2011) studied the effect of different sterilization modes and different lactose contents (a reducing saccharide – a reagent for the initial phase of the complex of Maillard´s reactions) on the quality of sterilized processed cheese. The crucial aim was to recommend the appropriate condition for sterilized processed cheese production for the industry that would minimize the negative effect of heat sterilization on the samples quality. In particular, the lactose content of 0%–2% (w/w) was tested. Moreover, the following sterilization modes were studied: 110°C for 100 min, 115°C for 32 min, 120°C for 10 min, and 125°C for 3.2 min. The sterilization modes were designed to achieve the practically same F-value that ranged from 7.8 to 7.9 min ($z = 10$°C). The C values for the above-mentioned sterilization modes were 44.3, 20.4, 9.2, and 4.3 min, respectively ($z_c = 31.4$°C). Therefore, from the Eq. 9.3 and the z value and the z_c value used, it could be implied that the usage of the higher sterilization temperatures applied for the shorter holding times could lead to the maintenance of a higher nutritional value of the sterilized foodstuffs.

All sterilization modes used led to samples where neither aerobic nor anaerobic spore-forming microorganisms were detected. Therefore, all products were evaluated as appropriate sterilized processed cheese. The results unambiguously pointed out that when the sterilization temperature rose (and the holding time adequately fell for reaching the constant F value) the deterioration processes decreased (especially losses of amino acids content, decreasing of bioavailable lysine content, increasing of firmness, and/or worsening of color and flavor). The intensity of the degradation processes related with protein and/or amino acid reactions (especially the complex of Maillard´s reaction and/or Strecker's degradation) also declined when the lactose content decreased.

Based on the results of the studies presented in section 9.6, the following recommendations for the practice (industry) can be summarized to obtain sterilized processed cheese of high quality:

- The mixture of the raw materials should contain components, in which the concentration of lactose (generally reducing saccharides) will be as minimal as possible, definitely under 1% w/w.
- An appropriate sterilization regime can be recommended at the level of 120°C for 10 min (based on the tested sterilization modes). Of course, the sterilization mode 125°C for 3.2 min provided even better results. On the other hand, from a point of view of practice, the holding time of 3.2 min represents a very short reaction time for the sterilization equipment with capacity of several tons.
- For long-term storage (in the order of months), the storage temperature should be set under 25°C.

9.7 Conclusion

Technological aspects are important in processed cheeses production that includes many factors. These aspects affect consistency of processed cheeses. Processed cheese blocks, spreads, or sauces need different producing conditions. Various type of agitation speed, heating rate, melting temperature, time under holding temperature, cooling rate, and further provide different final product. It is necessary to choose the right combination of processing parameters with respect of all used raw materials and requirement processed cheese consistency.

References

Acharya, M.R., Mistry, V.V., 2007. Influence of processing condition on Cheddar cheese meltability. Milchwissenschaft 62, 1–4.

Adamiec, J., Cejpek, K., Rössner, J., Velíšek, J., 2001. Novel strecker degradation products of tyrosine and dihydroxyphenylalanine. Czech J. Food Sci. 19, 13–18.

Agoda-Tandjawa, G., Le Garnec, C., Boulenguer, P., Gilles, M., Langendorff, V., 2017. Rheological behavior of starch/carrageenan/milk proteins mixed systems: role of each biopolymer type and chmeical characteristics. Food Hydrocoll. 73, 300–312.

Awad, R.A., Abdel–Hamid, L.B., El–Shabrawy, S.A., Singh, R.K., 2002. Texture and microtexture of block type processed cheese with formulated emulsifying salt mixtures. Lebensm. Wiss. Technol. 35 (1), 54–61.

Barth, A.P., Tormena, C.F., Viotto, W.H., 2017. pH influences hydrolysis of sodium polyphosphate in dairy matrices and the structure of processed cheese. J. Dairy Sci. 100 (11), 1–9.

Bayarri, S., Carbonell, I., Costell, E., 2012. Viscoelasticity and texture of spreadable cheeses with different fat contents at refrigeration and room temperatures. J. Dairy Sci. 95 (12), 6926–6936.

Belsito, P.C., Ferreita, M.V.S., Cappato, L.P., Cavalcanti, R.N., Vidal., V.A.S., Pimenetel, T.C., Esmerino, E.A., Balthazar, C.F., Neto, R.P.C., Tavares, M.I.B., Zacarchenco, P.B., Freitas, M.Q., Silva, M.C., Raices, R.S.L., Pastore, G.M., Pollonio, M.A.R., Cruz, A.G., 2017. Manufacture of Requeijão cremoso processed cheese with galactooligosaccharide. Carbohydr. Polym. 174, 869–875.

Bennett, R.J., Trivedi, D., Hemar, Y., Reld, D.C.W., Illingworth, D., Lee, S.K., 2006. The effect of starch addition on the rheological and microstructural properties of model processed cheese. Aust. J. Dairy Technol. 61 (2), 157–159.

Berger, W., Klostermeyer, H., Merkenich, K., Uhlmann, G., 2002. Process Cheese Manufacture – A JOHA Guide. BK Giulini Chemie GmbH & Co. OHC, Ladenburg, p. 239.

Bourriot, S., Garnier, C., Doublier, J.L., 1999. Micellar–casein–κ–carrageenan mixtures. I. Phase separation and ultrastructure. Carbohydr. Polym. 40 (2), 145–157.

Bowland, E.L., Foegeding, E.A., 1999. Factors determining large–strain (fracture) rheological properties of model processed cheese. J. Dairy Sci. 82, 1851–1859.

Brickley, C.A., Auty, M.A.E., Piraino, P., McSweeney, P.L.H., 2007. The effect of natural cheddar cheese ripening on the functional and textural properties of the processed cheese manufactured therefrom. J. Dairy Sci. 72 (9), 483–490.

Brickley, C.A., Govindasamy–Lucey, S., Jaeggi, J.J., Johnson, M.E., McSweeney, P.L.H.Lucey, J., A., 2008. Influence of emulsifying salts on the textural properties of nonfat process cheese made from direct acid cheese based. J. Dairy Sci. 91, 39–48.

Bubelová, Z., Tremlová, B., Buňková, L., Pospiech, M., Vítová, E., Buňka, F., 2015. The effect of long-term storage on the quality of sterilized processed cheese. J. Food Sci. Technol. 52, 4985–4993.

Buňka, F., Hrabě, J., Kráčmar, S., 2004. The effect of sterilisation on amino acid contents in processed cheese. Int. Dairy J. 14, 829–831.

Buňka, F., Pavlínek, V., Hrabě, J., Rop, O., Janiš, R., Krejčí, J., 2007. Effect of 1–monoglycerides on viscoelastic properties of processed cheese. Int. J. Food Prop. 10, 819–828.

Buňka, F., Štětina, J., Hrabě, J., 2008. The effect of storage temperature and time on the consistency and color of sterilized processed cheese. Eur. Food Res. Technol. 228, 223–229.

Buňka, F., Kříž, O., Veličková, A., Buňková, L., Kráčmar, S., 2009. Effect of acid hydrolysis time on amino acids determination in casein and processed cheeses with different fat content. J. Food Compos. Anal. 22, 224–232.

Buňka, F., Doudová, L., Weiserová, E., Kuchař, D., Michálek, J., Slavíková, S., Kráčmar, S., 2012. The effect of different ternary mixtures of sodium phosphates on hardness of processed cheese spreads. Int. J. Food Sci. Technol. 47 (10), 2063–2071.

Buňka, F., Doudová, L., Weiserová, E., Kuchař, D., Ponížil, P., Začalová, D., Nagyová, G., Pachlová, V., Michálek, J., 2013. The effect of ternary emulsifying salt composition and cheese maturity on the textural properties of processed cheese spreads. Int. Dairy J. 29 (1), 1–7.

Buňka, F., Doudová, L., Weiserová, Černíková, M., Kuchař, D., Slavíková, Š., Nagyová, G., Ponížil, P., Grůber, T., Michálek, J., 2014. The effect of concentration and composition of ternary emulsifying salts on the textural properties of processed cheese spreads. Lebensm. Wiss. Technol. 58, 247–255.

Bylund, G., 1995. Dairy Processing Handbook. Tetra Pak Processing Systems, Lund.

Carić, M., Gantar, M., Kaláb, M., 1985. Effect of emulsifying agents on the microstructure and other characteristics of process cheese – a review. Food Microstruct. 4, 297–312.

Chatziantoniou, S.E., Thomareis, A.S., Kontominas, M.G., 2015. Effect of chemical composition on physico–chemical, rheological and sensory properties of spreadable processed whey cheese. Eur. Food Res. Technol. 241 (6), 737–748.

Chatziantoniou, S.E., Thomareis, A.S., Kontominas, M.G., 2019. Effect of different stabilizers on rheological properties, fat globule size and sensory attributes of novel spreadable processed whey cheese. Eur. Food Res. Technol. 245, 2401–2412.

Contarini, G., Povolo, M., Leardi, R., Toppino, P.M., 1997. Influence of heat treatment on the volatile compounds of milk. J. Agric. Food Chem. 45, 3171–3177.

Cunha, C.R., Dias, A.I., Viotto, W.H., 2010. Microstructure, texture, colour and sensory evaluation of a spreadable processed cheese analogue made with vegetable fat. Food Res. Int. 43 (3), 723–729.

Černíková, M., Buňka, F., Pavlínek, V., Březina, P., Hrabě, J., Valášek, P., 2008. Effect of carrageenan type on viscoelastic properties of processed cheese. Food Hydrocoll. 22 (6), 1054–1061.

Černíková, M., Nebesářová, J., Salek, R. N., Říháčková, L., Buňka, F., 2017a. Microstructure, textural and viscoelastic properties of model processed cheese with different dry matter and fat in dry matter content. J. Dairy Sci. 100, 4300–4307.

Černíková, M., Salek, R. N., Kozáčková, D., Běhalová, H., Luňáková, L., Buňka, F., 2017b. The effect of selected processing parameters on viscoelastic properties of model processed cheese spreads. Int. Dairy J. 66, 84–90.

Černíková, M., Nebesářová, J., Salek, R. N., Popková, R., Buňka, F., 2018a. The effect of rework content addition on the microstructure and viscoelastic properties of processed cheese. J. Dairy Sci. 101, 2956–2962.

Černíková, M., Pachlová, V., Holas, O., Moudrá, K., Slintáková, K., Buňka, F., 2018b. The effect of dairy fat source on viscoelastic properties of full–fat processed cheese spreads. Eur. J. Lipid Sci. Technol. 120, 1700319.

Černíková, M., Salek, R. N., Kozáčková, D., Buňka, F., 2018c. The effect of different agitations and temperature maintainings on viscoelastic properties of full–fat processed cheese spreads. LWT – Food Sci. Technol. 89, 244–247.

Da Silva, D.F., De Souza Ferreira, S.B., Bruschi, M.L., Britten, M., Matumoto–Pintro, P.T., 2016. Effect of commercial konjac glucomannan and konjac flours on textural, rheological and microstructural properties of low fat processed cheese. Food Hydrocoll. 60, 308–316.

Dalgleish, D.G., 1997. Adsorption of protein and the stability of emulsions. Trends Food Sci. Technol. 8, 1–6.

Dimitreli, G., Thomareis, A.S., 2004. Effect of temperature and chemical composition on processed chesse apparent viscosity. J. Food Eng. 64, 265–271.

Dimitreli, G., Thomareis, A.S., 2007. Texture evaluation of block–type processed cheese as a function of chemical composition and in relation to its apparent viscosity. J. Food Eng. 79 (4), 1364–1373.

Dimitreli, G., Thomareis, A.S., 2008. Effect of chemical composition on the linear viscoelastic properties of spreadable–type processed cheese. J. Food Eng. 84 (3), 368–374.

Dimitreli, G., Thomareis, A.S., 2009. Instrumental textural and viscoelastic properties of processed cheese as affected by emulsifying salts and in relation to its apparent viscosity. Int. J. Food Prop. 12, 261–275.

Efigênia, M., Povoa, B., Moraes-Santos, T., 1997. Effect of heat treatment on the nutritional quality of milk proteins. Int. Dairy J. 7, 609–612.

El–Bakry, M., Duggan, E., O'Riordan, E.D., O'Sullivan, M., 2011. Effect of chelating salt type on casein hydration and fat emulsification during manufacture and post–manufacture functionality of imitation cheese. J. Food Eng. 102 (2), 145–153.

El–Nawawi, S.A., Heikal, Y.A., 1995. Factors affecting the production of low–ester pectin gels. Carbohydr. Polym. 26, 189–193.

Europa, 2008. Regulation (EC) No 1333/2008 of the European Parliament and of the Council of 16 December 2008 on food additives. Off. J. Eur. Union L 354, 16–33 31.12.2008 <https://eur-lex.europa.eu/legal-content/EN/ALL/?uri=celex%3A32008R1333>.

Ferrão, L.L., Ferreira, M.V., Cavalcanti, R.N., Carvalho, A.F.A., Pimentel, T.C., Silva, H.L.A., Silva, R., Esmerino, E.A., Netto, R.P.C., Tavares, M.I.B., Freitas, M.Q., Menezes, C.V., Cabral, L.M., Moraes, J., Silva, M.C., Mathisa, S.P., Raices, R.S.L., Pastore, G.M., Cruz, A.G., 2018. The xylooligosaccharide addition and sodium reduction in requeijão crema processed cheese. Food Res. Int. 107, 137–147.

Friedman, M., 1996. Food browning and its prevention. J. Agric. Food Chem. 44, 632–653.

Fu, J.T., Rao, M.A., 1999. The influence of sucrose and sorbitol on gel–sol transition of low–methoxyl pectin + Ca^{2+} gels. Food Hydrocoll. 13, 371–380.

Fu, W., Watanabe, Y., Inoue, K., Moriguchi, N., Fusa, K., Yanagisawa, Y., Mutoh, T., Nakamura, T., 2018. Effects of pre–cooked cheeses of different emulsifying condition on mechanical properties and microstructure of processed cheese. Food Chem. 245, 47–52.

Fu, W., Nakamura, T., 2018. Effect of starch on the mechanical properties and microstructure of processed cheeses with different types of casein network structures. Food Hydrocoll. 79, 587–595.

Gabriele, D., Cindio, de B., D'Antona, P., 2001. A weak gel model for foods. Rheol. Acta 40, 120–127.

Gaillard, S., Leguérinel, I., Savy, N., Mafart, P., 2005. Quantifying the combined effects of the heating time, the temperature and the recovery medium pH on the regrowth lag time of *Bacillus cereus* spores after a heat treatment. Int. J. Food Microbiol. 105, 53–58.

Garnier, C., Axelos, M.A.V., Thibault, J.F., 1994. Selectivity and cooperativity in binding of calcium ions by pectins. Carbohydr. Res. 256, 71–81.

Gaucheron, F., Mollé, D., Briard, V., Léonil, J., 1999. Identification of low molar mass peptides released during sterilization of milk. Int. Dairy J. 9, 515–521.

Genovese, D.B., Ye, A., Singh, H., 2010. High methoxyl pectin/apple particles composite gels: effect of particle size and particle concentration on mechanical properties and gel structure. J. Texture Stud. 41, 179–187.

Giri, A., Kanawjia, S.K., Singh, M.P., 2017. Effect of inulin on physico–chemical, senzory, fatty acid profile and microstructure of processed cheese spread. J. Food Sci. Technol. 54 (8), 2443–2451.

Görner, F., Valík, Ľ., 2004. Aplikovaná Mikrobiológia Poživatín. Malé centrum, Bratislava, p. 528 ISBN 80–967064–9–7.

Grosso, C.R.F., Rao, M.A., 1998. Dynamic rheology of structure development in low–methoxy pectin + Ca^{2+} + sugar gels. Food Hydrocoll. 12, 357–363.

Guinee, T.P., 2003. Pasteurized processed cheese products. In: Roginski, H., Fuquay, J.W., Fox, P.F. (Eds.), Encyklopedia of Dairy Science, 1. Elsevier, Amsterdam, pp. 411–418.

Guinee, T.P., Carić, M., Kaláb, M., 2004. Pasteureized processed cheese and substitute/imitation cheese products. In: Fox, P.F. (Ed.), Cheese: Chemistry, Physics and Microbiology. Elsevier, Amsterdam, pp. 349–394 ISBN 0–1226–3653–8.

Guinee, T.P., O'Callaghan, D.J., 2013. Effect of increasing the protein–to–fat ratio and reducing fat content on the chemical and physical properties of processed cheese product. J. Dairy Sci. 96 (11), 6830–6839.

Gupta, S.K., Karahadian, C., Lindsay, R.C., 1984. Effect of emulsifier salts on textural and flavour properties of processed cheese. J. Dairy Sci. 67, 764–778.

Gustaw, W., Mleko, S., 2007. The effect of polysaccharides and sodium chloride on physical properties of processed cheese analogs containing whey proteins. Milchwissenschaft 62 (1), 59–62.

Hladká, K., Randulová, Z., Tremlová, B., Ponížil, P., Mančík, P., Černíková, M., Buňka, F., 2014. The effect of cheese maturity on selected properties of processed cheese without traditional emulsifying agents. LWT – Food Sci. Technol 55, 650–656.

Hoffmann., W., Schrader, K., 2015. Dispersion analysis of spreadable processed cheese with low content of emulsifying salts by photocentrifugation. Int. J. Food Sci. Technol. 50, 950–957.

Hosseini–Parvar, S.H., Matia–Merono, L., Golding, M., 2015. Effect of basil seed gum (BSG) on textural, rheological and microstructural properties of model processed cheese. Food Hydrocoll. 43, 557–567.

Huang, H., Hallianan, R., Maleky, F., 2018. Comparison of different oleogels in processed cheese products formulation. Int. J. Food Sci. Technol. 53, 2525–2534.

Jooyandeh, H., 2009. Effect of fermented whey protein concentrate on texture of Iranian white cheese. J. Texture Stud. 40, 497–510.

Kaláb, M., Modler, W.H, 1985. Milk gel structure. XV. Electron microscopy of whey protein-based cream cheese spread. Milchwissenschaft 193–196.

Kaláb, M., Yun, J., Yiu, S.H., 1987. Textural properties and microstructure of process cheese food rework. Food Microstruct. 6, 181–192.

Kapoor, R., Metzger, L.E., Biswas, A.C., Muthukummarappan, K., 2007. Effect of natural cheese characteristic on process cheese properties. J. Dairy Sci. 90 (4), 1625–1634.

Kapoor, R., Metzger, L.E., 2008. Process cheese: scientific and technological aspects – a review. Compr. Rev. Food Sci. Food Saf. 7, 194–214.

Kasper, H., Burghardt, W., 2015. Výživa v medicíně a dietetika. Tisk Centrum, s.r.o., Moravany, Praha, pp. 64–66 ISBN 978–80–247–4533–6.

Kommineni, A., Amacharla, J., Metzger, L.E., 2012. Effect of xylitol on the functional properties of low–fat processed cheese. J. Dairy Sci. 95 (11), 6252–6259.

Kristensen, D., Hansen, E., Arndal, A., Trinderup, R.A., Skibsted, L.H., 2001. Influence of light and temperature on the colour and oxidative stability of processed cheese. Int. Dairy J. 11, 837–843.

Lauck, R. M., inventor; Stauffer Chemical Co., assignee. 1972. Using salvage cheese in preparing pasteurized process cheese. New York. US patent 3, 697, 292.

Lazárková, Z., Buňka, F., Buňková, L., Valášek, P., Kráčmar, S., Hrabě, J., 2010. Application of different sterilizing modes and the effect on processed cheese quality. Czech J. Food Sci. 28, 168–176.

Lazárková, Z., Buňka, F., Buňková, L., Holáň, F., Kráčmar, S., Hrabě, J., 2011. The effect of different heat sterilization regimes on the quality of canned processed cheese. J. Food Process Eng. 34, 1860–1878.

Lee, S.K., Klostermeyer, H., Schrader, K., Büchheim, W., 1996. Rheological properties and microstructure of model processed cheese containing low molecular weight emulsifiers. Nahrung – Food 40, 189–194.

Lee, S.K., Klostermeyer, H., 2001. The effect of pH on the rheological properties of reduced–fat model processed cheese spreads. Lebensm. Wiss. Technol. 34, 288–292.
Lee, S.K., Buvalda, R.J., Euston, S.R., Foegeding, E.A., McKenna, A.B., 2003. Changes in the rheology and microstructure of processed cheese during cooking. Lebensm. Wiss. Technol. 36, 339–345.
Lee, S.K., Klostermeyer, H., Anema, S.G., 2015. Effect of fat and protein–in–water concentration on the properties of model processed cheese. Int. Dairy J. 50, 15–23.
Lenze, S., Wolfschoon–Pombo, A., Schrader, K., Kulozik, U., 2019. Effect of the compositional factors and processing conditions on the creaming reaction during process cheese manufacturing. Food Bioproc. Technol. 12, 575–586.
Le Tohic, C., O´Sullivan, J.J., Drapala, K.P., Chartrin, V., Chan, T., Morrison, A.P., Kerry, J.P., Kelly, A.L., 2018. Effect of 3D printing on the structure and textural properties of processed cheese. J. Food Eng. 220, 56–64.
Löfgren, C., Hermansson, A.M., 2007. Synergistic rheological behaviour of mixed HM/LM pectin gels. Food Hydrocoll. 21, 480–486.
Lootens, D., Capel, F., Durand, D., Nicolai, T., Boulenguer, P., Langendorff, V., 2003. Influence of pH, Ca concentration, temperature and amidation on the gelation of low methoxyl pectin. Food Hydrocoll. 17, 237–244.
Lu, Y., Shirashoji, N., Lucey, J.A., 2007. Rheological, textural and melting properties of commercial samples of some of different types of pasteurized processed cheese. Int. J. Dairy Technol. 60 (2), 74–80.
Lucey, J.A, Fox, P.F., 1993. Importance if calcium and phosphate in cheese manufacture: a review. J. Dairy Sci. 76, 1714–1724.
Macků, I., Buňka, F., Pavlínek, V., Leciánová, P., Hrabě, J., 2008. The effect of pectin concentration on viscoelastic and sensory properties of processed cheese. Int. J. Food Sci. Technol. 43, 1663–1670.
Macků, I., Buňka, F., Voldánová, B., Pavlínek, V., 2009. Effect of addition of selected solid cosolutes and pectin on viscoelastic properties of model processed cheese. Food Hydrocoll. 23, 2078–2084.
Mafart, P., Couvert, O., Leguérinel, I., 2001. Effect of pH on the heat resistance of spores. Comparison of two models. Int. J. Food Microbiol. 63, 51–56.
Marchesseau, S., Gastaldi, E., Lagaude, A., Cuq, J.L., 1997. Influence of pH on protein interactions and microstructure of process cheese. J. Dairy Sci. 80 (8), 1483–1489.
May, C.D., 2000. Pectins. In: Phillips, G.O., Williams, P.A. (Eds.), Handbook of Hydrocolloids. Woodhead Publishing Limited and CRC Press, Boca Raton, pp. 169–188 ISBN 0–8493–0850–X.
McIntyre, I., O´Sullivan, M., O´Riordan, D., 2017. Monitoring the progresion of calcium and protein solubilisation as affected by calcium chelators during small–scale manufacture of casein–based food matrices. Food Chem. 237, 597–604.
Mehanna, N.S., Hassan, F.A.M., El–Messery, T.M., Mohamed, A.G., 2017. Production of functional processed cheese by using tomato juice. Int. J. Dairy Sci. 12, 155–160.
Meyer, A., 1973. Processed Cheese Manufacture. Food Trade Press Ltd., London, U. K., p. 329.
Mizuno, R., Lucey, J.A., 2005. Effects of emulsifying salts on the turbidity and calcium–phosphate–protein interactions in casein micelles. J. Dairy Sci. 88 (9), 3070–3078.
Mizuno, R., Lucey, J.A, 2007. Properties of milk protein gels formed by phosphates. J. Dairy Sci. 90 (10), 4524–4531.
Mohamed, A.G., Shalaby, S.M., Gafour, W.A., 2016. Quality characteristics and acceptability of an analogue processed spreadable cheese made with carrot paste (Daucus carota L.). Int. J. Dairy Sci. 11, 91–99.

Mohamed, A.G., Shalaby, S.M., 2016. Texture, chemical properties and sensory evaluation of a spreadable processed cheese analogue made with apricot pulp (Prunus armeniaca L.). Int. J. Dairy Sci. 11, 61–68.

Mohammadi, A., Fadaei, V., 2018. The effect of homogenization on texture of produced dry matter processed cheese. Food Sci. Technol. 38, 190–195.

Molins, R.A., 1991. Phosphates in food. CRC Press, Boca Raton.

Mounsey, J.S., O'Riordan, E.D., 2001. Characteristics of imitation cheese containing native starches. J. Food Sci. 66 (4), 586–591.

Mozuraityte, R., Berget, I., Mahdalová, M., Grønsberg, A., Øye, E.R., Greiff, K., 2019. Sodium reduction in processed cheese spreads and the effect on physicochemical properties. Int. Dairy J. 90, 45–55.

Muslow, B.B., Jaros, D., Rohm, H., 2007. Processed cheese and cheese analogues. In: Tamine, A.Y. (Ed.), Structure of Dairy Products. Blackwell Publishing Ltd., Oxford, pp. 210–235 ISBN 978-1-4051-2975-6.

Nagyová, G., Buňka, F., Salek, R.N., Černíková, M., Mančík, P., Grůber, T., Kuchař, D., 2014. Use of sodium polyphosphates with different linear length in the production of spreadable processed cheese. J. Dairy Sci. 97, 111–122.

Noronha, N., O'Riordan, E.D., O'Sullivan, M., 2008a. Influence of processing parameters on the texture and microstructure of imitation cheese. Eur. Food Res. Technol. 226 (3), 385–393.

Nastaj, M., Terpiłowski, K., Sołowiej, B.G., 2020. The effect of native and polymerised whey protein isolate addition on surface and microstructural properties of processed cheeses and their meltability determined by Turbiscan. Int. J. Food Sci. Technol 55 (5), 1–9.

Noronha, N., Duggan, E., Ziegler, G.R., O'Riordan, E.D., O'Sullivan, M., 2008b. Investigation of imitation cheese matrix development using light microscopy and NMR relaxometry. Int. Dairy J. 18, 641–648.

Olson, N.F., Vakaleris, D.G., Price, W., V., Knight, S., G., 1958. Acidity and age of natural cheese as factor affecting the body of pasteurized processed cheese spread. J. Dairy Sci. 41, 1005–1016.

Özilgen, M., 1998. Food Process Modeling and Control: Chemical Engineering Applications. Gordon and Breach Science Publishers, Sydney.

Piska, I., Štětina, J., 2004. Influence of cheese ripening and rate of cooling of the processed cheese mixture on rheological properties of processed cheese. J. Food Eng. 61, 551–555.

Pluta, A., Ziarno, M., Smolińska, A., 2000. Możliwości zastosowania hydrokoloidów w produkcji serów topionych. Przemysl Spożywczy 5, 42–44.

Přikryl, J., Hájek, T., Švecová, B., Salek, R.N., Černíková, M., Červenka, L., Buňka, F., 2018. Antioxidant properties and textural characteristics of processed cheese spreads enriched with rutin or quercetin: the effect of processing conditions. LWT – Food Sci. Technol. 87, 266–271.

Rafiq, S.M., Ghosh, B.C., 2017. Effect of potato incorporation on the physico–chemical, textural and sensory properties of processed cheese. Food Measure 11, 776–780.

Ramel, P.R., Marangoni, A.G., 2017. Characterization of the polymorphisms of milk fat within processed cheese products. Food Struct. 12, 15–25.

Ramel, P.R., Marangoni, A.G., 2018. Processed cheese as a polymer matrix composite: a particle tool kit for the replacement of milk fat with canola oil in processed cheese. Food Res. Int. 107, 110–118.

Ribeiro, K.O., Rodrigues, M.I., Sabadini, E., Cunha, R.L., 2004. Mechanical properties of acid sodium caseinate- κ–carrageenan gels: effect of co–solute addition. Food Hydrocoll. 18, 71–79.

Rufián-Henares, J.A., Guerra-Hernández, E., García-Villanova, B., 2004. Pyrraline content in enteral formula processing and storage and model systems. Eur. Food Res. Technol. 219, 42–47.

Rufián-Henares, J.A., Guerra-Hernández, E., García-Villanova, B., 2006. Available lysine and fluorescence in heated milk proteins/dextrinomaltose or lactose solutions. Food Chem. 98, 685–692.

Saadi, A.M., 2018. Effect of replacing milk fat with oil replacers on the quality of processed cheese. Biosci. Res. 15, 4374–4380.

Sádlíková, I., Buňka, F., Budínský, P., Voldánová, B., Pavlínek, V., Hoza, I., 2010. The effect of selected phosphate emulsifying salts on viscoelastic properties of processed cheese. LWT – Food Sci. Technol. 43, 1220–1225.

Sadek, Z.I., Refaat, B.M., Adb El–Shakour, E.H., Mehanna, N.S., Hassan, M.S., 2017. Biocontrol of processed cheese by incorporation of probiotic bacteria and its metabolites. Int. J. Dairy Sci. 12, 93–104.

Salek, R.N., Černíková, M., Nagyová, G., Kuchař, D., Bačová, H., Minarčíková, L., Buňka, F., 2015. The effect of ternary mixtures composition containing phosphate and citrate emulsifying salts on selected textural properties of spreadable processed cheese. Int. Dairy J. 44, 37–43.

Salek, R.N., Černíková, M., Maděrová, S., Lapčík, L., Buňka, F., 2016. The effect of different composition of ternary mixtures of emulsifying salts on the consistency of processed cheese spreads manufactured from Swiss–type cheese with different degrees of maturity. J. Dairy Sci. 99, 3274–3287.

Salek, R.N., Černíková, M., Pachlová, V., Bubelová, Z., Konečná, V., Buňka, F., 2017. Properties of spreadable processed Mozzarella cheese with divergent compositions of emulsifying salts in relation to the applied cheese storage period. LWT – Food Sci. Technol. 77, 30–38.

Salek, R.N., Černíková, M., Lorencová, E., Pachlová, V., Kůrová, V., Šenkýřová, J., Buňka, F., 2020. The impact of Cheddar or white brined cheese with various maturity degrees on the processed cheese consistency: a comparative study. Int. Dairy J. 111, 104816.

Schäffer, B., Lörinzy, D., Belágyi, J., 1999. DSC and electronmicroscopic investigation of dispersion–type processed cheeses made without peptization. J. Therm. Anal. Calorim. 56, 1211–1216.

Schär, W., Bosset, J.O., 2002. Chemical and physico–chemical changes in processed cheese and ready–mafe from fondue during storage. A review. LWT 35, 15–20.

Shirashoji, N., Jaeggi, J.J., Lucey, J.A., 2006. Effect of trisodium citrate concentration and cooking time on the physicochemical properties of pasteurized process cheese. J. Dairy Sci. 89 (1), 15–28.

Shirashoji, N., Jaeggi, J.J., Lucey, J.A., 2010. Effect of sodium hexametaphosphate concentration and cooking time on the physicochemical properties of pasteurized process cheese. J. Dairy Sci. 93 (7), 2827–2837.

Smewing, J., 1999. Hydrocolloids. In: Rosenthal, A.J. (Ed.), Food Texture: Measurement and Perception. Aspen Publishers, pp. 282–303 ISBN 978–0–8342–1238–1.

Smith, G.F., Rivera, E., 2016. Processed cheese without emulsifying salts. United States Patent. US 20090092730 A1. [2017-08-02]. <http://patft.uspto.gov/netacgi/nph–Parser?Sect2=PTO1&Sect2=HITOFF&p=1&u=/netahtml/PTO/search–bool.html&r=1&f=G&l=50&d=PALL&RefSrch=yes&Query=PN/9232808>.

Smith, G.F., Rivera, E., 2017. Processed cheese without emulsifying salts. United States Patent. US 20160081362 A1. [2017-08-02]. <http://patft.uspto.gov/netacgi/nph–Parser?Sect2=PTO1&Sect2=HITOFF&p=1&u=/netahtml/PTO/search–bool.html&r=1&f=G&l=50&d=PALL&RefSrch=yes&Query=PN/>.

Solhi, P., Azadmard–Damirchi, S., Hesari, J., Hamishehkar, H., 2020. Effect of fortification with asparagus powder on the qualitative properties of processed cheese. Int. J. Dairy Technol. 73 (1), 226–233.

Sołowiej, B., Glibowski, P., Muszyński, S., Wydrich, J., Gawron, A., Jeliński, T., 2015. The effect of fat replacement by inulin on the physicochemical properties and microstructure of acid casein processed cheese analogues with added whey protein polymers. Food Hydrocoll. 44, 1–11.

Spanguolo, P.A., Dalgliesh, D.G., Goff, H.D., Morris, E.R., 2005. Kappa–caffageenan interactions in systems containing casein micelles and polysaccharide stabilizers. Food Hydrocoll. 19 (3), 371–377.

Subramanian, R., Muthukumarappan, K., Gunasekaran, S., 2006. Linear viscoelastic properties of regular- and reduced-fat pasteurized process cheese during heating and cooling. Int. J. Food Prop. 9 (3), 377–393.

Sutheerawattananonda, M., Fulcher, R.G., Martin, F.B., Bastian, E.D., 1997. Fluorescence image analysis of process cheese manufactured with trisodium citrate and sodium chloride. J. Dairy Sci. 80 (4), 620–627.

Swenson, B.J., Wendorff, W.L., Lindsay, R.C., 2000. Effects of ingredients on the functioinality of fat–free process cheese spreads. J. Food Sci. 65, 822–825.

Syrbe, A., Bauer, W.J., Klostermeyer, H., 1998. Polymer science concepts in dairy systems – an overview of milk protein and food hydrocolloid interaction. Int. Dairy J. 8, 179–193.

Talbot–Walsh, G., Kannar, D., Selomulya, G., 2018. A review on technological parameters and recent advances in the fortification of processed cheese. Trends Food Sci. Technol. 81, 193–202.

Tamine, A.Y., 2011. Processed Cheese and Analogues. Wiley–Blackwell publishing, Oxford, UK, p. 350 ISBN 978–1–4051–8642–1.

Tohamy, M.M., Ali, M.A., Shaaban, H.A.G., Mohamad, A.G., Hasanain, A.M., 2018. Production of functional spreadable processed cheese using Chlorella vulgaris. Acta Sci. Pol. technol. Aliment. 17 (4), 347–358.

Tremlová, B., Štarha, P., Buňka, F, Gistingrová, Z., Hrabě, J., 2006. The effect of sterilization on size and shape of fat globules in model processed cheese samples. Acta Veterinaria Brno 75, 419–426.

Trivedi, D., Bennett, R.J., Hemar, Y., Reid, D.C.W., Lee, S.K., Illingworth, D., 2008a. Effect of different starches on the rheological and microstructural properties of (I) model processed cheese. Int. J. Food Sci. Technol. 43, 2191–2196.

Trivedi, D., Bennett, R.J., Hemar, Y., Reid, D.C.W., Lee, S.K., Illingworth, D., 2008b. Effect of different starches on the rheological and microstructural properties of (II) commercial processed cheese. Int. J. Food Sci. Technol. 43, 2197–2203.

Udyarajan, C.H.T., Horne, D.S., Lucey, J.A., 2007. Use of time–temperature superposition to study the rheological properties of cheese during heating and cooling. Int. J. Food Sci. Technol. 42, 686–698.

USA, 2020. Food and Drugs Administration. Electronic Code of Federal Regulations: Title 21: Food and Drugs, Part 133 Cheese and Related Cheese Products. [2018-02-20]. <https://www.ecfr.gov/cgi–bin/text–idx?SID=a7f0a0b592020b523fd3ade7718dc51f&mc=true&tpl=/ecfrbrowse/Title21/21cfr133_main_02.tpl>.

Vannucci, L., Masi, L., Gronchi, G., Fossi, C., Carossino, A.M., Brandi, M.L., 2017. Calcium intake, bone mineral density, and fragility fractures: evidence from an Italian outpatient population. Arch. Osteoporos. 12 (1), 1–7.

Vega, C., Dalgleish, D.G., Goff, H.D., 2005. Effect of κ–carrageenan addition to dairy emulsions containing sodium caseinate and locust bean gum. Food Hydrocoll. 19, 187–195.

Weiserová, E., Doudová, L., Galiová, L., Žák, L., Michálek, J., Janiš, R., Buňka, F., 2011. The effect of combinations of sodium phosphates in binary mixtures on selected texture parameters of processed cheese spreads. Int. Dairy J. 21 (12), 979–986.

Weiss, D., Stangierski, J., Baranowska, H.M., Rezler, R., 2018. Kinetic models of quality parameters of spreadable processed Gouda cheese during storage. Food Sci. Biotechnol. 27 (12), 1–8.

Williams, P.A., Phillips, G.O., 2009. Introduction to food hydrocolloids. In: Phillips, G.O., Williams, P.A. (Eds.), Handbook of Hydrocolloids. Woodhead Publishing Limited and CRC Press, Boca Raton, pp. 87–102 ISBN 978-1-4398-0820-7.

Winter, H.H., Chambon, F., 1986. Analysis of linear viscoelasticity of a crosslinking polymer at the gel point. J. Rheol. 30, 367–382.

Zanoni, B., Pagliarini, E., Giovanelli, G., Lavelli, V., 2003. Modelling the effects of thermal sterilization on the quality of tomato puree. J. Food Eng. 56, 203–206.

Zhang, X., Yang, Y., Zhai, D., 2011. Effect of blueberry on spreadable processed cheese. J. Northeast Agric. Univ. 18 (2), 73–78.

Packaging, cooling, and storage of processed cheese

10

Imène Felfoul and Hamadi Attia
University of Sfax, Analysis, Valuation and Food Safety Laboratory (LAVASA), National Engineering School of Sfax (ENIS), Route Soukra, Sfax, Tunisia

10.1 Introduction

Cheese has been present in the human diet for many years (Salque et al., 2013). Although, cheese is considered a nutritional energy source of protein and calcium, it is very well known for its poor stability over time. Therefore, to overcome this issue, processed cheese was developed, which is a derivative of natural cheese with higher stability over time and lesser requirement for refrigeration (Guinee et al., 2004).

Processed cheese was produced, for the first time, at the beginning of the 20th century. Today, a large variety of processed cheese products with different consistencies and flavors are manufactured all over the world. For instance, in the USA, processed cheese is subdivided into four main groups of cheese based on the ingredients type and relative amount in its production, namely processed cheese blocks, processed cheese foods, processed cheese spreads, and pasteurized blended cheese (Carić and Kaláb, 1993). In addition, processed cheese must legally contain more than 51% of natural cheese. Fig. 10.1 describes the factors that contribute to the popularity of the processed cheese products.

The manufacturing of processed cheese consists of blending shredded natural cheeses of different types and maturity degrees with emulsifying salts and water, and by heating the obtained mixture under partial vacuum with constant agitation until obtaining a homogeneous mass. All components, such as fat, protein, and other ingredients, should be distributed homogeneously throughout the product. In addition to natural cheeses, many other dairy and nondairy ingredients can be included in the mix (Guinee et al., 2017).

The quality of cheese products can be assessed by several functional properties which determine the cheese behavior during the preparation, processing, packaging, cooling, storage, and consumption (Fox et al., 2000). Flowability, meltability, slicing ability, and oiling-off during cooking describe the functionality of the cheese, and differ depending on the type of the cheese. For instance, melting ability, stretching ability, amount of free oil, and browning are substantial for Mozzarella cheese (Rowney et al., 1999).

The quality of processed cheese is derived from both the formulas and the processing conditions. The type and age of natural cheeses, pH, moisture, protein/fat ratio, amount and type of the emulsifying salts and whey protein concentration are possible factors that can influence the functional properties of the cheese (Shimp, 1985). With respect to processing conditions, an appropriate combination of mechanical (agitation time and

Processed Cheese Science and Technology: Ingredients, Manufacture, Functionality, Quality, and Regulations.
DOI: https://doi.org/10.1016/B978-0-12-821445-9.00013-3

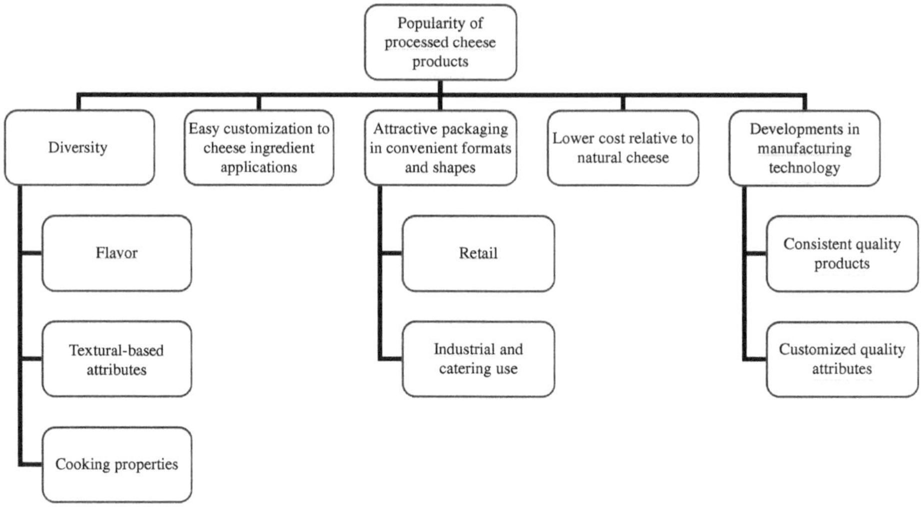

Figure 10.1 Factors determining the popularity of processed cheese.

rate), thermal (temperature and time), and cooling properties is required to produce a cheese product with desired functional properties (Cari and Kaláb, 1993; Fox et al., 2000; Glenn, 2003).

Processed cheese can be packaged in different formats. For example, retail products are available as triangle portions or foil-covered blocks or individually wrapped-slices, while industrial and catering products are available as blocks, slabs, or spreadable products filled into drums or buckets. The variety in appearance, flavor, texture, and packaging format is monitored by modifications in cheeses formulation, processing conditions, and composition (Guinee et al., 2004; Kapoor and Metzger, 2008; Tamime, 2011).

Moreover, cooling is the final step during the production of processed cheese. Consequently, the microstructure of the product developed during cooling affects the final quality of the cheese product. The cheese industry has noticed that faster cooling results in a softer product (Cari and Kaláb, 1993). Furthermore, complex biochemical changes, such as proteolysis, glycolysis, and lipolysis, continue to develop during cheese storage. Subsequently, the developments of the desired cheese products flavor and texture are influenced by moisture content, salt concentration, and microflora (Fox et al., 2000). Therefore, there is a need to evaluate the implications of packaging, cooling and storage effects on cheese functional properties and quality.

10.2 Production of processed cheese

Processed cheese is produced using a large range of dairy ingredients (cream, butter, anhydrous milk fat, curd, milk or milk powder, milk protein powder, whey powder, and caseinates) and nondairy ingredients and additives, such as coloring, hydrocolloids, and sensory active mixtures, which are used to modify the composition or the functional

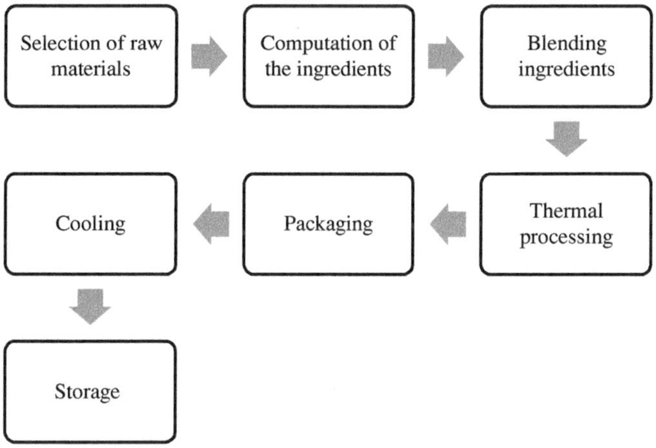

Figure 10.2 The manufacturing process for processed cheese.

properties (meltability and firmness) of the final cheese product. The key elements for producing processed cheeses are the emulsifying salts, which are usually sodium salts of phosphates, polyphosphates, or citrates. The consistency of the cheese product can be firm or spreadable, depending on the water content and the nature of the processed cheeses. The processed cheese making involves several operations carried out in the following order (Guinee et al., 2004; Mizuno and Lucey, 2005) as shown in Fig. 10.2:

1. Selection of raw materials according to the desired characteristics of the final product.
2. Computation of the ingredients in the different mixtures.
3. Weighing and blending the raw materials in the melting device.
4. Premixing the emulsifying salts with the amount of water needed.
5. Heat processing (heating with agitation in a water bath at a temperature from 80 to 105°C for many minutes).
6. Distribution of the hot blend in different volumes, then evacuation and sealing in various wrapping materials.
7. Cooling of the cheese products at room temperature.
8. Storage of the final products at 5°C.

10.2.1 Selection of raw materials

According to the European guide for the hygienic manufacture of processed cheese (2018), the quality of raw materials and ingredients (flavoring substances, salts, additives, gases and processing aids) is substantial to obtain a final product with higher quality meeting the consumers' nutritional and health requirements.

In some countries, processed cheeses made from only one cheese variety at different maturity degrees are very popular such as processed Cheddar, Gruyere and Mozzarella in the USA and Canada; Cheddar cheese in the UK and Australia; Emmental in Western Europe. Many processed cheeses are manufactured from natural cheese, with the type and amount of cheese varying between different processed cheese products (Guinee et al., 2004). Frequently, processed cheeses are produced from more than one type of

Table 10.1 Formulation of processed cheese.

References	Gouda cheese	Edam cheese	Ras cheese	Cheddar cheese	Butter	Skim milk powder	Emulsifying salts	Water
Mohamed et al. (2020)	–	–	47.58	15.86	7.61	–	2.13	26.82
Salek et al. (2020)	–	48	–	–	16	–	3.00	33.00
Cardoso et al. (2020)	30				28	1.5	2.10	38.4
Shalaby et al. (2018)	–	–	38.44	12.80	10.26	5.12	2.50	30.88
Přikryl et al. (2018)	–	45	–	–	14.25	–	2.30	38.45
Dimitreli and Thomareis (2008)	78.4	–	–	–	7.6	–	3.00	11.00
Dimitreli and Thomareis (2007)	81.8	–	–	–	–	8.7	3.00	6.50

cheese to promote a specific product flavor (Guinee et al., 2004). This results in easier processing and a better flavor balance. The most important criteria for the selection of cheese are type, flavor, age, composition, texture, and acidity, as they must be chosen carefully according to the required properties of the final cheese product.

Some defective cheeses can be used to produce processed cheese. Indeed, some physical defects can be corrected by skillful blending. Natural cheeses with microbial defects such as sporulation and gas production should not be selected for processing. However, proper selection of good quality natural cheese does not guarantee that the processed cheese will be of the desired high quality (Cariç and Kalab, 1987).

10.2.2 Computation of the ingredients

The processed cheese formulation is designed with target levels of moisture, fat and emulsifying salts to cheese casein ratio to provide a final product having the desired composition. In general, the weight ratio of emulsifying salts to casein protein in the cheese is kept constant to avoid the confounding effects of potential variations in protein hydration and physical properties associated with changing the level of emulsifying salts (Shirashoji et al., 2010; Guinee and O'Kennedy, 2012). The target protein to-fat ratios can be achieved by using cheese, butter oil, water, and emulsifying salts as ingredients in the processed cheese formulation. Table 10.1 summarizes the possible formulations of processed cheeses.

10.2.3 Blending ingredients in melting device

Blending of the ingredients aims to ensure a uniform final product quality (Guinee, 2009). Processed cheese is manufactured by mixing cheese in the presence of emulsifying salts as well as other dairy and nondairy ingredients followed by heat processing and continuous stirring to form a homogenous product with an extended shelf life. However, due to the heat processing applied and the extended shelf life, spore-forming bacteria may result in safety and stability issues.

Processed cheese may be considered as stable oil-in-water emulsions obtained from natural cheeses or dairy proteins and edible oils or fats with the addition of emulsifying agents as well as other dairy and nondairy ingredients, with natural cheeses as a basic mixture according to Meyer (1973). The type and level of the blended ingredients depend on the processed cheese category. These ingredients are subjected to heat treatment and continuous mixing process to homogenize and enhance the texture of the final cheese product (Hosseini-Parvar et al., 2015).

The formulation of processed cheeses differs mainly based on fat content of the natural cheese, which is the major ingredient, accounting for ≥ 51 g 100 g^{-1} regardless of the category of processed cheeses produced. Other dairy ingredients can be added at different levels, namely processed cheese preparations according to the *Codex Alimentarius* Commission (FAO/WHO, 2006). Based on their composition, water content and consistency, three major groups of processed cheese can be distinguished: processed cheese foods, processed cheese spreads and processed cheese blocks. Both block and spreadable type processed cheeses have similar ingredients but the proportions as well as the manufacturing process vary with respect to the final cheese product. Natural cheese of different ages and degrees of maturity are used in the production of processed cheeses in the presence of emulsifying salts and other dairy and nondairy ingredients.

10.2.4 Thermal processing

The heat treatment of the ingredients is performed by direct or indirect steam with constant agitation. Thermal processing extends the shelf life of the final product through destroying all pathogenic and spoilage microorganisms. Heat treatment also promotes the production of a final product with the desired physicochemical, functional, and microstructural properties as well as its stability over time. A partial vacuum can also be applied during cooking. Indeed, it may be useful to regulate moisture with the application of a direct steam injection. Partial vacuum can also be beneficial in removing the air, thus preventing the presence of air openings in the final cheese product.

10.3 Chemical composition of processed cheese

The composition of processed cheese, including pH, dry matter, fat, protein, and ash contents is determined according to the standard methods published by the International Dairy Federation (IDF), the Association of Official Analytical Chemists (AOAC, 2007) and the International Organization for Standardization (ISO). Nevertheless, many countries have their legal standards covering the composition of processed cheese. Table 10.2 summarizes the average chemical composition and pH values of various processed cheeses. Moisture and protein contents, emulsifying salts and processing conditions can have a major impact on the rheological characteristics as well as the sensory attributes of the final cheese product (Tamime, 2011). The pH of the molten mix is another factor that affects the properties of processed cheese. The

Table 10.2 Chemical composition and pH of processed cheese.

References	Dry matter (%)	Fat (%)	Protein (%)	Lactose (%)	Ash (%)	pH (–)
Mohamed et al. (2020)	45.00	22.50	13.52	0.95	4.96	5.82
Cardoso et al. (2020)	38.7	23.00	9.7	2.5	2.9	–
Shalaby et al. (2018)	44.64	22.73	12.74	3.49	4.01	5.73
Přikryl et al. (2018)	37	18.5	–	–	–	5.80
Tamime (2011)	41.1	16.5	10	13	–	–
Gliguem et al. (2009)	47.30	32.20	9.00	2.10	2.70	5.40
Gliguem et al. (2009)	43.30	22.50	10.50	6.80	3.70	5.52
Dimitreli and Thomareis (2008)	58.00	29.30	15.0	–	–	5.90
Piska and Stetina (2004)	48.5	37	–	–	–	5.58

pH of the processed cheese samples ranged from 5.40 to 5.90 for all processed cheese as shown in Table 10.2, which can be characterized as acceptable for processed cheese products. Moreover, the buffering capacity of the emulsifying salts used, could serve as possible explanations for the "narrow" variability observed in the pH values (Dimitreli and Thomareis, 2009; Lee and Klostermeyer, 2001; Lu et al., 2008; Salek et al., 2020).

10.4 Packaging of processed cheese

The hot molten blend is pumped directly to the filling machines or by gravity from the cooker through the closed pipeline system or homogenizer. Several packaging formats are possible using special filling machines (Buts and Mostert, 2011; Meyer, 1973; Zehren and Nusbaum, 1992). For instance, retail products are available as foil-covered blocks or triangle portions, individually wrapped- or stacked-slices in plastic film, fingers, tubes, tubs of spread, whereas products for the catering industrial trade are available as sliceable blocks, slabs, sausage-shapes wrapped in various types of plastic film, dips/sauce, cans, tubes, or pastes/spreadable products filled into drums or buckets. Recently, the slicing and packing of the cheese slices are newly developed, which is convenient for sandwiches mainly for young children in their lunch boxes. The slices of cheese can be obtained by mechanically slicing rectangular blocks of processed cheese. During the manufacture of cheese slices, the hot processed cheese is continuously pumped into an endless "tube" of plastic film, which is automatically flattened and crimped into a chain of individually wrapped slices by rotary crimping conveyors (Zehren and Nusbaum, 1992). Then, the chain of wrapped cheese slices is cooled to <10°C, dried, and cut into individual portions, stacked, and packed.

Pumping the hot molten cheese can be performed using a manifold with 8–12 nozzles. Consequently, the hot processed cheese is ejected as a continuous sheet on counter-rotating refrigerated rollers which allow it to be rapidly cooled from 80–70°C to <30°C. Then, the refrigerated sheet is automatically cut into parallel ribbons, and each ribbon is cut into slices. Then, these cheese slices are individually wrapped in plastic film, stacked, and packed (Guinee, 2017).

Sometimes, some problems occur such as a delay or a stoppage of the packing lines, resulting in extending the holding time of the hot molten cheese at 70–90°C, which cause the over creaming of the product. It is manifested as the development of a pudding-like consistency with a dull appearance of the cheese product. This defect becomes obvious when the product has cooled. Over-creaming manifests differently depending on the processed cheese packaging type. In fact, for block processed cheese and slices, over-creaming indicates the development of a coarse structure upon cooling. Thus, over creaming presents a real problem for industrials mainly in pumping the blend into the filling machines. It also causes the deterioration of the final product's properties, for instance, loss of cooking properties, loss of spreadability, loss of surface sheen, nonuniform greasy appearance in both slices and blocks processed cheeses (Guinee, 2017).

The variability in packaging format is mainly monitored by the changes in composition and processing conditions (Guinee et al., 2004; Kapoor and Metzger, 2008; Meyer, 1973; Tamime, 2011; Zehren and Nusbaum, 1992). Processed cheese can be stored at a low temperature or frozen without packaging. Then, they can be reprocessed as a blend ingredient in later batches of processed cheeses.

The design and type of packaging along with the quality and price policy are the main determining factors in the processed cheese sale. Product quality control in the production process, sensory, physicochemical, and microbiological parameters of processed cheeses packaged in polypropylene cups, briquettes, slices, smoked sausage processed cheese are carried out for the purpose of release of standard and safe production according to the requirements imposed on them by the relevant regulatory documents.

The packaging material can be considered as a potential source of contamination for processed cheeses. Thus, it must be properly handled, stored, and disinfected before it is filled with the processed cheese. It should be noted that even after the thermal treatment, the material may serve as a source of spore recontamination (Glass and Doyle, 2005). Spores can germinate and get too high in the final product whether the pH and water activity are adequate, which can lead to the product's spoilage.

The following indicators are studied in the literature:

- Organoleptic parameters: Taste, smell, consistency, view of the section, color - IDF (1997) « *Sensory Evaluation of Dairy Products by Scoring (Reference Method)* », Standard 99C, International Dairy Federation, Brussels.
- Total solid content - IDF (2004a) « *Cheese and Processed Cheese: Determination of Total Solid Content (Reference Method)* », Standard 4, International Dairy Federation, Brussels.
- Fat content - IDF (2004b) « *Cheese and Processed Cheese: Determination of Fat Content Gravimetric Method (Reference Method)* », Standard 5, International Dairy Federation, Brussels.
- Ash content – IDF (1964) « *Determination of the Ash Content of Processed Cheese* », Standard 27, International Dairy Federation, Brussels.
- Coliforms - IDF (2004c) « *Microbiology of Food and Animal Feeding Stuffs: Horizontal Method for the Detection and Enumeration of Coliforms* », Standard 4831, Most probable number technique, International Standards Organization, Geneva.
- Molds and yeast - ISO (2004) « *Milk and Milk Products: Enumeration of Colony-Forming Units of Yeasts and/or Moulds* », Standard 6611, Colony-count technique at 25°C, International Standards Organization, Geneva.

- Pathogenic microorganisms, including salmonella - IDF (2001) « *Milk and Milk Products: Detection of* Salmonella », Standard 93, International Dairy Federation, Brussels.

According to Shulga et al. (2021), organoleptic, physicochemical, and microbiological indicators of processed cheeses in polypropylene cups, in briquettes, slices and smoked bass, corresponded to the requirements imposed on them by the relevant normative documents. The highest score in the "sensory analysis" was attributed to processed cheese packaged in briquettes. The safety of cheeses is controlled by the content of residual amounts of heavy metals, antibiotics, radionuclides, toxic elements, pesticides, nisin, and hormonal preparations.

Processed cheeses in briquettes are tightly wrapped in foil, slices of processed cheeses are packed in a polyethylene shell, cheese bread with the taste and aroma of mushrooms is packed in polypropylene cups, and smoked cheese is wrapped in a paraffin layer. The packaging of all processed cheeses was not damaged. The surface of the cheese after removing the package is clean, smooth, not dried, not moldy. The taste and smell are pronounced, characteristic of each type of cheese. The consistency and color of the dough, characteristic of each type of processed cheese, are uniform throughout the mass of the cheese, there is no pattern and air voids (Shulga et al., 2021).

Processed cheese is packed at a temperature of ≥75°C to prevent the development of any potential microbiological contaminants from the environment or from the packaging material. At the stage of processing, the microbiological hazards can be caused from the environment. Thus, packaging materials must be sterilized to prevent the presence of yeasts and molds. Processed cheese may be contaminated with fragments from the packing materials used (e.g., foil, glue, labels, cardboard, clips) or metal fragments from the packing machine. The critical control points of the packaging processing depend on the packaging machine(s) used, and metal sieves (5.2-mm diameter holes) are fitted at the outlet of the filling nozzle of the packaging machine. However, the product could also be contaminated with residues of detergents and disinfectants.

10.5 Cooling of processed cheese

Cooling step must be as fast as possible for processed cheese spreads and relatively slow for processed cheese blocks since fast cooling softens the product. However, slow cooling can intensify Maillard reactions and promote the growth of spore-forming bacteria. Cooling of processed cheese is carried out slowly in a way that 20°C can be reached in almost 50 h. First, the molten cheeses were kept at room temperature (20–22°C) for 8 h. Then, they were moved to a cooling chamber at 10–15°C for 24 h. At last, the processed cheeses were kept at a relatively low temperature of 5–8°C in a cold room (Guinee, 2017).

The control of cooling process is exhibited in monitoring the temperature upon cooling the cheese product, which is usually set at 10°C, as well as the airflow in the cold room. The rate of cooling of the processed cheese affects the rheological properties of the final product (Piska and Stetina, 2004). Moreover, in the cooling chambers, the final processed cheese should be kept clean and free from pests.

The cooling phase contributes to the development of the final properties of the processed cheese as well as its related products. These properties depend on the blend formulation. In fact, thermal processing, and cooling rate depend on the processed cheese consistency, that is, firm or semi-soft. Many factors can contribute to the development of the structure of the final product during the cooling phase namely fat crystallization, incorporation of recombinant fat globules, and protein-protein interactions. With a better understanding of cheese physical and rheological properties and cheese microstructure during cooling, cheese manufacturers can control cooling timetables to help optimize quality attributes in processed cheese.

10.5.1 Effect of cooling on physical properties of processed cheese

Upon cooling, the processed cheeses seemed to be wet and fatty due to the presence of free fat and moisture release unlike molten cheeses heated at 60–100°C. These differences are related to the increase in protein aggregation induced by hydrophobic interactions of the protein network and to the coalescence of fat globules, as well as the release of free fat and water from the processed cheese matrix. These changes occurring during heating are influenced by numerous factors, namely the thermal processing of the cheese, the cheese composition, and the type of cheese coagulation (acid or enzymatic coagulation).

In some cases, the heat-treated processed cheeses were unstable, that is, undergoing moisture-exudation and oiling-off during cooling and storage. In 1991, Walter Gerber and Fritz Stettler developed a new processing method of processed cheese. They manufactured a stable heat-treated Emmental cheese type "Schachtelkäse" using sodium citrate as a "melting salt" to the powdered cheese before heating and shearing processing (Meyer, 1973). It has been reported afterwards that other cheeses, namely Cheddar cheese, could also be used to produce stable processed cheese products using different "melting salts," such as sodium phosphate or other emulsifying salts. The mechanism of heat stability's effect of emulsifying salts or "melting salts" is defined as follows: heated proteins were solubilized by removal of bound calcium and increasing pH, and emulsification of free fat by the hydrated proteins. Similarly, many food ingredients other than cheese were later included before processing (Price and Bush, 1974a, 1974b).

10.5.2 Effect of cooling on rheological properties

The monitoring of cooling stage is mostly limited to control the temperature as the finished product is being cooled to 10°C as well as the airflow in the cooling chamber. The rate of cooling of processed cheeses influences the rheological properties of the end products (Piska and Stetina, 2004).

A previous study performed by Piska and Štětina (2004) on the influence of the cooling rate on texture and rheological properties of processed cheese showed that a rapid cooling led to decreases of the complex modulus, yield stress, and the textural attributes, namely hardness, adhesiveness, and gumminess of processed cheese, but increases of the tangent of the phase shift. Consequently, the rapidly cooled samples showed more viscous behavior than those slowly cooled samples. The spreadability of the rapidly cooled samples was better, but very soft consistency and the stickiness to the aluminum foil were observed during the sensory evaluation.

10.5.3 Effect of cooling on functionality of processed cheese

Firmness and meltability are two important functional properties during the production of processed cheeses. These two criteria can be handled by changing formulations including appropriate types and ages of natural cheeses, pH, emulsifying salt types and concentrations, and protein/fat ratio (Shimp 1985). When the composition is not adjustable, changing processing parameters, namely cooling rate, provides another approach in modulating cheese texture.

In the industrial production, cheese blocks could reach around four hundred kilograms thus requiring cooling for approximately 7 days (Reinbold and Ernstrom, 1988; Reinbold et al., 1992a, 1992b). The rates of cooling in a vast block would have a large distribution, yielding texture differences inside the cheese block. At higher meltability, a faster cooling is achieved due to the shrunk of the block dimension, thus a better melt. Consequently, the cheese manufacturing will be accelerated, improving the production capacity, speeding the material flow, and saving the storage space.

A previous study conducted by Zhong et al. (2004) on the influence of cooling on processed cheese functionality revealed that using dynamic tests, the storage moduli during cooling at different rates showed the same trend as the industrial observations, i.e., a slower cooling contribute to a firmer cheese. Thus, to illustrate the cooling effects on cheese functionality, a heat transfer model was derived within cheese loaves during cooling at a constant temperature. The average rates of cooling were similar in a 5-pound cheese loaf, generating a comparable texture within the cheese block. To test the meltability of cheese loaves, cheese was cut with a wire and the measurement was carried out at the sliced position. No trend was observed for the Schreiber scores and the forces required for cutting. As for the forced convection treatment, that is, faster cooling rates, a smaller force was required for cutting, and a better melt was observed.

10.5.4 Effect of cooling on microstructure of processed cheese

Cooling is the final stage in processed cheese manufacturing. It started when protein network forms and fat crystallization occurs. During this stage, cheese microstructure is developed based on protein-protein interactions, fat crystallization and fat-protein interactions (Fox et al., 2000). These interactions have a great impact on the microstructure of processed cheese, which makes cooling stage extremely important. The microstructure of the processed cheese formed during cooling affects the final product's quality.

During the cooling step, the packaged processed cheese mixture circulates in a large refrigerator. Indeed, processed cheese blends enter in the refrigeration chamber at a temperature above 80°C and exit at 5°C. Nevertheless, the cooling rate of the cheese is not maintained at the same level from 80 to 5°C due to the nonuniformity of the applied temperature during cooling. Two cooling stages may be applied to cool down the processed cheese at different rates and temperatures. During the first cooling stage, the protein matrix is created. The fat crystallizes during the second cooling stage of the cheese product. The interactions and microstructure established during the two phase cooling have a substantial effect on the final cheese texture (Zhong, 2003; Caric and Kalab, 1993).

Furthermore, a faster cooling rate develops a higher number of smaller fat crystals, whereas slower cooling results in fewer numbers of larger crystals (DeMan, 1964; Herrera and Hartel, 2000; Campos et al., 2002). Faster cooling rates also result in an increase in solid fat content, which yields a firmer fat product (Lopez et al., 2000; DeMan, 1964; Herrera and Hartel, 2000; Campos et al., 2002, Tang and Marangoni, 2006). This result may also be attributed to differences in crystalline structure (crystal size, distribution of size, shape, polymorph, surface characteristics, the interaction between crystals and liquid oil) (Mulder, 1953).

Price (2007) used confocal laser scanning microscopy (CLSM) to investigate the effects of cooling rate on the microstructure of processed cheese analogues. Micrographs of cheese analogues revealed a protein ring surrounding fat droplet, which are formed during cooling from 80 to 5°C.

Zhong (2003) proposed a viable mechanism for casein gelation, which is a critical process in many dairy foods. The author found that slower cooling rates of protein gels lead to a firmer gel network. The mechanism was based on the number and size of casein flocs formed upon cooling. For slower cooling rates, milk fat crystals are larger sized, which blocks the behavior of the casein network. The casein structure develops before fat crystallization (80–40°C). Consequently, a more uniform casein structure and a more elastic gel are established upon slower cooling during the formation of protein network, thus must limit the fat crystal sizes to produce a firmer cheese. However, cheese is a combination of fat and protein and undergoes more complex two cooling phases (80°C–40°C and 40°C–5°C). Therefore, the protein-to-fat ratio is an important factor in cheese microstructure.

10.6 Storage of processed cheese

The final product should be stored at temperatures below 10°C, although such low temperatures may include crystal formation (calcium monophosphate, calcium diphosphate, calcium pyrophosphate, etc.). The relative humidity of air should also be controlled (around 85%) in storage rooms of processed cheese. Moreover, the storage rooms must be clean and free from pests. After processing, the cheese mixture is immediately transferred to packaging hoppers. Molten cheese is packaged to the product specifications (Zhong, 2003). For cheese slices, the melted mixture is spread uniformly by steel rollers and sectioned using rotary knives of the appropriate dimensions. Regardless of the cheese form, processed cheese is packaged in the absence of oxygen to inhibit molds growth (Zhong, 2003).

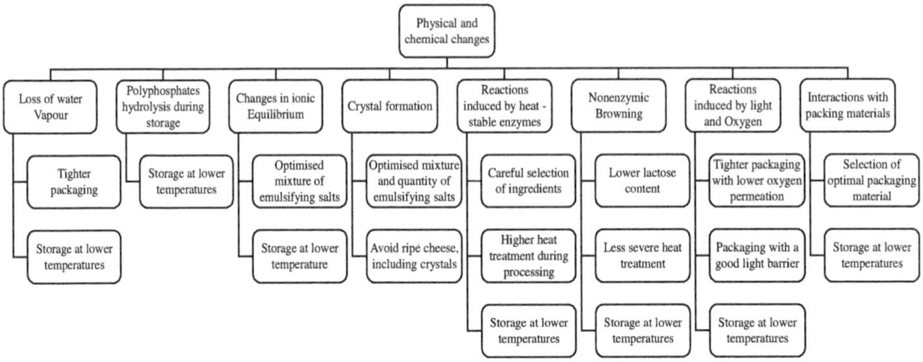

Figure 10.3 Physicochemical changes in processed cheese during storage.

10.6.1 Physical and chemical changes in processed cheese during storage

Berger et al. (1989) considered that the processed cheese is a "semi-preserved food" with a limited shelf life. Processed cheese packaged in plastic foils has a shelf life of not more than 3-4 months. Cheese products stored in metal cans or tubes may have longer shelf lives. Chambre and Daurelles (1997) revealed that processed cheese products generally retain their good quality for up to 6–12 months at room temperature. According to Sturm (1998), the shelf life of some cheese products stored at room temperature is as follows:

- 8 weeks for cheese slices;
- 20 weeks for small cheese portions;
- >1 year for cheese products packed in cans or tubes.

Fig. 10.3 summarizes the possible physicochemical changes and their relevance during the storage of processed cheese.

10.6.1.1 Loss of water vapor

In general, the packaging material for processed cheese does not provide an impervious barrier to water vapor. The storage temperature has a considerable effect on weight loss of processed cheese, which leads to a firmer texture of the final product (Schar and Bosset, 2002). According to Ney (1988), processed cheese in slices form, stored at 20°C for 1 month, presented a weight loss of about 2–5 g/kg.

10.6.1.2 Hydrolysis of polyphosphates

The hydrolysis of polyphosphate starts during the melting operation and continues until the storage period of processed cheese (Ney, 1988). According to Chambre and Daurelles (1997), the hydrolysis of polyphosphates to pyrophosphates (triphosphates and diphosphates) and orthophosphates establishes new acid functions, thus decreasing the pH of the cheese product and consequently inhibits its eventual increase. The buffering ability of polyphosphates depends on the chain length. Indeed, the higher the chain length, the lower the buffering capacity. Long-chain polyphosphates improve calcium binding (Berger et al., 1989; Caric and Kalab, 1997). Ney (1988) measured the rheological properties of numerous processed cheese slices during 1-month storage at 4, 13 and 20°C. Processed cheese slices were very hard to break. The breaking angles decreased during storage period. These results were related to the hydrolysis of polyphosphate (Ney, 1988).

10.6.1.3 Changes in ionic equilibria

pH and ionic balance stabilize slowly after processing and cooling stages of the processed cheese. At the storage stage, the ionic balance in the processed cheese is impacted by both the water vapor loss and the hydrolysis of polyphosphates. Furthermore, the interactions between protein, calcium, sodium and emulsifying salts affect the water binding property as well as the texture of the final cheese product. In general, processed cheeses become firmer during the storage period. Spreadable cheese products must be stored at 10–15°C to avoid the formation of a firmer and shorter texture (Chambre and Daurelles, 1997).

10.6.1.4 Formation of crystals

Crystals, such as calcium phosphate, calcium diphosphate, calcium lactate, lactose, and calcium tyrosinate, are due to compounds with low solubility. Many factors promote crystals formation such as excessive amounts of emulsifying salts or lactose, high calcium content in the natural cheese, high pH and long storage period of the processed cheese at low temperatures (Caric and Kalab, 1997). Moreover, mechanical treatments and ratio of phosphates/citrates in the emulsifying salt influence the crystallization on the surface of processed cheese slices (Uhlmann et al., 1983). Small white crystals (calcium citrate or tertiary sodium calcium citrate) appear on the surface of processed cheese slices (Berger et al., 1989).

10.6.1.5 Reactions induced by heat-stable enzymes

The manufacturing of processed cheese causes the denaturation and/or the deactivation of many enzymes mainly proteases and lipases. However, denatured proteases and heat-stable lipases can still show proteolytic and lipolytic activities, respectively. In 1994, Mulvihill and McCarthy investigated the proteolytic and rheological changes in cheese analogues during storage. Proteolysis was influenced significantly by plasmin content over the storage period (1–35 weeks). The hydrolysis of β–casein was more marked than that of α_{s1}–casein during storage. Moreover, processed cheese may establish a lipolytic activity over storage (Schar and Bosset, 2002).

10.6.1.6 Maillard reaction

The Maillard reaction plays a major role in products containing high amounts of lactose (10–80 g/kg) which is the case of processed cheese (Berger et al., 1989). During the storage phase and especially at higher temperatures, the formation of both a brownish color and an off-flavor limit the shelf life of the cheese product. Low lactose and protein contents as well as less temperature of processing and storage (approximately 5°C) may reduce the extent of Maillard reaction (Berger et al., 1989; Piergiovanni et al., 1989). The production and the composition of stirred-curd Cheddar cheese have great effects on the non-enzymic browning reaction of processed cheese (Bley et al., 1985). Faster cooling of processed cheese decreased also the intensity of brown color.

10.6.1.7 Reactions induced by light and oxygen

Light-induced reactions occur for cheese slices conditioned in transparent foil but do not happen in opaque packaging materials namely aluminum foils or steel cans. Usually, transparent foil is used for processed cheese under nitrogen/carbon dioxide atmosphere with oxygen content of less than 20 mL/L. The oxidative reactions lead to various modifications in the cheese product, namely color change (Kristensen et al., 2001) and off-flavor formation (methional and carbonyl compounds) as well as cholesterol oxidation. These reactions depend on the processed cheese composition, exposure to light and packaging properties. According to Goyal and Babu (1991), cheese products packaged in tin cans had the least chemical changes.

10.6.1.8 Interactions with packaging materials

Schwope et al. (1987) measured the migration of butylhydroxytoluene (BHT) and Irganox 1010 into processed cheese slices. The authors reported that the molecular weight strongly affect the migration of the materials. Indeed, BHT with a low molecular weight migrated faster than Irganox 1010. Furthermore, coating is required to prevent the undesirable migration of aluminum into the cheese. Indeed, the coating of aluminum foil is carried out using a protective polymer and a polyvinyl co-polymer at a sealing temperature of 65–75°C (Sturm, 1998).

10.6.2 Effect of storage on physicochemical properties of processed cheese

The pH values of processed cheeses gradually decreased during the storage period (3 months of storage) due to the hydrolysis of emulsifying salts and their interaction with proteins (Mohamed et al., 2016). Protein, fat, and calcium contents of initial raw milk affect the cheese matrix of the resulted cheese (Barłowska et al., 2011). The penetrometer reading values decreased gradually during processed cheese storage (Fawzia et al., 2008; Hussein et al., 2005).

The oil separation percent of processed cheese increased with the storage period. The oil separation depends on the state of milk protein and fat in the obtained emulsion of processed cheese, which is affected by the type and amount of the raw materials used in the preparation of processed cheese, pH, cooking time, and temperature (Mohamed et al., 2016). The increase in oil separation with storage can be due to the decrease in the pH values of final cheese products as well as the changes occurring in soluble nitrogen content (Abdel-Hamid et al., 2000; Omar et al., 2012). Moreover, melting index of processed cheese was reduced during the storage period (Mohamed et al., 2020).

10.6.3 Effect of storage on color parameters of processed cheese

A recent study performed by Mohamed et al. (2020) on the changes of the processed cheese color during 3 months storage showed lower values of whiteness degree and b* parameter, which reflects the yellowish color, for processed cheese during the storage period. According to Wijesinha-Bettoni and Burlingame (2013), this result could be mainly due to initial β-carotene content in milk.

10.6.4 Effect of storage on sensory properties of processed cheese

Schar and Bosset (2002) studied the ready-made fondue cheese packaged in multilayer aluminum pouches, and reported several changes in the mouth-feel of the cheese product. Indeed, the cheese structure is more homogeneous and the corresponding mouth perception is becoming creamier. An increasing "old" off-flavor limits its overall acceptance. Changes in fondue cheese depend on storage temperature. They are slower at 5°C than at 20°C.

10.6.5 Effect of storage on rheological properties of processed cheese

Interest in the mechanical properties of processed cheeses has increased (Konstance and Holsinger, 1992). Some dynamic tests were performed by Diefes et al. (1993) to compare rheological properties of Mozzarella cheese subjected to different storage procedures. These authors reported that the cheese product become hard and more elastic solid after freezing, and softer and more elastic liquid during refrigeration.

Ak and Gunasekaran (1999) investigated the changes in rheological properties of Mozzarella cheese during 1 month of refrigerated aging. These authors noticed a decrease of both loss (G'') and storage (G') moduli with aging. This effect was more noticeable at 20°C and in the first 2 weeks of storage than at 10°C. Hydrolysis of caseins in Mozzarella cheese during refrigerated storage has been reported (Ak et al., 1993; Ak and Gunasekaran, 1995; Farkye et al., 1991). The loss of protein network by proteolysis implies the decrease in G' during processed cheese aging. However, a decrease in viscous contribution (G") over storage was also observed by Ak and Gunasekaran (1999). This contribution is probably because proteolysis generates ionic groups (Creamer and Olson, 1982) that bind water, thus reducing viscous dissipation. Water binding property contributes to the elastic behavior of the cheese.

Salek (2017) confirmed that proteolysis is the major phenomenon that occurs during cheese aging along with glycolysis and lipolysis and affects the physical characteristics of nearly all cheeses. For processed cheese, rheological and textural properties are influenced by the age of the cheese used as well as the technological operation during cheese processing. Therefore, increasing cheese maturity level results in more pronounced proteolytic reactions. However, these properties are also affected by dry matter, solid fat content, pH, type and content of emulsifying salts added, thermal processing, and storage conditions (Brickley et al., 2007; Pachlova et al., 2011; Piska and Stetina, 2004).

10.7 Conclusion

Processed cheese products are supposed to be stable over time with a very long shelf life. However, even cheeses without any microbial contamination preserve their high quality attributes for only a few months at room temperature. The packaging material can be considered as a potential source of contamination for processed cheeses. The cooling phase contributes to the development of the final properties of the processed cheese as well as its related products. However, cooling must be as fast as possible. During storage, the flavor and structure of the processed cheese slowly change. The possible causes of these variations are examined in the present chapter. By investigating the changes during packaging, cooling, and storage stages, it will be useful for cheese makers to better predict and monitor the overall quality of the processed cheeses during all processing phases.

References

Abdel-hamid, L.B., El-Shabraway, S.A., Awad, R.A., Singh, R.K., 2000. Chemical properties of processed ras cheese spreads as affected by emulsifying salt mixtures. J. Food Process. Preserv. 24, 191–208.

Ak, M.M., Bogenrief, D., Gunasekaran, S., Olson, N.F., 1993. Rheological evaluation of Mozzarella cheese by uniaxial horizontal extension. J. Texture Stud. 24, 437–453.

Ak, M.M., Gunasekaran, S., 1995. Measuring elongational properties of Mozzarella cheese. J. Texture Stud. 26, 147–160.

Ak, M.M., Gunasekaran, S., 1999. Dynamic rheological properties of mozzarella cheese during refrigerated storage. J. Food Sci. 61, 566–569.

AOAC, 2007. In: Horowitz, W.E. (Ed.), Official Methods of Analysis. Association of Official Analytical Chemist, Arlington.

Barłowska, J., Szwajkowska, M., Litwinczuk, Z., Krol, J., 2011. Nutritional value and technological suitability of milk from various animal species used for dairy production. Compr. Rev. Food Sci. Food Safety 10, 291–302.

Berger, W., Klostermeyer, H., Merkenich, K., Uhlmann, G., 1989. Die Schmelzka Seherstellung. Benckiser-Knapsack GmbH, Ladenburg.

Bley, M.E., Johnson, M.E., Olson, N.F., 1985. Factors affecting nonenzymatic browning of processed cheese. J. Dairy Sci. 68, 555–561.

Brickley, C.A., Auty, M.A.E., Piraino, P., McSweeney, P.L.H., 2007. The effect of natural Cheddar cheese ripening on the functional and textural properties of the processed cheese manufactured therefrom. J. Food Sci. 72, 480–490.

Buts, E.M., Mostert, J.F., 2011. Packaging materials and equipment. In: Tamime, A.Y. (Ed.), Processed Cheese and Analogues. John Wiley & Sons, Ltd, Chichester, UK, pp. 199–218.

Campos, R., Narine, S.S., Marangoni, A.G., 2002. Effect of cooling rate on the structure and mechanical properties of milk fat and lard. Food Res. Int. 35, 971–981.

Cardoso, L.G., Bordignon Junior, I.J., da Silva, R.V., Mossmann, J., Reinehr, C.O., Briao, M.B., Colla, L.M., 2020. Processed cheese with inulin and microencapsulated chia oil (*Salvia hispanica*). Food Biosci 37, 100731.

Caric, M., Kaláb, M., 1987. Processed cheese products. In: Fox, P.F. (Ed.), Cheese: Chemistry, Physics, and Microbiology. Elsevier Applied Science Publishers Ltd., London, pp. 339–383.

Caric, M., Kaláb, M., 1993. Processed cheese products. In: Fox, P.F. (Ed.), Cheese, Chemistry, Physics and Microbiology, 2. Major Cheese Groups. Chapman and Hall, London, UK, pp. 467–505.

Caric, M., Kalab, M., 1997. Processed cheese products. In: Fox, P.F. (Ed.), Cheese: Chemistry, Physics and Microbiology. Elsevier Applied Science, London, New York, pp. 467–505.

Chambre, M., Daurelles, J., 1997. Le fromage fondu. In: Eck, A., Gillis, J.C. (Eds.), Le Fromage. Technique et documentation Lavoisier, Paris, pp. 691–708.

Creamer, L.K., Olson, N.F., 1982. Rheological evaluation of maturing Cheddar cheese. J. Food Sci. 42, 631–636 646.

DeMan, J.M., 1964. Physical properties of milk fat. J. Dairy Sci. 47, 1194–1200.

Diefes, H.A., Rizvi, S.S.H., Bartsch, J.A., 1993. Rheological behavior of frozen and thawed low-moisture, part-skim Mozzarella cheese. J. Food Sci. 58, 764–769.

Dimitreli, G., Thomareis, A.S., 2007. Texture evaluation of block-type processed cheese as a function of chemical composition and in relation to its apparent viscosity. J. Food Eng. 79, 1364–1373.

Dimitreli, G., Thomareis, A.S., 2008. Effect of chemical composition on the linear viscoelastic properties of spreadable type processed cheese. J. Food Eng. 84, 368–374.

Dimitreli, G., Thomareis, A.S., 2009. Instrumental textural and viscoelastic properties of processed cheese as affected by emulsifying salts and relation to its apparent viscosity. Int. J. Food Prop. 12, 262–275.

FAO/WHO, 2006. General standard for cheese CXS 283-1978 Formerly CODEX STAN A-6-1973. Adopted in 1973. Revised in 1999. Amended 2006. Available from: www.codexalimentarius.org/input/../standards/175/CXS_283e.pdf.

Farkye, N.Y., Kiely, L.J., Allshouse, R.D., Kindstedt, P.S., 1991. Proteolysis in Mozzarella cheese during refrigerated storage. J. Dairy Sci. 74, 1433–1438.

Fawzia, H., Abd Rabou, Abd El-Fattah, A.M., El-Sayed, M.M., Mohamed, A.G., 2008. Effect of some formulated emulsifying stats on processed cheese properties. Egypt. J. Dairy Sci. 39, 126–131.

Fox, P.F., Guinee, T.P., Cogan, T.M., McSweeney, P.L.H., 2000. Fundamentals of Cheese Science. Aspen Publishers, Inc, Gaithersburg, MD, USA.

Glass, K., Doyle, M.E., 2005. Safety of Processed Cheese, FRI Briefings. Food Research Institute, University of Wisconsin, Madison.

Glenn, T.A., Daubert, C.R., Farkas, B.E., Stefanski, L.A., 2003. Statistical analysis of creaming variables impacting process cheese melt quality. J. Food Qual. 26, 299–321.

Gliguem, H., Ghorbel, D., Lopez, C., Michon, C., Ollivon, M., Lesiure, P., 2009. Crystalisation and polymorphism of triacylglycerol contribute to rheological properties of processed cheese. J. Agr. Food Chem. 57, 3195–3203.

Goyal, G.K., Babu, K.E., 1991. Influence of packaging and storage on the chemical quality of processed cheese. Ind. J. Dairy Sci. 44, 274–279.

Guinee, T.P., 2017. Pasteurized Processed and Imitation Cheese Products. In: McSweeney, P.L.H., Fox, P.F., Cotter, P.D., Everett, D.W. (Eds.), Cheese. Elsevier Academic Press, Oxford, UK, pp. 1133–1184.

Guinee, T.P., 2009. The role of dairy ingredients in processed cheese products. In: Corredig, M. (Ed.), Dairy-Derived Ingredients. Woodhead Publishing, Oxford, UK, pp. 507–538.

Guinee, T.P., Caric, M., Kaláb, M., 2004. Pasteurized processed cheese and Substitute/Imitation products. In: Fox, P.F., McSweeney, P.L.H., Cogan, T.M., Guinee, T.P. (Eds.), Cheese Chemistry, Physics and Microbiology. Elsevier Academic Press, Oxford, UK, pp. 349–394.

Guinee, T.P., O'Kennedy, B.T., 2012. Reducing the level of added di-sodium phosphate alters the chemical and physical properties of processed cheese. Dairy Sci. Technol. 92, 469–486.

Herrera, M.L., Hartel, R.W., 2000. Effect of processing conditions on crystallization kinetics of a milk fat model system. J. Am. Oil Chem. Soc. 77, 1177–1187.

Hosseini-Parvar, S.H., Matia-Merino, L., Golding, M., 2015. Effect of basil seed gum (BSG) on textural, rheological and microstructural properties of model processed cheese. Food Hydrocoll 43, 557–567.

Hussein, G.A.M., Fathi, F.A., Mohamed, A.G., 2005. Quality and acceptability of processed cheese spreads made from total milk proteinate and casein co-precipitate. Egypt. J. Dairy Sci. 33, 261–277.

IDF, 1964. «*Determination of the Ash Content of Processed Cheese*». Standard 27, International Dairy Federation, Brussels.

IDF, 1997. «*Sensory Evaluation of Dairy Products by Scoring (Reference Method)*». Standard 99C, International Dairy Federation, Brussels.

IDF, 2001. «*Milk and Milk Products: Detection of* Salmonella». Standard 93, International Dairy Federation, Brussels.

IDF, 2004a. «*Cheese and Processed Cheese: Determination of Total Solid Content (Reference Method)*». Standard 4, International Dairy Federation, Brussels.

IDF, 2004b. «*Cheese and Processed Cheese: Determination of Fat Content Gravimetric Method (Reference Method)*». Standard 5, International Dairy Federation, Brussels.

IDF, 2004c. «ISO (2006f) *Microbiology of Food and Animal Feeding Stuffs: Horizontal Method for the Detection and Enumeration of Coliforms*». International Standards Organization, Geneva Standard 4831, Most probable number technique.

ISO, 2004. «*Milk and Milk Products: Enumeration of Colony-Forming Units of Yeasts and/or Moulds*». International Standards Organization, Geneva Standard 6611, Colony-count technique at 25°C.

Kapoor, R., Metzger, L.E., 2008. Process cheese: scientific and technological aspects—a review. Compr. Rev. Food Sci. Food Safety. 7, 194–214.

Konstance, R.P., Holsinger, V.H., 1992. Development of rheological test methods for cheese. Food Technol 1, 105–109.

Kristensen, D., Hansen, E., Arndal, A., Trinderup, R.A., Skibsted, L.H., 2001. Influence of light and temperature on the colour and oxidative stability of processed cheese. Int. Dairy J. 11, 837–843.

Lee, S.K., Klostermeyer, H., 2001. The effect of pH on the rheological properties of reduced-fat model processed cheese spreads. Lebensm. Wiss. Technol. 34, 288–292.

Lopez, C., Lesieur, P., Keller, G., Ollivon, M., 2000. Thermal and structural behavior of milk fat: 1. Unstable species of cream. J. Colloid Interface Sci. 229, 62–71.

Lu, Y., Shirashoji, N., Lucey, J.A., 2008. Effects of pH on the textural properties and meltability of pasteurized process cheese made with different types of emulsifying salts. J. Food Sci. 73, 363–369.

Meyer, A., 1973. Processed Cheese Manufacture. Food Trade Press Ltd, London, UK.
Mizuno, R., Lucey, J.A., 2005. Effects of emulsifying salts on the turbidity and calcium-phosphateeprotein interactions in casein micelles. J. Dairy Sci. 88, 3070–3078.
Mohamed, A.G., Abbas, H.M., Kassem, J.M., Gafour, W.A., Attalah, A.G., 2016. Impact of myrrh essential oil as a highly effective antimicrobial agent in processed cheese spreads. Int. J. Dairy Sci. 11, 41–51.
Mohamed, A.G., Ibrahim, O.A., Gafour, W.A.M.S., Farahat, E.S.A., 2020. Comparative study of processed cheese produced from sheep and cow milk. J. Food Process. Preserv. 45, 15003.
Mulder, H., 1953. Melting and solidification of milk fat. Neth. Milk Dairy J. 1, 149–174.
Ney, K.H., 1988. Gerat zur Messung des Biegebruchverhaltens von Schmelzka sescheiben. Alimenta 2, 31–36.
Omar, M.A.M., Mohamed, A.G., Ahmed, E.A.M., Hasanain, A.M., 2012. Production and quality evaluation of processed cheese containing legumes. J. Applied Sci. Res. 8, 5372–5380.
Pachlova, V., Bunka, F., Bunkova, L., Weiserova, E., Budinský, P., Zaludek, M., 2011. The effect of three different ripening/storage conditions on the distribution of selected parameters in individual parts of Dutch-type cheese: cheese ripened/stored under three different conditions. Int. J. Food Sci. Technol. 46, 101–108.
Piergiovanni, L., De Noni, I., Fava, P., Schiraldi, A., 1989. Nonenzymatic browning in processed cheeses. Kinetics of the Maillard reaction during processing and storage. Ital. J. Food Sci. 1, 11–20.
Piska, I., Stetina, J., 2004. Influence of cheese ripening and rate of cooling of the processed cheese mixture on rheological properties of processed cheese. J. Food Eng. 61, 551–555.
Price, K.M., 2007. Ph.D. dissertation. North Carolina State University, Raleigh, NC.
Price, W.V., Bush, M.G., 1974a. The process cheese industry, a review. 1. Industrial growth and problems. J. Milk Food Technol. 37, 135–152.
Price, W.V., Bush, M.G., 1974b. The process cheese industry, a review. II. Research and development. J. Milk Food Tech. 37, 179–198.
Přikryl, J., Hájek, T., Švecová, B., Salek, R.N., Černíková, M., Červenka, L., Buňka, F., 2018. Antioxidant properties and textural characteristics of processed cheese spreads enriched with rutin or quercetin: The effect of processing conditions. LWT - Food Sci. Technol. 87, 266–271.
Reinbold, R.S., Ernstrom, C.A., Hansen, C.F., 1992b. Effect of container materials on temperature profiles of large blocks of Swiss and Cheddar cheese. J. Dairy Sci. 75, 2637–2647.
Reinbold, R.S., Ernstrom, C.A., 1988. Effect of nonuniform cooling on moisture, salt, and pH distribution in 290-kilogram blocks of stirred-curd Cheddar cheese. J. Dairy Sci. 71, 1499–1506.
Reinbold, R.S., Ernstrom, C.A., Hansen, C.F., 1992a. Temperature, pH, and moisture profiles during cooling of 290-kilogram stirred curd Cheddar cheese blocks. J. Dairy Sci. 75, 2071–2082.
Rowney, M., Roupas, P., Hickey, M.W., Everett, D.W., 1999. Factors affecting the functionality of Mozzarella cheese. Aust. J. Dairy Technol. 54, 94–102.
Salek, R.N., Cerníkova, M., Pachlova, V., Bubelova, Z., Konecna, V., Bunka, F., 2017. Properties of spreadable processed Mozzarella cheese with divergent compositions of emulsifying salts in relation to the applied cheese storage period. LWT - Food Sci. Technol. 77, 30–38.
Salek, R.N., Lorencov, E., Mískova, Z., Lazarkova, Z., Pachlova, V., Adamek, R., Bezdekova, K., Bunka, F., 2020. The impact of Chios mastic gum on textural, rheological and melting properties of spread-type processed cheese during storage. Int. Dairy J. 109, 104755.

Salque, M., Bogucki, P., Pyzel, J., Sobkowiak-Tabaka, I., Grygiel, R., Szmyt, M., Evershed, R.P., 2013. Earliest evidence for cheese making in the sixth millennium BC in northern Europe. Nature 493, 522–525.

Schar, W., Bosset, J.O., 2002. Chemical and physico-chemical changes in processed Cheese and ready-made fondue during storage. A review. Lebensm.- Wiss. u.-Technol. 35, 15–20.

Schwope, A.D., Till, D.E., Ehntholt, D.J., Sidman, K.R., Whelan, R.H., Schwartz, P.S., Reid, R.C., 1987. Migration of BHTand Irganox 1010 from low density polyethylene (LDPE) to foods and food-simulating liquids. Food Chem. Toxicol. 25, 317–326.

Shalaby, S.M., Mohamed, A.G., Farahat, E.S., 2018. Preparation of functional and nutritional spreadable processed cheese fortified with vegetables and mushroom. Int. J. Current Res. 10, 74075–74082.

Shimp, L.A., 1985. Process cheese principles. Food Technol 39, 63–69.

Shirashoji, N., Jaeggi, J.J., Lucey, J.A., 2010. Effect of sodium hexametaphosphate concentration and cooking time on the physicochemical properties of pasteurized process cheese. J. Dairy Sci. 93, 2827–2837.

Shulga, L.V., Medvedeva, K.L., Lantzov, A.V., Valshonok, E.O., 2021. effect of packaging method on quality parameters of processed cheeses. Scientific notes of UO VGAVM 57, 2021. doi:10.52368/2078-0109-2021-57-1-4-7.

Sturm, W., 1998. Verpackung milchwirtschaftlicher Lebensmittel. Kempten: Edition IMQ. EAN: 9783000000348.

Tamime, A.Y., 2011. Processed cheese and analogues. In: Tamime, A.Y. (Ed.), Processed Cheese and Analogues. John Wiley & Sons Ltd, Chichester, West Sussex, UK.

Tang, D., Marangoni, A.G., 2006. Microstructure and fractal analysis of fat crystal networks. J. Am. Oil Chem. Society. 83, 377–388.

Uhlmann, G., Klostermeyer, H., Merkenich, K., 1983. Kristallisationserscheinungen in Schmelzka seprodukten - I. Phanomen und Ursachen. Milchwissenschaft. 38, 582–585.

Wijesinha-Bettoni, R., Burlingame, B., 2013. Milk and dairy product composition. In: Muehlhoff, E, Bennett, A, McMahon, D (Eds.), Milk and Dairy Products in Human Nutrition. Food and Agriculture Organization of the United Nations, Rome, Italy.

Zehren, V.L., Nusbaum, D.D., 1992. Process Cheese. Cheese Reporter Publishing Company, Inc, Madison, WI, USA.

Zhong, Q., 2003. Ph.D. dissertation. North Carolina State University, Raleigh, NC.

Zhong, Q., Daubert, C.R., Farkas, B.E., 2004. Cooling effects on processed cheese functionality. J. Food Process Eng. 27, 392–412.

Special processed cheeses, cheese spreads, and analogue cheeses

11

Oznur Cumhur[a] *and Meral Kilic-Akyilmaz*[b]
[a] Bilecik Seyh Edebali University, Faculty of Applied Science, Deparment of Tourism Management, Bilecik, Turkey, [b] Istanbul Technical University, Department of Food Engineering, Istanbul, Turkey

11.1 Introduction

Cheese is part of a healthy diet being rich in protein, calcium, and phosphorus. Processed or process cheese was developed for utilization and preservation of natural cheese. Reduction of food waste is overseen as a social, environmental and economical solution to use natural resources efficiently for the future of the world. Valorization of food waste is a current research field that serves the aim of efficient utilization of natural resources by derivation of new food products or supplements from these resources. Development of processed cheeses also serves to this aim by use of a variety of natural ingredients besides recycling unconsumed cheese.

Processed cheese was first developed to increase stability and shelf life of cheese at ambient temperatures and to utilize unmarketable improperly shaped and aged cheeses in 1916 in the USA (Tunick, 2013). Since then, different types of products have been developed based on the same basic formulation and manufacturing scheme. Analogue cheeses are manufactured from milk protein commonly rennet casein, milk fat, water, and emulsifying salts as substitutes of a certain type of cheese, such as Mozzarella or Cheddar. Imitation cheeses are produced similarly to analogue cheeses but milk protein, milk fat or both are partially or wholly replaced by non-dairy ingredients. They have a different composition compared to those of natural and processed cheeses. Starch has been used as a common replacement of protein and fat by the industry to lower the cost of milk-based ingredients in imitation cheese. The extra cost in production of imitation cheeses involved in preparation and reconstitution of ingredients, such as casein and milk fat is compensated by low-cost of starch and plant oil/fat (Fox et al., 2017). Special processed cheeses cannot be called cheese as they are not produced solely from milk, milk protein, milk fat, or other milk solids.

Recently, plant-based cheeses have been developed by the use of plant proteins and oils/fats for vegan/vegetarian consumers or individuals with milk protein allergy. There is also a need for production of cheese-like products using non-dairy components for some consumers that cannot tolerate or digest one or more components of cheese due to a specific disturbance or a metabolic disease. Furthermore, in countries where there is no sufficient supply of safe milk, analogue or imitation cheeses can be produced for consumption of cheese-like products by consumers. Functional processed

Processed Cheese Science and Technology: Ingredients, Manufacture, Functionality, Quality, and Regulations.
DOI: https://doi.org/10.1016/B978-0-12-821445-9.00009-1

Table 11.1 Types of special processed cheese.

Type	Special feature	Use
Analogue or imitation cheese	Made from dairy and/or nondairy ingredients	Cheese-like food As an ingredient in foods, such as pizza, toast, salad, sandwiches, cheeseburger, lasagne, pasta dishes, sauces, desserts, cheese cake, ready-to-prepare meals
Functional processed cheese	Made from dairy and/or nondairy ingredients Reduced calorie, cholesterol, protein, fat or salt Probiotic, prebiotic, or synbiotic Fortified with vitamin, mineral, dietary fiber, or bioactive compounds	Functional food Special diets
Plant-based processed cheese	Made from nondairy ingredients	Vegan cheese Vegetarian cheese Nondairy cheese

cheeses can also be developed as the composition and manufacture allow addition of functional ingredients. SPCs present oppurtunities for development of cheese-derived or cheese-like products with varying structural and nutritional composition for both dairy processors and consumers.

While traditional processed cheese is manufactured from cheese at different age or a mixture of cheese and curd, SPC can be formulated with dairy and/or non-dairy components for obtaining a specific composition, structure or functionality. Nondairy ingredients can result in products with poor texture, meltability, water holding and fat-emulsification that require optimization of formulation with various starches and stabilizers. Spreadability, stretchability, sliceability, shreddability, flowability, and melting without browning upon heating are some of the required structural functionalities from these cheeses depending on the final use. In this chapter, formulation, manufacture, and functionality of SPC varieties are reviewed.

11.2 Special processed cheese varieties

There are a wide variety of SPC depending on ingredients (dairy based, nondairy, or plant based), composition (low fat, low protein, or reduced salt), structural functionality (meltability, shreddability, or spreadability) and nutritional functionality (high fiber, prebiotic, probiotic, synbiotic, antioxidant) (Table 11.1). Each variety targets a different end use or consumer segment with special expectations or needs from cheese. In addition, cheese analogues or imitation cheeses have been developed with low protein and phenylalanine contents for the diets of patients with deficiencies in amino acid metabolism such as phenylketonuria (El-Shazly et al., 2010; Kiziloz et al., 2009; Hussein and Shalaby, 2018).

The largest commercial share of analogue cheeses belongs to pizza cheese which has similar functionalities to those of low moisture Mozzarella cheese. Analogues of various natural cheeses, such as Cheddar, Monterey Jack, Parmesan, Romano, Blue and Cream cheeses have also been produced by the industry. These cheeses are utilized as ingredients in pizza, hamburger, salad, sandwich, spaghetti, cheese sauce, dip, and ready-to-prepare meals due to their low cost and stability. Reduction of fat, cholesterol, sodium and enrichment with vitamins and minerals are also possible in these types of products. Addition of functional health-promoting ingredients, such as antioxidants, probiotics, prebiotics, and fiber can also create a different segment of processed cheese called functional processed cheese. All these types of cheeses are covered under the name of special processed cheese in this study.

11.3 Ingredients and formulation

The major ingredients of SPC are protein, fats/oils, emulsifying salts, stabilizers, acid and flavor. Proteins, fats/oils, and stabilizers are major structure forming ingredients in these products. Emulsifying salts are also important but they can be omitted in some formulations where rennet casein is not used. Various ingredients used in different types of SPC are listed in Table 11.2. The criteria for choosing these components depend on product type, local regulations, expected functionalities and nutritional content of the cheese-like product. Ingredients influence microstructure, texture, organoleptic, and functional properties of the resulting product (Mulsow et al., 2007).

Primary protein source used in SPC is rennet casein which is composed of particles of paracasein micelles as in cheese. Other casein-based ingredients, such as acid casein, Na- and Ca-caseinates can also be used depending on the formulation. In addition, milk protein concentrate obtained by ultrafiltration of milk and micellar casein concentrate can be used. Casein is unique in forming a firm structure which can flow upon heating compared with other proteins. Rennet casein also provides stringiness and stretchability expected from a pasta-filata type cheeses. Whey proteins are not preferred as they interfere with the structure formed by casein decreasing flowability and increasing firmness (Savello et al., 1989; Guinee, 2017). However, they can be used in products for burgers where high degree of flowability is not desired (Guinee, 2017).

Skim-milk powder and whey powder at a level below 10% can improve the creaming or emulsification process and spreadability in SPC (Gouda and Abou El-Nour, 2003). However, these ingredients will provide lactose that can cause sweetness, Maillard browning and crystallization defects. Nevertheless, lactose at a certain level has been used as a filler in cheese-like product formulations. Whey powder can also cause salty taste due to its high mineral content.

Plant flours or protein isolates including those from soybean, peanut, wheat, pea, lupine, and zein have been used as a protein source in SPC (Amar, 2012; Li et al., 2013; Rinaldoni et al., 2014; Awad et al., 2014; Schindler et al., 2012; Mattice and Marangoni, 2020). Plant proteins are mostly globular and large proteins and they lack

Table 11.2 Ingredients used in manufacture of special processed cheese (Guinee, 2011a; Guinee 2011b; Talbot-Walsh et al., 2018).

Component	Ingredients
Protein	*Milk proteins* Rennet casein, acid casein, sodium or calcium-caseinate, whey protein concentrate or isolate, skim milk powder, milk ultrafiltrate, milk protein concentrate, micellar casein powder, microfiltered milk retentate *Vegetable proteins* Soybean protein, wheat gluten, lupine protein, chickpea protein, pea protein, zein
Fat or oil	*Milk fat* Cream, butter, butter oil, anhydrous milk fat *Vegetable oil* Native or partially hydrogenated soybean oil, palm kernel oil, peanut oil, canola oil, coconut oil, corn oil
Stabilizer	*Starch* Native or modified forms of wheat, maize, rice, potato, corn, tapioca starches *Hydrocolloids* Gelatin, carrageenans, guar gum, xanthan gum, locust bean gum, gum arabic, basil seed gum, pectin, sodium alginate, gellan
Emulsifying salts	Sodium or potassium salts of citrates, phosphates, polyphosphates, pyrophosphate
Acid	Lactic acid, acetic acid, citric acid
Flavorings ingredients	Cheese, enzyme modified cheese, salt, starter distillate, smoke extract, yeast extract, spices, plant extracts, fruits, vegetables, nuts, meat, fish Sucrose, dextrose, corn syrup
Coloring agents	Annatto, β-carotene, paprika, curcumin, riboflavin, chlorophyll preparations
Preservatives	Potassium sorbate, calcium or sodium propionate, nisin, natamycin, phosphates, plant extracts
Functional ingredients	*Vitamins and minerals* Vitamin A, vitamin D, vitamin E, vitamin C, folic acid, magnesium oxide, zinc oxide, iron, riboflavin, thiamine *Bioactive compounds* Carotenoids, flavonoids, phenolic acids, phytosterols, long-chain polyunsaturated fatty acids, microalgae, plant extracts *Dietary fiber* *Probiotics* *Bacillus, Lactobacillus, Bifidibacterium* spp. *Prebiotic* Inulin, fructooligosaccharide, konjac glucomannan

flexibility of caseins. As a result, they cause poor texture with lesser elasticity, hardness, flowability and stretchability, an adhesive/ sticky body and poor flavor (Guniee, 2017). Therefore, the structure formed by these proteins can be supplemented by starch and/or stabilizers for a desired texture.

One of the major structuring ingredients of SPC is fat/oil. Fat/oil is a softening and filling agent in the cheese structure that also contributes to flowing or melting of the product along with casein. SPC can be produced by using different types and amounts of fat/oil. Milk fat, such as cream, butter, butter oil, or anhydrous milk fat can be used to improve texture, flavor, and flowability of the product. Plant oils or their hydrogenated forms from soybean, peanut, palm kernel, cottonseed, coconut, or corn have been used to replace milk. Plant fats with a melting profile similar to that of milkfat can be preferred for a cheese-like structure and meltability in the product. The use of plant oils also allows a reduction in saturated fat, trans fats, and cholesterol (Cunha et al, 2010; Huang et al., 2018; Rafiq and Ghosh, 2017a; Ramel and Marangoni, 2018). To improve the quality properties of the products with low-fat content, water, protein, starch, dietary fiber, or stabilizers are additionally used in formulation (Johnson et al., 2009; Bi et al., 2015).

Starch is a common substitute of protein or fat in formulation of SPC. Starch can form a firmer structure with lower water holding and flow compared to that of casein depending on the level of substitution. Native starch does not affect structure and function of the product if added at a level of less than 3% corresponding to substitution of 10%–15% of casein (Mounsey and O'Riordan, 1999, 2001). If used at a higher level, native starch can cause a firm and brittle structure with low flowability (Guinee, 2011a; Mounsey and O'Riordan, 1999, 2001). Native and modified starches at different levels can be used in formulation of imitation cheeses depending on targeted functionalities. For example, tapioca starch is preferred in plant-based cheeses for its elasticity (Fu and Nakamura, 2020).

Stabilizers are required for further stabilization of the processed cheese structure primarily by binding water. They are commonly used at a level less than 0.8% (Gouda and Abou El-Nour, 2003). Their use at excessive levels can impair textural properties and meltability (Swenson et al., 2000). Gum arabic, locust bean gum, gelatin, pectin, carboxymethyl cellulose, gellan, carrageenan, and agar or combination of these can provide a stable structure depending on the formulation.

Acids including lactic, citric and propionic acid are used for reducing pH, preventing microbial growth and improving texture and flavor (Tunick, 2013). Coloring ingredients such as annatto and β-carotene can be used in some types. Flavoring ingredients, including enzyme-modified cheese, spices, vegetables, and fruits can be added according to the consumer preferences. Sweeteners can be added in fruit-flavored products. Flavoring materials should be of good microbiological quality or this can be attained by thermal treatment before addition to the blend.

Preservatives are required in SPC since water activity and pH of the products could be high enough for germination and growth of pathogenic and spoilage microorganisms. Especially process cheese spread with a pH between 5.5-6.0 and a minimum water content of 50% support the growth of *Clostridium* spores causing gas and off-odor during storage (Borch and Lycken, 2007). Heat treatment applied in processing

(≥80°C) is sufficient to inactivate vegetative cells but not spores. Spore-formers *Bacillus, Geobacillus*, and *Clostridium* spp. are reported as common contaminants of processed cheeses and cheese spreads depending on microbial quality of ingredients, hygienic conditions during processing, packaging materials, and storage conditions (Lycken and Borch, 2006; Buňková and Buňka, 2017). Salt and emulsifying salts were found to have inhibitory effect on microbial growth. Phosphates were found to exert a bacteriostatic effect in processed cheese and their effect was strengthened with increasing phosphate content. Polyphosphate with P_2O_5 content of 68% at a level of 1.75% was found to inhibit growth of *Clostridium* species isolated from processed cheese spreads (Borch and Lycken, 2007). Nisin, a bacteriocin synthesized by some strains of *Lactococcus lactis,* is commonly used against *Clostridium* in processed cheeses and cheese spreads at 5–15 ppm (Muir et al., 1999; Williams and Delves-Broughton, 2003). Nisin was also found synergistic with lauryl arginine ester (1:3, w/w), a surfactant, against *Bacillus subtilis* and *Clostridium sporogenes* (Al-Nemr et al., 2016). Benzoic acid, sorbic acid, and natamycin are also used against mold and yeast growth. Another alternative preservation method could be addition of starter lactic acid bacteria even though there is no fermentation period in the manufacture (Chatziantoniou et al., 2015; Chatziantoniou et al., 2019). Sadek et al. (2017) showed that lactic acid bacteria were effective in decreasing the viability and survival of *Bacillus cereus* and *Clostridium perfringens* in processed cheese.

Emulsifying salts are citrates and phosphates which are required if casein-derived ingredients are used in processed cheese manufacture. They disperse and solubilize caseins by chelating calcium and stabilization of pH. Sodium salts of citrates, mono-, di- and polyphosphates are used. Calcium on caseins is replaced by sodium in emulsifying salts, this allows dispersion and hydration of casein in water. Emulsifying salts that can bind large amounts of calcium (sodium hexametaphosphate) produce a firmer process cheese than salts with lesser binding ability (sodium diphosphate) (Kaláb, 1995). Extent of emulsification of fat determines texture and melt or flow of the end product; therefore, degree of emulsification should be controlled. For example, some oiling off is desirable in pizza cheese where degree of emulsification is low (Fox et al., 2017). The amount of emulsifying salts is normally 1%–3% depending on the composition (Gouda and Abou El-Nour, 2003). Combination of emulsifying salts is commonly used as it results in a better structure with additional benefits from each.

Sodium present in emulsifying salts can increase sodium content of the resulting cheese as these are additive to the sodium from cheese ingredients. Potassium salts have been used to replace sodium salts in cheese but excessive levels can cause bitterness. Therefore, reduction of amount of sodium salts of phosphates or partial replacement of sodium salts by potassium have been practiced in the cheese industry. Production of analogue or imitation cheeses without emulsifying salts is possible if protein ingredients other than rennet casein are used and the formulation is changed (Guinee, 2017). For example, caseinates are water soluble and have the ability of water binding at pH 6.0 and emulsification of free fat yielding oil in water emulsion.

Water is required for dissolution of emulsifying salts, dispersion of ingredients, and formation of structure by the ingredients. Water influences the structure of proteins, the ionization of groups on the proteins and the emulsification of fat in processed cheese (Lee et al., 2004). Amount of water directly affects the textural properties of the cheese. Water can reduce hardness and stiffness and increase meltability by reducing protein-protein interactions and promoting protein-water interactions. Free water interrupt continuous protein network and weaken the structure (Hennelly et al., 2005). Water can be added at once or in portions at different stages of the production. Heating by direct injection of steam also contributes to the water content of the product.

Nutritional value of analogue cheeses can be increased by adding a variety of nutrients that provide consumers additional health benefits. Several researchers have investigated the fortification of processed cheese and its analogues with minerals (Baldwin et al., 2008), vitamins (Johnson et al., 2005; Kazmi et al., 2007; Upreti et al., 2002), algae (Mohamed et al., 2020; Tohamy et al., 2018), antioxidants (Přikryl et al, 2018), prebiotics, and dietary fiber (Giri et al., 2017; El-Shibiny et al., 2013; Montesinos-Herrero et al., 2006; Szafranska and Sołowiej, 2020). Another way to functionalize processed cheese is to add probiotic microorganisms. Some types of cheese have been shown to be a good medium for probiotic bacteria due to their low acidity, high buffering capacity, low oxygen content, and the presence of complex matrix with protein and fat that could protect the probiotics during their passage through the gastrointestinal tract (Rolim et al., 2020; Tamime et al., 2017). Several researchers have also supplemented process cheese with probiotics such as *Bacillus* and *Lactobacillus* strains (Ehsannia and Sanjabi, 2016a; Liu et al., 2006; Pivetta et al., 2020; Soares et al., 2019).

Vegetables, fruits, meats, spices and nuts have been added to SPC. These ingredients contribute to appearance and taste which allow development of a new variety. They also supply bioactive components with potential health benefits. Vegetables, fruits and meat are added to the blend of processed cheeses as cooked, dry, and canned. Carrot, lupine, soybean, celery, dill, leek, mushroom, pea, potato, pumpkin, parsley, and tomato have been added to processed cheese in recent years (Abbas et al., 2019a; Awad et al., 2014; Kassem et al., 2017; Khider et al., 2017; Mehanna et al., 2017; Mohamed et al., 2016; Rafiq and Ghosh, 2017b; Shalaby et al., 2018; Solhi et al., 2020). Similarly, fruits such as blueberries, kaki, and mandarin were used as ingredients (Abbas et al., 2019b; El-Messery et al., 2019; Zhang et al., 2011).

11.4 Manufacture of special processed cheeses

SPC may be produced either as a block or a spread by the manufacturing method of processed cheese. Rennet casein is commonly used in making processed cheese analogue as it provides a cheese-like structure and functionality. Manufacturing process include formulation and preparation of ingredients, mixing, heating, and packaging (Fig. 11.1). Typical formulation of SPC contain protein source, fat source, stabilizers, emulsifying salts depending on protein type, acid, salt and water (Table 11.3). There are two phases in SPC forming the structure. Dry protein ingredients with or without emulsifier salts, starch and stabilizers are hydrated sufficiently by mild heating and mixing forming the aqueous phase. Fat is melted separately at appropriate temperature forming the fat phase. A minimum heat treatment of 65.5°C for 30 s is recommended

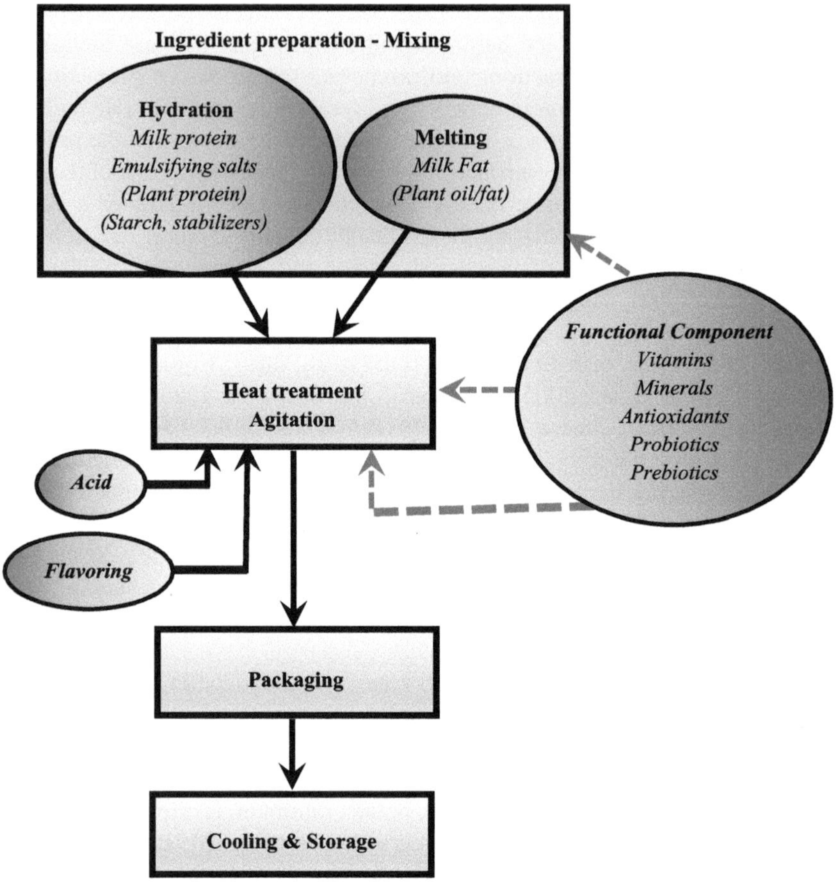

Figure 11.1 Flowchart for manufacture of special processed cheese.

Table 11.3 Composition of special processed cheeses (Liu et al., 2008; Kiziloz et al., 2009; Bergsma, 2019; Butt et al., 2019; Masotti et al., 2018; Mattice, and Marangoni, 2020).

Ingredient	**Cheese analogue**	**Imitation cheese with starch**	**Plant-based cheese analogue**
		(%, w/w)	
Casein, caseinates	15–25	5–25	–
Plant protein or flour			2–30
Fat/oil	10–26	10–26	1.5–25
Starch	–	3–25	3–17
Emulsifying salts	0.5–3.0	0.5–3.0	–
Stabilizer (gums)	<1	<1	<1
Acid (citric acid)	<1	<1	–
NaCl	<2	<2	<2

for processed cheese by FDA (Kapoor and Metzger, 2008). The water in oil emulsion structure is formed by mixing, shearing and heating the two phases at a temperature between 70–95°C, typically 80–85°C, for 5–10 min. Degree of emulsification of fat is controlled during the mixing and heating step according to expected functionality from the product as it affects firmness, spreadability and flowability, oiling-off and juiciness as in real cheese. For example, low degree of emulsification yields large fat droplets and flow and some oiling-off after heating which is desired in pizza cheese (Guinee, 2017). Acid is added after heating to allow sufficient hydration of proteins and emulsification of fat by the proteins. Flavoring ingredients, nutrients and bioactive components are also added after heating to protect their activity and functionality in cheese. Hot fluid mass obtained at heating step is packaged in a suitable container and the final structure of the product is attained after cooling.

Functional ingredients are sensitive to heat which requires special attention during their addition to the blend of SPC. Probiotic microorganisms must survive in the product during processing and storage (Castro et al., 2015; Ehsannia and Sanjabi, 2016b; Talbot-Walsh et al., 2018). They can be added after heat treatment but adding probiotics after heat treatment can lead to re-contamination of the product (Al Khalaileh and Ajo, 2013). Encapsulated probiotics could be added before heating. Another alternative could be use of spore-forming and low pH resistant probiotic species such as *Bacillus* instead of those with low heat resistance such as *Lactobacillus* and *Bifidobacterium* (Cutting, 2011; Ehsannia and Sanjabi, 2016a; Ehsannia and Sanjabi, 2016b; Soares et al, 2019). Other limiting factors for probiotics in cheese included high emulsifying salt concentration, low water activity, inadequate acidity, presence of oxygen and low storage temperature (Ehsannia and Sanjabi, 2016b).

In case of vitamin fortification of SPC, use of appropriate form of vitamin without loss in activity during processing and storage is important in terms of health-effect and claim. Fat-soluble vitamins A, D, and E exhibit good heat stability, but some water-soluble vitamins, including vitamin C, thiamine and folate, are unstable during thermal treatment and storage (Belitz et al., 2009; Talbot-Walsh et al., 2018). Mineral fortification of SPC is also possible without affecting quality and functionality. Cheddar cheese-like cheese was iron-supplemented with Fe-casein, Fe-whey protein, or $FeCl_3$ (Zhang and Mahonay, 1991). Baldwin et al. (2008) developed Mg-fortified processed cheese that had salty taste and nutritional benefits.

Fruits, vegetables, meat and peanut are added before heating (Abbas et al., 2019a, Awad et al., 2014; El-Messery et al., 2019; El-Sayed, 2017; Kassem et al., 2017; Mohamed et al., 2016; Zhang et al., 2011; Mohamed and Shalaby, 2016) or during heating process (Krumov et al., 2010; Shalaby et al., 2018) to ensure even distribution in cheese mass and to provide pasteurization. Alternatively, low-contamination or pre-treated flavoring ingredients can be added after heat treatment (Alimardanova et al., 2015; Khider et al., 2017), but mixing is essential for a homogeneous blend at this stage.

Processed cheese spreads are produced from the same ingredients as in processed cheese and SPC. However, they contain more water to have soft and spreadable structure at room temperature. The moisture content varies between 44 and 60% and the minimum fat content is 20% (Tunick, 2013). In the manufacture, the ingredients

Table 11.4 Quality and functional attributes of special processed cheeses (Drake et al., 1999; Muir et al., 1999; Adhikari et al., 2009).

Physical properties before cooking	Sliceability, shreddability, smoothness, hardness, cohesiveness, springeness, gumminess, chewiness, adhesiveness, elasticity, sweating, stickiness
Physical properties after cooking	Meltability, flowability, oiling-off, stretchability, blistering, crispiness, browning, apparent viscosity
Physical properties of spreadable types	Spreadability, graininess, mouth coating, melt-in-mouth, stickiness, greasiness
Microbial quality	Total viable bacteria count Molds and yeast count
Sensory properties	Color, odor, flavor, cheese taste, sourness, creamy, milky, buttery, saltiness, bland taste, rancidity, bitterness, foreign odor, foreign flavour, cardboard, oxidized/fatty, minty, cooked
Nutritional properties	Macro and micronutrient composition, activity of probiotics, antioxidants

are heated at a temperature in the range of 85–95°C and pH is adjusted between 5.7 and 5.9, which is higher than that (5.4–5.6) of block type processed cheese. Cheese spreads can be manufactured from acid casein, Na- or Ca-caseinates. Na-caseinate is especially preferred due to its high water-binding capacity and efficient emulsifying properties required for creaming (Guinee, 2017). Stabilizers may be added to improve water retention (Gouda and Abou El-Nour, 2003).

11.5 Quality and functionality of special processed cheeses

Each specialty cheese has characteristic quality and functional attributes determined by type or targeted end use (Table 11.4). In general, processed cheese should have a shiny appearance, a good emulsion without oil or water on surface and elastic structure that is easily separated from the packaging film. Pizza cheese is expected to be shredded easily and uniformly, flow or melt with some oil-off and no browning upon heating. Processed cheese for toast and burger should also be meltable without excessive spread by heating.

Quality and functionality of special processed cheeses depend on ingredients, agitation speed, temperature, and time of heat treatment. These factors influence protein hydration and degree of emulsification that determine structure and stability of the end product. As a general rule, as temperature, time, and agitation speed are increased, there will be a stronger network due to more protein-protein and protein-water interactions creating a compact, firm and less meltable structure with high degree of fat emulsification (Noronha et al., 2008b). Rigidity of processed cheese was found to be directly proportional to dry matter content and inversely proportional to fat in dry matter content (Černíková et al., 2017). Ingredients especially protein, fat, starch and stabilizers have large impact on quality and functionality of SPC. Therefore, effects of the ingredients on quality and functionality of SPC are discussed below.

11.5.1 Protein

Rennet casein and caseinates impart a firm, cohesive and elastic texture with flow in processed cheese. Rennet casein also provides a fibrous structure with stretchability compared to those of Na- or Ca-caseinate and acid casein. Na-caseinate formed an imitation cheese with higher pH, hardness, and degree of fat emulsificaton compared to Ca-caseinate (Cavalier-Salou and Cheftel, 1991). Although addition of whey proteins besides rennet casein increased water holding and supported gel structure of processed cheese analogue (Mleko and Foegeding, 2000), they caused a reduction in meltability, some free oil and sticking to the package (Savello et al., 1989; Gustaw and Mleko, 2007). Addition of hydrocolloids such as xanthan gum, κ-carrageenan and locust bean gum lessened stickiness to package (Savello et al., 1989). Sołowiej et al. (2014) reported that acid casein at levels of 10-13% with 1-3% whey protein concentrate or isolate yielded meltable but adhesive processed cheese analogue. Nastaj et al. (2020) reported that meltability of processed cheese made from acid casein (10%) decreased by addition of 2%–6% of whey protein isolate or whey protein polymers. While whey protein polymers increased hydrophobicity, whey protein isolates increased hydrophilicity of the cheese surface.

11.5.2 Fat

Fat has an impact on the physical and sensory characteristics of cheese and it especially affects the cheese flavor and mouthfeel (Banks, 2007). Fat source or type can also influence structure of the product. Hanáková et al. (2013) found that coconut fat and butter caused higher firmness compared to that by palm oil in a processed cheese analogue. Reduction of fat causes an increase in moisture content and textural and sensory defects in cheese. Low and reduced-fat cheese analogues have fracturable, firm, and rubbery texture and a bland taste (Johnson et al., 2009). They also have lower meltability and less flexible and fluid structure compared to full fat counterpart (Johnson, 2011; Li et al., 2019; Chatli et al., 2017; Kommineni et al., 2012; Li et al., 2019; Sołowiej et al., 2015). Reduction in total fat and cholesterol levels in processed cheese was achieved by use fat mimetics or replacers (Tamime et al., 1999). Fat is replaced by fat mimetics such as resistant starch and dietary fiber that also provide health benefits (O'Riordan et al., 2011; Bi et al., 2015; Giri et al., 2017; Sołowiej et al., 2015). Use of soy lecithin at a level of 0.05% improved textural properties of reduced fat processed cheese (Drake et al., 1999). Rinaldoni et al. (2014) produced a spreadable reduced-fat cheese-like product with soy protein concentrate (5–15 g/L) and skimmed cow milk by acid coagulation method.

Nutritional status of analogue cheese has been improved by partially replacing the fat with functional ingredients that provide certain health benefits, e.g. inulin and resistant starch (Hennelly et al., 2006; Montesinos et al., 2006). Hennelly et al. (2006) used inulin gels to partially replace the fat in analogue cheese. Although meltability was found to be unaffected by the inclusion of inulin, hardness of the cheese increased. The authors suggested increase in the moisture content to offset the hardness. Montesinos et al. (2006) showed that resistant starch can be used to replace up to 50% fat in analogue cheese without impacting cheese meltability. Noronha et al. (2007) replaced 90% of the fat in an analogue cheese with resistant starch without an adverse effect on functionality. The authors stated that the crucial role played by fat as a plasticiser in full-fat analogue cheese may be fulfilled by water in low-fat and high-moisture products. Duggan et al. (2008) also noted that in low-fat analogue cheeses, the resistant starch does not compete with protein for water molecules but rather it entraps the water and allows it to remain available for the plasticisation of the cheese.

Replacement of butter by peanut cream fat in a SPC formula with starch reduced firmness and cohesiveness with an increase in fat globule size and decrease in their number and replacement by 50% received satisfactory sensory scores from panelists (El-Wahed and Hassanien, 2019). Kommineni et al. (2012) substituted fat with xylitol (2%–4%) in a low fat (5%) processed cheese formula. The cheese was softened with a lower elastic modulus upon addition of xylitol without an effect on melt temperature. Kim et al. (2009) developed a cholesterol-reduced processed Cheddar cheese by using β-cyclodextrin-treated Cheddar or butter in the formulation. In a recent study, oleogels prepared from soybean oil and rice bran or sunflower wax were used at a level of 10% to replace milk fat in a processed cheese made from Swiss cheese (Huang et al., 2018). The samples with oleogel had lesser agglomeration of oil and oil loss while rheological properties and meltability were similar to those of control sample with vegetable oil and commercial ones with milk fat. Benaouadj et al. (2017) reported that both fat (10%–20%) and starch content (1%–3%) and modification were effective on viscosity of SPC.

11.5.3 Emulsifying salts

Ability of emulsifying salts to bind calcium, crosslink casein and disperse casein during cooking have been found important in terms of their effects on textural properties and meltability of processed cheese (Lu et al., 2008). Combination of various phosphate salts have been shown to affect structural properties (Awad et al. 2004; Lu et al., 2008; Dimitrelli and Thomareis, 2009; Buňka et al., 2012). Degree of emulsification and meltability were found to be inversely related in processed cheese analogues made with acid and rennet casein (Savello et al., 1989). Increase in emulsifying salt concentration reduced melting and increased hardness of an imitation cheese with waxy maize starch, hard cheese and and κ-carrageenan (Beykont and Kilic-Akyilmaz, 2014). Lucey (2008) reported development of a process cheese base where citric acid, not emulsifying salts, was used to reduce cross-linkages between casein.

Cunha et al. (2012) manufactured processed cheese spread from Ca-caseinate with 37% dry matter and 64% fat in dry matter content with various phosphate salts. Sodium hexametaphosphate was found to be an efficient dispersion agent for caseins allowing a weak gel structure with good emulsification even though measured degree of calcium dissociation and pH were low. Reduction of emulsifying salts from 3% to 2% was possible by using 0.15% κ-carrageenan along with 1.34% Na-citrate and 0.66% DSP (Ahmad et al., 2016). Sołowiej et al. (2020) successfully replaced 50% of emulsifying salts with polymerized whey protein in an analogue processed cheese made with acid casein.

Hydrolysis of emulsifying polyphosphate salts in a spreadable processed cheese with Ca-caseinate was found to be dependent on calcium concentration and pH, being increased as the pH was lowered with lactic acid (Barth et al., 2017). There was a concamitant reduction in hardness, adhesiveness, and gumminess with a more fluid and continuous structure as the pH was increased from 5.2–5.6 to 6.0–6.8.

11.5.4 Starch and stabilizers

Starch has been used as a substitute for high cost ingredients, protein and fat, in imitation cheese. Type and amount of starch have a large impact on structure, functionality, and storage stability of imitation cheeses. Starch had no adverse effect if used at a level less than 3% (Mounsey and O'Riordan, 2008a; Trivedi et al., 2008; Sołowiej et al., 2016). However, when the starch content was increased over 8%, the imitation cheese became brittle, tough and less cohesive and stretchable with no meltability (Noronha et al., 2008a; Ye et al., 2009; Bi et al., 2015). Trivedi et al. (2008) reported that starch inclusion caused a gummy texture sticking to the packaging film.

High amylose starches, including wheat, potato, and corn starch caused a hard texture with low meltability and cohesiveness while rice and waxy maize starches led to a soft texture with some level of melting (Mounsey and O'Riordan, 2001; Ye et al., 2009). The functionality of rice starch was explained by low-water binding and smaller starch granules that do not disrupt protein structure excessively. In another study, waxy rice starch outperformed modified rice starch in achieving cheese-like properties (Mounsey and O'Riordan, 2008b). Waxy rice starch with an amylose content of 2.5% and used at a level 2.5% led to a desirable structure due to low degree of fat emulsification and retrogradation during storage (O'Riordian et al., 2011).

Adverse effects of starch on the structure and functionality of imitation cheese were explained by molecular interactions of starch or protein with water. Starch was found to reduce free water available in the structure (Noronha et al., 2008a). Starch and rennet casein were found to compete for water that impaired hydration of protein hence emulsification of fat (Mounsey and O'Riordan, 2008a; Ye et al., 2009). In addition, starch and protein were found to form separate networks in imitation cheese (Mounsey and O'Riordan, 2008a; Ye et al., 2009). Sufficient water should be present for hydration and dispersion of protein and dissolving and gelatinization of starch. Temperature and time of heat treatment should be adjusted for proper network development.

In an effort for reducing adverse effect of starch on the structure, Butt et al. (2019) replaced 10% of fat and/or 15% of rennet casein in an imitation cheese formula with cold water-soluble starches obtained by citric or lactic acid treatment of rice starch. Authors noted that modified starch citrate or lactate had high water binding with a low tendency towards retrogradation. Hardness and elasticity of the cheese were increased by lactic acid-modified starch and meltability was reduced by replacement of both fat and rennet casein but reduction of only casein resulted in similar meltability to that of full protein and fat cheese. Benaouadj et al. (2017) also reported that modified starches increased viscosity of processed cheese more compared with native starches. Tapioca starch at a level of 2.5% enhanced viscosity without affecting creaming and physical properties of processed cheese compared with potato starch (Fu and Nakamura, 2020).

Starch can be used at higher levels than 10% when the structure is supported by stabilizers. Thermoreversible gel forming hydrocolloid stabilizers such as gelatin and κ-carrageenan are preferred when high levels of rennet casein is to be replaced. Enzyme- or acid-treated, cross-linked starch and starch derivatives have been used with or without hydrocolloid stabilizers to partially or completely replace rennet casein in imitation cheeses (Zallie and Chiu, 1990; Carpenter et al., 1998; Yoder et al., 1996). Kiziloz et al. (2009) developed an imitation cheese with partially α-amylase-hydrolysed waxy maize starch and κ-carrageenan that allowed 80% replacement of rennet casein and meltability, hardness, and cohesiveness in imitation cheese close to those of a high protein equivalent.

Use of starch as a substitute for fat or rennet casein reduced storage stability of imitation cheeses. Retrogradation and gelation during storage caused hardening and reduction in shreddability and meltability (Mounsey and O'Riordan, 1999, 2001; Beykont and Kilic-Akyilmaz, 2014). Shrinkage of starch network during storage for 1 mo at 4°C caused release of fat and moisture drying and forming a crust after cooking (Ennis and Mulvihill, 1997; O'Riordan et al., 2011).

Imitation cheese products with a low protein hence phenylalanine contents have been developed for patients with phenylketonuria who have to consume a phenylalanine-limited diet. Protein can be replaced by starch and stabilizers to have safe concentrations of phenylalanine when cheese was consumed in a typical daily diet. Kiziloz et al. (2009) achieved reduction of 80% of rennet casein by waxy maize starch and κ-carrageenan. Partial hydrolysis of waxy maize starch together with κ-carrageenan allowed flow or melt in the imitation cheese upon heating. El-Shazly et al. (2010) substituted protein with branched-chain amino acids with improved protein quality and low phenylalanine content in a SPC. The product with 10% protein substitution allowed the minimum content of phenylalanine and was shown to be applicable in diet management. While protein substitution by 10% yielded the highest solubility and penetration, the best flavor and appearance were obtained by 5% substitution of protein. Hussein and Shalaby (2018) developed an imitation cheese with acceptable sensory and quality properties by using non-dairy ingredients including modified potato starch, palm oil and oat, or wheat germ. pH was found to be an important factor affecting functional properties of analogue processed cheeses with starch. Beykont and Kilic-Akyilmaz (2014) reported that imitation cheese made from a blend of waxy maize starch, a hard cheese and and κ-carrageenan became firm and unmeltable when pH was reduced from 5.2 to 4.6 by using citric acid explained by increased protein-protein interactions. Sołowiej et al. (2016) also found that imitation cheese made from acid casein and acetylated distarch adipate (ADA) or hydroxypropyl distarch phosphate had more elastic and viscous properties at pH values of 4.5–5.5 and 6.0–7.0, respectively, with satisfactory melting at pH between 5.0 and 7.0. Talbot-Walsh et al. (2019) reported that imitation cheese network with 5% potato, waxy maize or corn starch can be improved by decreasing or increasing pH outside the range of 4.8–5.3. They explained that increase or decrease in pH caused the continuous network of starch to break which enhanced the intermolecular bonds between starch and protein molecules in the structure. Dispersion of starch in the structure led to a reduction in hardness and improved meltability with better sensory properties at pH 6.

Stabilizers can be used with or without starch to strengthen and stabilize the structure of SPC. Hanáková et al. (2013) reported that κ-carrageenan was the best stabilizer as it has a low enough gelation temperature for processing and desirable consistency with lower adhesiveness compared to other carrageenans, arabic gum, and locust bean gum at a level of 1% in processed cheese analogue. Usage of pectin at a level between 0.2%–0.8% increased rigidity and reduced spreadability in a process cheese made from rennet casein with 40% dry matter and 50% fat in dry matter contents without affecting flavor (Macků et al., 2008). A pectin gel prepared from low-methoxylated pectin at low calcium level counteracted the structural defects in 50% reduced-fat processed cheese analogue (Liu et al., 2008). Pectin acted as a linkage point compacting the structure of low fat cheese as that of full fat cheese. Chatziantoniou et al. (2019) produced spreadable processed cheeses from a whey cheese by addition of xanthan gum, guar gum, locust bean gum, and κ-carrageenan at a level of 0.8 added individually or in 1:1 mixtures. The mixtures of locust bean gum with xanthan gum or κ-carrageenan acted synergistically and formed the firmest and the least spreadable cheese. The partial substitution of κ-carrageenan with xanthan or guar gums led to a decrease in firmness and increase in spreadability and smoothness. Xanthan gum resulted in the largest fat globules and the mixture of κ-carrageenan with locust bean gum gave the smallest ones.

11.5.5 Functional components

Upreti et al. (2002) supplemented pasteurized processed cheese with vitamin D_3 using water- and fat-dispersible forms. The vitamin solution was added before heat treatment (79°C/1 min). There was no detectable degradation of vitamin D_3 in process cheese stored at room (21–29°C) and refrigeration (4–6°C) temperature during 9 months of storage. However, when pasteurized cheese was heated in the oven (232°C/5 min) for culinary applications, 25% to 30% of vitamin D_3 was lost in both cheeses in which water and oil dispersible forms of vitamin D_3 were used. This low recovery of vitamin D after baking was attributed to destruction or a lack of recovery due to protein binding or entrapment in processed cheese by the authors. Kazmi et al. (2007) also supplemented Cheddar-like cheese with vitamin D_3 by adding it to milk in pre-dissolved crystalline and emulsified forms. While there was no loss in emulsified vitamin D3, 75–9% of the crystalline form was lost to whey and 7% of it was degraded after three months of ripening.

Various spices and their extracts have been used to improve the microbiological, functional and sensory properties of processed cheeses. El-Sayed (2017) reported that addition of thyme powder, which has antioxidant and antimicrobial properties, to processed cheese at a ratio between 0.1% and 0.4% improved the structure, flavor and organoleptic quality of the cheese. Ground black pepper (*Piper nigrum* L.), a dry savory herb (*Satureja hortensis* L.) and extracts of these spices were added to different batches of processed cheese at 5 and 10 min of heating (85°C/15 min). Extracts of black pepper and the herb improved the microbiological quality and sensory properties of the product (Krumov et al., 2010). Aqueous extracts of dried rosemary, lemongrass, cinnamon, cloves, garden cress, oregano and sage at a level of 10% (w/v) had significant effects on the sensory properties of processed cheese. While lemongrass and garden cress extracts improved color of fortified cheeses, lemongrass, cinnamon, oregano and rosemary extracts positively affected odor and taste (Tayel et al., 2015).

Solhi et al. (2020) reported that addition of asparagus powder, which is rich in phenolic compounds and dietary fiber, upto 1.5% increased antioxidant activity, proteolysis and rigidity but it reduced spreadability of SPC with 43% dry matter, 12% protein, and 54% fat in dry matter. Peanuts also provide plant-based proteins along with unsaturated fatty acids and fiber (Arya et al., 2016). Peanut paste prepared from roasted and ground peanuts added to SPC formula at a level of 10% before heat treatment did not have negative effects on textural properties, appearance, meltability and acceptability of the product (Rafiq and Ghosh, 2017a).

Some microalgae species such as *Spirulina maxima* (Mohamed et al., 2020) and *Chlorella vulgaris* (Mohamed et al, 2013; Tohamy et al, 2018) were used as a source of nutrients, antioxidants, and color for development of a functional processed cheese. The algae were added after heat treatment not to destroy activity of its components. *Chlorella vulgaris* and *Spirulina maxima* at a concentration of 2% (w/w) were shown to be compatible with cheese structure and did not affect its sensory acceptability and functional properties (Mohamed et al., 2013; Tohamy et al., 2018; Mohamed et al., 2020). Radical scavenging activity of the cheese improved as the concentraton of *Spirulina maxima* was increased (Mohamed et al., 2020).

11.5.6 Plant-based ingredients

Use of plant-based ingredients can cause a foreign taste which is actually a characteristic taste for the plant (Bachmann, 2001). Green or off-flavor formation by legume proteins during storage limits their application in process cheese (Schindler et al., 2012). Use of lupine paste as a substitute for cheese at levels of 25–100% in a processed cheese with dry matter content of 42% and fat in dry matter content of 50% of cheese base increased firmness and oil separation index and reduced cohesiveness, springiness and meltability (Awad et al., 2014). While storage at 5°C enhanced these effects, substitution by 25% was found acceptable in terms of sensory properties (Awad et al., 2014). Szafrańska et al. (2020) developed a processed cheese sauce from acid casein, WPC80 and rapeseed oil fortified with dietary fiber from bamboo, acacia, potato and citrus at a level of 1%–4% to lower calories and cholesterol. Potato fiber increased viscosity the most with more water holding while the samples with orange and bamboo fibers were found more acceptable in sensory analysis.

Li et al. (2013) developed cheese spreads from soy protein by fermentation using lactic acid bacteria, acidification with glucono-delta-lactone and/or papain treatment to prevent the unwanted soybean taste. The combined use of fermentation, glucono-delta-lactone and enzymatic hydrolysis exhibited more acceptable sensory features and better spreadability compared to individual applications. Similarly, the aroma profile of pea protein preparations changed significantly and undesirable flavor of peas decreased by fermentation by lactic acid bacteria (Schindler, et al, 2012). Amar and Surono (2012) showed that flavor of soy protein-based Gouda analogue cheese can be improved by using butter, mixed cheese starter cultures and sodium citrate.

A probiotic cheese-like product was developed by fermenting soy milk with starter cultures and probiotic *Lactobacillus rhamnosus*. Probiotic organism withstood the process and storage at 10°C in the product and it did not have a negative effect on the fermentation and sensory properties of cheese (Liu et al., 2006). Sharma et al. (2018) developed a peanut-based cheese analogue, using a probiotic strain of *Lactobacillus rhamnosus*. In this vegan cheese analogue, a decrease in the viable count of probiotic bacteria was detected after 15 days of storage at 4°C which was ascribed to low pH (<4.2).

Plant-based cheeses have been developed by using plant flour or protein isolate and oil. Bergsma (2019) used different types and amounts of root and tuber starches in a vegan cheese-like formula containing potato protein, sunflower oil and salt. The shredability and stretchability of the product was reported to be the best at a starch content of 17%. The analogue exhibited a soft structure at low starch concentrations (10%–16%), while cheese shreds became dry and crumbled at high starch concentrations (18%–25%). The meltability of the samples were changed according to starch type in descending order, waxy potato > rice > tapioca starch > potato starch > corn > wheat.

A plant-based cheese was developed with zein as a partial substitute for fat and starch at a level upto 30% (w/w) by Mattice and Marrangoni (2020). Their formula also contained corn-tapioca starch and sunflower-coconut oil at a starch:oil ratio of 2:1 and xanthan gum where corn starch was required for zein's functionality. Plant-based cheeses had similar textural, rheological, and melt-stretch properties to those of Cheddar cheese. Zein was found to be a potential ingredient in plant-based cheeses due to formation of networks via non-covalent interactions in aqueous environments and weakening of these noncovalent bonds at high temperatures (Mattice and Marrangoni, 2020).

11.6 Future trends in special processed cheeses

The advantages of special processed cheeses compared to natural or processed cheeses for both producers and consumers are (Guinee, 2017):

1. Adjustment of composition that suits special dietary needs with a wide variety of ingredients
2. Production of functional cheeses without affecting cheese structure and functionality
3. Low cost of manufacture depending on ingredients used (usage of starch or plant oil/fat, no ripening period and related costs)
4. Ingredients do not depend on milk supply and have longer shelf life which allows production in countries with insufficient supply of milk
5. Simplicity of production procedures and equipments
6. Use as ingredient in other food products to impart color, texture, and taste

These advantanges and flexibility in manufacture of SPC continue to drive the development of cheese-like products with low-cost ingredients and sustainable manufacturing. While processed cheese sales have shown a decline due to an unjustified unhealthy image, plants-based cheese is a fast-growing category in the market in recent years (Boisseau, 2019). Even though plant proteins were not preferred in the past for cheese making, now, there is a large potential of plant proteins due to unsustainability, cost, and environmental pollution associated with processing of animal foods and increasing trend toward vegan/vegetarian foods among consumers.

11.7 Conclusion

SPCs are versatile products with variable composition, functionality, sensory properties and nutritional value. This group of processed cheese extends from analogue to plant-based cheese-like products. They can be developed with cost-effective ingredients and simple manufacturing procedure by using diverse flavoring and nutritional components for specific consumer groups.

SPC production from nondairy ingredients with or without dairy ingredients is a developing sector in the food industry. SPC products open up new oppurtunities for both manufacturers and consumers. They permit to adjust composition, reduce cost, help in efficient use of natural resources, lowering environmental pollution by the industry and reach to food by countries with insufficient supply of natural resources. However, they cannot totally replace the nutritional value, flavor and functionality of the products that they imitate. Therefore, they should be allowed to be produced when there is a justified need but not as a competitor of the products they imitate. Strict regulation and inspection are needed for prevention of their use for adulteration of natural cheeses and unjustified benefit.

Microbial safety and storage stability of SPC need to be evaluated before marketing as they are produced from various nontraditional ingredients that can contain pathogens with a potential to grow in a cheese-like medium. Sensory properties are also critical for acceptability and consumption of SPC by all consumer groups. Furthermore, functional SPC could be an appealing product to health-conscious consumers. New ingredients, including different types of protein, starch, stabilizers, new flavorings and bioactive components, and processing technologies need to be investigated to improve quality, nutritional value, and functionality of SPC in future studies.

References

Abbas, H.M., Mohamed, A.G., Bayiomi, H.M., 2019a. Assessment the quality of imitated pumpkin processed cheese during cold storage. J. Biol. Sci. 19, 51–56.

Abbas, H.M., Zaky, W.M., Hassan, L.K., Shahein, N.M., Mohamed, A.G., Samy, N.M., Farahat, E.S.A., 2019b. Impact of kaki (*Diospyros kaki*) juice on the rheological, sensory and color properties of spreadable processed cheese analogue. J. Biol. Sci. 19, 231–236.

Adhikari, K., Cole, A., Grün, I., Heymann, H., Hsieh, F.H., Huff, H., 2009. Physical and sensory characteristics of processed cheeses manufactured by extrusion technology. J. Sci. Food Agric. 89, 1428–1433.

Ahmad, S., Butt, M.S., Pasha, I., Sameen, A., 2016. Quality of processed Cheddar cheese as a function of emulsifying salt replaced by κ-carrageenan. Int. J. Food Prop. 19, 1874–1883.
Al Khalaileh, N.A., Ajo, R.Y., 2013. Production of processed spread cheese supplemented with probiotic bacteria and studying growth and viability under different storage conditions. Carpathian J. Food Sci. Technol. 5 (1), 13–22.
Alimardanova, M.K., Tlevlesova, D.A., Simov, Z., Dimitrov, D., Matibayeva, A., 2015. Incorporating Allium odorum as a vegetable ingredient of processed cheeses. Res. J. Pharm. Biol. Chem. Sci. 6 (3), 330–338.
Al-Nemr, T.M., Mohamed, S.E., Barbabosa, A., Salem, A.Z.M., 2016. Influence of nisin and lauryl arginine ester against some foodborne pathogens in recombined feta and processed spread cheese. J. Food Saf. 36 (2), 172–179.
Amar, A., Surono, I.S., 2012. Physico-chemical, and sensory properties of soy based Gouda cheese analog made from different concentration of fat, sodium citrate and various cheese starter cultures. Makara J. Technol. 16 (2), 149–156.
Arya, S.S., Salve, A.R., Chauhan, S., 2016. Peanuts as functional food: A review. J. Food Sci. Technol. 53 (1), 31–41.
Awad, R.A., Abdel-Hamid, L.B., El-Shabrawy, S.A., Singh, R.K., 2004. Physical and sensory properties of block processed cheese with formulated emulsifying salt mixtures. Int. J. Food Prop. 7 (3), 429–448.
Awad, R.A., Salama, W.M., Farahat, A.M., 2014. Effect of lupine as cheese base substitution on technological and nutritional properties of processed cheese analogue. Acta Sci. Pol. Technol. Aliment. 13, 55–64.
Bachmann, H.P., 2001. Cheese analogues: a review. Int. Dairy J. 11, 505–515.
Baldwin, C.J., Smith, G.F., Akashe, A., 2008. Methods of fortifying process cheese with magnesium phosphate and products thereof. U.S. Patent No. 7,547,457.
Banks, J., Weimer, B., 2007. Producing low fat cheese. In: Weimer, B.C. (Ed.), Improving the Flavour of Cheese. CRC Press, Boca Raton, pp. 520–536.
Barth, A.P., Tormena, C.F., Viotto, W.H., 2017. pH influences hydrolysis of sodium polyphosphate in dairy matrices and the structure of processed cheese. J. Dairy Sci. 100 (11), 8735–8743.
Belitz, H.D., Grosch, W., Schieberle, P., 2009. Food Chemistry. Springer, Berlin.
Benaouadj, F., Ziane-Zafour, A.H., Rebiha, M., 2017. Effects of modified starch and fat on the rheological characteristics of newly formulated processed cheese: Use of experimental design method. J. Dispersion Sci. Technol. 38 (5), 693–698.
Bergsma, J., 2019. Vegan cheese analogue. U.S. Patent No. 16,079,182.
Beykont, E., Kilic-Akyilmaz, M., 2014. Physical properties of an imitation cheese as affected by emulsifying salts and citric acid. J. Food Process. Preserv. 38 (4), 1918–1925.
Bi, W., Zhao, W., Li, D., Li, X., Yao, C., Zhu, Y., Zhang, Y., 2015. Effect of resistant starch and inulin on the properties of imitation mozzarella cheese. Int. J. Food Prop. 19 (1), 159–171.
Boisseau, 2019. Natural cheese retail sales stay put. https://www.dairyfoods.com/articles/93517-natural-cheese-retail-sales-stay-put. Accessed 3 August, 2020.
Borch, E., Lycken, L., 2007. Influence of long-chain polyphosphate and heat treatment on *Clostridium cochlearium* and *Clostridium sporogenes* isolated from processed cheese spread. J. Food Prot. 70 (3), 744–747.
Buňka, F., Doudová, L., Weiserová, E., Kuchař, D., Michálek, J., Slavíková, Š., Kráčmar, S., 2012. The effect of different ternary mixtures of sodium phosphates on hardness of processed cheese spreads. Int. J. Food Sci. Technol. 47 (10), 2063–2071.
Buňková, L., Buňka, F., 2017. Microflora of processed cheese and the factors affecting it. Crit. Rev. Food Sci. Nutr. 57 (11), 2392–2403.

Butt, N.A., Ali, T.M., Awan, S., Hasnain, A., 2019. Use of alcoholic–alkaline treated starch citrates and lactates as fat and protein replacer in imitation Mozzarella cheese. J. Food Process 42, e13297.

Carpenter, R.N., Finnie, K.J., Olsen, R.L., 1998. Imitation cheese composition and products containing starch. U.S. Patent No. 5,807,601.

Castro, J.M., Tornadijo, M.E., Fresno, J.M., Sandoval, H., 2015. Biocheese: a food probiotic carrier. Biomed. Res. Int. 1–11, 723056.

Cavalier-Salou, C., Cheftel, J.C., 1991. Emulsifying salts influence on characteristics of cheese analogs from calcium caseinate. J. Food Sci. 56, 1542–1547.

Černíková, M., Nebesářová, J., Salek, R.N., Řiháčková, L., Buňka, F., 2017. Microstructure and textural and viscoelastic properties of model processed cheese with different dry matter and fat in dry matter content. J. Dairy Sci. 100 (6), 4300–4307.

Chatli, M.K., Gandhi, N., Singh, P., 2017. Efficacy of sodium alginate as fat replacer on the processing and storage quality of buffalo mozzarella cheese. Nutr. Food Sci. 47 (3), 381–397.

Chatziantoniou, S.E., Thomareis, A.S., Kontominas, M.G., 2019. Effect of different stabilizers on rheological properties, fat globule size and sensory attributes of novel spreadable processed whey cheese. Eur. Food Res. Technol. 245, 2401–2412.

Chatziantoniou, S.E., Thomareis, A.S., Kontominas, M.G., 2015. Effect of chemical composition on physico-chemical, rheological and sensory properties of spreadable processed whey cheese. Eur. Food Res. Technol. 241 (6), 737–748.

Cunha, C.R., Alcantara, M.R., Viotto, W.H., 2012. Effect of the type of emulsifying salt on microstructure and rheological properties of "*requeijão cremoso*" processed cheese spreads. J. Food Sci. 77 (8), 176–181.

Cunha, C.R., Dias, A.I., Viotto, W.H., 2010. Microstructure, texture, colour and sensory evaluation of a spreadable processed cheese analogue made with vegetable fat. Food Res. Int. 43 (3), 723–729.

Cutting, S.M., 2011. *Bacillus* probiotics. Food Microbiol. 28 (2), 214–220.

Dimitreli, G., Thomareis, A.S., 2009. Instrumental textural and viscoelastic properties of processed cheese as affected by emulsifying salts and in relation to its apparent viscosity. Int. J. Food Prop. 12 (1), 261–275.

Drake, M.A., Truong, V.D., Daubert, C.R., 1999. Rheological and sensory properties of reduced-fat processed cheeses containing lecithin. J. Food Sci. 64 (4), 744–747.

Duggan, E., Noronha, N., O'Riordan, E.D., O'Sullivan, M., 2008. Effect of resistant starch on the water binding properties of imitation cheese. J. Food Eng. 84 (1), 108–115.

Ehsannia, S., Sanjabi, M.R., 2016a. Physicochemical, microbiological and spoilage analysis of probiotic processed cheese analogues with reduced emulsifying salts during refrigerated storage. J. Food Sci. Technol. 53 (2), 996–1003.

Ehsannia, S., Sanjabi, M.R., 2016b. Quality characterization of processed cheese inoculated by *Bacillus coagulans* during cold storage: compositional and sensorial attributes and probiotic microorganism viability. J. Food Process. Preserv. 40 (4), 667–674.

El-Messery, T.M., El-Said, M.M., Farahat, E., 2019. Production of functional processed cheese supplemented with nanoliposomes of mandarin peel extract. Pak. J. Biol. Sci. 22 (5), 247–256.

El-Sayed, H.A., 2017. Effect of thyme on the quality and shelf life of block processed cheese. J. Food Dairy Sci. 8 (8), 335–340.

El-Shazly, H.A.M., Awad, R.A., Essawy, E.A.Y., Kamal, T.M., Salama, W.M., 2010. Protein substitution to produce a processed cheese with high branched-chain amino acids of medical and genetic importance. Egypt. J. Med. Hum. Genet. 11 (2), 121–133.

El-Shibiny, S., Abd El-Gawad, M.A., Assem, F.M., Seleet, F.L., Abou Dawood, S.A., Elaaser, M., 2013. Preparation, composition and microbiological and rheological properties of functional processed cheese supplemented with rice bran. J. Appl. Sci. Res. 9, 4927–4934.

El-Wahed, E.M., Hassanien, M.F.R., 2019. Chemical, rheological and sensory characteristics of processed cheese spread analogues. Carpathian J. Food Sci. Technol. 11 (1), 135–148.

Ennis, M.P., Mulvihill, D.M., 1997. Cheese analogues. In: Cogan, T.M., Fox, P.F., Ross, R.P. (Eds.), Proceedings of 5th Cheese Symposium, Teagasc, Dublin, pp. 1–15.

Fox, P.F., Guinee, T.P., Cogan, T.M., McSweeney, P.L.H., 2017. Processed cheese and substitute/imitation cheese products. Fundamentals of Cheese Science. Springer, NY, pp. 589–627.

Fu, W., Nakamura, T., 2020. Moisture content impact creaming effect and microstructure of processed cheese containing different textural starches. Int. Dairy J. 105, 104685.

Giri, A., Kanawjia, S.K., Singh, M.P., 2017. Effect of inulin on physico-chemical, sensory, fatty acid profile and microstructure of processed cheese spread. J. Food Sci. Technol. 54 (8), 2443–2451.

Gouda, A., Abou El-Nour, A., 2003. Processed cheese. In: Caballero, B., Finglas, P., Toldra, F. (Eds.), Encyclopedia of Food Sciences and Nutrition. Academic Press, San Diego, pp. 1108–1115.

Guinee, T.P., 2011a. Cheese analogues. In: Fuguay, J.F., Fox, P.F., Mcsweeney, P.L.H. (Eds.), Encyclopedia of Dairy Sciences. Academic Press, London, pp. 814–821.

Guinee, T.P., 2011b. Pasteurized processed cheese products. In: Fuguay, J.F., Fox, P.F., Mcsweeney, P.L.H. (Eds.), Encyclopedia of Dairy Sciences. Academic Press, London, pp. 805–813.

Guinee, T.P., 2017. Pasteurized processed and imitation cheese products. In: McSweeney, P.L.H., Fox, P.F., Cotter, P.D., Everett, D.W. (Eds.), Cheese: Chemistry, Physics and Microbiology. Academic Press, London, pp. 1133–1184.

Gustaw, W., Mleko, S., 2007. The effect of polysaccharides and sodium chloride on physical properties of processed cheese analogs containing whey protein. Milchwissenschaft 62, 59–62.

Hanáková, Z., Buňka, F., Pavlínek, V., Hudečková, L., Janiš, R., 2013. The effect of selected hydrocolloids on the rheological properties of processed cheese analogues made with vegetable fats during the cooling phase. Int. J. Dairy Technol. 66, 1–6.

Hennelly, P.J., Dunne, P.G., O'Sullivan, M., O'Riordan, D., 2005. Increasing the moisture content of imitation cheese: effects on texture, rheology and microstructure. Eur. Food Res. Technol. 220 (3–4), 415–420.

Hennelly, P.J., Dunne, P.G., O'Sullivan, M., O'Riordan, D., 2006. Textural, rheological and microstructural properties of imitation cheese containing inulin. J. Food Eng. 75 (3), 388–395.

Huang, H., Hallinan, R., Maleky, F., 2018. Comparison of different oleogels in processed cheese products formulation. Int. J. Food Sci. Technol. 53, 2525–2534.

Hussein, G.A.M., Shalaby, S.M., 2018. Properties of imitation cheese products prepared with non-dairy ingredients. Saudi J. Life Sci. 3 (9), 578–587.

Johnson, J.L., Mistry, V.V., Vukovich, M.D., Hogie-Lorenzen, T., Hollis, B.W., Specker, B.L., 2005. Bioavailability of vitamin D from fortified process cheese and effects on vitamin D status in the elderly. J. Dairy Sci. 88 (7), 2295–2301.

Johnson, M.E., 2011. Cheese: Low-fat and reduced-fat cheese. In: Fuguay, J.F., Fox, P.F., Mcsweeney, P.L.H. (Eds.), Encyclopedia of Dairy Sciences. Academic Press, London, pp. 833–842.

Johnson, M.E., Kapoor, R., McMahon, D.J., McCoy, D.R., Narasimmon, R.G., 2009. Reduction of sodium and fat levels in natural and processed cheeses: scientific and technological aspects. Compr. Rev. Food Sci. Food Saf. 8 (3), 252–268.

Kaláb, M., 1995. Practical aspects of electron microscopy in cheese research. In: Malin, E., Tunick, M.H. (Eds.), Chemistry of Structure-Function Relationships in Cheese. Advances in Experimental Medicine and Biology, vol. 367. Springer, Boston, MA, pp. 247–276.

Kapoor, R., Metzger, L.E., 2008. Process cheese: Scientific and technological aspects-A review. Comp. Rev. Food Sci. Food Safe. 7 (2), 194–214.

Kassem, J.M., Abbas, H.M., Mohamed, A.G., Bahgaat, W.K., El-Messery, T.M., 2017. Sweet processed cheese spread analogue as a novel healthy dairy product. Int. J. Dairy Sci. 12 (5), 331–338.

Kazmi, S.A., Vieth, R., Rousseau, D., 2007. Vitamin D3 fortification and quantification in processed dairy products. Int. Dairy J. 17, 753–759.

Khider, M, Seoudi, O., Abdelaliem, Y.F., 2017. Functional processed cheese spreads with high nutritional value as supplemented with fresh and dried mushrooms. Int. J. Nutr. Food Sci. 6 (1), 45–52.

Kim, S.Y., Hong, E.K., Ahn., J., Kwak, H.S., 2009. Chemical and sensory properties of cholesterol-reduced processed cheese spread. Int. J. Dairy Technol. 62 (3), 348–353.

Kiziloz, M.B., Cumhur, O., Kilic, M., 2009. Development of the structure of an imitation cheese with low protein content. Food Hydrocoll. 23, 1596–1601.

Kommineni, A., Amamcharla, J., Metzger, L.E., 2012. Effect of xylitol on the functional properties of low-fat process cheese. J. Dairy Sci. 95 (11), 6252–6259.

Krumov, K., Ivanov, G., Slavchev, A., Nenov, N., 2010. Improving the processed cheese quality by the addition of natural spice extracts. Adv. J. Food Sci. Technol. 2 (6), 335–339.

Lee, S.K., Anema, S., Klostermeyer, H., 2004. The influence of moisture content on the rheological properties of processed cheese spreads. Int. J. Food Sci. Technol. 39 (7), 763–771.

Li, H., Yu, H., Liu, Y., Wang, Y., Li, H., Yu, J., 2019. The use of inulin, maltitol and lecithin as fat replacers and plasticizers in a model reduced-fat mozzarella cheese-like product. J. Sci. Food Agric. 99 (12), 5586–5593.

Li, Q., Xia, Y., Zhou, L., Xie, J., 2013. Evaluation of the rheological, textural, microstructural and sensory properties of soy cheese spreads. Food Bioprod. Process. 91 (4), 429–439.

Liu, D.M., Li, L., Yang, X.Q., Liang, S.Z., Wang, J.S., 2006. Survivability of *Lactobacillus rhamnosus* during the preparation of soy cheese. Food Technol. Biotechnol. 44 (3), 417–422.

Liu, H., Xu, X.M., Guo, S.D., 2008. Comparison of full-fat and low-fat cheese analogues with or without pectin gel through microstructure, texture, rheology, thermal and sensory analysis. Int. J. Food Sci. Technol. 43 (9), 1581–1592.

Lu, Y., Shirashoji, N., Lucey, J.A., 2008. Effects of pH on the textural properties and meltability of pasteurized process cheese made with different types of emulsifying salts. J. Food Sci. 73 (8), E363–E369.

Lucey, J.A., 2008. Advances in nonfat/lowfat process cheese for melting and ingredient use. J. Dairy Sci. 91 (E-Suppl. 1), 153–154.

Lycken, L., Borch, E., 2006. Characterization of *Clostridium* spp. isolated from spoiled processed cheese products. J. Food Prot. 69 (8), 1887–1891.

Macků, I., Buňka, F., Pavlínek, V., Leciánová, P., Hrabě, J., 2008. The effect of pectin concentration on viscoelastic and sensory properties of processed cheese. Int. J. Food Sci. Technol. 43 (9), 1663–1670.

Masotti, F., Cattaneo, S., Stuknytė, M., De Noni, I., 2018. Status and developments in analogue cheese formulations and functionalities. Trends Food Sci. Technol. 74, 158–169.

Mattice, K.D., Marangoni, A.G., 2020. Physical properties of plant-based cheese products produced with zein. Food Hydrocoll., 105746.

Mehanna, N.S., Hassan, F.A.M., El-Messery, T.M., Mohamed, A.G., 2017. Production of functional processed cheese by using tomato juice. Int. J. Dairy Sci. 12, 155–160.

Mleko, S., Foegeding, E.A., 2000. Physical properties of rennet casein gels and processed cheese analogs containing whey proteins. Milchwissenschaft 55, 513–516.

Mohamed, A.G., Abo-El-Khair, B.E., Shalaby, S.M., 2013. Quality of novel healthy processed cheese analogue enhanced with marine microalgae Chlorella vulgaris biomass. World Appl. Sci. J. 7, 914–925.

Mohamed, A.G., El-Salam, B., Gafour, W., 2020. Quality characteristics of processed cheese fortified with *Spirulina* powder. Pak. J. Biol. Sci. 23 (4), 533–541.

Mohamed, A.G., Shalaby, S.M., 2016. Texture, chemical properties and sensory evaluation of a spreadable processed cheese analogue made with apricot pulp (*Prunus armeniaca* L.). Int. J. Dairy Sci. 11 (2), 61–68.

Mohamed, A.G., Shalaby, S.M., Gafour, W.A., 2016. Quality characteristics and acceptability of an analogue processed spreadable cheese made with carrot paste (Daucus carota L.). Int. J. Dairy Sci. 11 (3), 91–99.

Montesinos-Herrero, C., Cottell, D.C., O'Riordan, E.D., O'Sullivan, M., 2006. Partial replacement of fat by functional fibre in imitation cheese: Effects on rheology and microstructure. Int. Dairy J. 16 (8), 910–919.

Mounsey, J.S., O'Riordan, E.D., 1999. Empirical and dynamic rheological data correlation to characterize melt characteristics of imitation cheese. J. Food Sci. 64, 701–703.

Mounsey, J.S., O'Riordan, E.D., 2001. Characteristics of imitation cheese containing native starches. J. Food Sci. 66, 586–591.

Mounsey, J.S., O'Riordan, E.D., 2008a. Alteration of imitation cheese structure and melting behaviour with wheat starch. Eur. Food Res. Technol. 226 (5), 1013–1019.

Mounsey, J.S., O'Riordan, E.D., 2008b. Characteristics of imitation cheese containing native or modified rice starches. Food Hydrocoll. 22 (6), 1160–1169.

Muir, D.D., Tamime, A.Y., Shenana, M.E., Dawood, A.H., 1999. Processed cheese analogues incorporating fat-substitutes 1. Composition, microbiological quality and flavour changes during storage at 5°C. LWT Food Sci. Technol. 32 (1), 41–49.

Mulsow, B.B., Jaros, D., Rohm, H., 2007. Processed cheese and cheese analogues. In: Tamime, A.Y. (Ed.), Structure of Dairy Products. Blackwell Publishing, Oxford, pp. 210–235.

Nastaj, M., Terpiłowski, K., Sołowiej, B.G., 2020. The effect of native and polymerised whey protein isolate addition on surface and microstructural properties of processed cheeses and their meltability determined by Turbiscan. Int. J. Food Sci. Technol. 55, 2179–2187.

Noronha, N., Duggan, E., Ziegler, G.R., O'Riordan, E.D., O'Sullivan, M., 2008a. Inclusion of starch in imitation cheese: its influence on water mobility and cheese functionality. Food Hydrocoll. 22 (8), 1612–1621.

Noronha, N., O'Riordan, D.E., O'Sullivan, M., 2007. Replacement of fat with functional fibre in imitation cheese. Int. Dairy J. 17, 1073–1082.

Noronha, N., O'Riordan, E.D., O'Sullivan, M., 2008b. Influence of processing parameters on the microstructure and texture of imitation cheese. Eur. Food Res. Technol. 226, 385–393.

O'Riordan, E.D., Duggan, E., O'Sullivan, M., Noronha, N., 2011. Production of analogue cheeses. In: Tamime, A.Y. (Ed.), Processed Cheeses and Analogues. Blackwell Publishing, UK, pp. 219–244.

Pivetta, F.P., Silva, M.N., Tagliapietra, B.L., Richards, N.S.S., 2020. Addition of green banana biomass as partial substitute for fat and encapsulated *Lactobacillus acidophilus* in requeijão cremoso processed cheese. Food Sci. Technol. 40 (2), 451–457.

Přikryl, J., Hájek, T., Švecová, B., Salek, R.N., Černíková, M., Červenka, L., Buňka, F., 2018. Antioxidant properties and textural characteristics of processed cheese spreads enriched with rutin or quercetin: The effect of processing conditions. LWT Food Sci. Technol. 87, 266–271.

Rafiq, S.M, Ghosh, B.C., 2017a. Effect of peanut addition on the fatty acid profile and rheological properties of processed cheese. J. Food Process Technol. 8, 690–695.

Rafiq, S.M., Ghosh, B.C., 2017b. Effect of potato incorporation on the physico-chemical, textural and sensory properties of processed cheese. J. Food Meas. Charact. 11 (2), 776–780.

Ramel, P.R., Marangoni, A.G., 2018. Processed cheese as a polymer matrix composite: a particle toolkit for the replacement of milk fat with canola oil in processed cheese. Food Res. Int. 107, 110–118.

Rinaldoni, A.N., Palatnik, D.R., Zaritzky, N., Campderrós, M.E., 2014. Soft cheese-like product development enriched with soy protein concentrates. LWT Food Sci. Technol. 55 (1), 139–147.

Rolim, F.R.L., Freitas Neto, O.C., Oliveira, M.E.G., Oliveira, C.J.B, Queiroga, R.C.R.E., 2020. Cheeses as food matrixes for probiotics: In vitro and in vivo tests. Trends Food Sci. Technol. 138–154.

Sadek, Z.I, Refaat, B.M., Abd El-Shakour, E.H., Mehanna, N.S., Hassan, M.S., 2017. Biocontrol of processed cheese by incorporation of probiotic bacteria and its metabolites. Int. J. Dairy Sci. 12, 93–104.

Savello, P.A., Ernstrom, C.A., Kalab, M., 1989. Microstructure and meltability of model process cheese made with rennet and acid casein. J. Dairy Sci. 72, 1–11.

Schindler, S., Zelena, K., Krings, U., Bez, J., Eisner, P., Berger, R.G, 2012. Improvement of the aroma of pea (*Pisum sativum*) protein extracts by lactic acid fermentation. Food Biotechnol. 26 (1), 58–74.

Shalaby, S.M., Mohamed, A.G., Farahat, E.S., 2018. Preparation of functional and nutritional spreadable processed cheese fortified with vegetables and mushroom. Int. J. Curr. Res. 10 (10), 74075–74082.

Sharma, P., Sharma, D., Amin, A., 2018. Development of a functional fermented peanut-based cheese analog using probiotic bacteria. J. Biotechnol. Comput. Biol. Bionanotechnol. 99 (4), 435–441.

Soares, M.B., Almada, C.N., Almada, C.N., Martinez, R.C.R., Pereira, E.P.R., Balthazar, C.F., et al., 2019. Behavior of different *Bacillus* strains with claimed probiotic properties throughout processed cheese ("requeijão cremoso") manufacturing and storage. Int. J. Food Microbiol. 307, 108288.

Solhi, P., d-Damirchi, S.A., Hesari, J., Hamishehkar, H., 2020. Production of the processed cheese containing tomato powder and evaluation of its rheological, chemical and sensory characteristics. J. Food Sci. Technol. 57 (6), 2198–2205.

Sołowiej, B.G., Nastaj, M., Szafrańska, J.O., Muszyński, S., Gustaw, W., Tomczyńska-Mleko, M., Mleko, S., 2020. Effect of emulsifying salts replacement with polymerised whey protein isolate on textural, rheological and melting properties of acid casein model processed cheeses. Int. Dairy J. 105, 104694.

Sołowiej, B., Cheung, I.W.Y., Li-Chan, E.C.Y., 2014. Texture, rheology and meltability of processed cheese analogues prepared using rennet or acid casein with or without added whey proteins. Int. Dairy J. 37 (2), 87–94.

Sołowiej, B., Dylewska, A., Kowalczyk, D., Sujka, M., Tomczyńska-Mleko, M., Mleko, S., 2016. The effect of pH and modified maize starches on texture, rheological properties and meltability of acid casein processed cheese analogues. Eur. Food Res. Technol. 242 (9), 1577–1585.

Sołowiej, B., Glibowski, P., Muszyński, S., Wydrych, J., Gawron, A., Jeliński, T., 2015. The effect of fat replacement by inulin on the physicochemical properties and microstructure of acid casein processed cheese analogues with added whey protein polymers. Food Hydrocoll. 44, 1–11.

Swenson, B.J., Wendorff, W.L., Lindsay, R.C, 2000. Effect of ingredients on the functionality of fat-free process cheese spreads. J. Food Sci. 65, 822–825.

Szafrańska, J., Sołowiej, B.G., 2020. Effect of different fibres on texture, rheological and sensory properties of acid casein processed cheese sauces. Int. J. Food Sci. Technol. 55, 1971–1979.

Talbot-Walsh, G., Kannar, D., Selomulya, C., 2018. A review on technological parameters and recent advances in the fortification of processed cheese. Trends Food Sci. Technol. 81, 193–202.

Talbot-Walsh, G., Kannar, D., Selomulya, C., 2019. pH effect on the physico-chemical, microstructural and sensorial properties of processed cheese manufactured with various starches. LWT Food Sci. Technol. 111, 414–422.

Tamime, A.Y., Muir, D.D., Shenana, M.E., Kalab, M., Dawood, A.H., 1999. Processed cheese analogues incorporating fat-substitutes 2. Rheology, sensory perception of texture and microstructure. LWT Food Sci. Technol. 32 (1), 50–59.

Tamime, A.Y., Saarela, M., Wszolek, M., Ghoddousi, H., Linares, D.M., Shah, N.P., 2017. Production and maintaining viability of probiotic microorganism. In: Tamie, A.Y., Thomas, L.V. (Eds.), Dairy Products. Probiotic Dairy Products. Wiley, New Jersey, pp. 67–164.

Tayel, A.A., Hussein, H., Sorour, N.m., El-Tras, W.F., 2015. Foodborne pathogens prevention and sensory attributes enhancement in processed cheese via flavoring with plant extracts. J. Food Sci. 80, M2886–M2891.

Tohamy, M.M., Ali, M.A., Shaaban, H.A.G., Mohamad, A.G., Hasanain, A.M., 2018. Production of functional spreadable processed cheese using Chlorella vulgaris. Acta Sci. Pol. Technol. Aliment. 17 (4), 347–358.

Trivedi, D., Bennet, R.J., Hemar, Y., Reid, D.C.W., Lee, S.K., Illingworth, D., 2008. Effect of different starches on rheological and microstructural properties of (I) model processed cheese. Int. J. Food Sci. Technol. 43, 2191–2196.

Tunick, M., 2013. The Science of Cheese. Oxford University Press, Inc., ProQuest Ebook Central, pp. 159–173. http://ebookcentral.proquest.com/lib/itup/detail.action?docID=1538395.

Upreti, P, Mistry, V.V, Warthesen, J.J., 2002. Estimation and fortification of vitamin D3 in pasteurized process cheese. J. Dairy Sci. 85 (12), 3173–3181.

Williams, G.C., Delves-Broughton, J., 2003. Nisin. In: Caballero, B., Finglas, P., Toldra, F. (Eds.), Encyclopedia of Food Sciences and Nutrition. Academic Press, San Diego, CA, pp. 4128–4135.

Ye, A., Hewitt, S., Taylor, S., 2009. Characteristics of rennet–casein-based model processed cheese containing maize starch: Rheological properties, meltabilities and microstructures. Food Hydrocoll. 23 (4), 1220–1227.

Yoder, D., Chang, S.G., Xu, A., Domoras, T., 1996. Caseinate replacement composition for imitation cheese and process form making imitation cheese. U.S. Patent No 5 (486), 375.

Zallie, J.P., Chiu, C.W., 1990. Imitation cheeses containing enzymatically debranched starches in lieu of caseinates. U.S. Patent No. 4,937,091.

Zhang, D., Mahoney, A.W., 1991. Iron fortification of process cheddar cheese. J. Dairy Sci. 74, 353–358.

Zhang, X., Yang, Y., Zhao, D., 2011. Effect of blueberry on spreadable processed cheese. J. Northeast Agric. Univ. 18 (2), 73–78.

Techniques used for processed cheese characterization

12

Apostolos S. Thomareis and Georgia Dimitreli
Department of Food Science and Technology, International Hellenic University, Thessaloniki, Greece

12.1 Introduction

The main techniques used to characterize processed cheeses both in industry and in research, distinguished into nine categories, will be discussed. All the techniques can also be employed for natural cheeses which are the main raw materials for the production of processed cheeses. Several techniques are horizontal and can also be applied in the case of other raw materials used.

As is well known, the appropriate taking of a representative sample is required before the measurements. ISO (2008a) gives guidance on methods of sampling milk and milk products for chemical, microbiological, physical and sensory analysis. In particular, it sets out the requirements for general arrangements (sampling personnel, sealing and labeling of samples, replicate samples, sampling report), apparatus (sampling equipment and containers), sampling, preservation, storage and transport of samples.

12.2 Compositional analyses

Knowledge of the chemical composition is essential for both the production of processed cheeses and their quality control.

Most of the methods to be examined are recommended by the International Organization for Standardization (ISO) in International Standards or Technical Specifications, and by the Association of Official Analytical Chemists (AOAC) International as Official Methods (OM) of Analysis.

12.2.1 *Total solids/moisture content*

Commonly used thermogravimetric methods are listed there.

a) *Oven method*: A weighed cheese test portion mixed with sand is dried by heating it in a drying oven at 102°C ($\geq$ 3 h). The dried test portion is then weighed to determine the loss of mass (ISO, 2004b).

b) *Vacuum oven method*: Cheese sample is dried in vacuum oven (pressure $\leq$100 mm Hg) at 100°C to constant weight (about 4 h). Loss in weight is expressed as moisture (OM 926.08, AOAC, 2016).

Processed Cheese Science and Technology: Ingredients, Manufacture, Functionality, Quality, and Regulations.
DOI: https://doi.org/10.1016/B978-0-12-821445-9.00007-8

c) *Rapid methods*: As oven drying is slow, rapid commercial thermogravimetric methods have been developed with a more efficient heating mode (Anonymous, 2002, 2019):
  - Microwave drying: heating by absorption of microwaves (2–5 min),
  - Infrared drying: heating by absorption of infrared radiation (5–15 min), and
  - Halogen drying: heating by absorption of infrared radiation with halogen radiator (2–10 min). This method is most commonly used in commercial devices. It is also applied in combination with microwave drying.

12.2.2 Fat content

a) *Gravimetric method (Reference method)*: A cheese test portion is digested with hydrochloric acid, and then ethanol is added. The acid-ethanolic solution is extracted with diethyl ether and light petroleum and the solvents are removed by distillation or evaporation. The mass of the substances extracted is determined. This is usually known as the Schmid-Bondzynski-Ratzlaff principle (ISO, 2004a).
b) *Van Gulik method*: The protein is dissolved in sulfuric acid, and then the fat of the cheese is separated in a Van Gulik butyrometer by centrifuging, the separation being assisted by the addition of a small quantity of *iso*-amyl alcohol. The fat content is then read directly from the butyrometer scale (ISO, 2008b,c).
c) *NMR spectroscopy*: NMR (nuclear magnetic resonance) is based on the fact that certain nuclei (protons) absorb and reemit radio frequency energy over a narrow band of frequencies when placed in a static magnetic field. On this basis, a rapid commercial method has been developed where the fat content is determined by the proton transverse relaxation time (T2) of fat molecules in previously dried food, in 30 s (Anonymous, 2019).

12.2.3 Total nitrogen and nitrogenous fractions content

a) *Total nitrogen (Kjeldahl method)*: A test portion of milk or milk product is digested with a mixture of concentrated sulfuric acid and potassium sulfate. Copper(II) sulfate is used as a catalyst to thereby convert organic nitrogen present to ammonium sulfate. The function of the potassium sulfate is to elevate the boiling point of the sulfuric acid and to provide a stronger oxidizing mixture for digestion. Excess sodium hydroxide is added to the cooled digest to liberate ammonia. The liberated ammonia is steam distilled into excess boric acid solution and titrated against a hydrochloric acid standard volumetric solution. The total nitrogen content (w_N) is calculated from the amount of ammonia produced and the crude protein content (w_{cp}) from the nitrogen content obtained ($w_{cp} = 6.38w_N$) (ISO, 2014a; OM 2001.14, AOAC, 2016).
b) *Soluble nitrogenous fractions (SN, TCA-SN, PTA-SN)*: They are obtained from a citrate solution of cheese after separation by precipitation of the insolubles in a medium having a pH of 4.4 (SN) or in 12% trichloroacetic acid (TCA-SN) or in 5% phosphotungstic acid (PTA-SN). The nitrogen content in each fraction is determined by Kjeldahl method (ISO, 2011b).
c) *Protein*: Knowing the cheese contents in total nitrogen, w_N (or *TN*), and soluble nitrogen in 12% trichloroacetic acid, $w_{TCA\text{-}SN}$, which is the nonprotein nitrogen content, w_{np} (or *NPN*), the protein content, w_p, of cheese is (Lee et al., 1979):

$$w_p = 6.38(w_N - w_{TCA-SN}) = 6.38(TN - NPN)$$

d) *Water-soluble nitrogen*: A test portion of ground cheese is suspended in water. The suspension is homogenized, centrifuged at 3000 × *g* and then filtered. The nitrogen content in the filtrate determined using the Kjeldahl method is the water-soluble nitrogen content (*WSN*) (Bütikofer et al., 1993).

e) *Noncasein nitrogen (NCN)*: Its content is determined by the Kjeldahl method in the filtrate of a citrate solution of cheese brought to pH 4.6, with sodium acetate and acetic acid, as casein is precipitated at pH 4.6 (Lee et al., 1979).

f) *Peptization coefficient (PC)*: A test portion of ground cheese is suspended in water. After ultracentrifugation at 300,000 × *g*, the nitrogen determined by the Kjeldahl method in the supernatant is the nonsedimentable nitrogen (*NSN*). The peptization coefficient is (Lee et al., 1979):

$$PC(\%) = \frac{NSN - NCN}{100 - NCN} \times 100$$

j) *Protein (dye-binding method)*: This rapid commercial method (3 min test) is based on protein-tagging technology where the sample is mixed with an anionic dye that binds to cationic groups of the basic amino acid residues (histidine, arginine, and lysine) and to free amino terminal groups. The amount of unbound dye remaining in solution, after the insoluble protein has been removed by filtration, is determined by measuring its absorbance. The amount of dye bound is proportional to the protein content (Anonymous, 2019).

12.2.4 Chloride and salt content

a) *Potentiometric titration method*: A test portion of ground cheese is suspended in water. The suspension is acidified with nitric acid, and then the chloride ions are titrated potentiometrically with a silver nitrate standard solution (ISO, 2006b; OM 983.14, AOAC, 2016).

b) *Volhard method*: A test portion of ground cheese is suspended in water. The suspension is acidified with nitric acid and silver nitrate, more than enough to combine with all the chlorides, is added. The excess silver nitrate is titrated with potassium thiocyanate using ferric alum as indicator (OM 935.43, AOAC, 2016).

12.2.5 Titratable acidity and pH

a) *Titratable acidity*: A cheese test portion is suspended in water. Filtrate is titrated with standard 0.1M NaOH using phenolphthalein. Results are expressed as lactic acid (OM 920.124, AOAC, 2016).

b) *pH*: Ground cheese is put in close contact with the combined electrode of a pH-meter, previously calibrated with two buffer solutions at pH 4.01 and 7.00. The pH value is recorded. For semihard and hard cheeses, addition of a small amount of water is needed to get proper contact (Ardö and Polychroniadou, 1999).

12.2.6 Ash and mineral content

a) *Ash (Gravimetric method)*: A test portion of ground cheese is dried and then ignited with a burner. Ignition is completed in a furnace at ≤550°C. Residue is ash (OM 935.42, AOAC, 2016).

b) *Phosphorus*: Cheese is digested by concentrated sulfuric acid and hydrogen peroxide. Molybdenum blue is formed by addition of a sodium molybdate-ascorbic acid solution. Molecular absorption of the blue color formed is measured spectrometrically at a wavelength of 820 nm (ISO, 2010; OM 990.24, AOAC, 2016).

c) *Calcium, magnesium and phosphorus*: Cheese test sample is dried and ashed. Residue is dissolved and diluted in acidified aqueous solution. Portions are diluted for colorimetric determination of phosphorus content at 400 nm, and atomic absorption spectrophotometric determination of calcium content at 422.7 nm and magnesium content at 285.2 nm (OM 991.25, AOAC, 2016).

d) *Calcium, sodium, potassium, and magnesium*: The organic matter of milk or milk product is decomposed by dry ashing or by wet digestion using nitric acid either in an open microwave-assisted wet digestion system or in a pressurized microwave-assisted wet digestion system or in a pressurized polytetrafluoroethylene (PTFE) decomposition vessel or any appropriate instrumentation in wet digestion. The ash containing calcium, sodium, potassium and magnesium is dissolved in a nitric acid solution in the case of dry ashing, or the digests diluted in the case of wet digestion. The test and calibration solutions are atomized into an air–acetylene flame of an atomic absorption spectrometer and their absorption is measured at appropriate wavelengths (422.7 nm for calcium, 589.6 nm for sodium, 766.5 nm for potassium, and 285.2 nm for magnesium) (ISO, 2007b).

e) *Calcium, copper, iron, potassium, magnesium, manganese, sodium, phosphorus, and zinc*: The test portion of milk or milk product is heated at 200°C either with nitric acid in a closed-vessel microwave digestion system or with a combination of hydrogen peroxide, nitric acid, and hydrochloric acid in an open-vessel microwave digestion system. Digested test solution or an appropriate dilution is presented to the inductively coupled plasma–atomic emission spectrometer. The instrument is calibrated with acid(s)-matched standard calibration solutions. An ionization buffer (cesium) and internal standard(s) (chromium, indium, strontium, and yttrium) are used with or without appropriate in-line addition to minimize easily ionizable element effects and to correct for physical and chemical interferences. The following is a list of wavelengths (in nanometers) for each element that have been found acceptable for main foodstuffs, given in priority order: Ca, 317.933; Cu, 324.754; Fe, 259.94; K, 766.491; Mg, 285.213; Mn, 257.610; Na, 589.592; P, 213.618; and Zn, 213.857 (OM 2011.14, AOAC, 2016; ISO, 2018d).

12.2.7 Lactose content

a) *Enzymatic method utilizing the glucose moiety of the lactose*: A solution or suspension of the processed cheese test portion is deproteinated to obtain a pure extract which is reacted with the following enzymes and biochemical substances: (i) β-galactosidase, to split the lactose into glucose and galactose; (ii) hexokinase and adenosine triphosphate (ATP), to phosphorylate the glucose (both that originally present and that liberated by the β-galactosidase) to glucose 6-phosphate (G6P); (iii) glucose 6-phosphate dehydrogenase (G6P-DH) in the presence of nicotinamide adenine dinucleotide phosphate ($NADP^+$), to catalyze the oxidation of G6P to 6-phosphogluconate, the $NADP^+$ being reduced to NADPH. The amount of NADPH is determined from the absorbance of the test solution at 340 nm and is proportional to the lactose content (ISO, 2002a).

b) *Enzymatic method utilizing the galactose moiety of the lactose*: A solution or suspension of the processed cheese test portion is deproteinated to obtain a pure extract which is reacted with the following enzymes and biochemical substances: (i) β-galactosidase, to split the lactose into glucose and galactose; (ii) β-galactosidase dehydrogenase in the presence of nicotinamide adenine dinucleotide (NAD^+) to catalyze the oxidation of galactose into galactonic acid, the NAD^+ being reduced to NADH. The amount of NADH is determined from the absorbance of the test solution at 340 nm and is proportional to the lactose content (ISO, 2002b).

12.2.8 Citric acid content

a) *Enzymatic method*: An extract of the cheese sample is treated with the following enzymes and biochemical substances: (i) citrate lyase to convert citric acid to oxalacetate and acetate; (ii) malate dehydrogenase and lactate dehydrogenase, in the presence of reduced nicotinamide adenine dinucleotide (NADH), to catalyze the reduction of oxalacetate and its decarboxylation product pyruvate to L-malate and L-lactate, respectively, with the subsequent conversion of NADH to its oxidized form (NAD^+). The decrease in concentration of NADH is determined from the absorbance of the test solution at 340 nm and is proportional to citric acid content (ISO, 2006e).
b) *Colorimetric method:* Cheese is dispersed in water and protein is precipitated by trichloroacetic acid. Filtrate is treated with pyridine and acetic anhydride and absorbance of yellow color of citric acid complex is measured at 428 nm (OM 976.15, AOAC, 2016).

12.2.9 Nitrate and nitrite contents

a) *Method using cadmium reduction and spectrometry*: A test portion of milk or milk product is dispersed in warm water, with precipitation of the fat and proteins, then filtration. The nitrate ions are reduced to nitrite ions in a portion of the filtrate by means of copperized cadmium. A red color is developed in portions of both unreduced filtrate and the reduced solution, by addition of sulfanilamide and N-1-naphthyl ethylenediamine dihydrochloride. Spectrometric measurements are carried out at a wavelength of 538 nm. The nitrite content of the sample and the total nitrite content after reduction of nitrate ions are calculated by comparing the measured absorbances with those of a set of sodium nitrite calibration solutions. The nitrate content is calculated from the difference between these two contents (ISO, 2004e).
b) *Method using segmented flow analysis (Routine method)*: For nitrate determination, a test portion of milk or milk product is suspended in water. Part of the suspension is transferred to the analyzer for dialysis. The nitrate ions are reduced to nitrite. The nitrite content is determined by a spectrometric method. Standard nitrate solutions are determined by the same procedure. The nitrate content is calculated by comparing the reading obtained from the test portion with the readings from the standard solutions.

   For nitrite determination, a test portion of milk or milk product is suspended in a solution of ammonium and sodium chloride. Part of the suspension is transferred to the analyzer for dialysis. The nitrite content is determined by a spectrometric method. Standard nitrite solutions are determined by the same procedure. The nitrite content is calculated by comparing the reading obtained from the test portion with that from the standard solution (ISO, 2004f).
c) *Method using cadmium reduction and flow injection analysis with in-line dialysis (Routine method)*: A test portion of milk or milk product is suspended in a warm extraction buffer solution. Fat is separated by centrifuging and rapid cooling. Analyses are made of small portions of the defatted solution by flow injection analysis (FIA). Inline dialysis is used to remove

protein and remaining fat. The nitrate ions are reduced to nitrite ions by cadmium. The nitrite ions are reacted with sulfanilamide and N-1-naphthyl ethylenediamine dihydrochloride to give a red-colored azo dye. The color is measured in a flow cell at maximum absorption of the dye at 540 nm with reference to the absorption measured at 620 nm. The nitrite and nitrate contents of the test sample are calculated with reference to the measured absorbances for a series of standard solutions of nitrite and nitrate, respectively (ISO, 2004g).

12.2.10 Total solids, moisture, fat, nonfat solids, protein, lactose, and salt contents (NIR spectroscopy)

The near-infrared (NIR) spectrometer is based on diffuse reflectance or transmittance measurement in the whole near infrared wavelength region of 700–2500 nm or segments of this or at selected wavelengths. On this basis, rapid commercial methods (10–60 s) for determining various food constituents have been developed.

According to ISO (2006c), the milk product is pretreated to obtain a homogeneous test sample representing the chemical composition of the sample material. It is loaded into the sample holder of the NIR spectrometer. The absorbance at wavelengths in the near infrared region is measured and the spectral data are transformed to constituent concentrations by calibration models developed on representative samples ($\geq$120) from the population to be tested.

The calibration may be performed using different techniques: multiple linear regression (MLR) or multivariate algorithms such as partial least-squares regression (PLS), locally weighted regression (LWR) or artificial neural networks (ANN).

When calibration equations have been developed, they should be validated on an independent test set, preferably sampled after the calibration period. The test set should cover all variations in the sample population and should contain at least 25 samples.

12.2.11 Peroxide value and fatty acids composition of milk fat

a) *Peroxide value*: A test portion of anhydrous milk fat is dissolved in a mixture of methanol/1-decanol/n-hexane, then iron(II) chloride and ammonium thiocyanate are added. The peroxides oxidize the iron(II) which forms a red iron(III) complex with the ammonium thiocyanate. The amount of substance is calculated from a photometric determination (500 nm) of the red iron(III) complex, after a fixed period of reaction (ISO, 2006a).
b) *Fatty acids*: Fatty acid methyl esters (FAME) of milk fat are prepared by transesterification. They are separated and determined by capillary gas-liquid chromatography. Individual FAME are quantified by reference to a milk fat of known composition (ISO, 2002c, 2002d).
c) *Fatty acids*: FAME are prepared by direct transesterification with methanolic sodium methoxide for liquid milk samples. The same transesterification procedure is applied to milk powder sample after dissolution (i.e., reconstitution) in water, and to fat after extraction from various milk products (e.g., low fat products, cheeses).

   FAME are separated using capillary gas-liquid chromatography, identified by comparison with the retention time of pure standards and quantified as fatty acids by reference to an internal standard (C11:0 FAME), added to the sample, and instrument response factors.

The transesterification performance is verified using tritridecanoin (C13:0 TAG) as second internal standard (ISO, 2015b; OM 2012.13, AOAC, 2016).

12.2.12 Natamycin content

A known quantity of sample is extracted with methanol. The extract is diluted with water followed by cooling to between −15°C and −20°C to precipitate most of the fat, followed by filtration. The natamycin content or surface-area-related natamycin mass is determined in the filtrate (after concentration, if necessary) by

a) *UV molecular absorption spectrometry*: Absorbance is measured at 317 nm, using natamycin standard solutions between 311 nm and 329 nm (ISO, 2018a).
b) *High-performance liquid chromatography (HPLC)*: The mass of natamycin in the injected aliquot portion of the test solution can be found by interpolation on a calibration graph made with standard solutions of natamycin (ISO, 2018b).

12.2.13 Nisin A content

The cheese sample is grated and extracted with dilute formic acid at 80°C. After ultracentrifugation, interfering proteins are separated by means of filtration through an ultrafiltration (UF) membrane. In the purified extract, nisin A is separated using a polymeric stationary phase and detected using mass spectrometry (LC-MS or LC-MS-MS) (ISO, 2009d).

12.2.14 Benzoic and sorbic acid contents

Fats and proteins of milk and milk products are removed from a slightly alkaline solution of the product by Carrez precipitation. Following dilution of the resultant solution with methanol, the supernatant liquid is filtered. The benzoic acid and sorbic acid are separated by high-performance liquid chromatography (HPLC) on a reversed-phase C_{18} column, measuring the absorbance at 227 nm and 250 nm (ISO, 2008e).

12.3 Physicochemical measurements

Physicochemical properties are related to the stability and shelf life of processed cheeses. They include emulsion stability, water activity, color, and fat particle size distribution. Although pH is a physicochemical property, it is discussed in subchapter 12.2 because it is closely related to titratable acidity.

12.3.1 Emulsion stability

Emulsion stability of processed cheeses is evaluated under the accelerated conditions of free oil formation tests. According to Kindstedt and Fox (1991), free oil formation, also called oiling off or fat leakage, is the tendency of liquid fat to separate from melted cheese and accumulate as oil pockets, particularly at the surface.

a) *Filter paper test*: Cheese disks are placed on filter paper on a sheet of asbestos for 5 min in an oven at 110°C. The disks are then lifted from the paper and the total area covered by melted fat is measured. Free oil formation is expressed as a function of original disk area (Breene et al., 1964; Kindstedt and Rippe, 1990). A variant is suggested by Thomas (1973): Cylindrical samples are gently pressed between two sheets of filter paper and incubated at 45°C for 2 h. The diameters of the fat spread before and after heating are used in the calculation of oil separation index.

b) *Modified Gerber test*: Grated cheese sample is accurately weighed in a screw-cap Pyrex tube, which is then immersed in boiling water for 4 min. After addition of acidified water (pH 2.2 with HCl) onto the melted cheese, the sample tube is centrifuged. After addition of aqueous methanol, the sample tube is again centrifuged, and placed in a water bath at 60°C. The layer of butter oil released at the surface is quantitatively transferred to a Gerber bottle. Finally, the bottle is centrifuged, tempered at 60°C, and the butter oil is measured. Free oil is expressed as percentage in cheese or in cheese fat (Kindstedt and Fox, 1991; Chatziantoniou et al., 2015).

c) *Photocentrifugation*: Dispersion analysis is carried out using an apparatus which combines the centrifugation of the sample with the continuous measurement of the light transmission through it. The processed cheese containing vials are heated in a 65°C water bath for 20 min. Then, thin cuvettes are filled with the sample and centrifuged at 1200 × *g* for up to 150 min at 60°C. Every 30 s, transmission profiles are captured at 870 nm throughout the full length of the cell. The measured instability index is quantified by the clarification at a given separation time, divided by the maximum clarification. The value of the index is a dimensionless number between 0 (very stable) and 1 (very unstable) (Hoffmann, 2016).

12.3.2 Water activity

a) *Microclimate method*: The water activity (a_w) of a food is estimated by determining the equilibrium relative humidity (ERH) which does not lead to any variation of the sample mass (a_w = ERH/100). Grated cheese is weighted in a Petri dish. The dish is then placed over of a saturated salt solution of known ERH (or a_w) poured in an airtight container, at 20°C. Solutions of three salts are used for the sample: potassium dichromate ($a_w = 0.980$), potassium dihydrogen phosphate ($a_w = 0.965$), and ammonium dihydrogen phosphate ($a_w = 0.935$). Petri dishes are thoroughly weighted after 24 h. Variation of the cheese weight (ΔW) is recorded. The differences are plotted on graph $\Delta W = f(a_w)$. The a_w of the cheese is the value at the intercept of the straight on the a_w axis, where $\Delta W = 0$ (Ardö and Polychroniadou, 1999).

b) *Resistive electrolytic hygrometer*: It uses a sensing element in the form of a liquid electrolyte held between two small glass rods by capillary force. The electrolyte changes resistance if it absorbs or loses water vapor. The resistance is directly proportional to relative air humidity, and also to the water activity of the sample (once vapor-liquid equilibrium is established) (Mathew et al., 2019).

c) *Capacitance hygrometer*: It consists of two charged plates separated by a polymer membrane dielectric. As the membrane adsorbs water, its ability to hold a charge increases and the capacitance is measured. This value is roughly proportional to the water activity of the sample (Mathew et al., 2019).

d) *Dew-point hygrometer*: The sample is placed in a hermetic measurement chamber that contains a mirror of which the temperature can be made to vary. The mirror is cooled until water-vapor condensation (dew) appears on its surface. This temperature (dew point) is then used to find the relative humidity of the chamber using psychrometric charts (ISO, 2017e). This method is the most accurate and the fastest (often $\leq$5 min).

12.3.3 Color

For color measurement of food products, colorimeters and spectrophotometers are used, based on the recommendations of International Commission on Illumination (CIE). Colorimeters quantify color by measuring three primary color components of light seen by the human eye, specifically red, green, and blue. This tristimulus color measurement provides data on the proportion of these three components present in the light reflected (mainly solids) or transmitted (typically liquids) by a food product. Colorimeters can be used for food production and inspection. In contrast, spectrophotometers measure the spectral reflectance or transmittance of an object over the entire spectrum of human visible light wavelengths (400–700 nm). They offer greater specificity, precision and accuracy than colorimeters, and can also be used for food research and development.

Color measurements are related to light source, spectral characteristics of the object, and the spectral sensitivity characteristics of the human eye.

The light source simulates the commonly used CIE standard illuminant D65 corresponding to daylight with a correlated color temperature of approximately 6500 K (CIE, 2018).

Regarding the type of illuminating and viewing geometry, there are two common types of directional illumination systems: An instrument with 45°:0° geometry that illuminates an object from a 45° angle and detects the reflected light at 0°, and an instrument with 0°:45° geometry that illuminates an object at 0° and detects the reflected light at a 45° angle. There are also two common types of diffused illumination systems: An instrument with d:8° geometry that illuminates an object from scattered directions and detects the reflected light at an 8° angle, and an instrument with 8°:d geometry that illuminates an object at an 8° angle and detects the reflected light from scattered directions (Butts, 2015; CIE, 2018).

The reflected light is detected by sensors reproducing CIE 1931 2° or CIE 1964 10° (more commonly) standard colorimetric observers. These correspond to color-matching functions for visual field size of 2° and 10°, respectively, representing the human population having normal color vision (CIE, 2018; ISO/CIE, 2019).

In food technology, the color is commonly expressed in the CIE 1976 L*a*b* (or CIELAB) color space (Fig. 12.1). In this three-dimensional space, color is described as a point by three rectangular coordinates: L* (lightness) from black (0) to white (100), a* from green (negative values) to red (positive values), and b* from blue (negative values) to yellow (positive values).

Approximate correlates of the perceived attributes of lightness, chroma and hue are also introduced in the same diagram, but in cylindrical coordinates. They are calculated as follows (CIE, 2018; ISO/CIE, 2019):

CIE 1976 lightness: L* (as defined above)
CIE 1976 a,b (CIELAB) chroma: $C^*_{ab} = (a^{*2} + b^{*2})^{1/2}$
CIE 1976 a,b (CIELAB) hue angle: $h_{ab} = \arctan(b^*/a^*)$

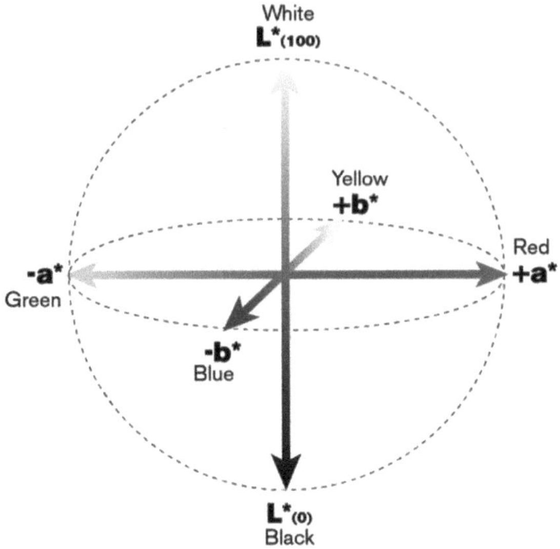

Figure 12.1 Illustration of CIE 1976 L*a*b* (or CIELAB) color space.

To compare colors, CIE 1976 a,b (CIELAB) color difference, ΔE^*_{ab}, is calculated as the Euclidean distance between the points representing two colors in the space (CIE, 2018; ISO/CIE, 2019):

$$\Delta E^*_{ab} = [(\Delta L^*)^2 + (\Delta a^*)^2 + (\Delta b^*)^2]^{1/2}$$

An observer sees the difference between two colors as follows (Mokrzycki and Tatol, 2011; Chudy and Gierałtowska, 2020):

- $0 < \Delta E^*_{ab} \leq 1$: observer does not notice the difference
- $1 < \Delta E^*_{ab} \leq 2$: only experienced observer can notice the difference
- $2 < \Delta E^*_{ab} \leq 3.5$: inexperienced observer also notices the difference
- $3.5 < \Delta E^*_{ab} \leq 5$: clear difference in color is noticed
- $5 < \Delta E^*_{ab}$: observer notices two different colors

These data represent experimentally verified statistics.

12.3.4 Fat particle size distribution

Fat particle size distribution of processed cheese is usually determined using a laser light scattering particle size analyzer, equipped with a wet cell dispersion unit having a pump, stirrer and ultrasonicator. Sample preparation involves its dispersion in distilled or reverse osmosis water containing EDTA to chelate Ca^{++} and dissociate casein, and Tween 20 to stabilize fat particles, at a pH value adjusted to 10 using NaOH (Walstra, 1965). After dispersion for 1 h, sample is allowed to rest at refrigerated conditions (4–5°C) overnight. Prior to measurement, sample is equilibrated at room temperature for 1 h. The resulting sample suspension is added to the dispersion unit until an obscuration

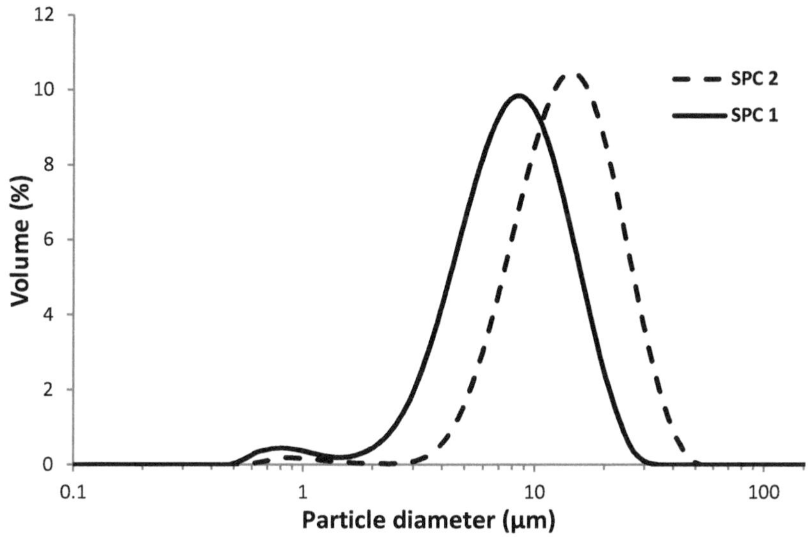

Figure 12.2 Fat particle size distribution of two spreadable processed cheeses (SPC) (Chatziantoniou, 2020).

rate between 10% and 20% is reached. The refractive indices used are 1.456 and 1.330 for fat and water, respectively (Lee et al., 2004; Trivedi et al., 2008; Chen and Liu, 2012; Lee et al., 2015; Chatziantoniou et al., 2019).

Fat particle size distribution is presented as volume (%) versus particle diameter (μm) (Fig. 12.2), and also as cumulative volume (%) versus particle diameter (μm). The fat particle size distribution width or dispersion index or span is calculated as follows:

$$\text{Span} = (d_{0.9} - d_{0.1})/d_{0.5}$$

where $d_{0.1}$, $d_{0.5}$ and $d_{0.9}$ are particle diameters at 10%, 50% and 90% cumulative volume, respectively. The average size of the fat particles measured is expressed as the volume-to-surface mean diameter, d_{32} (μm), or the volume-weighted mean diameter, d_{43} (μm), as follows:

$$d_{32} = \sum n_i d_i^3 / \sum n_i d_i^2 \qquad d_{43} = \sum n_i d_i^4 / \sum n_i d_i^3$$

where n_i is the number of particles with diameter d_i.

12.4 Rheological measurements

Processed cheese as hot cheese melt, during processing or subsequent melting applications, as well as processed cheese sauces are liquid materials. Also, processed cheese spreads can be considered semiliquid or semisolid materials. The fundamental methods used for the rheological evaluation of processed cheese in liquid or semiliquid state are tube and rotational viscometry, lubricated squeezing flow viscometry, and dynamic small-amplitude oscillatory rheology (Steffe, 1996; Tamime et al., 2011).

12.4.1 Tube viscometry

Tube viscometers establish a pressure difference to create a flow through a capillary tube. By measuring the pressure drop and volumetric flow rate in steady laminar flow through a uniform circular tube of known length and diameter, the viscosity of a fluid can be determined (Darby, 1996). Thus, the apparent viscosity (η_{app}) of a fluid at any specific shear stress or shear rate can be determined by calculating the true shear rate ($\dot{\gamma}$) and the shear stress at the wall (τ_w) from the raw data ($\eta_{app} = \tau_w / \dot{\gamma}$) (Darby, 1984). There are many measurement errors that may occur in using tube viscometers, such as kinetic energy losses, end effects, elastic energy, turbulence, pressure losses prior to the capillary, heat effects, wall effects, effect of time-dependent properties. As these errors affect rheological measurements and make the results unreliable, data corrections are required (Whorlow, 1992; Steffe, 1996).

A pneumatic tube viscometer was used for the evaluation of processed cheese apparent viscosity just after emulsification (Dimitreli and Thomareis, 2004; Dimitreli et al., 2005; Dimitreli and Thomareis, 2007, 2009).

12.4.2 Rotational viscometry

Rotational viscometers used in processed cheese include concentric cylinder, cone and plate, mixer, and Brookfield viscometers.

a) *Concentric (coaxial) cylinder viscometers*: The fluid sample is placed in the narrow gap between two cylinders, one of which is rotating at a fixed angular velocity and the other is stationary. There is a balance of torques. Thus, the torque acting on the rotating cylinder balances the torque due to the shear stress acting on its surface from the fluid. The shear stress and shear rate in the gap are calculated from the torque and angular velocity, respectively (Darby, 1996; Steffe, 1996). This type of instrument can be used for measuring the viscosity of non-Newtonian liquids (McClements, 1999).

Kirchmeier et al. (1978), Kirchmeier (1980), and Lee et al. (2003) studied the flow properties of processed cheese using concentric cylinder viscometers.

b) *Cone and plate viscometers*: The fluid is held in the gap between a shallow cone (usually $<4°$ from horizontal) and a flat plate or disk, with the apex of the cone almost touching the plate. One member is rotated at a fixed angular velocity and the developed torque is measured, as in concentric cylinder, and these values are used to determine the shear stress and shear rate in the gap. This geometry is very popular because (i) measurements may be done on very small volumes, (ii) the measured values of torque and angular velocity may be related directly to shear stress and shear rate in the gap with no complex data manipulation if the gap is small, and (iii) corrections are not usually required (Darby, 1984). The cone and plate viscometers are suitable for non-Newtonian liquids (McClements, 1999).

A cone and plate viscometer was applied by Lee et al. (2004) to determine apparent viscosity of processed cheese spreads at 70°C.

c) *Mixer (or paddle) viscometers*: This type of viscometer consists of two or more paddles attached to a shaft that is caused to rotate at one or more speeds while the torque resistance is measured. They may resemble a flag or a star-shaped geometry. The stirring action makes them particularly useful for food suspensions containing particulates that settle out (Bourne, 2002).

The Rapid Visco Analyzer (RVA) is a useful example of mixer viscometry. This is a computer-integrated instrument capable of rapidly measuring the apparent viscosity of products over a range of temperature and agitation conditions. In this test, a sample of processed cheese is placed in a disposable RVA canister, a stirring paddle is inserted and the cheese is analyzed using a 14-min test. During the test, the temperature is raised from 25°C to 90°C in 5 min, held for 3 min at 90°C, and then cooled from 90°C to 25°C in 6 min. In the first 2 min of the test, the stirring speed is gradually increased from 0 to 300 rpm, and then maintained at 300 rpm for the remainder of the test. The apparent viscosity of the processed cheese is collected during the entire test (Prow and Metzger, 2005; Tamime et al., 2011).

Another application of mixer viscometer is the vane method. In this test, a viscometer equipped with a 4-bladed stainless steel vane is used to determine the yield stress of cheese spread sample at a constant rotational speed and at refrigeration or room temperature. The measurement is carried out in the original container with minimal disturbance during the introduction of the vane into the sample (Breidinger and Steffe, 2001; Plog, 2014).

d) *Brookfield viscometers*: This type of instrument may be held in the hand or supported on a stand. A motor gives a series of speeds of rotation that are constant. Various spindles that take the form of cylinders, disks, and T bars are attached to a small chuck. When the spindle is immersed in the liquid and the motor switched on, the viscous drag of the fluid on the spindle is registered as torque on a dial (Bourne, 2002). The viscometer is often used with a Helipath Stand. This is a stand designed to slowly lower or raise the viscometer so that its rotating shearing element will describe a helical path through the sample while torque data are obtained. In this way, channeling effect is eliminated and problems associated with slip are minimized (Steffe, 1996).

The apparent viscosity of hot cheese melts after emulsification (Kalab et al., 1987) and processed cheese analogues (Sołowiej et al., 2016) was determined using viscometers equipped with T-bar type spindles and Helipath Stand.

12.4.3 Lubricated squeezing flow viscometry

Lubricated squeezing flow is often used to determine flow properties of highly viscous materials such as semiliquid or semisolid food products (e.g., processed cheese, melted cheese, etc.) (Tamime et al., 2011). The sample is squeezed out between two parallel lubricated plates, where biaxial extensional flow is achieved, at either controlled force (or stress) or controlled speed (or strain rate) by the use of a texture analyzer. Quite often the upper plate is moving at a constant speed, while the lower plate is stationary. The main advantage of this method is that the presence of slip, as a result of self-lubrication, is not only acknowledged but is actually a prerequisite for a proper test. In the case where steady flow is achieved, that is, stress and strain rate reach constant values then the elongational viscosity can be determined (Campanella et al., 1987; Rüegg et al., 1991). If the stress and strain are a function of time then the biaxial stress growth coefficient can be calculated (Shukla et al., 1995), as in Fig. 12.3. Power law

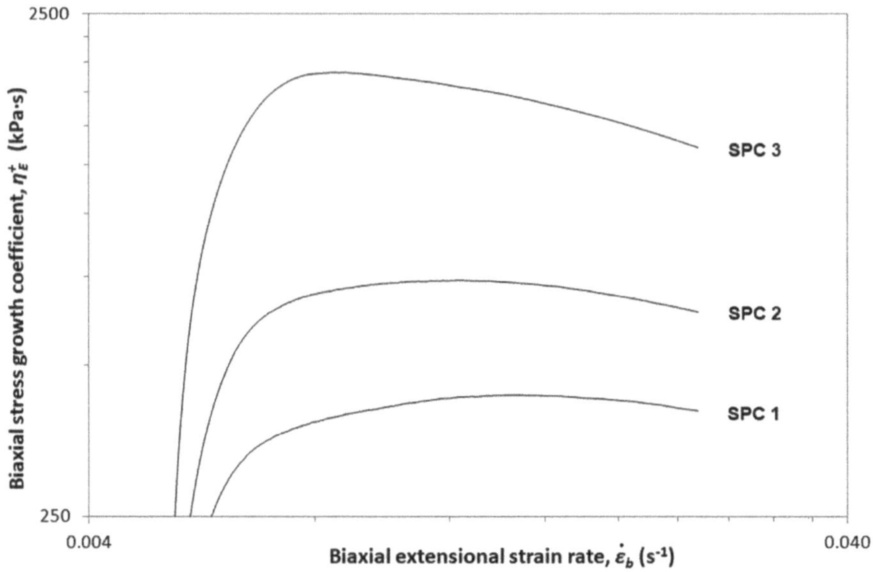

Figure 12.3 Variation of biaxial stress growth coefficient of three spreadable processed cheese (SPC) samples in relation to the biaxial extensional strain rate (deformation up to 80%, compression speed 6 mm/min) (Chatziantoniou, 2020).

fluid parameters (consistency coefficient, flow behavior index) may be determined from linear regression of experimental data. However, only the linear portion of the curve, where steady flow is present, should be used in analysis (Steffe, 1996).

Casiraghi et al. (1985) determined the elongational viscosity of processed cheese spread. The elongational viscosity of melting processed American cheeses was also determined at various temperatures in the range 36–62°C (Campanella et al., 1987). Lubricated squeezing flow viscometry was used by Chatziantoniou et al. (2015, 2019) to study the rheological properties of spreadable processed whey cheeses.

12.4.4 Dynamic small-amplitude oscillatory rheology

In oscillatory tests, the viscoelastic properties are determined. The material is subjected to deformation or stress, which varies harmonically with time by the use of a rheometer. When a sinusoidal oscillating stress or strain at a specific frequency is applied to the material, the phase angle (δ) between the stress and strain, as well as the amplitude ratio is measured. From treating the obtained data, the storage or elastic modulus (G') that is a measure of the energy stored and recovered per cycle, and loss or viscous modulus (G'') that is a measure of the energy dissipated or lost as heat per cycle of sinusoidal deformation, can be determined (Ferry, 1980; Tabilo-Munizaga and Barbosa-Cánovas, 2005). A useful parameter in describing viscoelastic behavior of materials is the loss tangent or tangent of the phase angle (tan δ). It is dimensionless and is directly related to the energy lost per cycle divided by the energy stored per cycle:

$$\tan\delta = G''/G'$$

For an ideal elastic material, the stress and strain are in phase ($\delta = 0$ and $\tan\delta = 0$), while for an ideal viscous material, the stress and strain are 90° out of phase ($\delta = 90°$ and $\tan\delta = \infty$). Thus, if a viscoelastic material has $\delta < 45°$ ($\tan\delta < 1$ or $G' > G''$), it is more elastic than viscous (solid-like behavior). On the contrary, if the material has $\delta > 45°$ ($\tan\delta > 1$ or $G'' > G'$), it is more viscous than elastic (liquid-like behavior) (Ferry, 1980; Steffe, 1996). However, the overall response of the sample may be characterized by the complex modulus (G^*) and complex viscosity (η^*):

$$G^* = \sqrt{(G')^2 + (G'')^2} \qquad \eta^* = G^*/\omega$$

where ω is the angular frequency. Complex modulus describes the total resistance to deformation of a material that it is considered as an elastic solid and it is expressed in Pa. In contrast, complex viscosity describes the total resistance to flow of a material that it is considered as a viscous liquid and it is measured in Pa s (Dimitreli and Thomareis, 2008).

Oscillatory testing provides data over a wide range of frequencies. However, if fundamental parameters are required, it must be restricted to linear viscoelastic behavior. A strain or stress sweep, conducted by varying the amplitude of oscillation (input signal) at a constant frequency and noting the magnitude of phase lag and the amplitude ratio, is used to determine the limits of linear viscoelastic behavior by identifying a critical value of the sweep parameter. In the linear region, rheological properties are not strain or stress dependent. Strain and stress sweeps can be conducted with a controlled rate and a controlled stress, respectively. The frequency sweep is probably the most common mode of oscillatory testing because it shows how the viscous and elastic behavior of the material changes with the rate of application of strain or stress. In this test, the frequency is increased, while the amplitude of the input signal (stress or strain) is held constant (Fig. 12.4). Materials usually exhibit more solid-like characteristics at higher frequencies (Steffe, 1996). Dynamic rheological analysis can also be performed at a constant stress or strain, within the linear viscoelastic region, at a range of temperatures (typically 25–90°C). The data collected in the analysis can be used to characterize the physical properties of melted processed cheese (Tamime et al., 2011).

For liquid samples, the cone and plate or concentric cylinder geometry are appropriate, whilst for semisolid materials, the cone and plate or parallel plate geometry may be used (Darby, 1984; Steffe, 1996).

Viscoelastic properties of processed cheeses or processed cheese spreads by the use of dynamic rheology have been studied extensively by many researchers (Sutheerawattananonda and Bastian, 1998; Reparet and Noël, 2003; Dimitreli and Thomareis, 2008; Buňka et al., 2012; Černíková et al., 2018).

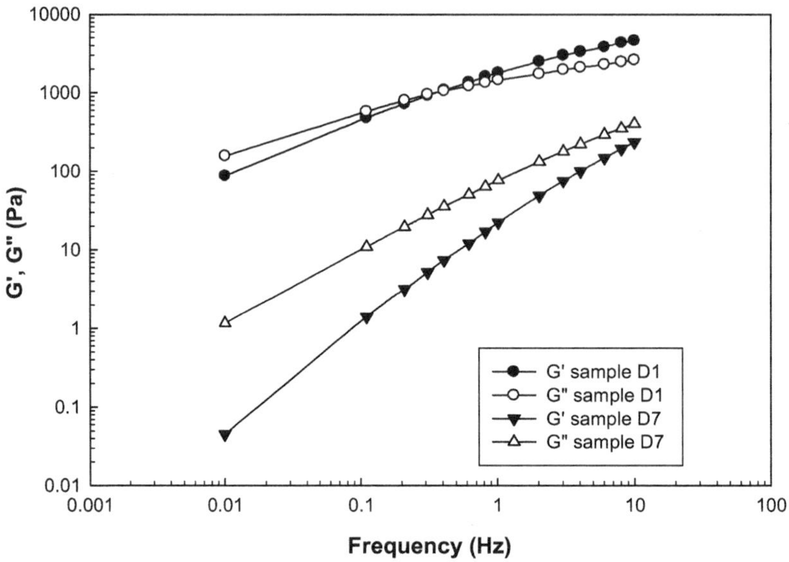

Figure 12.4 Frequency sweep for spreadable-type processed cheese samples differing in their chemical composition (D1, D7) (shear stress 3 Pa, temperature 28°C). G': elastic modulus; and G'': viscous modulus (Dimitreli, 2006).

12.5 Instrumental texture evaluation

A variety of tests have been developed to characterize textural properties of processed cheese in solid state (block-type processed cheese). Most of these tests provide fundamental rheological parameters and others are empirical in nature. These tests mainly include uniaxial compression, texture profile analysis, stress relaxation, and penetration tests.

12.5.1 Uniaxial compression test

Uniaxial compression test is commonly used for determining the textural properties of foods, including processed cheese. It is performed using a texture analyzer. A specimen of known shape (usually cylindrical) and size is placed between two parallel rigid plates of the instrument, and often the upper plate is moved downward at a constant crosshead speed (i.e., constant deformation rate) (ISO, 2006f), while recording the corresponding force as a function of time (Gunasekaran and Ak, 2003). The resulting force – time data pairs are converted into corresponding stress (σ) and strain (ε) values from which other rheological quantities can be calculated (Gunasekaran and Ak, 2003).

If the magnitudes of the applied force and deformation are small, the body may be assumed to be elastic and the test may be considered fundamental. The modulus of elasticity or Young's modulus (E or Y) can be computed as σ/ε (Rao, 1992). Instead of the Young's modulus, the modulus of deformability (M) can be determined, which is defined as σ_T/ε_T, where σ_T and ε_T are the true stress and strain, respectively. The modulus of deformability is representative of a material's overall resistance to deformation and may include irreversible and rate dependent deformations, both small and large. It depends on the test conditions, especially on the specimen dimensions and the deformation rate. In practice, cheese is subjected to large stresses and strains, which usually result in visible fracture. Compression tests usually involve deformation up to 80% that cause complete breakdown of the cheese structure. Incidentally, these strains simulate the mastication of cheese, where deformation of the cheese between the molars is typically about 70%. Other mechanical properties commonly determined from uniaxial compression tests on cheese include, fracture stress σ_f, fracture strain ε_f, and work to fracture W_f (Gunasekaran and Ak, 2003; ISO, 2006f). The large strain compression test is useful in that it provides a measure of the stresses and strains at which cheese is perfectly elastic, at which it fractures and at which it flows (Fox et al., 2000).

Marshall (1990) and Stampanoni and Noble (1991) applied uniaxial compression up to high deformations (80 and 84%, respectively), so as to study textural properties of processed cheeses. Pereira (2000) studied textural properties of processed cheese analogs using uniaxial compression up to 70% deformation.

12.5.2 Texture profile analysis

Texture profile analysis (TPA) is a technique that has been used widely for the textural evaluation of cheese in general and particularly of processed cheese. TPA is applied either by uniaxial compression or penetration of the samples using of a texture analyzer. The test is carried out by compressing a sample typically cylindrical in shape, twice, usually to 80% of its original height. Compression is achieved using parallel plates where one plate is fixed and the other plate moves with a reciprocating linear cyclical motion. Since this test is intended to reflect the human perception of texture, the first and second compression cycles are referred to as the first and second bite, respectively. A typical TPA curve is shown in Fig. 12.5.

Various textural parameters may be determined from this curve (van Vliet, 1991; Rao, 1992; Steffe, 1996): Brittleness or fracturability (B) is the force at the first major decrease in force curve or the force at which the material fractures (not shown in Fig. 12.5). Hardness 1 (H_1) is the peak force (maximum force) or the force necessary to attain a given deformation during the first compression. Work done on the sample during the first compression (A_1) is the area under the solid line up to the dashed line in the first compressive cycle or the work required to complete the first compression. Adhesiveness (A_3) is the area under the zero force line representing the work, due to a tensile force, needed to pull food apart and separate it from the compression plates during the first decompression (a similar adhesiveness characteristic is present in the second compressive cycle). Stringiness (S_2) is the extension (distance) of the sample during the first decompression. Hardness 2 (H_2) is the maximum force during the second compression. Work done on the sample during the second compression (A_2) is the area under the solid line up to the dashed line in the second compressive cycle. Springiness or elasticity (S_1) is the distance (or the strain) a food recovers during the

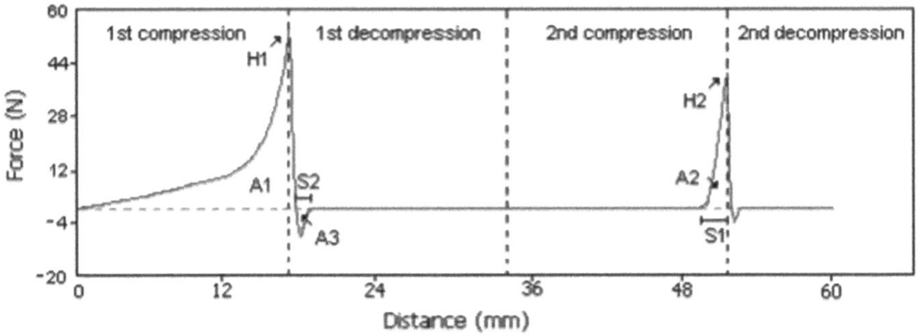

Figure 12.5 Typical texture profile analysis curve for processed cheese. H_1: hardness 1 (peak: force), H_2: hardness 2 (peak: force), A_1: work done on the sample during the first bite (area: work), A_2: work done on the sample during the second bite (area: work), A_3: adhesiveness (area: work), S_1: springiness (distance), S_2: stringiness (distance) (Dimitreli, 2006).

time between the end of the first compression cycle and the beginning of the second compression cycle or the rate at which a deformed material goes back to its undeformed condition after the deforming force is removed. Cohesiveness (C) is the ratio of A_2/A_1 or the quantity that simulate the strength of the internal bonds making up the body of the material. Chewiness (K) is the product of hardness 1, cohesiveness and springiness or the quantity of energy required to masticate a solid food product to a state ready for swallowing. Gumminess (G) is the product of hardness 1 and cohesiveness or the quantity of energy required to disintegrate a semisolid food product to a state ready for swallowing.

Evaluation of the texture of processed cheese, by the use of TPA in compression mode, was performed by many researchers (Harvey et al., 1982; Mulvihill and McCarthy, 1994; Piska and Štětina, 2004; Dimitreli and Thomareis, 2007; Chatziantoniou et al., 2015). The application of penetration by two sequential cycles using a texture analyzer was also reported by many researchers (Abou El-Nour et al., 1996; Buňka et al., 2013; Sołowiej et al., 2016; Černíková et al., 2017).

12.5.3 Stress relaxation test

One of the fundamental methods to study viscoelastic response is stress relaxation test, which generally involves the instantaneous application of a low constant deformation (usually $\leq 10\%$) to a material and maintaining this deformation constant throughout the test (Fox et al., 2000; Guinee, 2002). Stress relaxation test is commonly achieved via the use of a texture analyzer. The test duration is limited (on the order of minutes, e.g., 15 min or less) because degradation of sample may occur before the test is completed as a result of physical changes (e.g., moisture exchange with environment), microbial activity, etc. (Gunasekaran and Ak, 2003).

As long as the deformation is maintained constant, the manner in which the sample attempts to alleviate the imposed stress is recorded (Rao, 1992). A plot of the rate at

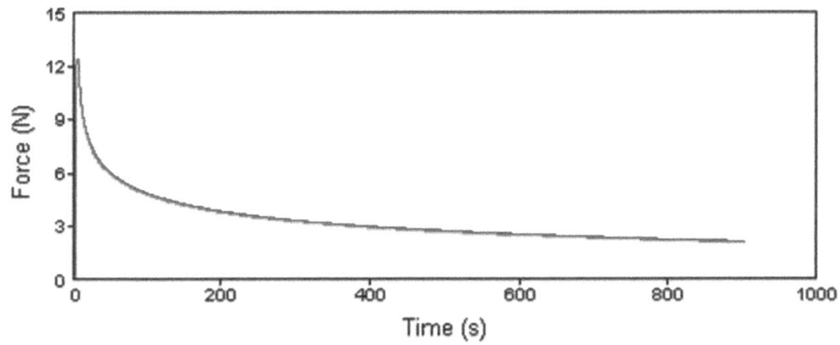

Figure 12.6 Typical stress relaxation curve for processed cheese (sample deformation 10%) (Dimitreli, 2006).

which the stress is dissipated as a function of time is known as a stress relaxation curve (Fig. 12.6). These curves are fitted to exponential decay equations (Guinee, 2002).

The model most often used to represent viscoelastic materials subjected to stress relaxation is the generalized Maxwell model. A single Maxwell model is represented by a spring and a dashpot in series. Spring expresses elasticity (solid behavior) and dashpot expresses viscosity (liquid behavior). The generalized Maxwell model is represented by n Maxwell models in parallel with each other and with an isolated spring, and described by the following equation (Dimitreli and Thomareis, 2007):

$$E(t) = E_1 \exp(-t/\tau_1) + E_2 \exp(-t/\tau_2) + \ldots + E_n \exp(-t/\tau_n) + E_\infty$$

where $E(t)$ is the modulus of elasticity of the entire model at any time t and has units of Pa, E_i and τ_i are the moduli of elasticity and the relaxation times, respectively, of the n Maxwell models, and E_∞ is the modulus of elasticity of the spring corresponding to the residual stress. Relaxation time is the time required for the stress to decay by 63%, and $\tau_1 > \ldots > \tau_n$. Coefficients of viscosity (η_i) of the samples in the dashpot of each Maxwell model are calculated as follows (van Vliet, 1991; Rao, 1992):

$$\eta_i(\text{Pa s}) = E_i \tau_i$$

Many materials show stress relaxation curves that decay to zero values, so there is no residual stress and $E_\infty = 0$. These materials are considered viscoelastic liquids. In contrast, the materials which present a residual (unrelaxed) stress, having $E_\infty > 0$, are considered viscoelastic solids.

An alternative model to describe stress relaxation curves of viscoelastic solids is suggested by Peleg and Normand (1983):

$$P_0 t/[P_0 - P(t)] = k_1 + k_2 t$$

where $P(t)$ is the decaying parameter (force, stress or modulus) at time t, P_0 is its initial value (at $t = 0$), and both k_1 and k_2 are constants.

Joshi et al. (2004) studied viscoelastic properties of processed cheese as a function of time. To analyze the data obtained by the stress relaxation experiment, both rheological models, namely the generalized Maxwell model and the Peleg and Normand model, were considered for representation. Viscoelastic parameters of block-type processed cheeses have been determined using stress relaxation test and a generalized Maxwell model with four Maxwell models and a spring (Dimitreli and Thomareis, 2007, 2009).

12.5.4 Penetration test

One of the simplest and fastest techniques that can be used to measure the firmness of foods is the penetration test. This empirical test has been used widely in the evaluation of processed cheese. There are different kinds of penetrometers that are equipped with either a cylindrical, conical or spherical head (plunger) that is plunged into the sample to a fixed depth and the corresponding force is recorded. Conversely, it is also possible to move the probe under a predetermined force and measure the corresponding deformation. In either case, the force, deformation and area under the force-deformation curve (energy or work) are measured and used as an indication of the textural quality of the food (Steffe, 1996; Fox et al., 2000; Tamime et al., 2011). Many researchers used different kinds of penetrometers so as to study processed cheese firmness (Thomas et al., 1980; Gupta and Reuter, 1993; Abou El-Nour et al., 1998; Morsy et al., 2014). Apart from penetrometers, texture analyzers equipped with plungers of various shapes have also been used by many researchers for the performance of the penetration tests (Brummel and Lee, 1990; Tamime et al., 1999; Buňka et al., 2012; Ghorbel et al., 2016).

12.6 Microstructural evaluation

Microscopy techniques, such as confocal laser scanning microscopy (CLSM), scanning electron microscopy (SEM) and transmission electron microscopy (TEM) are used to study the microstructure of foods, and processed cheeses in particular. The distribution of fat, polysaccharides and protein can be effectively characterized using CLSM, while SEM and TEM reveal details of the protein matrix and fat–protein interface (Tamime et al., 2011).

12.6.1 Confocal laser scanning microscopy

CLSM employs laser diode arrangements containing semiconductor materials for obtaining emission wavelengths in the range 300–460 nm or krypton, argon, helium and neon lasers for wavelengths in the range 450–633 nm. Incident photons from the laser beam are absorbed by fluorochromes within the sample and emitted at a longer wavelength via a dichroic mirror to a photon detector such as a photomultiplier tube. The signal is amplified and converted into pixels. The beam is scanned to produce a two-dimensional image. A z-stepping motor on the sample stage facilitates consecutive xy-planes of focus through the z-axis to produce three-dimensional data. In practice, xy-resolution of 200–250 nm and z-resolution of approximately 750 nm can easily be achieved in food materials (Auty, 2018).

The image formation in CLSM does not depend on transmitting light through the specimen, and, therefore, bulk or thick specimens can be used without specific sample preparation, as thin sectioning, fixation or dehydration. The main advantage of the CLSM is its ability to produce optical sections through a three-dimensional specimen (Rincón Cardona et al., 2013). This is because lasers provide a coherent high-intensity excitation illumination, whilst confocal optics enables imaging of undisturbed regions of the sample approximately 10 μm below its surface (Tamime et al., 2011). CLSM also possess the possibility to follow in situ the dynamics of different processes. Specially designed stages, which allow heating, cooling or mixing of the sample, give the possibility to simulate food processing under the microscope (Rincón Cardona et al., 2013). However, the drawback of using CLSM is related to the high costs of the equipment used and their maintenance required (El-Bakry and Sheehan, 2014).

In systems like cheese, one of the advantages of this technique is that it can both visualize and chemically differentiate cheese components through the use of protein and lipid specific stains. Acridine orange, Nile blue, fluorescein isothiocyanate (FITC), rhodamine B, and fast green FCF are added as extrinsic fluorochromes to stain protein, and oil red O and Nile red to stain fat. An even more detail analysis of food structure may be achieved by the simultaneous labeling of two or more components of foods with stains which are specific for each component (Rincón Cardona et al., 2013). CLSM can also be applied to processed cheese as a quality control tool for monitoring the degree of emulsification over time, the structural changes in fat and protein phases that occur during creaming action, the presence of nondairy biopolymers, such as starch and the presence of crystalline inclusions (Tamime et al., 2011).

A typical procedure for microstructural evaluation of processed cheese with CLSM involves the following: Cheese samples are prepared to approximate dimensions of 5 × 5 × 2 mm for visualization, and are deposited on a Petri dish with a cover glass at the bottom. Fat and protein phases are stained with Nile red and fast green FCF, respectively. Krypton/argon and helium/neon lasers are used to elicit the emission of the Nile red and fast green FCF dyes at excitation wavelengths of 488 and 633 nm, respectively. Emission filters of short pass 514 nm and band pass 670–810 nm are used for Nile red and fast green FCF, respectively. Micrographs are obtained as shown in Fig. 12.7 (Le Tohic et al., 2018).

CLSM has been used in several studies for the evaluation of processed cheese microstructure (Sutheerawattananonda et al., 1997; Lee et al., 2003; Sołowiej et al., 2016; Le Tohic et al., 2018).

12.6.2 Electron microscopy

Both scanning electron microscopy (SEM) and transmission electron microscopy (TEM) use electron microscopes. These are bulky and high cost devices that emit and accelerate electron beams, which with the help of electromagnetic and electrostatic fields can be focused on the sample surface (Clarke and Eberhart, 2002). Their main

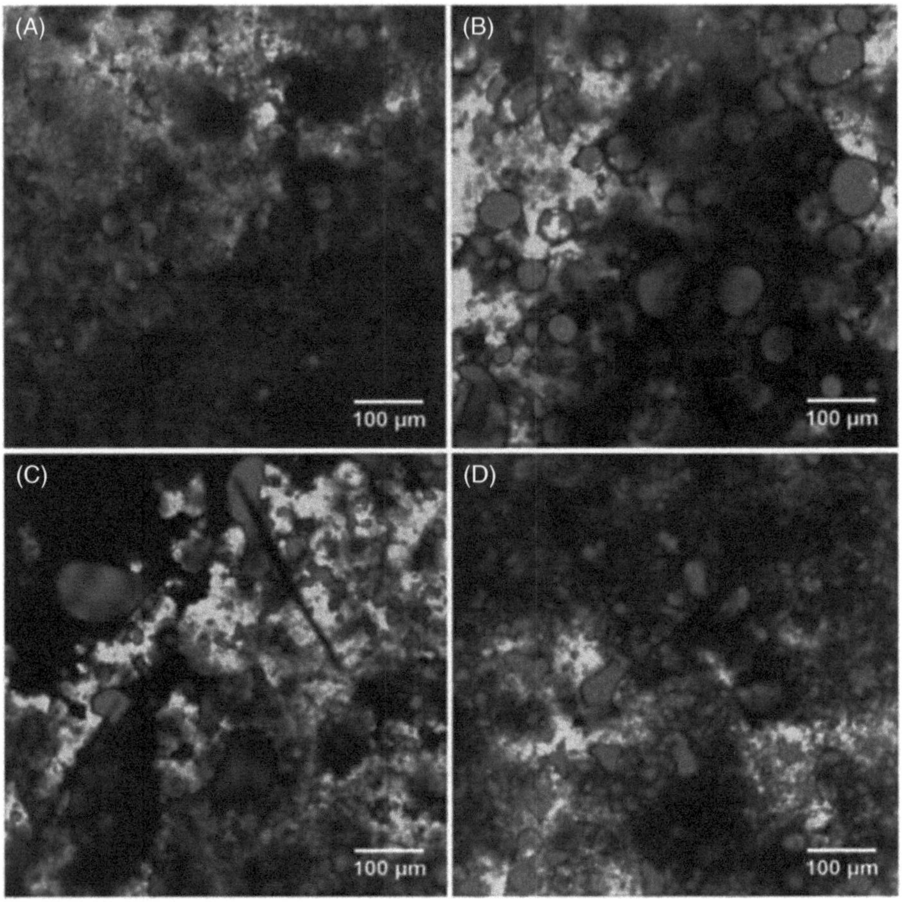

Figure 12.7 Confocal laser scanning micrographs of four different processed cheese samples. Red = fat globules; green = protein phase (Le Tohic et al., 2018).

components are the same: (i) an electron gun (electron source), (ii) a series of electromagnetic and electrostatic lenses to control the shape and trajectory of the electron beam, and (iii) electron apertures. All of these components live inside a chamber, which is under high vacuum to avoid electron deviation by air molecules (Elhassan, 2016; Nanakoudis, 2018).

12.6.2.1 Scanning electron microscopy

In SEM, the accelerated incident (primary) electrons scan the surface of the sample, and the electrons reemitted (secondary and back scattered) from the surface are captured by detectors. A three-dimensional image of the sample surface is thus formed on a computer screen. Scanning electron microscopes use an accelerating voltage of 1–30 kV, have large field of view, their maximum magnification is up to 1–2 million times, and their optimal spatial resolution is about 0.5 nm. The preparation of the sample is quite simple, while there is no limit to the thickness of the specimen. In general, the whole process of use is considered easy (Nanakoudis, 2018).

There are three main types of SEM, which are also used in processed cheeses: Conventional scanning electron microscopy, cryoscanning electron microscopy, and environmental scanning electron microscopy.

a) *Conventional scanning electron microscopy*: As this technique requires the use of a high vacuum, the sample to be tested must be completely dry. In addition, in order to obtain a high number of secondary and back scattered electrons allowing imaging, it is necessary to coat the sample with a conductive layer to avoid charging (El-Bakry and Sheehan, 2014).

In general, in the case of dairy products, various preparatory steps may be involved, such as sampling, fixation, dehydration, defatting, freezing, freeze-fracturing, critical-point drying, freeze-drying, dry-fracturing, mounting, and metal coating. Artifacts can occur due to changes in the microstructure of the sample because of inappropriate preparatory steps. Distortion of the sample images during microscopic examination can also cause artifacts (Kalab, 1984).

A typical cheese sample preparation for microstructural evaluation with conventional SEM involves the following: Cheese is cut with a razor blade into rectangular strips followed by the fixation, dehydration and defatting by the use of glutaraldehyde solutions (2%, buffered near the pH value of the cheese), ethanol and chloroform, respectively. The cheese samples are then freeze fractured in liquid nitrogen, thawed in ethanol, and finally critical-point dried with liquid carbon dioxide. The dried pieces are mounted on aluminum stubs using silver paint or 2-sided sticky tape and coated with a thin layer (20–30 nm in thickness) of gold–palladium in a sputter coater under an argon atmosphere. If fat is the component of interest, cheese samples that are fixed in glutaraldehyde are postfixed with imidazole buffered osmium tetroxide and not defatted. Furthermore, sample preparation is carried out at low temperatures to prevent any damage to the fat globules. Finally, after sample preparation, the microstructure examination of the cheese using conventional SEM of the prepared specimens is undertaken immediately, at an accelerating voltage of about 5–15 kV and a magnification of 500–10,000 times (El-Bakry and Sheehan, 2014). SEM micrographs of processed cheeses are shown in Fig. 12.8.

Conventional SEM has been used in several studies as a tool for the evaluation of processed cheese microstructure (Kalab et al., 1987; Marshall, 1990; El-Shibiny et al., 2007; Černíková et al., 2018).

b) *Cryo-scanning electron microscopy (cryo-SEM)*: In this technique, the sample is frozen quickly to minimize the size of ice crystals. A cryo-chamber or cooling stage is attached to the conventional SEM to ensure temperatures below −80°C during the whole sample examination protocol (Hassan and Awad, 2005; El-Bakry and Sheehan, 2014).

In cryo-SEM, cheese sample of approximately 1 × 3 × 5 mm is mounted on a specimen holder followed by cryofixation by rapid plunging into nitrogen slush (−210°C). Sample is then transferred (under vacuum at −180°C) to the cryo-chamber and its interior is exposed by slicing off the apex using a scalpel. The sample is sublimed (partially freeze dried), sputter-coated with gold and then introduced in the microscope chamber where it is examined at similar accelerating voltage and magnification to the conventional SEM technique (Hassan and Awad, 2005; El-Bakry and Sheehan, 2014).

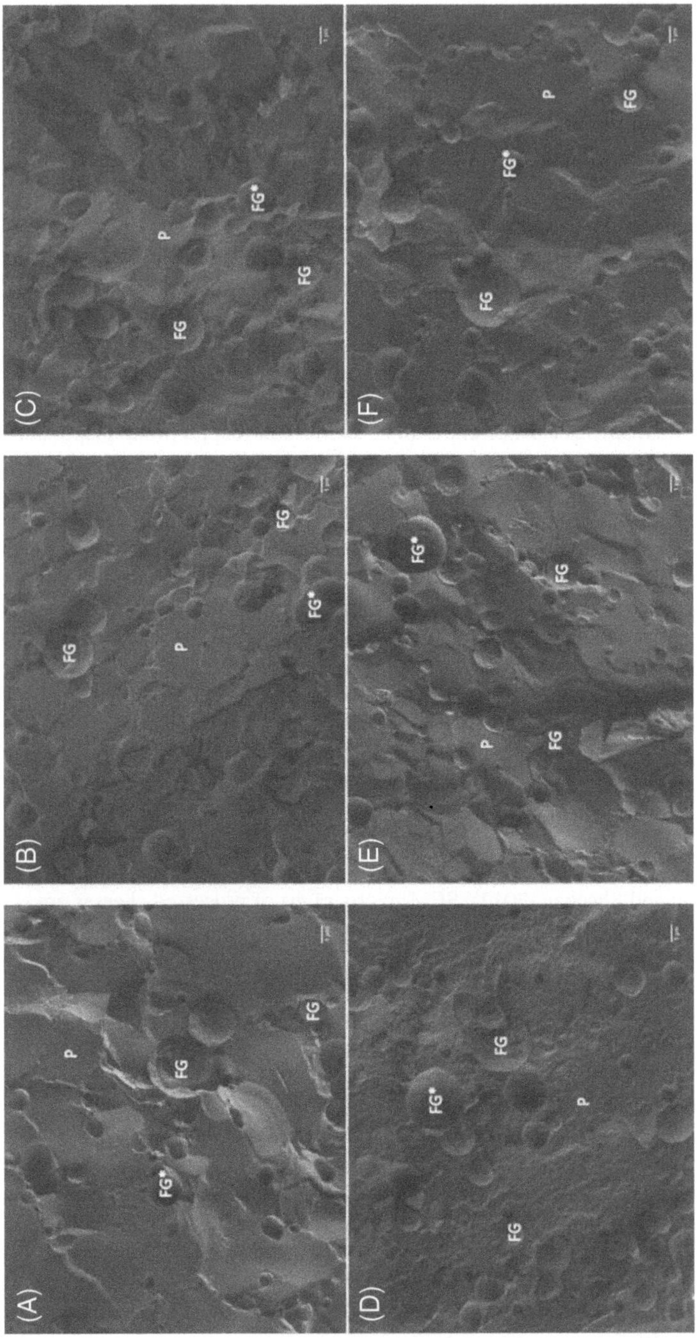

Figure 12.8 Scanning electron microscopy images of six different processed cheese samples. FG = fat globules; FG* = removed fat globules; P = protein phase. Scale bar = 1 μm (Černíková et al., 2018).

Cryo-SEM is often used for the microstructural analyses of cheese due to its advantages. First, the fat and almost all of the water are retained in the sample. In addition, the protein matrix is largely unchanged and the fat globules remain intact. Finally, sample preparation and consumption of microscopic grade chemicals are drastically reduced (Hassan and Awad, 2005; El-Bakry and Sheehan, 2014).

Cryo-SEM was used by many researchers (Hennelly et al., 2005; Montesinos-Herrero et al., 2006; Noronha et al., 2008; El-Bakry et al., 2011), to evaluate processed cheese microstructure.

c) *Environmental scanning electron microscopy (ESEM):* In this technique, sample is directly cut and mounted on the cooling stage (5°C), where no excessive sample preparation is needed. A low vacuum (1–20 Torr) gaseous environment is applied within the imaging chamber. The gas may be inert or water vapor (El-Bakry and Sheehan, 2014; Auty, 2018).

ESEM has many advantages over other SEM techniques. It allows the sample to be examined at a relatively higher pressure, so the sample does not need to be completely dry. This means that if water vapor is the gas used, hydrated samples (e.g., cheeses) can be imaged in their native state, that is, with no alterations due to necessary sample preparations. In addition, the use of ESEM has another major advantage, in that the sample not subjected to very low temperatures, during their preparation before microscopic examination. Finally, the sample is not damaged by the high energy electrons used to visualize it. The disadvantage of using the ESEM technique for microstructural analyses of cheese is the lower resolution of the micrographs obtained (El-Bakry and Sheehan, 2014; Auty, 2018).

ESEM was used by Noronha et al. (2008) and Gliguem et al. (2009) for the study of processed cheese microstructure.

12.6.2.2 Transmission electron microscopy

In transmission electron microscopy (TEM), the accelerated electrons pass through the sample and the transmitted electrons are detected. A two-dimensional image of the inner structure is thus directly shown on a fluorescent screen or, via a charge-coupled device (CCD) camera, onto a computer screen. Transmission electron microscopes use an accelerating voltage of 60–300 kV, have limited field of view, their maximum magnification is more than 50 million times, and their optimal spatial resolution is <50 pm. The preparation of the sample is laborious, while the thickness of the specimen is typically <150 nm. In general, extensive user training is required (Nanakoudis, 2018). As in the case of SEM, artifacts can occur due to inappropriate preparatory steps. Nevertheless, TEM allows the highest spatial resolution and the most consistent and accurate results of all microscopy techniques. It can provide a basis for imaging and the use of software for image analysis (El-Bakry and Sheehan, 2014).

TEM techniques applied most frequently in natural and processed cheeses are the thin sectioning and replica methods.

In the thin sectioning method, cheese cubes (1 mm^3) are fixed, dehydrated, embedded and sectioned, followed by the microscopic examination. More precisely, the fixation of the cheese sample is carried out using aqueous solutions of glutaraldehyde and osmium tetroxide, for the sequential fixation of protein and fat components. This is followed by a dehydration using ethanol mixed with acetone and embedding of the sample by a low-viscosity epoxy resin. The acetone increases the miscibility with the embedding agent used. Thin sections of 60–90 nm are cut using ultramicrotome with a diamond knife. The sections are shed in a copper grid, stained with aqueous solutions of uranyl acetate and lead citrate, and microscopically examined at an accelerating voltage of 60 kV (Geng et al., 2011; El-Bakry and Sheehan, 2014).

In the replica method, cheese cubes (1 mm^3) are soaked in glycerol overnight before being plunged into liquid nitrogen, followed by freeze-fracturing, sublimation and coating with platinum and carbon to create the replica. A carbon-platinum replica of the surface of the cheese is prepared by evaporating platinum at an angle of 20° to the surface followed by the evaporation of a carbon layer from directly above the surface. This replica is cleaned in a chromic acid solution for 2 days. Finally, the replica is washed in water, then in acetone, and mounted on copper/rhodium grids in order to be immediately examined by TEM at an accelerating voltage of 60 kV (Lee et al., 2003; El-Bakry and Sheehan, 2014).

TEM has been used in several studies as a tool for the evaluation of processed cheese microstructure (Kalab et al., 1987; Tamime et al., 1990; Awad et al., 2002; Lee et al., 2003, 2013).

12.6.3 Other microscopy techniques

In addition to the important techniques mentioned above, other older ones are used, and many newer ones can be successfully applied in the study of the microstructure of processed cheeses, and provide valuable information, such as: Bright field and polarized light microscopy, fluorescence light microscopy, cryo-transmission electron microscopy (Cryo-TEM), scanning transmission electron microscopy (STEM), Fourier transform infrared imaging (FT-IR), atomic force microscopy (AFM), fluorescence recovery after photo-bleaching (FRAP), fluorescence lifetime imaging (FLIM), confocal Raman spectroscopy, X-ray nano/microtomography, and superresolution microscopy.

12.7 Assessment of functional properties

Functional properties are defined as the food properties that determine the end use of a food product. The main functional properties are spreadability for spreadable-type processed cheeses, and sliceability, shreddability, meltability, and stretchability for block-type processed cheeses. The last three properties concern the cheeses used as topping in the pizza and other dishes that are baked in the oven. There are both subjective and objective methods for determining functional properties.

12.7.1 Spreadability

Spreadability is a measure of how easily and uniformly a food can be deformed and spread at end-use temperatures (Plog, 2014).

a) *Sensory assessment*: Processed cheese is spread on a slice of bread or a cracker by the panelists with a knife. Spreadability is rated for intensity or acceptance using a descriptive or hedonic scale, respectively (Coutouly et al., 2014; Abd El-Wahed and Hassanien, 2019).

b) *Spreadability rig*: This is an attachment for measuring spreadability using a texture analyzer. A processed cheese spread sample is filled into a 90° female cone. After tempering at 8°C, sample is displaced from the female cone by penetration with a precisely matched 90° male cone. Force is measured for the duration of the test, and spreadability is equated to the area under the curve. More easily spread samples require smaller forces to be displaced from the female cone (Swenson et al., 2000).

c) *Vane method*: A viscometer equipped with a vane is used to determine the yield stress of cheese spread sample. Low yield stress indicates a very spreadable product (Breidinger and Steffe, 2001; Plog, 2014).

d) *Squeezing flow viscometry*: A cheese sample disk is compressed between two lubricated parallel metal disks of the same diameter as the specimen. The compression is carried out, at room temperature and at constant speed, to 80% deformation, using a texture analyzer. Apparent stress at the end of compression (2 mm final specimen height) is determined and is a measure of the sample's consistency (Corradini and Peleg, 2005). It can be used for the instrumental evaluation of spreadability, since the test conditions correspond to spreading of sample in a thin layer. That is, the lower the apparent stress of the sample, the higher its spreadability (Chatziantoniou et al., 2019).

12.7.2 Sliceability

Sliceability is defined as a cheese's ability to be cut cleanly into thin slices, resist breakage or fracture at slices edges, and undergo a high level of bending before breaking (Guinee and Kilcawley, 2004).

a) *Slicer test*: The thickness of the thinnest slice without deformation that could be cut from loaves of processed cheeses, using an electric meat-cheese slicer with circular knife blade, is recorded as an index of sliceability (Gupta et al., 1984).

b) *Wire cutting test*: The force required to slice the processed cheese sample is measured with a texture analyzer in the compression mode. The crosshead attachment is a harp-type fixture with a single strand of piano wire (Heiland et al., 1988). The steady-state cutting energy can also be obtained (Goh et al., 2005).

12.7.3 Shreddability

Shreddability is defined as a cheese's ability to cut cleanly into long thin uniform strips, have low susceptibility to form curd fines, and resist sticking, matting or clumping when loosely packed (Guinee and Kilcawley, 2004). Before the assessment of shreddability, the cheese block is shred in a food processor with a stainless-steel shredding blade, under a constant load (Apostolopoulos and Marshall, 1994).

a) *Sensory assessment*: Shreddability is assessed visually by the panelists. Three important factors determine shreddability of a cheese: the length of the shreds, the amount of fragments present and the degree of stickiness between the shreds (Apostolopoulos and Marshall, 1994).

b) *Sieving test*: The shredded cheese is shaken by hand through two sieves to divide the shreds into three size classes (long shreds, short shreds, and fines). The percentages of shred classes and the percentages of cheese that adhered to the blade and to the top of the food processor are calculated and used to evaluate shreddability (Childs et al., 2007).

c) *Image Analysis*: The sample of shreds is laid over a black background, spread over a rectangular area assuring that no pieces are touching each other and then photographed. The prints are developed in black and white for maximum contrast. The images are then captured by a video camera connected to an image analyzer. All of the image processing and measurement routines are software-based allowing the identification and quantification of the image (Apostolopoulos and Marshall, 1994).

12.7.4 Meltability

Meltability can be defined as the ease with which cheese will flow or spread upon heating (Wang et al., 1998).

a) *Arnott test*: This test is based on the measurement of changes in the height of processed cheese cylinders heated in an oven at 100°C for 15 min. Meltability of processed cheese is quantified as the percent reduction in the height of cheese cylinder after heating (Arnott et al., 1957; Ha et al., 2013).
b) *Olson and Price test*: Processed cheese cylinder is adjusted in one end of a Pyrex tube. This melting tube is placed horizontally in an oven at 110°C for 8 min. The total distance in millimeters covered by the melted cheese is called "cheese-flow" and is taken as an indication of the cheese meltability (Olson and Price, 1958). Modifications to the method have been proposed by other researchers (Mounsey and O'Riordan, 1999; Everard et al., 2005).
c) *Schreiber test*: Thick disks of cheese samples are placed in a Petri dish and then heated at 232°C for 5 min in an oven, followed by cooling of the melted cheese for 30 min. The meltability of processed cheese is assessed with the Schreiber melt test score sheet, which has 11 concentric rings (Kosikowski, 1977; Ha et al., 2013). A modification to the method, with a melting temperature of 90°C, has been proposed by Muthukumarappan et al. (1999).
d) *USDA test*: Similar to the Schreiber test, a circular cheese slice is heated at 205°C for 10 min in a Petri dish. Cheese expansion is measured on an official meltability scale after cooling (U.S. Department of Agriculture, 1980).
e) *UW meltmeter*: This is a modified lubricated squeezing flow rheometer developed to evaluate melt/flow characteristics of cheeses. A cylindrical cheese sample is placed on the sample platform under a circular top plate that is attached to a linear variable differential transformer. The device can be operated either at constant-force or at constant-rate in conjunction with a texture analyzer, at various melting temperatures (e.g., 40 and 60°C). The biaxial stress growth coefficient can be used as an objective measure of cheese meltability (Wang et al., 1998).
f) *Rapid Visco Analyzer (RVA)*: The apparent viscosity of a disk of processed cheese during the heating time from 25°C to 90°C, the holding time at 90°C, and the cooling time from 90°C to 25°C gives information on its meltability. The features of interest obtained include melt time, hot viscosity, time at 5000 mPas during cooling, and solidification time (Prow and Metzger, 2005).
g) *Temperature-sweep dynamic shear test*: Meltability of processed cheese is determined with a dynamic stress rheometer using parallel plate geometry. The test combines low shear with thermal treatment of a cheese disk. Measurements are performed at constant stress, frequency and strain. Sample temperature is increased from 25°C to 90°C at a constant rate. As melting predictor is considered the transition temperature, that is, the lowest temperature at which the elastic and viscous properties are equivalent ($G' = G''$ or $\tan\delta = 1$) (Sutheerawattananonda and Bastian, 1998; Reparet and Noël, 2003).

12.7.5 Stretchability

Stretchability is the tendency of the heated cheese to form strings when extended (Kapoor and Metzger, 2008).

a) *Fork test*: Grated cheese is placed on a pizza crust containing pizza sauce, and after the pizza is baked in preheated oven at 287°C for 10 min, a fork is inserted manually into the melted cheese and raised vertically until all cheese strands break. Strand length at the time the strands break is used as a measure of stretchability of the cheese (U.S. Department of Agriculture, 1980; Fife et al., 2002).

b) *Tensile tests*: A circular plate is used as a template to hold a pizza crust, and a smaller circular piece (with a vertical rod attached) is cut out, allowing the center of the template to be raised independently of the edge. A similarly cut pizza crust is placed on the template and a standard weight of cheese sprinkled on top of the crust. The complete apparatus is heated in a microwave oven for 15 s to melt the cheese. On removal from the oven, the vertical rod is attached to the head of a tensile testing machine and pulled vertically, stretching the melted cheese. The extensibility of the cheese, defined as the distance of travel of the cross-head until all the cheese strings failed, is used as a measure of stretchability (Apostolopoulos, 1994; Fife et al., 2002).

   This method was modified by Guinee and O'Callaghan (1997), such that the two halves of a cheese covered pizza were pulled horizontally, rather than vertically, after cooking in an oven at 280°C for 4 min. To do this, the pizza pie is placed on the platform unit of the stretching apparatus, consisting of a fixed element and a rolling one. Stretch was defined as the distance travelled by the rolling element to the point of complete strand breakage (Fife et al., 2002).

c) *Compressive elongation test*: A cylindrical cheese sample melted at 65°C is compressed between two parallel plates and forced to elongate horizontally, using a texture analyzer. The elongational viscosity of the melted cheese is calculated; this could be used as an index of the ability of the cheese to form strings when it is stretched. This test is complementary to the tensile tests (Apostolopoulos, 1994).

d) *Utah State University (USU) stretch test*: Cheese plugs are placed into stainless steel cups and heated in a water bath for 30 min at 60, 70, 80 or 90°C. An immersed three-pronged hook-shaped probe pulls vertically the melted cheese. Stretch parameters (melt strength, stretch length and stretch quality) are determined from a force-distance graph produced by the probe, attached to a tensile tester (Fife et al., 2002).

e) *Ring-and-ball method*: A thin cheese slice held within a horizontal ring is forced downward under the weight of a steel ball as the sample is heated in a mineral oil bath, at 65°C. The extensional viscosity of melted cheese estimated from the stretch length versus time data and the stretch length at failure are taken as a measure of stretchability (Hicsasmaz et al., 2004).

12.8 Microbiological analyses

In processed cheese, microorganisms arise either from raw materials and when their thermal destruction is incomplete during processing, or from contamination after processing. Microbiological methods are used for the detection as well as for the enumeration of microorganisms. These are horizontal methods, that is, they are used for all foods independently, even for animal feeding stuffs.

To determine the microorganisms, it is necessary to cultivate them under suitable culture conditions. The culture media used can be selective (specific) or nonselective, in liquid form (broths) or, more frequently, in solid form (solidified by the addition of agar). In the case of solid culture media, two techniques are commonly used: the pour plate technique and the surface plating technique. In the first, the test sample is dispensed into empty Petri dishes and mixed with the molten agar culture medium of 44–47°C to form poured plates. In the second, the test sample is surface plated on the solid agar culture medium contained in Petri dishes.

If the test sample is liquid, the culture medium is inoculated directly with the test portion. However, if the test sample is solid, a dilution with a nine-fold quantity of a suitable diluent precedes the inoculation (initial suspension or 10^{-1} dilution). If the number of microorganisms per unit volume is large, further decimal dilutions are required (10^{-2}, 10^{-3}, etc.).

The incubation which follows is carried out according to the microorganisms to be determined under appropriate conditions of temperature (psychrotrophic, mesophilic, thermophilic), time, and atmosphere (aerobic, anaerobic).

The number of viable microorganisms or colony-forming units (CFU) per gram or per milliliter of the test sample is calculated from the number of colonies obtained in the plates containing fewer than 300 colonies.

General requirements and guidance for microbiological examinations are detailed in ISO (2007a). Regarding culture media, their preparation, production, storage, and performance testing are recommended in ISO (2014b).

Microbiological methods are distinguished into reference methods and alternative methods.

12.8.1 Reference methods

These are international methods recommended by ISO or AOAC. In this subchapter, ISO methods are mentioned.

12.8.1.1 Total aerobic mesophilic microorganisms

a) *Enumeration*: Plate count agar is inoculated with the test sample and its decimal dilutions by the pour plate technique. The plates are incubated under aerobic conditions at 30°C for 72 h. The number of microorganisms or total plate count per gram or per milliliter of the test sample is calculated (ISO, 2013a).

b) *Enumeration*: Plate count agar is inoculated with the test sample and its decimal dilutions by the surface plating technique. Then the same procedure as above is followed (ISO, 2013b).

12.8.1.2 Psychrotrophic microorganisms

For the enumeration of psychrotrophic microorganisms, the surface plating technique is used as in the case of total aerobic mesophilic microorganisms, with the difference that the incubation is done at 6.5°C for 10 days (ISO, 2019).

12.8.1.3 Yeasts and molds

a) *Enumeration*: In products with water activity greater than 0.95, surface-inoculated plates are prepared using dichloran-Rose Bengal chloramphenicol agar (DRBC agar), as selective culture medium. The plates are then aerobically incubated at 24–26°C for 5 days and colonies/propagules are counted. For the identity of any doubtful colonies, a binocular magnifier or microscope can be used. The number of yeasts and molds per gram or per milliliter of the test sample is calculated. Molds and yeasts are counted separately, if necessary (ISO, 2008f).

b) *Enumeration*: In products with water activity less than or equal to 0.95, the selective culture medium used is dichloran 18% w/w glycerol (DG18) agar, and the incubation period at 24–26°C can be extended up to 7 days (ISO, 2008g).

12.8.1.4 Mesophilic lactic acid bacteria

For the enumeration of mesophilic lactic acid bacteria, de Man, Rogosa and Sharpe (MRS) agar at pH 5.7 is inoculated with the test sample and its decimal dilutions by pour plate or surface plating technique. The dishes are incubated at 30°C for 69–75 h. The number of mesophilic lactic acid bacteria is calculated per gram or per milliliter of test sample (ISO, 1998).

12.8.1.5 Spore-forming bacteria

These are *Bacillus* spp., *Geobacillus* spp., and *Clostridium* spp. According to Palmas et al. (1999), vegetative forms are initially eliminated from the test sample by heat-treatment at 80°C for 15 min.

a) *Enumeration of* Bacillus *spp.*: Tryptic soy agar (TSA) is inoculated by surface plating technique with the test sample and its decimal dilutions, and then incubated aerobically at 37°C for 24–48 h (Palmas et al., 1999). Other culture media, such as sheep blood agar, plate count agar or nutrient agar, may also be used in aerobiosis at 35–37°C for 24–48 h (Thomareis, 2020).

b) *Enumeration of* Geobacillus *spp.*: The same culture media, technique and conditions are used as in the case of *Bacillus* spp., except for the temperature (55°C) (Thomareis, 2020).

c) *Enumeration of* Clostridium *spp.*: Reinforced clostridium medium (RCM) is inoculated by the pour plate technique with the test sample and its decimal dilutions, and then incubated anaerobically at 37°C for 48–72 h (Palmas et al., 1999). Other culture media, such as tryptic soy agar, sheep blood agar, plate count agar or nutrient agar, may also be used in anaerobiosis at 35–37°C for 48–72 h (Thomareis, 2020).

12.8.1.6 Bacillus cereus

a) *Enumeration*: The test sample and its decimal dilutions are surface plated on mannitol egg yolk polymyxin (MYP) agar. The plates are incubated under aerobic conditions at 30°C for 18–48 h. Colonies of *B. cereus* are confirmed by the hemolysis reaction, after subculturing on sheep blood agar and incubation at 30°C for 22–26 h. The number of *B. cereus* per gram or per milliliter of sample is calculated from the number of confirmed colonies (ISO 2004c).

b) *Detection of low numbers*: Single or double-strength tryptone soya polymyxin broth (TSPB) is inoculated with the test sample and incubated at 30°C for 44–52 h. After subculturing on polymyxin pyruvate egg yolk mannitol bromothymol blue agar (PEMBA) or MYP agar and incubation for 18–48 h at 37°C or 30°C, respectively, suspected colonies are confirmed by means of hemolysis test or by microscopic examination. The results are given as the "presence" or "absence" of *B. cereus* in grams or milliliters of product (ISO, 2006d).

c) *Enumeration of low numbers*: The whole of the above procedure takes place with the test sample and its decimal dilutions. The most probable number of *B. cereus* per gram or per milliliter of sample is calculated from confirmed dilutions by reference to most probable number tables (ISO, 2006d).

12.8.1.7 *Sulfite-reducing anaerobic bacteria*

For the enumeration of sulfite-reducing bacteria growing under anaerobic conditions, iron sulfite agar of 44–47°C is mixed in plates (or tubes) with the test sample and its decimal dilutions. The plates (or tubes) are incubated under anaerobic conditions for 24–48 h at 36–38°C, or at 49–51°C if thermophilic bacteria are suspected. Typical black colonies are counted. The black color of the colonies and the surrounding zone is due to the formation of iron(II) sulfide as a result of the reaction between sulfide ions and trivalent iron [Fe(III)] present in the medium.

The number of sulfite-reducing bacteria per milliliter or per gram of sample is calculated. If the number of their spores is requested, an elimination of the vegetative forms must be initially applied to the test sample by pasteurization (75°C for 20 min to boiling for several minutes) (ISO, 2003b).

12.8.1.8 *Clostridium perfringens*

For the enumeration of *Clostridium perfringens* by the pour plate technique, the test sample and its decimal dilutions are dispensed into empty Petri dishes and mixed with sulfite-cycloserine (SC) agar of 44–47°C to form poured plates. After solidification, an overlay of the same medium is added. The plates are incubated anaerobically at 37°C for 18–22 h. The characteristic colonies are enumerated and then confirmed using the following techniques: (i) Thioglycollate broth is inoculated with colonies and incubated under anaerobic conditions at 37°C for 18–24 h. From this culture, lactose sulfite (LS) broth is inoculated and incubated aerobically at 46°C for 18–24 h. (ii) Nitrate motility medium is inoculated with colonies and incubated under anaerobic conditions at 37°C for 24 h. (iii) Lactose-gelatine medium is inoculated with colonies and incubated under anaerobic conditions at 37°C for 24 h.

Bacteria that produce black colonies in SC agar, are nonmotile, usually reduce nitrate to nitrite, produce acid and gas from lactose, and liquefy gelatine in 48 h are considered to be *C. perfringens*. The number of *C. perfringens* per milliliter or per gram of sample is calculated (ISO, 2004d).

12.8.1.9 Salmonella *spp*

a) *Detection*: Buffered peptone water (BPW) is inoculated with the test portion and incubated at 34-38°C for 18 h. Rappaport-Vassiliadis medium with soya (RVS) broth or modified semisolid Rappaport-Vassiliadis (MSRV) agar and Muller-Kauffmann tetrathionate-novobiocin (MKTTn) broth are inoculated with the culture obtained. The RVS broth or the MSRV agar is incubated at 41.5°C for 24–48 h and the MKTTn broth at 37°C for 24–48 h. From the cultures obtained, xylose lysine deoxycholate (XLD) agar is inoculated and incubated at 37°C for 24 h. A second selective agar is also used. Colonies are subcultured and their identity is confirmed by means of biochemical and serological tests (ISO, 2017a).

b) *Enumeration*: In the most probable number (MPN) format, the initial suspension in BPW is added to the first (empty) row of three wells of a 12-well microtiter plate. To achieve dilutions, the second row of three wells containing BPW is inoculated from the first row, and so on for the third and fourth well. The 12-well microtiter plates are then incubated at 37°C for 18 h. Each well obtained is subcultured in a well containing a MSRV agar with incubation at 41.5°C for 24–48 h. From the cultures obtained, XLD agar is inoculated and incubated at 37°C for 24h. Colonies of *Salmonella* spp. obtained are confirmed by means of biochemical and serological tests. From the number of confirmed positive wells (dilutions), the MPN of *Salmonella* spp. per gram or per milliliter of the test sample is calculated by reference to most probable number tables (ISO, 2012).

c) *Serotyping*: The following antigens are determined for isolates biochemically confirmed as *Salmonella* spp.: O-antigens (somatic), H-antigens (flagellar) and Vi-antigens (capsular) (ISO, 2014c).

12.8.1.10 Listeria monocytogenes *and* Listeria *spp*

a) *Detection*: A selective primary enrichment medium containing half the concentrations of acriflavine and nalidixic acid (half-Fraser broth), which is also used as a dilution fluid for the test portion, is inoculated with the initial suspension and incubated at 30°C for 24–26 h. A full-strength secondary liquid enrichment medium (Fraser broth) is inoculated with the culture obtained and incubated at 37°C for 24 h. The cultures obtained are plated out on agar *Listeria* according to Ottaviani and Agosti (ALOA), and incubated at 37°C for 24–48 h. A second solid selective medium is also used. The colonies of *L. monocytogenes* or *Listeria* spp. are subcultured by plating out and confirmed by means of morphological and/or biochemical tests (ISO, 2017c).

b) *Enumeration*: The initial suspension is prepared in an appropriate diluent (buffered peptone water, half-Fraser broth, etc.). The test sample and/or decimal dilutions are surface plated on ALOA. The Petri dishes are incubated at 37°C for 24–48 h. Colonies of *L. monocytogenes* and/or *Listeria* spp. are confirmed by means of morphological and/or biochemical tests. From the number of confirmed colonies, the number of *L. monocytogenes* and/or *Listeria* spp. is calculated per gram, per milliliter, per square centimeter or per sampling device (ISO, 2017d).

12.8.1.11 Staphylococcus aureus *and other staphylococci*

a) *Enumeration*: The surface of Baird-Parker agar medium is inoculated with the test sample and its decimal dilutions. The plates are incubated aerobically at 35°C or 37°C and examined after both 24 h and 48 h. The number of staphylococci per milliliter, or per gram, of sample is calculated from the number of colonies obtained on plates, and confirmed by a positive coagulase test result (ISO, 1999a)

b) *Enumeration*: Poured plates of rabbit plasma fibrinogen agar medium are inoculated with the test sample and its decimal dilutions. The plates are incubated at 35°C or 37°C for 18–24 h, and a further 24 h if necessary. From the number of typical colonies per Petri dish, the number of coagulase-positive staphylococci is calculated per gram or per milliliter of test sample (ISO, 1999b).

c) *Detection of stressed cells and in low numbers*: Tubes with modified Giolitti and Cantoni broth are inoculated with the test sample and incubated at 37°C, anaerobically, for 24 h and 48 h. The presence of coagulase-positive staphylococci is indicated by the reduction of potassium tellurite. The surface of solid Baird-Parker medium is inoculated from positive tubes after 24 h and all the remaining tubes after 48 h. The plates are incubated at 37°C for 24 h and 48 h. The presence of coagulase-positive staphylococci is indicated by the reduction of potassium tellurite and an egg yolk reaction. Colonies are confirmed by a coagulase reaction. Alternatively, the surface of rabbit plasma fibrinogen agar is inoculated and, after the same incubation conditions, the presence of coagulase-positive staphylococci is indicated by colonies showing a specific rabbit plasma fibrinogen reaction. The results are given as the "presence" or "absence" of coagulase-positive staphylococci in x g or x ml of product (ISO, 2003a).

d) *Enumeration of stressed cells and in low numbers*: The whole of the above procedure takes place with serial decimal dilutions of the test sample. The most probable number of coagulase-positive staphylococci per gram or per milliliter of sample is calculated by reference to most probable number tables for confirmed dilutions (ISO, 2003a).

12.8.1.12 Escherichia coli

a) *Enumeration*: Cellulose membranes overlaid on mineral-modified glutamate agar (MMGA) are inoculated with the test sample and its decimal dilutions and then incubated at 37°C for 4 h. For isolation, the membranes from the resuscitation stage on the MMGA are transferred to tryptone-bile-glucuronide agar (TBX), and then incubated at 44°C for 18–24 h. The number of colony-forming units (CFU) of β-glucuronidase-positive *Escherichia coli* per gram or per milliliter of sample is calculated from the number of typical blue colonies (ISO, 2001a).

b) *Enumeration*: TBX agar is inoculated with the test sample and its decimal dilutions. The dishes are incubated for 18–24 h at 43–45°C then examined to detect the presence of colonies of β-glucuronidase-positive *Escherichia coli*. The number of colony-forming units (CFU) of β-glucuronidase-positive *Escherichia coli* per gram or per milliliter of sample is calculated (ISO, 2001b).

c) *Detection and enumeration of stressed cells*: Tubes of double and single-strength liquid selective enrichment medium (MMGM) are inoculated with the test sample and its decimal dilutions and incubated at 36–38°C for 22–26 h. The tubes are examined for acid production, indicating lactose fermentation. Each positive tube is subcultured to TBX agar and incubated at 43–45°C for 18–24 h. The TBX agar is examined for the presence of blue or blue-green colonies, indicating the presence of β-glucuronidase positive *Escherichia coli*. The most probable number of β-glucuronidase-positive *Escherichia coli* is determined according to the number of positive tubes (ISO, 2015a).

12.8.2 Alternative methods

These are easy and rapid proprietary methods for the detection or quantification of microorganisms, mainly food-borne pathogens. They are validated by compared to reference methods. Their certification is performed by an independent third party (AFNOR, MicroVal, NordVal), in accordance with the protocol specified by ISO (2016). In general, they can be distinguished into chromogenic, immunoenzymatic, molecular biology, and flow cytometry methods.

12.8.2.1 Chromogenic methods

These are methods of classical microbiology on selective culture medium and involving metabolic or enzymatic reactions specific to a given bacterial genus or species. The final compound resulting from these reactions is colored thus allowing specific coloring of the colonies (Garry, 2010).

12.8.2.2 Immunoenzymatic methods

These methods are based on the recognition of specific antigens of the microorganism sought. They are used within the framework of qualitative analyzes (presence/absence). For the performance of microbiological analyzes, ELISA (enzyme-linked immunosorbent assay) and ELFA (enzyme-linked fluorescence assay) methods, on the one hand, and immunoconcentration method, on the other hand, are applied (Garry, 2010).

a) *ELISA and ELFA methods*: From a practical standpoint, a first antibody coupled to the wall of a tube (or well) recognizes the antigen of the target microorganism and binds it. A second antibody labeled with a colored or fluorescent enzyme recognizes the antigen of the microorganism and also binds it (sandwich technique). These methods are applied after enrichment with specific broth for the desired germ. After rinsing, the presence of the germ is revealed by colorimetry (ELISA) or fluorometry (ELFA). The immunoenzymatic detection replaces the isolation on a dish practiced within the framework of the reference methods and makes it possible to gain 24–48 hours to obtain the analysis result (Garry, 2010).

b) *Immunoconcentration (or immunocapture) method*: The method is based on the use of antibodies, specific for the desired bacterial genus or species, coupled to magnetic beads and used to capture the microorganism. Once the microbial antigens are bound by the antibodies, the microbial cells are concentrated, using a magnet. The presence of the microorganism is then detected by inoculation on an agar dish or else by ELISA or ELFA. Immunoconcentration reduces the time required for the enrichment step (Garry, 2010).

12.8.2.3 Molecular biology methods

a) *Polymerase chain reaction (PCR) method*: The method is based on the amplification of DNA fragment(s) specific of the desired microorganism. A PCR reaction corresponds to the succession of around thirty cycles each comprising 3 steps: denaturation, annealing and elongation (extension). If in the same test several different zones are amplified, this is called multiplex PCR (Garry, 2010).

All the elements necessary for the reaction are grouped together in a tube which is subjected to the different temperatures corresponding to each step. First, the DNA is denatured (separation of the double strands) by raising the temperature (>90°C). Second, the temperature is quickly lowered to 50–65°C. The primers (DNA fragments specific for the desired bacterial species) recognize their complementary sequence on the target DNA strands. They each hybridize to their respective strands (annealing). Third, temperature of the tube is increased to 72°C, which allows Taq polymerase to add nucleotides. The nucleotides are not incorporated randomly but as a function of the target sequence (complementary nucleotides). A new DNA strand, the sequence of which is complementary to that of the target strand, has just been synthesized (elongation). The cycle is repeated about thirty times. The sequence thus amplified is revealed either by hybridization with a labeled probe (colorimetric reaction) or by migration on electrophoresis gel (Garry, 2010).

b) *Hybridization method*: Molecular hybridization is based on the complementary base pairing properties of nucleic acids. This principle is used to reveal the presence of target nucleic acid molecules, specific for a germ. 16S ribosomal RNA (16S rRNA) sequences are very frequently used for identification, as this RNA is highly conserved within the same species (Garry, 2010).

ISO (2011a) specifies the minimum requirements of performance characteristics for the detection of nucleic acid sequences (DNA or RNA), by molecular methods used for the detection and quantification of food-borne pathogens.

12.8.2.4 Flow cytometry method

Flow cytometry is a technique that allows particles, molecules or cells to scroll at high speed in the beam of a laser. The reemitted light allows to visualize and therefore to count these particles. The signals measured are linked either to the intrinsic optical properties of the particles corresponding to the light scattering phenomena, or to the fluorescence obtained by specific labeling of the particles. In the second case, by using fluorescent markers specific to a bacterial species, it is then possible to count the corresponding cells (Garry, 2010).

12.9 Sensory evaluation

Sensory evaluation is a scientific discipline used to measure, analyze, evoke, and interpret the reactions to those characteristics of foods and materials as perceived by the senses of sight, smell, taste, touch, and hearing (Stone and Sidel, 2004). Sensory evaluation or sensory analysis is the ultimate process of estimating the quality of a food which will lead to its acceptance or rejection by the consumer.

12.9.1 Types of sensory tests

There are two types of sensory tests: analytical and hedonic.

First, analytical or objective tests provide objective data on the sensory properties of foods and are carried out primarily by trained assessors. There are three classes of analytical tests: (i) discrimination or difference tests to determine whether there are sensory differences between samples (ii) threshold tests to determine threshold, the lowest concentration at which a sensory stimulus is detectable, and (iii) descriptive tests to identify the nature of a sensory difference and/or the magnitude of the difference.

Second, hedonic tests are known as subjective, affective or consumer tests. They provide subjective data on acceptability, liking or preference, and are carried out by untrained assessors (Drake, 2007; Kemp et al., 2009).

12.9.2 Analytical tests

12.9.2.1 Discrimination tests

The main discrimination tests are the following (ISO, 2005; Kemp et al., 2009; Lawless and Heymann, 2010; ISO, 2017b):

a) *Paired comparison test*: Assessors are presented with two coded samples and asked to determine if the samples are the same or different. They may be asked to describe any differences and determine which of the two samples has the greatest intensity of a specified attribute.

b) *Triangle test*: Assessors are presented with three coded samples and told that two samples are the same and one is different. They are asked to determine which sample is the odd. They may also be asked to describe the difference.

c) *Duo-trio test*: Assessors are presented with three samples, two coded and one labeled as a reference. They are asked to determine which is the most similar (or different) to the reference.

d) *Tetrad test*: Assessors are presented with four coded samples, two of which are of one type and two of another. They are asked to group the two sets of samples.

e) *Two-out-of-five test*: Assessors are presented with five coded samples, two of which are of one type and three of another. They are asked to group the two sets of samples.

f) *"A - not A" test*: Assessors are presented with several coded samples, which may be "A" or "not A". They are asked to indicate which sample is "A", after they have learnt to recognize sample "A".

g) *Ranking test*: Assessors are presented with several coded samples. They are asked to place the samples in order of intensity for a specified attribute.

Typically, a discrimination test is conducted with 25–40 participants who have been screened for their sensory acuity to common product differences and who are familiar with the test procedures. Statistical tests used in discrimination tests include binomial, chi-square, *t*-test, etc. (Lawless and Heymann, 2010).

12.9.2.2 Threshold tests

These tests are used to determine absolute threshold (lowest concentration at which a sensory stimulus can be detected), recognition threshold (lowest concentration at which a sensory stimulus can be recognized), difference threshold (lowest concentration at which differences in stimuli can be detected), terminal threshold (concentration above which there is no perceived increase in sensory stimulus), orthonasal threshold (threshold of volatile compound detected orthonasally), and retronasal threshold (threshold of volatile compound detected retronasally) (Drake, 2007). Data on sensory thresholds to stimuli are used in sensory studies in two main ways: as measures of the sensitivity of assessors or groups of assessors to specific stimuli; and as measures of the ability of substances to evoke sensory responses in assessors. In the first, the value of the threshold is taken as a description of an assessor's performance; in the latter, as a measure of a property of the substance (ISO, 2018c). Seventy-five to 100 individuals need to be tested to obtain a reliable threshold (Drake, 2007). The two commonly used statistical methods for determining a threshold are geometric means and logistic regression (Gallagher and Cuppett, 2007).

12.9.2.3 Descriptive tests

These tests can be applied to one or more samples in order to characterize, both qualitatively and quantitatively, one or more sensory properties. These are simple to very complex methods, the main ones of which are summarized below, although

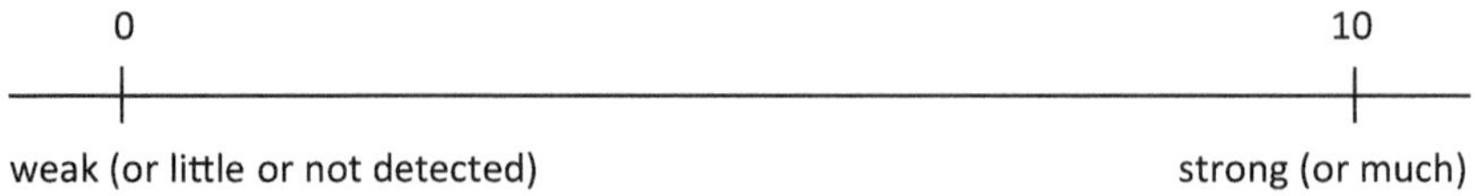

Figure 12.9 A 15-cm line scale for rating the intensity of an attribute, with indented anchors at 1.3 cm from the ends.

many sensory laboratories use modified versions tailored to meet their specific needs.

a) *Consensus profiling*: Assessors work as a group to agree on attributes and intensity ratings. It is relatively quick, but may be subject to bias, as assessors with stronger personalities may dominate the agreement process. No statistical analysis can be carried out (Kemp et al., 2009; ISO, 2017b).

b) *Flavor profile method*: Aroma, flavor and mouth-feel are assessed in terms of quality, intensity of attributes on a 5-point category scale (0, threshold, 1, 2, 3), order of perception, aftertaste and overall impression. A panel of 4–6 selected and trained assessors is asked to evaluate the samples individually and then discuss their evaluations as a group to determine a consensus score. The method does not lend itself to statistical analysis. Profile attribute analysis, a modification to the method, includes the assessment of visual, tactile and auditory attributes and the use of individual assessments, category/line scales, and data evaluation using analysis of variance (ANOVA) (Neilson et al., 1988; Kemp et al., 2009).

c) *Texture profile method*: Texture and mouth-feel properties of foods, including mechanical, geometric, fat and moisture content-related attributes, are assessed in terms of (i) quality, using a predetermined list of attributes, that is, hardness, fracturability, chewiness, gumminess, adhesiveness, viscosity and geometric structure, (ii) intensity, using a 5-point to 9-point scale, and (iii) order of perception, from first bite to complete mastication. A panel of 6–10 assessors is selected for their ability to discriminate textural differences on the product type to be assessed. They are then trained using predetermined scales and work in consensus, which means the data cannot be processed statistically. In later modifications, however, they work individually using other types of scales (Brandt et al., 1963; Szczesniak et al., 1963; Kemp et al., 2009).

d) *Quantitative descriptive analysis*: This method can produce a full qualitative and quantitative sensory description. Assessors (8–15), selected for their ability to discriminate and describe products in the category to be studied, agree on a list of qualitative attributes and then work individually to rate the attributes on a line scale with indented anchors (Fig. 12.9). Assessors receive limited training and the primary aim for assessors is to be consistent within themselves rather than with the rest of the panel. Assessments are made in replicates of 2-6 repeat evaluations, data are translated into mean scores and statistically analyzed using ANOVA, individual assessor performance is monitored and compared to that of the panel, and results are presented graphically in spider plots (Kemp et al., 2009; Sidel et al., 2018), as shown in Fig. 12.10.

e) *Spectrum method*: A full qualitative and quantitative description can be produced. Sensory qualities are assessed using a predefined, standardized lexicon from which terms are selected. The lexicon includes technical terms and terms that are applied across all products. Perceived intensity is assessed with a usually 15-point numbered absolute scale that can be universal or product specific and is anchored at multiple points with well-defined references. Selected assessors (12–15) receive in-depth training on attributes and intensity references so that they

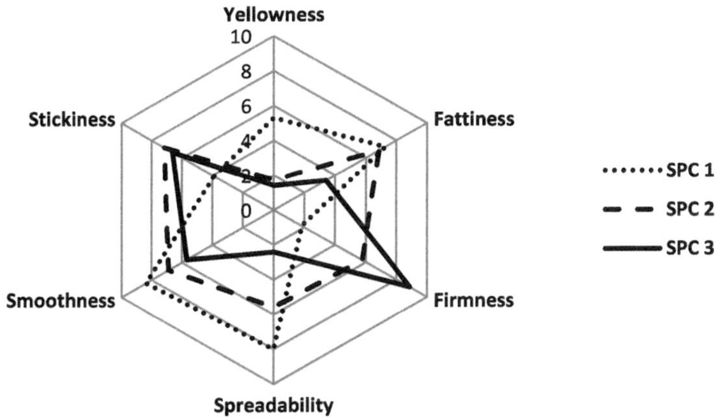

Figure 12.10 Spider plot of six sensory attributes for three samples of spreadable processed cheese (SPC) (Chatziantoniou, 2020).

all make assessments and score similarly. Assessors agree on the attributes and order of assessment, and then rate intensity individually. ANOVA and principal component analysis (PCA) are used for data evaluation (Kemp et al., 2009; Lawless and Heymann, 2010; Dus et al., 2018).

f) *Free choice profiling*: Untrained or minimally trained assessors evaluate products using their own individual sets of descriptive terms, and then rate intensity of product attributes. The individual sensory maps are treated with generalized Procrustes analysis (GPA) to create a consensus configuration (ISO, 2005, 2017b; Kemp et al., 2018).

g) *Flash profiling*: Untrained assessors evaluate products using their own individual sets of descriptive terms, and then rank the products on an ordinal scale for each attribute. GPA is also used for data treatment (ISO, 2008d; Kemp et al., 2018).

h) *Time-intensity profiling*: It is a quantitative and dynamic sensory method aiming to record continuously over a period of time the development and decline of the perceived intensities of specified sensory attributes. ANOVA is used for data analysis (ISO, 2008d; Lawless and Heymann, 2010; Kemp et al., 2018).

12.9.3 Hedonic tests

Hedonic tests are mainly used (i) to compare a product with competitor products and (ii) to optimize a product to be liked by a large number of consumers. These tests are divided into two categories: preference tests and acceptance tests.

12.9.3.1 Preference tests

The main preference tests are the following (Kemp et al., 2009; Lawless and Heymann, 2010):

a) *Paired preference tests*: Assessors are presented with two coded samples and asked to identify which sample is preferred. A no-preference option is sometimes included in a paired preference test.

Degree of liking/disliking		Sample
9	Like extremely	----------
8	Like very much	----------
7	Like moderately	----------
6	Like slightly	----------
5	Neither like nor dislike	----------
4	Dislike slightly	----------
3	Dislike moderately	----------
2	Dislike very much	----------
1	Dislike extremely	----------

Figure 12.11 The 9-point hedonic scale.

Degree of intensity/liking		Sample
5	Much too salty	----------
4	Too salty	----------
3	Just about right	----------
2	Not salty enough	----------
1	Not nearly salty enough	----------

Figure 12.12 The 5-point just-about-right scale for processed cheese saltiness.

b) *Preference ranking test*: Assessors are presented with several coded samples. They are asked to rank the samples in either descending or ascending order of preference or liking.
c) *Best-worst scaling*: Assessors are presented with more than two coded samples. They are asked to choose the one they like most and the one they like least.

Statistical tests used in preference tests include binomial, chi-square, *t*-test, etc.

12.9.3.2 Acceptance tests

These are also known as acceptability scaling and are used to assess the attractiveness of consumers to food. Assessors are asked to assess several samples using a rating scale for the degree of liking or disliking for each of the characteristics to be evaluated. The most common scale is the 9-point category hedonic scale (Fig. 12.11), also known as the liking scale. The use of the bipolar 5-point category just-about-right (JAR) scale that combines intensity and hedonic judgments is also frequent (Fig. 12.12). Their main disadvantage is that the perception of what is just-about-right varies across individuals. Statistical tests used in acceptance tests include binomial, chi-square, *t*-test, ANOVA, etc. (Kemp et al., 2009; Lawless and Heymann, 2010).

Typically, for hedonic testing, the assessors are 75–150 consumers, regular users of the product to be evaluated (Lawless and Heymann, 2010).

12.9.4 International standards on sensory evaluation of milk and milk products

Among the many international standards published by ISO on sensory analysis, three of them concern milk and milk products. More precisely, ISO (2009a) specifies general guidance for the recruitment, screening, selection, training, monitoring, and regular performance checking of selected assessors and expert sensory assessors. Particular emphasis is given to the criteria and procedures applied in screening and training, where various sensory methods are used to evaluate odors and the basic tastes of aqueous solutions, as well as selected odor, flavor and texture attributes of suggested dairy products.

General practices are also specified for the sensory evaluation of dairy products by trained assessors (ISO, 2009b). In particular, requirements for the panel leader, panel preparation, documents, test room, and method used are recommended. The method for each dairy product includes sampling and sample preparation, apparatus and materials, assessment of appearance, odor/flavor and consistency, and sensory attributes listed in international tables.

In order to evaluate compliance with product specifications for sensory properties, ISO (2009c) gives guidance on a general method based on sensory scoring. Sensory properties of individual samples of milk and milk products are analyzed under standardized conditions by a panel of expert milk and milk product assessors. Each assessor evaluates the samples independently and uses a discrete 5-point interval scale to estimate the magnitude of a possible deviation in a product from a pre-established sensory specification (1: very considerable deviation-5: no deviation). The method is especially applicable in process and quality control.

12.10 Statistical techniques

The elaboration of the experimental measurements requires statistical methods to be applied to the experimental data depending on the experimental design used and the expected outcomes. The first stages of analysis are often those that summarize the data, such as descriptive statistics. Then parametric or nonparametric tests are applied, so as to establish the significance of treatment effects. Correlation and regression analysis elucidate the relation, if any, between the different variables under evaluation. Further analysis, by using principal component analysis, may be carried out to explain the main variance within the attribute ratings. Experimental designs in food product development are also applied, such the optimization technique of the response surface methodology.

12.10.1 Descriptive statistics

The easier way to present the experimental data is by the use of descriptive statistics, which comprise methods used to summarize the characteristics of a sample, for example, the mean (average value), median, variance, standard deviation, standard error, etc. The variance and the standard deviation (SD) give the degree of variation in the sample data. The higher these measures are, the greater the variation judged relative to the mean. A minimum value for any is zero (i.e., no variation in the sample values – they are all the same). The standard error (SE) takes into account not only the variation in the sample data, but also the sample size (Bower, 2009).

The experimental data from the evaluation of processed cheese properties have been presented as mean ± SD (Dimitreli and Thomareis, 2004, 2008; Chen and Liu, 2012; Černíková et al., 2017, 2018) or mean ± SE (Giri et al., 2017; Černíková et al., 2017, 2018).

12.10.2 Parametric tests

These are comparison tests for interval and ratio data. Parametric tests should be used when data are normally distributed and there is homogeneity of variances. These tests compare the mean values of the measure. The null hypothesis (H_0) is that the mean values do not differ; the alternative hypothesis (H_a) is that they differ. The appropriate test depends on the number of groups being compared and whether or not they are independent. These tests are commonly used in analysis of instrumental and sensory data (Bower, 2009; Granato et al., 2014). The most usual tests are the *t*-test and analysis of variance (ANOVA):

a) *t-test*: It can be used in a number of forms. One-sample *t*-test compares a sample mean to a standard value (hypothesized mean). Two-sample *t*-test compares the means of two independent samples. Paired *t*-test compares the means of two related (paired) samples. The *t*-test is used for small samples (<30) (Bower, 2009).

A *t*-test was applied by Sádlíková et al. (2010) for the evaluation of processed cheese properties.

b) *ANOVA*: It is used to compare three or more groups (thus being an independent sample test, that is, there are three or more separate groups). In the simplest form, it is an extension of the *t*-test. A dependent variable (usually continuous and parametric) and an independent variable (usually categorical) need to be identified. If an overall significance is established then the location of specific differences can be indicated by examination of the mean values, but a multiple comparison test (see in Section 12.10.4) is required for location of significant differences (Bower, 2009).

Depending on the number of factors to be analyzed, ANOVA can be distinguished in one-way ANOVA, two-way ANOVA and factorial ANOVA. In one-way ANOVA only one factor is assessed while, two-way ANOVA is used for two factors in which only the main effects are analyzed. The two-way ANOVA determines the differences and possible interactions when response variables are from two or more categories. The use of two-way ANOVA enables comparison and contrast of variables resulting from independent or joint actions. Factorial ANOVA is used for *n* factors and analyzes the main ones and their interaction effects (Granato et al., 2014).

Analysis of variance has been used in several studies as a tool for determining the significance of main effects in processed cheese variables. One-way ANOVA was applied to the experimental data by Dimitreli and Thomareis (2008, 2009), Ghorbel et al. (2016), Giri et al. (2017), and Motevalizadeh et al. (2018). Two-way ANOVA was used by Adhikari et al. (2009), Bayarri et al. (2012), Chen and Liu (2012), and Sołowiej et al. (2016). Three-way ANOVA was applied by Adhikari et al. (2009).

12.10.3 Nonparametric tests

Nonparametric tests do not require that the data have a normal distribution and that there is homogeneity of variances. They are distinguished in comparison tests for nominal and ordinal data (Bower, 2009).

Nominal data are nonparametric in nature and make fewer assumptions regarding the parent population from which the data originate. Data are in the form of category groups or classifications (two or more). The tests have a variety of applications in food studies, mainly in consumer and sensory testing. These data are often generated in consumer survey tests where demographic variables are gathered (Bower, 2009). Nominal data are usually tested using chi-square (χ^2) methods and binomial test:

a) *Chi-square tests*: They compare the proportions of observed occurrences with the theoretical expected ones. If the null hypothesis is true, then the proportions will be equal. This test is very commonly used in analysis of questionnaire or difference data as a test of proportions or distribution in that it tests the "goodness-of-fit" of experimental data to expected data (Bower, 2009).

b) *Binomial test*: Based on the general binomial theorem, this test calculates the probability of getting the result by chance in a choice between two objects (product sample/methods of production, etc.). The total number of occurrences of the choice is viewed as the trials. The null hypothesis for this test is that experimental results do not differ significantly from what is expected. In food science, the main application is for sensory difference and consumer preference tests (Bower, 2009).

Ordinal data are generated by ranking tests and when groups of consumers or sensory assessors rate items on a scale that has some logical order (Bower, 2009). The data are ranked from the lowest to the highest and each value is assigned, in order, the integer values from 1 to n. These tests compare the median of samples and some take into account the symmetry and shape of the parent distribution. The null hypothesis (H_0) is that the median values do not differ; the alternative (H_a) is that they differ (Bower, 2009; Granato et al., 2014). A number of tests for ordinal data are available which can also be applied to higher level of measurement scales (Bower, 2009; Granato et al., 2014):

a) *Sign and Wilcoxon tests*: When two related samples are rated by the same assessors, there is choice between the sign test and the Wilcoxon signed rank test. These tests are analogous to paired-sample t-test.

b) *Mann-Whitney test*: In order to compare two independent samples, Mann-Whitney U test is used to examine the data in a manner analogous to the use of two-sample t-test.

c) *Friedman test*: This test is used to determine differences in central location (median) for analysis of trials with one-way repeated measures having three or more related samples based on ranked data.

d) *Kruskal-Wallis test*: This test is used for comparison of rating between three or more independent samples. It essentially provides a one-way ANOVA for three or more independent samples based on ranked data.

Nonparametric tests have been used in several studies on processed cheese, such as chi-square test (Subramanian and Vivekanandan, 2020), binomial test (Bubelová et al., 2015), sign test (Frühbauerová, 2020), Mann-Whitney test (Černíková et al. 2010), Friedman test (Lazárková et al., 2010), Wilcoxon and Kruskal-Wallis tests (Černíková et al., 2008; Buňka et al., 2012, 2013; Černíková et al., 2017, 2018).

12.10.4 Multiple comparison tests

Multiple comparison tests are performed only if a significant difference for ANOVA is obtained. There are many possible tests as Fisher's least significant difference (LSD), Tukey's honest significant difference (HSD), Duncan's multiple range test (MRT), etc. The tests are performed at the same significance level as ANOVA. Choice of test depends on assumptions regarding the data as for application of a parametric test. The more powerful ones risk committing the Type I error (accepting equality of means when they are actually different) and conversely the weaker ones the Type II (rejecting equality when they are not different) (Bower, 2009; Granato et al., 2014). In some, the number of comparisons causes the significance value to rise directly with the number (Fisher's test). Tukey's test avoids this but has less power for detection of true differences as numbers increase. Duncan's test calculates a range value, which varies according to the number of comparisons (Bower, 2009). Fisher's test is used where sample sizes are small and when the distribution of the residues is normal. The test is much more robust than Tukey's test but it is the most sensitive to Type I errors (yet it provides an important tool for comparing means after an ANOVA procedure) (Granato et al., 2014). Choosing a test depends essentially on the confidence in the data (Bower, 2009), as well as on the investigator's experience (Granato et al., 2014). Instrumental data will usually qualify for the more powerful tests. For sensory and consumer data, it may be more advisable to use a more conservative alternative. Consideration of the error types, for example, in cases where missing a true difference is of high consequence, a more powerful test is appropriate. A compromise is a "medium test" such as Tukey's test (Bower, 2009).

Significant differences between processed cheese samples were determined by Tukey's test in several researches (Chen and Liu, 2012; Sołowiej et al., 2016; Giri et al., 2017). Fisher's test (Bayarri et al., 2012; Ghorbel et al., 2016), and Duncan's test (Morsy et al., 2014; Motevalizadeh et al., 2018) have also been used.

12.10.5 Correlation

Correlation is a method of analysis used to study the possible association between two continuous variables. This parametric test requires both data sets to consist of independent random variables that conform to normal distribution. A measure of the correlation is the correlation coefficient (r) that measures the degree of linear association between the two sets of data, and its value lies between -1 and $+1$. The closer the absolute value, $|r|$, is to 1, the stronger the correlation between the data values. The correlation between two variables is positive if high values for one variable are associated with high values for the other variable and negative if one variable is low when the other is high. A correlation coefficient close to 0 indicates that there is no linear relation, or at best a very weak correlation, between the two variables (Granato et al., 2014). To achieve an adequate assessment of the effect in correlation studies, certain features must be present: sufficient range of values, independent readings, and linear relationship. If the relationship is not linear then the coefficient will not be valid. Correlation analysis is commonly applied in sensory and instrumental data (Bower, 2009).

Thomareis et al. (2013), Chatziantoniou et al. (2015), Ghorbel et al. (2016), and Chatziantoniou et al. (2019) used correlation analysis in order to determine the strength of association between data sets of processed cheese.

12.10.6 Regression analysis

Regression analysis is used to examine the relationship between sets of dependent and independent variables. Regression is a natural extension of the results of a correlation analysis. Regression describes the efficacy of fit of a model, while correlation estimates the linear relationship of two variables (Bower, 2009; Granato et al., 2014).

a) *Simple regression*: Regression analysis calculates a mathematical expression of predictive ability and allows an accurate plot of a line on the graph. In regression, a clear distinction must be made on which variable is predicting the other. The predictor variable (or explanatory variable) is designated as the independent variable and the other as the dependent or response variable. In simple linear regression, the equation is of the form $Y = a + b\,X$, where X is the predictor variable and Y is the dependent variable. There are two coefficients in the equation: a is the intercept and b represents the slope of the line. These coefficients are determined during the regression analysis along with the plotting of the best-fit line using the method of least squares. This ensures that the experimental points are at a minimum distance from the line. An overall measure of the strength of the regression is given by the coefficient of determination (R^2). The closer this is to 1, the stronger the effect (Bower, 2009).

   A simple linear regression was performed by Kycia et al. (2006) and Thomareis et al. (2013), in order to analyze the experimental data from the evaluation of processed cheese.
b) *Multiple regression*: This method is similar to simple regression, it involves one dependent variable, but two or more independent variables. The relationship between these and the dependent variable is established. On analysis, prediction of the dependent variable is possibly based on several measures and thus the model can be more effective. Linearity is assumed in the basic multiple regression model $Y = a + b_1\,X_1 + b_2\,X_2 + \cdots + b_k\,X_k$, with k independent variables. Each independent variable is analyzed for its contribution to variation in the dependent variable. Any that contribute little can be removed and the model reassessed for the fit to improve predictive ability. An overall measure of the strength of the regression is given by the coefficient of multiple determination (R^2). The closer this is to 1, the better the regression equation that fits the data (Bower, 2009).

   Dimitreli and Thomareis (2008), Shirashoji et al. (2010), Buňka et al. (2012, 2013), and Ghorbel et al. (2016) used linear multiple regression to evaluate experimental data of processed cheeses analysis.

12.10.7 Principal component analysis

Principal component analysis (PCA) is a form of multivariate analysis. This method has received extensive application in food science because of its ability to reduce large data matrices to more readable formats. It also provides a useful platform on which to base understanding of the multivariate concept. The procedure is based on the fact that when there are many measures on a particular object then some of these are likely to be correlated. A composite variable derived from inter-correlated variables can constitute the first component, explaining the highest percentage of the total variation in the original variables. A second component, uncorrelated with the first, can then be derived to examine more variation (the next highest percentage), and so on. The analysis is often based on derivation of measures from a correlation matrix, which displays the inter-correlation of all the variables. The component matrix gives the loading of each variable on the components. A high magnitude (close to +1 or −1) for loading means that the variable is highly correlated with the component, but >0.3 can be enough for importance (Bower, 2009).

PCA has been used by Sunesen et al. (2002) and Bayarri et al. (2012) for the evaluation of processed cheese properties.

12.10.8 Response surface methodology

Response surface methodology (RSM) includes optimization procedures for the settings of factorial variables, such that the response reaches a desired maximum or minimum value. The response is in effect modeled by factorial techniques and ANOVA, but these are extended for more detailed modeling of the effects. RSM is based on the factorial study results (screening, then three-level factorial), and is a type of augmentation where extra treatments are added to focus the effects and improve the predictive power of the model. The additional treatments are located within the factorial space (center point) and outwith it (star points) – such a structure is known as a central composite design. On analysis by multiple regression, the enhanced model is produced and the equation can be used to plot the response surface. Contour plots reveal not only optimum levels for factors, but also regions where a response shows the same magnitude. RSM is suitable for the formulation of food products in food development (Bower, 2009).

Mixture designs are a type of response surface methodology where the combination of all the quantitative factors must add up to 100%. All foods are mixtures and this would seem an ideal model for formulation. A difficulty arises in that in a mixture design each quantity can be present in amounts from 0 to 100%, which cannot apply to all food ingredients in some formulae, thus more complex analysis is required (Bower, 2009).

RSM was used by Shirashoji et al. (2006, 2010) as a tool for the investigation of the effect of emulsifying salt concentration and cooking time on the physicochemical and rheological properties of processed cheese. RSM was also applied to optimize a pizza cheese (Motevalizadeh et al., 2018).

12.11 Concluding remarks and further perspectives

These are the main techniques already applied to processed cheeses and their raw materials. They were categorized according to their relevance or purpose. Others are very simple methods and use rudimentary equipment or almost unskilled personnel, while others are complex methods and require sophisticated devices or highly skilled operators. These are methods necessary for production, quality control, quality improvement, new product development, as well as scientific research. Most of these techniques are horizontal and apply to all types of food. Many of them are used to meet the requirements of food legislation, while all are aimed at consumer satisfaction.

The implementation of HACCP in the food industry, along with the general need to save time, increasingly requires the use of rapid analytical methods. To date, many methods of this type, mainly chemical and microbiological, have been developed, in particular proprietary methods. However, still other rapid and automated methods and techniques, of all the categories described above, need to be developed. These techniques must be quantitative and inexpensive and, at the same time, of high precision and accuracy.

References

Abd El-Wahed, E.M., Hassanien, M.F.R., 2019. Carpathian J. Food Sci. Technol. 11 (1), 135–148.

Abou El-Nour, A., Scheurer, G.J., Buchheim, W., 1998. Milchwissenschaft 53, 686–690.

Abou El-Nour, A., Scheurer, G.J., Omar, M.M., Buchheim, W., 1996. Milchwissenschaft 51, 684–687.

Adhikari, K., Cole, A., Grün, I., Heymann, H., Hsiehd, F.H., Huff, H., 2009. J. Sci. Food Agric. 89, 1428–1433.

Anonymous, 2002. Methods of Moisture Content Determination - Application Brochure. Mettler Toledo GmbH, Greifensee, Switzerland, p. 67.

Anonymous, 2019. Compositional Analysis - Dairy Industry. CEM Corporation, Matthews, North Carolina, USA, p. 18.

AOAC, 2016. Official Methods of Analysis of AOAC International, 20th Edition Association of Official Analytical Chemists International, Rockville, Maryland, USA Latimer, G.W. Jr (Ed.).

Apostolopoulos, C., 1994. J. Dairy Res. 61 (3), 405–413.

Apostolopoulos, C., Marshall, R.J., 1994. J. Food Qual. 17 (2), 115–128.

Ardö, Y., Polychroniadou, A., 1999. Laboratory Manual for Chemical Analysis of Cheese. European Communities, Luxemburg.

Arnott, D.R., Morris, H.A., Combs, W.B., 1957. J. Dairy Sci. 40 (8), 957–963.

Auty, M.A.E., 2018. Microscopy techniques for dairy products – An introduction. Ch. 1. In: El-Bakry, M., Sanchez, A., Mehta, B.M. (Eds.), Microstructure of Dairy Products. John Wiley & Sons Ltd, Sussex, UK, pp. 1–32.

Awad, R.A., Abdel-Hamid, L.B., El-Shabrawy, S.A., Singh, R.K., 2002. LWT - Food Sci. Technol. 35 (1), 54–61.

Bayarri, S., Carbonell, I., Costell, E., 2012. J. Dairy Sci. 95, 6926–6936.

Bourne, M.C., 2002. Food Texture and Viscosity - Concept and Measurement, 2nd Edition. Academic Press, San Diego, California, USA, p. 423.

Bower, J.A., 2009. Statistical Methods for Food Science. Introductory Procedures for the Food Practitioner, 1st Edition. Blackwell Publishing Ltd, Chichester, Oxford, Iowa, p. 307.

Brandt, M.A., Skinner, E.Z., Coleman, J.A., 1963. J. Food Sci. 28 (4), 404–409.

Breene, W.M., Price, W.V., Ernstrom, C.A., 1964. J. Dairy Sci. 47 (11), 1173–1180.

Breidinger, S.L., Steffe, J.F., 2001. J. Food Sci. 66 (3), 453–456.

Brummel, S.E., Lee, K., 1990. J. Food Sci. 55 (5), 1290–1292.

Bubelová, Z., Tremlová, B., Buňková, L., Pospiech, M., Vítová, E., Buňka, F., 2015. J. Food Sci. Technol. 52 (8), 4985–4993.

Buňka, F., Doudová, L., Weiserová, E., Kuchař, D., Michálek, J., Slavíková, S., Kráčmar, S., 2012. Int. J. Food Sci. Technol. 47 (10), 2063–2071.

Buňka, F., Doudová, L., Weiserová, E., Kuchař, D., Ponížil, P., Začalová, D., Nagyová, G., Pachlová, V., Michálek, J., 2013. Int. Dairy J. 29, 1–7.

Bütikofer, U., Rüegg, M., Ardö, Y., 1993. Lebensmittel - Wissenschaft und - Technologie 26, 271–275.

Butts, K., 2015. Qual. Mag. 54 (2), 18–19.

Campanella, O.H., Popplewell, M.L., Rosenau, J.R, Peleg, M., 1987. J. Food Sci. 52 (5), 1249–1251.

Casiraghi, E.M., Bagley, E.B., Christianson, D.D., 1985. J. Texture Stud. 16 (3), 281–301.

Černíková, M., Buňka, F., Pavlínek, V., Březina, P., Hrabě, J., Valašek, P., 2008. Food Hydrocolloids 22, 1054–1061.

Černíková, M., Buňka, F., Pospiech, M., Tremlová, B., Hladkác, K., Pavlínek, V., Březina, P., 2010. Int. Dairy J. 20 (5), 336–343.

Černíková, M., Nebesářová, J., Salek, R.N., Popková, R., Buňka, F., 2018. J. Dairy Sci. 101, 2956–2962.

Černíková, M., Nebesářová, J., Salek, R.N., Řiháčková, L., Buňka, F., 2017. J. Dairy Sci. 100, 4300–4307.

Chatziantoniou, S.E., 2020. Development and Study of Processed Whey Cheeses. Ph.D. Thesis. University of Ioannina, Ioannina, Greece, p. 276.

Chatziantoniou, S.E., Thomareis, A.S., Kontominas, M.G., 2015. Eur. Food Res. Technol. 241, 737–748.

Chatziantoniou, S.E., Thomareis, A.S., Kontominas, M.G., 2019. Eur. Food Res. Technol. 245, 2401–2412.

Chen, L., Liu, H., 2012. J. Dairy Sci. 95 (9), 4823–4830.

Childs, J.L., Daubert, C.R., Stefanski, L., Foegeding, E.A., 2007. J. Dairy Sci. 90 (5), 2163–2174.

Chudy, S., Gierałtowska, U., 2020. Int. J. Dairy Sci. 15 (2), 108–113.

CIE, 2018. Colorimetry, Technical Report CIE 015:2018, 4th Edition. International Commission on Illumination, Vienna, Austria, p. 111.

Clarke, A.R., Eberhart, C.N., 2002. Microscopic Techniques for Materials Science. CRC Press, Boca Raton, Boston, New York, Washington, p. 424 and Woodhead Publishing Limited, Cambridge.

Corradini, M.G., Peleg, M., 2005. J. Texture Stud. 36, 605–629.

Coutouly, A., Riaublanc, A., Axelos, M., Gaucher, I., 2014. Dairy Sci. Technol. 94, 125–144.

Darby, R., 1984. Emulsion rheology. In: Lissant, K.J. (Ed.), Emulsion and Emulsion Rheology, Part III. Marcel Dekker, New York, Basel.

Darby, R., 1996. Chemical Engineering Fluid Mechanics. Marcel Dekker, New York, Basel, Hong Kong.

Dimitreli, G., 2006. Ph.D. Thesis. Lincoln University, UK, p. 279.

Dimitreli, G., Thomareis, A.S., 2004. J. Food Eng. 64, 265–271.

Dimitreli, G., Thomareis, A.S., 2007. J. Food Eng. 79, 1364–1373.

Dimitreli, G., Thomareis, A.S., 2008. J. Food Eng. 84, 368–374.

Dimitreli, G., Thomareis, A.S., 2009. Int. J. Food Prop. 12, 261–275.

Dimitreli, G., Thomareis, A.S., Smith, P.G., 2005. Int. J. Food Eng. 1 (4), 1–17 Article 2.

Drake, M.A., 2007. J. Dairy Sci. 90 (11), 4925–4937.

Dus, C., Stapleton, L., Trail, A., Retiveau-Krogmann, A., Civille, G.V., 2018. Spectrum method. In: Kemp, S.E., Hort, J., Hollowood, T. (Eds.), Descriptive Analysis in Sensory Evaluation, Ch. 9. Wiley-Blackwell, Hoboken, New Jersey, USA, p. 319.

El-Bakry, M., Duggan, E., O'Riordan, E.D., O'Sullivan, M., 2011. J. Food Eng. 103 (2), 179–187.
El-Bakry, M., Sheehan, J., 2014. J. Food Eng. 125, 84–96.
El-Shibiny, S., Shenana, M.E., El-Nagar, G.F., Abdou, S.M., 2007. Int. J. Dairy Sci. 2 (1), 13–22.
Elhassan, M.S.M., 2016. Ph.D. Thesis. University of Pretoria, South Africa.
Everard, C.D., O'Donnell, C.P., Fagan, C.C., Sheehan, E.M., Delahunty, C.M., O'Callaghan, D.J., 2005. Int. J. Food Prop. 8, 267–275.
Ferry, J.D., 1980. Viscoelastic Properties of Polymers, 3rd Edition. John Wiley & Sons, New York, Chichester, Brisbane, Toronto, Singapore, p. 672.
Fife, R.L., McMahon, D.J., Oberg, C.J., 2002. J. Dairy Sci. 85 (12), 3539–3545.
Fox, P.F., Guinee, T.P., Cogan, T.M., McSweeney, P.L.H., 2000. Fundamentals of Cheese Science. Aspen Publishers, Gaithersburg, Maryland, p. 588.
Frühbauerová, M., Červenka, L., Hájek, T., Salek, R.N., Velichová, H., Buňka, F., 2020. Potravinarstvo Slovak J. Food Sci. 14, 230–238.
Gallagher, D., Cuppett, J., 2007. Water Sci. Technol. 55 (5), 69–75.
Garry, P., 2010. Méthodes Rapides pour la Recherche ou le Dénombrement des Bactéries Pathogènes. IFIP, Rennes, Toulouse, Maisons-Alfort, France, p. 25.
Geng, X.L., van den Berg, F.W.J., Bager, A.N., Ipsen, R., 2011. Int. Dairy J. 21 (9), 711–717.
Ghorbel, D., Ben Bettaïeb, N., Ghrib, F., Ben Slema, M., Attia, H., 2016. Int. J. Food Prop. 19, 1513–1521.
Giri, A., Kanawjia, S.K., Singh, M.P., 2017. J. Food Sci. Technol. 54 (8), 2443–2451.
Gliguem, H., Ghorbel, D., Grabielle-Madelmont, C., Goldschmidt, B., Lesieur, S., Attia, H., Ollivon, M., Lesieur, P., 2009. J. Therm. Anal. Calorim. 98, 73–82.
Goh, S.M., Charalambides, M.N., Williams, J.G., 2005. Eng. Fract. Mech. 72 (6), 931–946.
Granato, D., de Araújo Calado, V.A., Jarvis, B., 2014. Food Res. Int. 55, 137–149.
Guinee, T.P., 2002. Cheese rheology. In: Roginski, H., Fuquay, J.W., Fox, P.F. (Eds.), Encyclopedia of Dairy Sciences. Academic Press, Elsevier Science, London, UK.
Guinee, T.P., Kilcawley, K.N., 2004. Cheese as an ingredient. In: Fox, P.F., McSweeney, P.L.H., Cogan, T.M., Guinee, T.P. (Eds.), Cheese: Chemistry, Physics and Microbiology, 3rd Edition 2. Elsevier Academic Press, Amsterdam, Netherlands, pp. 395–428.
Guinee, T.P., O'Callaghan, D.J., 1997. J. Food Eng. 31, 147–161.
Gunasekaran, S., Ak, M.M., 2003. Cheese Rheology and Texture. CRC Press, Boca Raton, FL, USA.
Gupta, S.K., Karahadian, C., Lindsay, R.C., 1984. J. Dairy Sci. 67 (4), 764–778.
Gupta, V.K., Reuter, H., 1993. Lait 73, 381–388.
Ha, H.K., Lee, M.R., Noh, C.W., Kim, J.W., Lee, W.J., 2013. J. Agric. Life Sci. 47 (6), 219–225.
Subramanian, H., Vivekanandan, K., 2020. Int. J. Market. Human Resour. Manage. 11 (1), 16–26.
Harvey, C.D., Morris, H.A., Jenness, R., 1982. J. Dairy Sci. 65, 2291–2295.
Hassan, A.N., Awad, S., 2005. J. Dairy Sci. 88 (12), 4214–4220.
Heiland, W.K., Konstance, R.P., Flanagan, J.F., Craig, J.C., Holsinger, V.H., 1988. J. Dairy Sci. 71 (9), 2388–2394.
Hennelly, P.J., Dunne, P.G., O'Sullivan, M., O'Riordan, D., 2005. Eur. Food Res. Technol. 220 (3-4), 415–420.
Hicsasmaz, Z., Shippelt, L., Rizvi, S.S.H., 2004. J. Dairy Sci. 87 (7), 1993–1998.
Hoffmann, W., 2016. Dairy Sci. Technol. 96, 251–259.
ISO, 1998. International Standard ISO 15214, 1st Edition International Organization for Standardization, Geneva, Switzerland.

ISO, 1999a. International Standard ISO 6888-1, 1st Edition International Organization for Standardization, Geneva, Switzerland.
ISO, 1999b. International Standard ISO 6888-2, 1st Edition International Organization for Standardization, Geneva, Switzerland.
ISO, 2001a. International Standard ISO 16649-1, 1st Edition International Organization for Standardization, Geneva, Switzerland.
ISO, 2001b. International Standard ISO 16649-2, 1st Edition International Organization for Standardization, Geneva, Switzerland.
ISO, 2002a. International Standard ISO 5765-1|IDF 79-1, 1st Edition International Organization for Standardization, Geneva, Switzerland.
ISO, 2002b. International Standard ISO 5765-2|IDF 79-2, 1st Edition International Organization for Standardization, Geneva, Switzerland.
ISO, 2002c. International Standard ISO 15884|IDF 182, 1st Edition International Organization for Standardization, Geneva, Switzerland.
ISO, 2002d. International Standard ISO 15885|IDF 184, 1st Edition International Organization for Standardization, Geneva, Switzerland.
ISO, 2003a. International Standard ISO 6888-3, 1st Edition International Organization for Standardization, Geneva, Switzerland corrected version.
ISO, 2003b. International Standard ISO 15213, 1st Edition International Organization for Standardization, Geneva, Switzerland.
ISO, 2004a. International Standard ISO 1735|IDF 5, 3rd Edition International Organization for Standardization, Geneva, Switzerland.
ISO, 2004b. International Standard ISO 5534|IDF 4, 2nd Edition International Organization for Standardization, Geneva, Switzerland.
ISO, 2004c. International Standard ISO 7932, 3rd Edition International Organization for Standardization, Geneva, Switzerland.
ISO, 2004d. International Standard ISO 7937, 3rd Edition International Organization for Standardization, Geneva, Switzerland.
ISO, 2004e. International Standard ISO 14673-1|IDF 189-1, 2nd Edition International Organization for Standardization, Geneva, Switzerland.
ISO, 2004f. International Standard ISO 14673-2|IDF 189-2, 2nd Edition International Organization for Standardization, Geneva, Switzerland.
ISO, 2004g. International Standard ISO 14673-3|IDF 189-3, 2nd Edition International Organization for Standardization, Geneva, Switzerland.
ISO, 2005. International Standard ISO 6658, 2nd Edition International Organization for Standardization, Geneva, Switzerland.
ISO, 2006a. International Standard ISO 3976|IDF 74, 2nd Edition International Organization for Standardization, Geneva, Switzerland.
ISO, 2006b. International Standard ISO 5943|IDF 88, 4th Edition International Organization for Standardization, Geneva, Switzerland.
ISO, 2006c. International Standard ISO 21543|IDF 201, 1st Edition International Organization for Standardization, Geneva, Switzerland.
ISO, 2006d. International Standard ISO 21871, 1st Edition International Organization for Standardization, Geneva, Switzerland.
ISO, 2006e. Technical Specification ISO/TS 2963|IDF/RM 34, 1st Edition International Organization for Standardization, Geneva, Switzerland.
ISO, 2006f. Technical Specification ISO/TS 17996|IDF/RM 205, 1st Edition International Organization for Standardization, Geneva, Switzerland.

ISO, 2007a. International Standard ISO 7218. 3rd Edition International Organization for Standardization, Geneva, Switzerland.
ISO, 2007b. International Standard ISO 8070|IDF 119, 2nd Edition International Organization for Standardization, Geneva, Switzerland.
ISO, 2008a. International Standard ISO 707|IDF 50, 3rd Edition International Organization for Standardization, Geneva, Switzerland.
ISO, 2008b. International Standard ISO 3432|IDF 221, 2nd Edition International Organization for Standardization, Geneva, Switzerland.
ISO, 2008c. International Standard ISO 3433|IDF 222, 2nd Edition International Organization for Standardization, Geneva, Switzerland corrected version.
ISO, 2008d. International Standard ISO 5492. 2nd Edition Amendment 1:2016 International Organization for Standardization, Geneva, Switzerland.
ISO, 2008e. International Standard ISO 9231|IDF 139, 1st Edition International Organization for Standardization, Geneva, Switzerland.
ISO, 2008f. International Standard ISO 21527-1, 1st Edition International Organization for Standardization, Geneva, Switzerland.
ISO, 2008g. International Standard ISO 21527-2, 1st Edition International Organization for Standardization, Geneva, Switzerland.
ISO, 2009a. International Standard ISO 22935-1|IDF 99-1, 1st Edition International Organization for Standardization, Geneva, Switzerland.
ISO, 2009b. International Standard ISO 22935-2|IDF 99-2, 1st Edition International Organization for Standardization, Geneva, Switzerland.
ISO, 2009c. International Standard ISO 22935-3|IDF 99-3, 1st Edition International Organization for Standardization, Geneva, Switzerland.
ISO, 2009d. Technical Specification ISO/TS 27106|IDF/RM 217, 1st Edition International Organization for Standardization, Geneva, Switzerland.
ISO, 2010. International Standard ISO 2962|IDF 33, 3rd Edition International Organization for Standardization, Geneva, Switzerland.
ISO, 2011a. International Standard ISO 22118. 1st Edition International Organization for Standardization, Geneva, Switzerland.
ISO, 2011b. International Standard ISO 27871|IDF 224, 1st Edition International Organization for Standardization, Geneva, Switzerland.
ISO, 2012. Technical Specification ISO/TS 6579-2, 1st Edition International Organization for Standardization, Geneva, Switzerland.
ISO, 2013a. International Standard ISO 4833-1, 1st Edition International Organization for Standardization, Geneva, Switzerland.
ISO, 2013b. International Standard ISO 4833-2, 1st Edition International Organization for Standardization, Geneva, Switzerland.
ISO, 2014a. International Standard ISO 8968-1|IDF 20-1, 2nd Edition International Organization for Standardization, Geneva, Switzerland.
ISO, 2014b. International Standard ISO 11133, 1st Edition International Organization for Standardization, Geneva, Switzerland corrected version.
ISO, 2014c. Technical Report ISO/TR 6579-3, 1st Edition International Organization for Standardization, Geneva, Switzerland.
ISO, 2015a. International Standard ISO 16649-3, 1st Edition International Organization for Standardization, Geneva, Switzerland corrected version.
ISO, 2015b. International Standard ISO 16958|IDF 231, 1st Edition International Organization for Standardization, Geneva, Switzerland.

ISO, 2016. International Standard ISO 16140-2, 1st Edition International Organization for Standardization, Geneva, Switzerland.

ISO, 2017a. International Standard ISO 6579-1, 1st Edition International Organization for Standardization, Geneva, Switzerland.

ISO, 2017b. International Standard ISO 6658, 3rd Edition International Organization for Standardization, Geneva, Switzerland.

ISO, 2017c. International Standard ISO 11290-1, 2nd Edition International Organization for Standardization, Geneva, Switzerland.

ISO, 2017d. International Standard ISO 11290-2, 2nd Edition International Organization for Standardization, Geneva, Switzerland.

ISO, 2017e. International Standard ISO 18787, 1st Edition International Organization for Standardization, Geneva, Switzerland.

ISO, 2018a. International Standard ISO 9233-1|IDF 140-1, 2nd Edition International Organization for Standardization, Geneva, Switzerland.

ISO, 2018b. International Standard ISO 9233-2|IDF 140-2, 2nd Edition International Organization for Standardization, Geneva, Switzerland.

ISO, 2018c. International Standard ISO 13301, 2nd Edition International Organization for Standardization, Geneva, Switzerland.

ISO, 2018d. International Standard ISO 15151|IDF 229, 1st Edition International Organization for Standardization, Geneva, Switzerland.

ISO, 2019. International Standard ISO 17410, 2nd Edition International Organization for Standardization, Geneva, Switzerland.

ISO/CIE, 2019. International Standard ISO/CIE 11664-4, 1st Edition International Organization for Standardization, Geneva, Switzerland.

Joshi, N.S., Jhala, R.P., Muthukumarappan, K., Acharya, M.R., Mistry, V.V., 2004. Int. J. Food Prop. 7 (3), 519–530.

Kalab, M., 1984. Food Microstruc. 3, 95–111.

Kalab, M., Yun, J., Yiu, S.H., 1987. Food Microstruc. 6, 181−192.

Kapoor, R., Metzger, L.E., 2008. Compr. Rev. Food Sci. Food Saf. 7, 194–214.

Kemp, S.E., Hollowood, T., Hort, J., 2009. Sensory Evaluation - A Practical Handbook. Wiley-Blackwell, Chichester, West Sussex, UK, p. 196.

Kemp, S.E., Ng, M., Hollowood, T., Hort, J., 2018. Introduction to descriptive analysis. In: Kemp, S.E., Hort, J., Hollowood, T. (Eds.), Descriptive Analysis in Sensory Evaluation, Ch. 1. Wiley-Blackwell, Hoboken, New Jersey, USA, pp. 3–39.

Kindstedt, P.S., Fox, P.F., 1991. J. Food Sci. 56 (4), 1115–1116.

Kindstedt, P.S., Rippe, J.K., 1990. J. Dairy Sci. 73 (4), 867–873.

Kirchmeier, O., 1980. Milchwissenschaft 35, 75–77.

Kirchmeier, O., Weiss, G., Kiermeier, F., 1978. Zeitschrift für Lebensmittel-Untersuchung und Forschung 166, 212–220.

Kosikowski, F.V., 1977. Process cheese and related types. In: Kosikowski, F.V., Mistry, V. (Eds.), Cheese and Fermented Milk Foods, 1. Edwards Brothers Inc, Ann Arbor, Michigan, USA, pp. 382–406.

Kycia, K., Pluta, A., Zmarlicki, S., 2006. Polish J. Food Nutr. Sci. 56 (special issue 1s), 113–118.

Lawless, H.T., Heymann, H., 2010. Sensory Evaluation of Food - Principles and Practices, 2nd Edition. Springer, New York, USA, p. 596.

Lazárková, Z., Buňka, F., Buňková, L., Valášek, P., Kráčmar, S., Hrabě, J., 2010. Czech J. Food Sci. 28 (3), 168–176.

Le Tohic, C., O'Sullivan, J.J., Drapala, K.P., Chartrin, V., Chan, T., Morrison, A.P., Kerry, J.P., Kelly, A.L., 2018. J. Food Eng. 220, 56–64.

Lee, B.O., Paquet, D., Alais, C., 1979. Lait 59, 589–596.
Lee, S.K., Anema, S., Klostermeyer, H., 2004. Int. J. Food Sci. Technol. 39 (7), 763–771.
Lee, S.K., Buwalda, R.J., Euston, S.R., Foegeding, E.A., McKenna, A.B., 2003. Lebensmittel - Wissenschaft und - Technologie 36 (3), 339–345.
Lee, S.K., Huss, M., Klostermeyer, H., Anema, S.G., 2013. Int. Dairy J. 32, 79–88.
Lee, S.K., Klostermeyer, H., Anema, S.G., 2015. Int. Dairy J. 50, 15–23.
Marshall, R.J., 1990. J. Sci. Food Agric. 50, 237–252.
Mathew, S., Raman, M., Parameswaran, M.K., Rajan, D.P., 2019. Fish and Fishery Products Analysis - A Theoretical and Practical Perspective. Springer, Singapore.
McClements, D.J., 1999. Food Emulsions - Principles, Practice and Techniques. CRC Press LLC, Boca Raton, London, New York, Washington.
Mokrzycki, W.S., Tatol, M., 2011. Mach. Graph. Vis. 20 (4), 383–411.
Montesinos-Herrero, C., Cottell, D.C., O'Riordan, E.D., O'Sullivan, M., 2006. Int. Dairy J. 16 (8), 910–919.
Morsy, T.A., Mohamed, A.G., Kholif, S.M., Gafour, W.A., 2014. Int. J. Dairy Sci. 9 (3), 74–81.
Motevalizadeh, E., Mortazavi, S.A., Milani, E., Hooshmand-Dalir, M.A.R., 2018. Food Sci. Nutr. 6 (2), 356–372.
Mounsey, J.S., O'Riordan, E.D., 1999. J. Food Sci. 64 (4), 701–703.
Mulvihill, D.M., McCarthy, A., 1994. Int. Dairy J. 4, 15–23.
Muthukumarappan, K., Wang, Y.C., Gunasekaran, S., 1999. J. Dairy Sci. 82, 1068–1071.
Nanakoudis, A., 2018. SEM and TEM? What's the difference?. Lambda News https://www.lambdaphoto.co.uk/news/2018/02/21.
Neilson, A.J., Ferguson, V.B., Kendall, D.A., 1988. Profile methods - Flavor profile and profile attribute analysis. In: Moskowitz, H.R. (Ed.), Applied Sensory Analysis of Foods, Vol. 1, Ch. 2, CRC Press, Boca Raton, Florida, USA, pp. 21–41.
Noronha, N., Duggan, E., Ziegler, G.R., Stapleton, J.J., O'Riordan, E.D., O'Sullivan, M., 2008. Food Res. Int. 41 (5), 472–479.
Olson, N.F., Price, W.V., 1958. J. Dairy Sci. 41 (7), 999–1000.
Palmas, F., Cosentino, S., Fadda, M.E., Deplano, M., Mascia, V., 1999. Lait 79, 607–613.
Peleg, M., Normand, M.D., 1983. Rheol. Acta 22 (1), 108–113.
Pereira, R.B., 2000. Ph.D. Thesis. Massey University, New Zealand, p. 325.
Piska, I., Štětina, J., 2004. J. Food Eng. 61, 551–555.
Plog, J.P., 2014. Appl. Rheol. 24 (4), 10–11.
Prow, L.A., Metzger, L.E., 2005. J. Dairy Sci. 88 (4), 1277–1287.
Rao, V.N.M., 1992. Classification, description and measurement of viscoelastic properties of solid foods. In: Rao, M.A., Steffe, J.F. (Eds.), Viscoelastic Properties of Foods, Ch. 1. Elsevier Applied Science London, New York, pp. 3–47.
Reparet, J.M., Noël, Y., 2003. Lait 83, 321–333.
Rincón Cardona, J.A., Huck Iriart, C., Herrera, M.L, 2013. Applications of confocal laser scanning microscopy (CLSM) in foods. In: Lagali, N. (Ed.), Confocal Laser Microscopy - Principles and Applications in Medicine, Biology, and the Food Sciences, Ch. 11. IntechOpen, London, UK, pp. 203–234.
Rüegg, M., Eberhard, P., Popplewell, L.M., Peleg, M., 1991. Melting properties of cheese. In: Rheological and Fracture Properties of Cheese. Bulletin of the International Dairy Federation No 268. FIL-IDF, Brussels.
Sádlíková, I., Buňka, F., Budinský, P., Barbora, V., Pavlínek, V., Hoza, I., 2010. LWT - Food Sci. Technol. 43, 1220–1225.
Shirashoji, N., Jaeggi, J.J., Lucey, J.A., 2006. J. Dairy Sci. 89 (1), 15–28.

Shirashoji, N., Jaeggi, J.J., Lucey, J.A., 2010. J. Dairy Sci. 93 (7), 2827–2837.
Shukla, A., Rizvi, S.S.H., Bartsch, J.A., 1995. J. Texture Stud. 26, 313–323.
Sidel, J.L., Bleibaum, R.N., Tao, K.W.C., 2018. Quantitative descriptive analysis. In: Kemp, S.E., Hort, J., Hollowood, T. (Eds.), Descriptive Analysis in Sensory Evaluation, Ch. 8. Wiley-Blackwell, Hoboken, New Jersey, USA, pp. 287–318.
Sołowiej, B., Dylewska, A., Kowalczyk, D., Sujka, M., Tomczyńska-Mleko, M., Mleko, S., 2016. Eur. Food Res. Technol. 242, 1577–1585.
Stampanoni, C.R., Noble, A.C., 1991. J. Texture Stud. 22, 381–392.
Steffe, J.F., 1996. Rheological Methods in Food Process Engineering, 2nd Edition. Freeman Press, East Lansing, p. 418.
Stone, H., Sidel, J.L., 2004. Sensory Evaluation Practices, 3rd Edition. Elsevier Academic Press, San Diego, London, p. 408.
Sunesen, L.O., Lund, P., Sørensen, J., Hølmer, G., 2002. LWT - Food Sci. Technol. 35 (2), 128–134.
Sutheerawattananonda, M., Bastian, E.D., 1998. J. Texture Stud. 29, 169–183.
Sutheerawattananonda, M., Fulcher, R.G., Martin, F.B., Bastian, E.D., 1997. J. Dairy Sci. 80 (4), 620–627.
Swenson, B.J., Wendorff, W.L., Lindsay, R.C., 2000. J. Food Sci. 65 (5), 822–825.
Szczesniak, A.S., Brandt, M.A., Friedman, H.H., 1963. J. Food Sci. 28 (4), 397–403.
Tabilo-Munizaga, G., Barbosa-Cánovas, G.V., 2005. J. Food Eng. 67, 147–156.
Tamime, A.Y., Kalab, M., Davies, G., Younis, M.F., 1990. Food Struct. 9, 23–37.
Tamime, A.Y., Muir, D.D., Shenana, M.E., Kalab, M., Dawood, A.H., 1999. Lebensmittel Wissenschaft und – Technologie 32, 50–59.
Tamime, A.Y., Muir, D.D., Wszolek, M., Domagala, J., Metzger, L., Kneifel, W., Durrschmid, K., Domig, K.J., Hill, A., Smith, A., Guinee, T.P., Auty, M.A.E, 2011. Quality control in processed cheese manufacture. In: Tamime, A.Y. (Ed.), Processed Cheese and Analogues, 1st Edition. Ch. 10. Blackwell Publishing Ltd, Oxford, Chichester, Iowa, pp. 245–340.
Thomareis, A.S., 2020. Unpublished data. International Hellenic University, Thessaloniki, Greece.
Thomareis, A.S., Chatziantoniou, S., Kotsoglou, N., Zagana, D., 2013. International Conference on Food and Biosystems Engineering (FaBE 2013). Skiathos Island, Greece.
Thomas, M.A., 1973. Austr. J. Dairy Technol. 28, 77–80.
Thomas, M.A., Newell, G., Abad, G.A., Turner, A.D., 1980. J. Food Sci. 45, 458–459.
Trivedi, D., Bennett, R.J., Hemar, Y., Reid, D.C.W., Lee, S.K., Illingworth, D., 2008. Int. J. Food Sci. Technol. 43 (12), 2191–2196.
U.S. Department of Agriculture, 1980. USDA Specifications for Mozzarella Cheeses. Agric. Marketing Service, USDA, Washington, USA.
van Vliet, T., 1991. Terminology to be used in cheese rheology. In: Rheological and Fracture Properties of Cheese. Bulletin of the International Dairy Federation No 268. FIL-IDF, Brussels, pp. 5–15.
Walstra, P., 1965. The Netherlands Milk Dairy J. 19 (2), 93–109.
Wang, Y.C., Muthukumarappan, K., Ak, M.M., Gunasekaran, S., 1998. J. Texture Stud. 29, 43–55.
Whorlow, R.W., 1992. Rheological Techniques, 2nd Edition. Ellis Horwood, New York, London, Toronto, Sydney, Tokyo, Singapore, p. 460.

Quality and testing of processed cheese: defects, QA, and QC

13

Apostolos S. Thomareis and Soumela E. Chatziantoniou
Department of Food Science and Technology, International Hellenic University, Thessaloniki, Greece

13.1 Introduction

The production of safe and high-quality processed cheese is essential for the health of consumers as well as for the reliability of manufactures and the continuation of consumer trust. In this direction, manufacture of processed cheese has to rely on the implementation of quality assurance systems, based on the requirements of good manufacturing and hygiene practices (GMP, GHP), as well as of Hazard Analysis and Critical Control Point (HACCP) system.

Apart from the safety of finished products, manufacturers should also secure that processed cheese is manufactured with particular specifications and should not have defects. Defects in processed cheese, which are of physical, chemical and microbial origin, are caused by poor quality raw materials, improper processing conditions, postmanufacture handling or a combination of the above. Inconsistencies in processed cheese quality may result in downgrading of product and a consequential economic loss to manufacturers.

The objective of this chapter is to review the scientific and technological aspects of processed cheese quality assurance and control. The implementation of prerequisite programs and HACCP principles, the sampling, as well as the defects of processed cheese and solutions for their avoidance will be discussed.

13.2 Food quality and safety

Quality is the degree to which a set of inherent characteristics of an object fulfills requirements. The term "quality" can be used with adjectives, such as poor, good or excellent. Examples of what is referred to as an object may be a product, service, process, person, organization, system, and resource (ISO, 2015a). In this section, the "object" will usually be the product, that is, processed cheese. In general, a food product should be safe for human consumption and organoleptically acceptable. It should also meet customer and food law requirements. Finally, it should be chemically and microbiologically stable throughout its defined shelf life. Therefore, the inherent characteristics of a food product, such as all chemical, physical, microbiological, sensory, functional, and nutritional attributes should meet the needs and expectations of the consumer.

In order to achieve good quality and be able to continually improve it, an organization should apply quality management. International Organization for Standardization

Processed Cheese Science and Technology: Ingredients, Manufacture, Functionality, Quality, and Regulations.
DOI: https://doi.org/10.1016/B978-0-12-821445-9.00008-X

(ISO) has specified requirements for a quality management system (QMS) when an organization: (i) needs to demonstrate its ability to consistently provide products and services that meet customer and applicable statutory and regulatory requirements, and (ii) aims to enhance customer satisfaction through the effective application of the system, including processes for improvement of the system and the assurance of conformity to customer and applicable statutory and regulatory requirements. All of these requirements are generic and are intended to be applicable to any organization, regardless of its type or size, or the products, and services it provides (ISO, 2015b).

Food safety is the assurance that food will not cause an adverse health effect for the consumer when it is prepared and/or consumed in accordance with its intended use. Food safety is related to the occurrence of food safety hazards in end products (ISO, 2018). Since the safety of a food is the most important aspect of its quality, ISO has specified requirements for a food safety management system (FSMS), which is a set of interrelated or interacting elements to establish policy and objectives and to achieve those objectives, used to direct and control an organization with regard to food safety. FSMS can help an organization improve its overall performance in food safety. The potential benefits are: (i) the ability to consistently provide safe foods and products and services that meet customer and applicable statutory and regulatory requirements, (ii) the addressing risks associated with the objectives, and (iii) the ability to demonstrate conformity to specified FSMS requirements (ISO, 2018).

Quality management can include establishing quality policies and quality objectives, and processes to achieve these quality objectives through quality planning, quality assurance, quality control, and quality improvement (ISO, 2015a).

Although they are closely related and both encompass vital aspects of quality management, quality assurance and quality control are fundamentally different in their focus. Quality assurance (QA) is the part of quality management focused on providing confidence that quality requirements will be fulfilled, while quality control (QC) is the part of quality management focused on fulfilling quality requirements (ISO, 2015a).

More precisely, quality assurance is a set of activities used for ensuring quality in the processes by which products are manufactured. That is, it is process-oriented. QA is also proactive because it aims to prevent defects before they occur. It involves the prevention aspect of quality management and is usually the responsibility of the entire team in the organization.

On the contrary, quality control is a set of activities used for ensuring quality in the products during and after manufacturing. That is, it is product-oriented. QC is also reactive because it aims to identify defects after they appear. It involves the inspection aspect of quality management and is usually the responsibility of a specific product testing team. However, quality control is an integral part of quality assurance, as it is the confirmation of the effectiveness of the latter, by examination of the product.

Food businesses fulfill their responsibilities for food quality and safety by implementing QMS (e.g., ISO 9001) and FSMS (e.g., ISO 22000) throughout the entire food chain (primary production, manufacturing and processing, wholesale, retail and markets, food service, and households). These management systems can be supported by various quality and safety assurance systems such as Good Agricultural Practices (GAP), Good Manufacturing Practices (GMP), Good Storage and Distribution Practices (GSDP), Good Hygiene Practices (GHP), and HACCP. However, it is absolutely necessary to apply a series of selected quality controls at the various stages of the food chain.

Broadly defined, GAPs apply available knowledge to addressing economic, social and environmental sustainability for on-farm production and postproduction processes, resulting in safe and healthy food and nonfood agricultural products (FAO, 2003). GAPs for dairy farmers are about implementing sound practices on dairy farms, collectively called Good Dairy Farming Practices. These practices should ensure that the milk and milk products produced are safe and suitable for their intended use, and also that the dairy farm enterprise is viable into the future, from the economic, social, and environmental perspectives. Dairy farmers should apply good practice in the following areas: animal health, milking hygiene, animal feeding and water, animal welfare, environment, and socioeconomic management (FAO/IDF, 2011).

GMPs are the combination of manufacturing and quality control procedures aimed at ensuring that products are consistently manufactured to their designated specifications (IFST, 2018). They include measures such as the correct dosage of ingredients, the appropriate processing temperature, etc. (EU, 2016).

GHPs are the fundamental measures and conditions applied at any step within the food chain to provide safe and suitable food (CAC, 2020a). They include measures such as the appropriate cleaning and disinfection, the personal hygiene, etc. (EU, 2016). GMPs (IFST, 2018; LII, 2021b) and GHPs (EU, 2004a; CAC, 2020a) are complementary, but they have many common measures, and sometimes no differentiation is made between these two systems of preventive practices (EU, 2016).

Finally, HACCP is a systematic approach to the identification, evaluation, and control of food safety hazards (NACMCF, 2017). HACCP is a food safety assurance system in which safety is addressed through the analysis and control of biological, chemical, and physical hazards from raw material production, procurement and handling, to manufacturing, distribution, and consumption of the finished food product. Prior to the application of the HACCP system, the food business operator should have implemented prerequisite programs (PRP). These are programs comprising GAPs, GMPs and GHPs, as well as other practices and procedures such as training and traceability, which establish the basic environmental and operational conditions that lay the foundation of the implementation of HACCP (CAC, 2020a).

13.3 Prerequisite programs

In this section, the GHPs of the Codex Alimentarius Commission (CAC, 2020a) will be taken as prerequisite programs for the implementation of the HACCP system in processed cheese manufacture. They will be briefly mentioned, but will be supported and supplemented by other relevant guidance.

The development, implementation and maintenance of GHPs provide the conditions and activities that are necessary to support the production of safe and suitable food at all stages of the food chain, from primary production (milk) through to handling of the final product (processed cheese). However, if hazards occur or are present at levels which demonstrate that GHP procedures are not sufficient to provide safe food, it will be necessary to develop and implement a HACCP system.

13.3.1 Primary production

In dairy farms, measures can be taken to minimize the occurrence and levels of hazards in the further food chain by applying prerequisite programs.

Milk production should not be carried out in polluted areas by toxic or offensive odors or sources of contaminated water (CAC, 2001b, 2009). During rearing, milking, as well as during handling, storage and transport of milk, producers should implement measures to control contamination of milk from feedstuffs, pesticides, veterinary drugs, fecal matter, waste, pest, and other objectionable agents (CAC, 2009; FAO/IDF, 2011; ISO, 2011, 2016; ACS, 2017).

After milking, milk should be cooled immediately to not more than 8°C in the case of daily collection, or not more than 6°C if collection is not daily. During transport, the temperature of the milk should not be more than 10°C (EU, 2004c,d). Appropriate facilities and procedures should be in place to ensure that cleaning, maintenance and personal hygiene do not compromise milk safety.

13.3.2 Dairy establishment - design of facilities and equipment

As a dairy establishment, the production unit of natural cheese (raw material), as well as the production unit of processed cheese (finished product) is designated. Dairy establishment should not be located where there is a threat to food safety or suitability, such as pollution, flooding, pests and wastes, and hazards cannot be controlled by reasonable measures. The grounds of the dairy establishment should be kept in a condition that will protect against the contamination of food (LII, 2021b).

The size, construction, design and layout of the dairy establishment should permit adequate maintenance and cleaning, and prevent cross-contamination (LII, 2021b). The surfaces of walls, partitions and floors should be constructed of nontoxic and impervious materials that are easy to clean and disinfect. Walls and partitions should have a smooth surface up to a height appropriate to the operation. Floors should be constructed to allow adequate drainage and cleaning. For easier cleaning, new construction should have rounded cove at the juncture of the wall and floor in all receiving, pasteurizing, manufacturing, packaging and storage rooms. Ceilings and overhead fixtures (e.g., lighting) should be constructed to be shatterproof where appropriate, and finished to minimize condensation, the buildup of dirt, and the shedding of particles (EU, 2004b; ACS, 2017; LII, 2021a).

Windows should be easy to clean, constructed to minimize the buildup of dirt and fitted with removable and cleanable insect-proof screens. On new construction, window sills should be slanted downward at approximately a 45° angle. Doors should have smooth and nonabsorbent surfaces, and be easy to clean and disinfect. Work surfaces that come into direct contact with food should be in sound condition, durable, and easy to clean, maintain, and disinfect. They should be made of smooth, nonabsorbent materials, and inert to the food, to detergents, and to disinfectants (EU, 2004b; ACS, 2017; LII, 2021a).

Adequate drainage and waste disposal systems and facilities should be provided and well maintained. They should be designed and constructed so that contamination of food or water supply is avoided (IFST, 2018).

Suitably designated facilities should be provided for cleaning utensils and equipment. Such facilities should have an appropriate supply of hot and/or cold water, where required. Adequate hand washing, changing and toilet facilities should be available so that an appropriate degree of personal hygiene can be maintained and to avoid personnel contaminating food (IFST, 2018). Suitable facilities should be available for heating, cooling, cooking, refrigerating and freezing food, and, when necessary, controlling ambient temperatures to ensure the safety and suitability of food.

Product grading or inspection is carried out in a clean, dry room or area designated specifically for this purpose at a temperature of at least 15.5°C (LII, 2021a). Adequate means of natural or mechanical ventilation should be provided to minimize airborne contamination of food and to control odors, ambient temperature and humidity. Air filtration should be used to eliminate dirt and dust from the incoming air (ACS, 2017).

Appropriate natural or artificial lighting which does not alter food color should not be less than 540 lux in inspection areas, 220 lux in work areas and 110 lux in other areas. Where contamination of product by broken glass is possible, light bulbs and fluorescent tubes should be protected against breakage (FAO, 1998; LII, 2021a).

Separate facilities for the safe and hygienic storage of food products, food ingredients, food packaging materials and nonfood chemicals (including cleaning materials, lubricants, and fuels) should be provided. Food storage facilities should be maintained, cleaned, protected from pests and contaminations, and should have regulated food storage conditions (ISO, 2019).

Consistent with the size and type of plant and the volume of dairy products manufactured, an adequately equipped laboratory should be maintained and properly staffed with qualified and trained personnel for quality control and analytical testing (LII, 2021a). Equipment and containers coming into contact with food should be made of nontoxic and corrosion-resistant materials suitable for food contact. They should be designed, constructed and located to ensure that they can be adequately cleaned and disinfected (EU, 2004e; LII, 2021b).

Equipment used to cook, heat, cool, store, or freeze food should be designed to achieve the required food temperatures as rapidly as necessary in the interests of food safety and suitability, and to maintain food temperatures effectively.

13.3.3 Training and competence

Personnel should have the knowledge and skills necessary to enable them to handle food hygienically. Training programs should include the nature of hazards associated with the food, the manner in which the food is produced, processed, handled and packed, and the GHPs. Periodic assessments of the effectiveness of training and instruction programs should be made. Training programs should be routinely reviewed and updated where necessary (FAO, 1998).

13.3.4 Dairy establishment maintenance, cleaning and disinfection, and pest control

Dairy establishment and equipment should be maintained in an appropriate condition to facilitate all cleaning and disinfection procedures, function as intended and prevent contamination of food, such as from pests, metal shards, flaking plaster, debris, chemicals, wood, plastic, glass, and paper (ACS, 2017).

Cleaning can be carried out by the separate or the combined use of physical methods, such as heat, scrubbing, turbulent flow, and vacuum cleaning (or other methods that avoid the use of water), and chemical methods using solutions of detergents, alkalis or acids, followed by rinsing with water. For the cleaning of equipment, both methods can be used: (i) cleaning in place (CIP), that is, system where equipment is cleaned by impingement or circulation of flowing chemical solutions, cleaning liquids and water rinses into, onto and over surfaces in equipment or systems without dismantling and designed for the purpose, and (ii) cleaning out of place (COP), that is, system where equipment is disassembled and cleaned in a tank or in an automatic washer by circulating a cleaning solution and maintaining a minimum temperature throughout the cleaning cycle (ISO, 2009).

Where necessary, cleaning should be followed by chemical disinfection with subsequent rinsing, unless the manufacturer's instructions indicate that rinsing is not required. Application of cleaning and disinfection procedures should be monitored for effectiveness and periodically verified by means such as visual inspections and audits, so as to ensure the procedures have been applied properly. The type of monitoring will depend on the nature of the procedures, but could include pH, water temperature, conductivity, cleaning agent concentration, disinfectant concentration, contact microbiological swabs, ATP bioluminescence techniques, etc. (IFST, 2018).

Pests (e.g., birds, rodents, insects, etc.) pose a major threat to the safety and suitability of food. Pest infestations can occur where there are breeding sites and a supply of food. Dairy establishment should be kept in good repair and condition to prevent pest access and to eliminate potential breeding sites and harborage. Dairy establishment and surrounding areas should be regularly examined for evidence of infestation. Detectors and traps (e.g., insect light traps, bait stations) should be designed and located so as to prevent potential contamination of raw materials, products or facilities. Pest infestations should be addressed immediately by a qualified person or company and appropriate corrective action should be taken. Domestic animals should also be excluded from manufacturing and storage areas (IFST, 2018).

Food waste should be collected and stored in covered containers and should not be allowed to accumulate and overflow in food handling, food storage, and other working areas. Containers are to be of an appropriate construction, kept in sound condition, be easy to clean and, where necessary, to disinfect. All waste is to be eliminated in a hygienic and environmentally friendly way (EU, 2004b).

13.3.5 Personal hygiene

Symptoms of illnesses, for which personnel could be considered for possible exclusion from food handling and/or medical examination, include jaundice, diarrhea, vomiting, fever, sore throat with fever, visibly infected skin lesions (boils, cuts, etc.) and discharge from the ears, eyes or nose. Personnel should maintain a high degree of personal cleanliness and, where appropriate, wear suitable protective clothing, head and beard covering, gloves and footwear (LII, 2021b). When engaged in food handling activities, personnel should refrain from smoking or vaping, spitting, chewing, eating, drinking, touching the mouth, nose or other places of possible contamination, and sneezing or coughing over unprotected food. Personal effects such as jewelry, watches, pins or other items such as false nails/eye lashes should not be worn or brought into food handling areas if they pose a threat to the safety and suitability of food (LII, 2021b). Visitors to food businesses, in particular to food manufacturing, processing or handling areas, should be instructed and supervised, wear protective clothing, and adhere to the other personal hygiene provisions for personnel.

13.3.6 Control of operation

Products (processed cheese) should be described in such a way as not to compromise awareness of the hazards or other factors, such as the suitability of the products for their intended use. Processes used should be described. It may be helpful to develop a flow diagram, which shows the sequence and interaction of all processing steps in the operation, including where raw materials, ingredients and intermediate products enter the flow and where end products, byproducts and waste are released or removed. Hygiene procedures and practices relevant to the business should be effectively implemented and monitored. When monitoring results indicate a deviation, a corrective action should be taken. Verification procedures are essential.

The temperature and time of cooking, cooling, processing, and storage should be controlled. The microbiological, chemical, physical and allergen specifications used for food safety or suitability should be based on sound scientific principles and state, where appropriate, sampling parameters, methods of analysis, acceptable limits, and monitoring procedures (FAO, 1998; FDA, 2005; CAC, 2013b; ACS, 2017; CAC, 2019, 2020b).

Systems should be in place to prevent or minimize microbiological contamination of foods occurring by direct contact or indirectly by food handlers, by contact with surfaces, from cleaning equipment, by splashing or by airborne particles. Systems should be in place throughout the food chain to prevent contamination of foods by extraneous materials, such as personnel belongings, especially any hard or sharp object, for example, jewelry, glass, metal shards, bone, plastic, wood fragments, that could cause injury or present a choking hazard. Systems should be in place to prevent or minimize contamination of foods by harmful chemicals, for example, cleaning materials, nonfood grade lubricants, chemical residues from pesticides and veterinary drugs such as antibiotics. Systems should be in place to take into account the allergenic nature of some foods, as appropriate for the food business. Presence of allergens, for example, milk and dairy products, derivatives of gluten-containing cereals, etc., should be identified in raw materials, other ingredients and products.

Only raw materials and other ingredients that are fit for purpose should be used. There should be adequate stock rotation, based on the principle of "first in, first out" (FIFO). The rework (reworking product) should be stored, handled and used in such a way that safety, quality, traceability and regulatory compliance are maintained (ISO, 2009; IFST, 2018). Packaging design and materials should be safe and suitable for food use, provide adequate protection for products to minimize contamination, prevent damage, and accommodate proper labeling (ISO, 2013a).

Water and steam used as an ingredient in processed cheese as well as in contact with food and food processing equipment (e.g., for rinsing) should be potable water and not contaminate the cheese (EU, 2004b; ACS, 2017; WHO, 2017; FAO/WHO, 2019). Tables 13.1 and 13.2 show the maximum acceptable concentrations of microbiological and chemical contaminants, respectively, in water intended for human consumption (EU, 2020; Health Canada, 2020).

Appropriate records for the food business operation should be retained for a period that exceeds the shelf life of the product or as determined by the competent authority. Withdrawal procedures should be applied for the rapid and efficient removal of an unsafe food from the market before it has reached the consumer. Recall procedures should also be applied for the rapid and effective removal of an unsafe food from the market when it may have reached the consumer and the notification of the consumer. Appropriate corrective actions should be taken (EU, 2001, 2002).

Table 13.1 Maximum acceptable concentrations of microbiological contaminants in drinking water (EU, 2020; Health Canada, 2020).

Contaminant	Concentration (cfu/100 mL)
Intestinal enterococci	0
Escherichia coli	0
Coliforms	0
Clostridium perfringens including spores	0

Table 13.2 Maximum acceptable concentrations of chemical contaminants in drinking water (EU, 2020; Health Canada, 2020).

	Concentration	
Contaminant	Value	Unit
Arsenic	10	μg/L
Cadmium	5	μg/L
Lead	5	μg/L
Mercury	1	μg/L
Nitrate	50	mg/L
Nitrite	0.5	mg/L
Pesticides total	0.5	μg/L

13.3.7 Product information and consumer awareness

Lot identification or other identification strategies are essential in product recall and also help effective stock rotation (FIFO). Each container of food should be permanently marked to identify the producer and the lot. A traceability/product tracing system should be designed and implemented throughout the entire food chain to facilitate the withdrawal and/or recall of products (EU, 2002; CAC, 2006; ISO, 2007; EU, 2011a). All products should be labeled with clear instructions to enable the consumer to handle, prepare, display, store, and/or use the product safely and correctly. Food allergens contained in the product should be identified (EU, 2011c; CAC, 2018, 2020b). Consumer education programs should also cover general food hygiene (WHO, 2006).

13.3.8 Transportation

Conveyances and bulk containers should be designed and constructed in such a way as not to contaminate foods or packaging, to be effectively cleaned and disinfected, to allow separation of different foods or foods from nonfood items, to provide protection against environmental contamination, and to maintain the temperature, humidity and other conditions necessary for food. Conveyances and containers for transporting food should be kept in an appropriate state of cleanliness, repair and condition. Where the same conveyance or container is used for transporting different foods or nonfoods, effective cleaning and, where necessary, disinfection, and drying should take place between loads (CAC, 2001a; EU, 2004b; ISO, 2019).

13.3.9 Food defense

Dairy establishments should assess the hazard to products posed by potential acts of sabotage, vandalism or terrorism and should put in place relative protective measures (use of locks, electronic card key or alternative systems) (ISO, 2009).

13.4 Hazard analysis and critical control point

The HACCP system is based on seven principles (EU, 2016; NACMCF, 2017):

Principle 1. Conduct a hazard analysis*:* identifying any hazards that must be prevented, eliminated or reduced to acceptable levels;

Principle 2. Determine the critical control points (CCP)*:* identifying the CCPs at the step or steps at which control is essential to prevent or eliminate all relevant hazards or to reduce them to acceptable levels;

Principle 3. Establish critical limits*:* establishing critical limits at CCPs, which separate acceptability from unacceptability for the prevention, elimination or reduction of identified hazards;

Principle 4. Establish monitoring procedures*:* establishing and implementing effective monitoring procedures at CCPs;

Principle 5. Establish corrective actions*:* establishing corrective actions when monitoring indicates that a CCP is not under control;

Principle 6. Establish verification procedures*:* establishing procedures, which shall be carried out regularly, to verify that the measures outlined in principles 1 to 5 are working effectively;

Principle 7. Establish record-keeping and documentation procedures*:* establishing documents and records commensurate with the nature and size of the food business to demonstrate the effective application of the measures outlined in principles 1 to 6.

The seven principles of HACCP are part of the next twelve steps necessary for its implementation.

13.4.1 Assembling the HACCP team and identifying the scope (Step 1)

The food business operator (FBO) should ensure that the appropriate knowledge and expertise are available for the development of an effective HACCP system. This may be achieved by assembling a multidisciplinary team responsible for different activities within the operation, for example, production, maintenance, quality control, cleaning, and disinfection. The HACCP team is responsible for developing the HACCP plan. The HACCP team should identify the scope of the HACCP system and applicable prerequisite programs. The scope should describe which food products and processes are covered (CAC, 2020a).

13.4.2 Description of the end product (Step 2)

All ingredients used for the production of processed cheese shall be adequately described, with regard to their microbiological, chemical or physical aspects which may be important to identify their food safety status and possible hazards (e.g., composition, origin, shelf life, previous processing and handling, any acceptance criteria and specifications applicable). Raw materials include cheeses, milk fat products (cream, butter, butter oil and ghee), other milk products (milk concentrates, buttermilk, milk powders, milk proteins, whey powders, lactose), and vegetable fat (only in the case of processed cheese analogues). Food ingredients are sodium chloride, potassium chloride, water, gelatin and starches, vinegar, spices, etc. Food additives are emulsifying agents (e.g., phosphates, citrates), acidity regulators (e.g., citric, lactic or acetic acid or sodium hydrogen carbonate and/or calcium carbonate), colors (e.g., annatto, beta-carotene, chlorophyll, riboflavin), preservatives (e.g., potassium or sodium sorbate, potassium or sodium propionate, nisin), stabilizers and thickeners, and anticaking agents. In addition, the primary packaging material shall be described (CAC, 2008a,b,c; ASSIFONTE, 2018).

For processed cheese, the typical intrinsic factors shall be described: pH (5.2 to 5.8), melting salts (up to 3%), any preservatives, for example, nisin (up to the maximum levels specified by legislation). Organoleptic aspects such as appearance, texture, taste and odor shall also be specified (ASSIFONTE, 2018). In addition, the dry matter, fat, protein and salt contents shall be reported. The name of product, the lot number, and the minimum shelf life shall be indicated on the package: 6 months for pasteurized processed cheese stored at 0–4°C (Anonymous, 2009) and 24 months for sterilized processed cheese stored at 6–23°C (Bubelová et al., 2015).

13.4.3 Identification of intended use (Step 3)

The HACCP team should describe the intended use of the product by the consumer (processed cheese for spreading, for sandwich or pizza making, etc.). In specific cases, vulnerable groups of the population (infants, the elderly, and the infirm or immunocompromised individuals) may have to be considered. When products are manufactured specifically for a vulnerable population, it may be necessary to enhance process controls, monitor control measures more frequently, and verify that the controls are effective by testing the products.

13.4.4 Construction of a flow diagram (Step 4)

A flow diagram that covers all steps in the production of a specific product, including any applicable rework, should be constructed. The flow diagram should indicate all inputs, including those of ingredients and food contact materials, water and air, if relevant (CAC, 2020a).

Fig. 13.1 shows a typical flow diagram including the basic steps involved in processed cheese manufacture. The steps are described as follows.

13.4.4.1 Natural cheese selection

Proper selection of natural cheese, as raw material, is of the utmost importance for the successful production of processed cheese. The most important criteria for cheese selection are: type, flavor, maturity, consistency, texture, acidity (pH) and microbial quality. A single variety of natural cheese or a mixture of different types of cheese can be used (Carić and Kaláb, 1993; Guinee et al., 2004).

The European Dairy Association (EDA) and the European Association of the Dairy Trade (EUCOLAIT) have developed guidance that assist in deciding whether a cheese material is suitable for further food processing, directly or pending further treatment, or must be disposed of and used in accordance with the animal by-products legislation (EU, 2009). Table 13.3 summarizes the minimum criteria for suitability of natural cheese as a raw material in the manufacture of food products (processed cheese, etc.) and the necessary specific treatments, as defined in the guidance (EDA/EUCOLAIT, 2018).

13.4.4.2 Blending of the raw cheese

The selected raw material determines the type and flavor of the finished product. In order to produce a processed cheese of regular consistency and unchanged quality and to introduce it to the market as a specified item, it is recommended to combine together at least four to six different lots of natural cheese of the same or various types (Meyer, 1973).

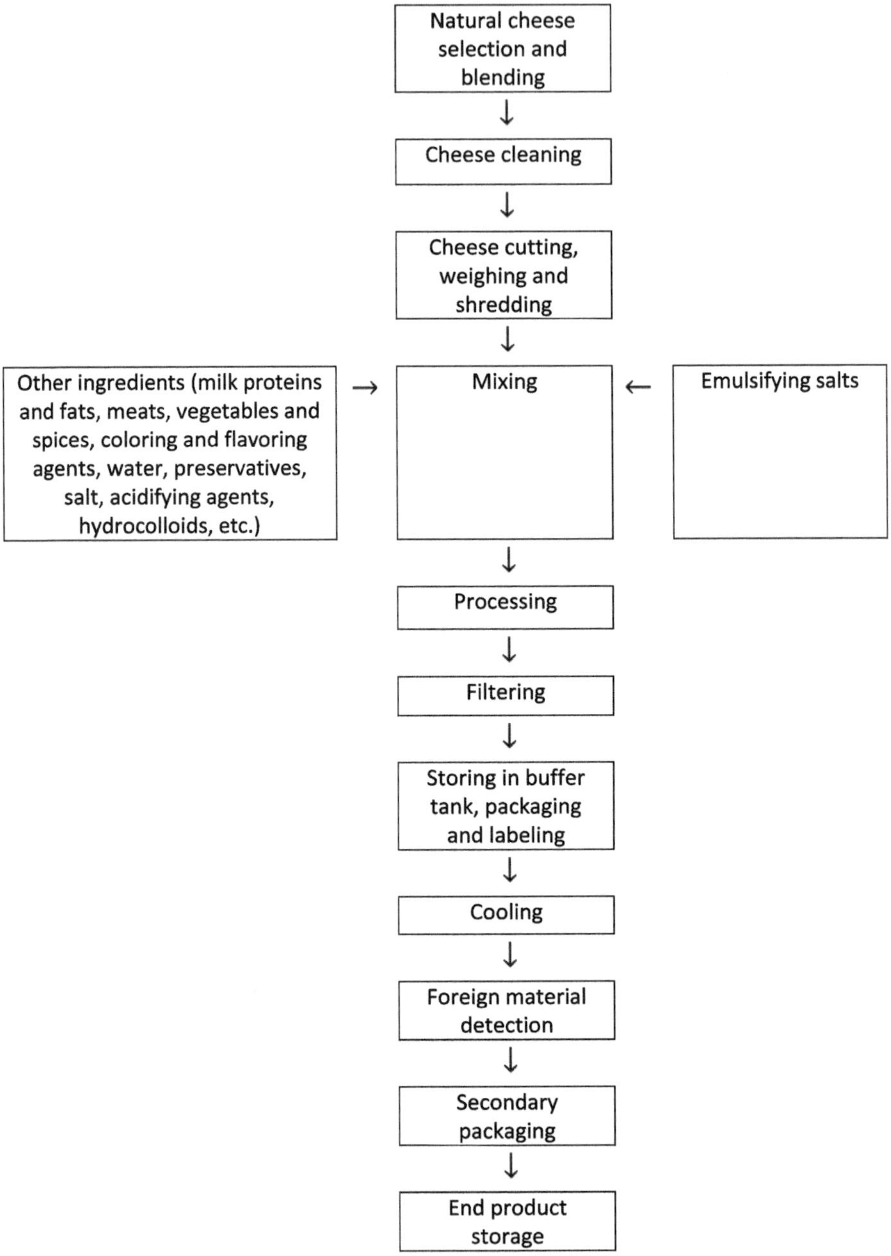

Figure 13.1 Typical flow diagram of processed cheese manufacture.

Table 13.3 Suitability criteria for natural cheese as a raw material for further food processing (adapted from EDA/EUCOLAIT, 2018).

Type of raw material	Type of deviation/defect	Assessment of suitability for usage	Specific treatment
1. Cheeses intended for direct consumption	1.1 Noncoated	Suitable	Remove packaging material prior to use. Remove unclean spots.
	1.2 Coated	Suitable	Remove coating material prior to use (peel, brush or scrub off, or apply a heat-based removal process).
2. Prepacked cheese and cheese cuts returned from the market or retailers	2.1 From approved wholesale enterprises and retailers	Suitable	Remove packaging and coating material prior to use.
	2.2 From other retailers (outlets)	Not suitable	Disposal and use as animal by-product (EU, 2009).
3. Samples intended for testing and analysis	3.1 Unopened shelf life samples that are held under controlled chilled storage within the manufacturing facilities	Suitable	Remove packaging, unclean spots and coating material prior to use. Keep at temperature and within time limits specified on the label or accompanying documents.
	3.2 Unopened accelerated shelf life samples that are held under elevated controlled storage conditions within the manufacturing facilities	Suitable, if supported by a full food safety evaluation	
	3.3 Leftovers of samples used for professional sensory testing		
	3.4 Unopened laboratory samples held in controlled chilled storage	Suitable	
	3.5 Leftovers of laboratory samples that have been opened in laboratory facilities	Not suitable	Disposal and use as animal by-product.

(*continued on next page*)

Table 13.3 Suitability criteria for natural cheese as a raw material for further food processing (adapted from EDA/EUCOLAIT, 2018)—cont'd

Type of raw material	Type of deviation/defect	Assessment of suitability for usage	Specific treatment
4. Cheese not complying with quality specifications	4.1 Erroneous texture 4.2 Structure defects (e.g., eyes) 4.3 White (crystallized) surface 4.4 Deviating taste 4.5 Compositional noncompliance 4.6 Mechanical damage or deformation of packaging or cheese 4.7 Mislabeled	Suitable	Remove packaging, unclean spots and coating material prior to use. Keep at temperature and within time limits specified on the label or accompanying documents.
5. Physically contaminated cheeses	5.1 Unspecified foreign material 5.2 Glass or hard plastic	Not suitable	Disposal and use as animal by-product.
	5.3 Metal	Suitable	Remove effectively the fragments before or during further processing by accepted methods (e.g., metal detector).
6. Chemically contaminated cheeses	Maximum levels (ML) for chemical contaminants or maximum residue limits (MRL) for pesticides and/or veterinary drugs exceeded	Not suitable	Disposal and use as animal by-product.
7. Cheese contaminated with yeast		Suitable	Specific treatment not required.
8. Cheese exceeding microbiological process hygiene criteria	8.1 Microbiological criteria for coagulase-positive staphylococci exceeded 8.2 Microbiological criteria for other hygiene indicators exceeded	Suitable, if free of staphylococcal enterotoxins (coagulase-positive staphylococci < 1×10^5 cfu/g)	Heat treatment (or equivalent treatment) that delivers at least 8 log reductions of coagulase-positive staphylococci and/or *L. monocytogenes*.

(*continued on next page*)

Table 13.3 Suitability criteria for natural cheese as a raw material for further food processing (adapted from EDA/EUCOLAIT, 2018)—cont'd

Type of raw material	Type of deviation/defect	Assessment of suitability for usage	Specific treatment
9. Cheese exceeding microbiological food safety criteria		Suitable	Heat treatment (or equivalent treatment) that delivers at least 8 log reductions of the pathogens in question.
10. Cheese with undesired mold colonies	10.1 Visible mold colonies that most likely are species typically used in the manufacture of mold-ripened cheeses	Suitable	Heat treatment at least 75°C for at least 1 minute.
	10.2 Visible mold colonies where it cannot be substantiated that they are likely to be species typically used in the manufacture of mold-ripened cheese (i.e., not covered by 10.1)	Suitable	Visible mold should be removed up to 2-3 cm in depth. Heat treatment at least 75°C for at least 1 minute.
	10.3 Cheese combined with flavoring foods (e.g., herbs, spices and fruits)	Suitable, if free of mycotoxins	Heat treatment at least 75°C for at least 1 minute.
	10.4 Cheese not complying with 10.1, 10.2 or 10.3	Not suitable	Disposal and use as animal by-product.
11. Line recoveries	11.1 Cheese edges off-cuts	Suitable	Specific treatment not required.
	11.2 Cheese curd		
	11.3 Floor sweepings	Not suitable	Disposal and use as animal by-product.
12. Cheese exceeding age specifications	12.1 Exceeding pre-established best before date	Suitable	Specific treatment not required.
	12.2 Exceeding pre-established expiration date	Not suitable	Disposal and use as animal by-product.

(*continued on next page*)

Table 13.3 Suitability criteria for natural cheese as a raw material for further food processing (adapted from EDA/EUCOLAIT, 2018)—cont'd

Type of raw material	Type of deviation/defect	Assessment of suitability for usage	Specific treatment
13. Cheese contaminated with cheese mites		Suitable the noninfested parts	Remove contaminated material. Heat treatment of noninfested parts at least 75°C for at least 1 minute.
14. Deteriorated cheese	14.1 Cheese contamination with other pests (e.g., vermin)	Not suitable	Disposal and use as animal by-product.
	14.2 Protein or fat decomposition	Suitable	Specific treatment not required.
	14.3 Nauseous taste and odor	Not suitable, although not unsafe	Disposal and use as animal by-product.

13.4.4.3 *Cleaning of the raw cheese*

The selected and blended raw material should be subjected to a cleaning treatment before it can be processed, since the rind of the cheese may be too hard and heavily contaminated. Wax or linen bandage should also be removed. The selected cheeses still retaining their labels are brought into the cleaning room by various methods, for example, transporting trucks, pallets, fork lift trucks or usually conveyor belts (Meyer, 1973). Conveyors should be constructed of material which can be properly cleaned, will not rust, or otherwise contaminate the cheese, and should be maintained in good repair (LII, 2021a). The cleaning department should be equipped with storage shelves to hold the cheese, machines for washing it, working and rind-removing tables, drying shelves and steam installation. This room should be well ventilated and lighted, and cheese waste should be removed regularly (Meyer, 1973).

13.4.4.4 *Cutting, weighing and shredding of the raw cheese*

The larger cheeses, such as Emmentaler, Cheddar, Cantal, Gouda, etc., are usually precut into smaller pieces. Cutting can be carried out by machines or by hand. Then the cheese is weighed in plastic containers using a platform or table scale (Meyer, 1973). After weighing, cheese is shredded. Applied in the form of grinding or milling, this operation increases the homogeneity of the blend, enables better contact between the emulsifying agent and the other blend ingredients and facilitates heat transfer during processing (Carić and Kaláb, 1993; Guinee et al., 2004). The shredders shall be of corrosion-resistant material and of such construction as to prevent contamination of the cheese and to allow thorough cleaning of all parts and product contact surfaces (LII, 2021a).

13.4.4.5 Addition of other ingredients

In order to adjust the nonfat solid content of processed cheese and improve its properties (emulsion stability, structure, spreadability, etc.), milk protein ingredients are used, such as skim milk powder, whey powder, whey protein concentrate, casein and caseinates, casein-whey protein coprecipitates, etc. Milk fat products (cream, butter and butter oil, etc.) are used to standardize the fat content and improve texture and flavor. Previously processed cheese (rework) is commonly used as an ingredient in processed cheese spreads to intensify the creaming properties of the blend (Carić and Kaláb, 1993). Nondairy ingredients, such as meats, vegetables and spices, coloring and flavoring agents, water, preservatives, salt, acidifying agents, hydrocolloids, etc., are also added to improve sensory properties and/or shelf life.

13.4.4.6 Addition of emulsifying agents

This is the last step in preparing the blend for processing. Emulsifying agents (melting salts) are responsible for the unique features of processed cheese production and the type of the resulting product. Emulsifying agents are of major importance in processed cheese production where they are used to provide a uniform structure during the melting process, and also of the products. Phosphates, polyphosphates and citrates are most common but sodium potassium tartrate, complex sodium phosphates, trihydroxyglutaric acid or diglycolic acid are sometimes used. These compounds are not emulsifiers in the strict chemical sense (i.e., they are not surface-active compounds), and since emulsification is not their only purpose, melting salts are conditionally termed "emulsifying agents". However, true emulsifiers may be included in commercially produced emulsifying agents (Carić and Kaláb, 1993).

The essential role of emulsifying agents in the manufacture of processed cheese is to supplement the emulsifying capability of cheese proteins. This is accomplished by: (i) sequestering (removing) calcium from the casein system, (ii) peptizing (solubilizing) and dispersing caseins, (iii) hydrating and swelling caseins, (iv) emulsifying the fat and stabilizing the emulsion, (v) adjusting pH and stabilizing it, and (vi) forming an appropriate structure of the product after cooling (Thomareis, 1986; Guinee et al., 2004).

13.4.4.7 Mixing

The mixing of the shredded raw cheese material with the simultaneous addition of other ingredients and emulsifying agents leads to obtaining some homogeneity and producing better breakdown of the casein. Mixing is obtained with various mixing and kneading machines (Meyer, 1973). However, shredding and mixing are not necessary when the processing machine has a high-speed rotating knife, for example, a Stephan Universal Machine.

13.4.4.8 Processing

This process step is specific to the manufacture of processed cheese. This term implies the conversion of a cheese blend into a smooth and uniform heat-treated mass of cheese. It is distinguished in melting and emulsification, and heat treatment (ASSIFONTE, 2018):

a) *Melting and emulsification*: Natural cheese consists of a paracasein gel with occluded fat globules and moisture. Emulsifying agents function indirectly, by promoting (together with heat and shearing) a number of physicochemical changes in the blend which convert insoluble paracasein into sodium paracaseinate, which binds water and emulsifies the free oil during processing.

b) *Heat treatment*: It is carried out using a batch cheese cooker or a continuous cheese cooker (scraped surface heat exchanger, tubular heat exchanger, and direct injection UHT system). In the manufacture of processed cheese, the design of the heat treatment varies considerably depending on the raw materials used, the type of processed cheese (e.g., spread or block), and the flavoring agents. However, all heat processes applied shall include at heating step of above 80°C for more than 1 minute.

Some examples of applied heat processing are the following (ASSIFONTE, 2018):

1. 135°C for not less than 5 minutes (a commercial sterilization), followed by hot filling,
2. 130°C for 5 or 8 s (a UHT treatment), followed by hot filling,
3. 106°C for not less than 2 minutes (a pasteurization), followed by cooling to 90°C and storage (at least 30 minutes) with natural cooling until filling at above 60°C,
4. 90°C for not less than 1 minute (a pasteurization), followed by storage with natural cooling until filling at above 60°C.

Examples 1 and 3 are typically applied where control of lactobacilli is desired.

The cheese cookers shall be the steam jacketed or direct steam type. They shall be constructed of stainless steel or other equally corrosion-resistant material. All product contact surfaces shall be readily accessible for cleaning. Each cooker shall be equipped with an indicating thermometer, and shall be equipped with a temperature recording device. If direct steam is applied to the product, only culinary steam shall be used (LII, 2021a).

13.4.4.9 Filtering

In-line filters are used to remove any solid cheese, burnt particles, calcium lactate crystals, and foreign bodies from the molten cheese blend (ASSIFONTE, 2018). The hoppers of all filters shall be covered but the cover may have sight ports. If necessary, the hopper may have an agitator to prevent buildup on side wall (LII, 2021a).

13.4.4.10 Homogenization

This is an optional step that however improves the stability of the emulsion, as well as the consistency, structure, appearance and flavor of the processed cheese (Carić and Kaláb, 1993). Various machines are suitable for this purpose: meat cutter, colloid mill, homogenizer and ultrasonic homogenizer (Meyer, 1973).

13.4.4.11 Storing in buffer tank, packaging and labeling

In cold fill production, the heat-treated cheese is naturally cooled down in the buffer tank below the pasteurization temperature before entering the primary packaging. All typical ubiquitous germs and spore-like yeasts and molds present in a normal production environment may therefore recontaminate the product surface (ASSIFONTE, 2018).

Another cold process technology involves extrusion of hot (e.g., 80°C) molten cheese onto a cold stainless-steel roller from which it is delivered in a flat strip on a conveyor to a slicer and packaging unit. During this period, the product is exposed to ambient air for a period of approximately 1 minute (ASSIFONTE, 2018).

On the contrary, in hot fill production, the processed cheese is filled into the primary packaging material at a temperature that minimizes the consequences of any recontamination (contact or airborne) due to the pasteurization effect resulting from the temperature of the product itself (ASSIFONTE, 2018). The filler valves and head shall be kept in good repair and capable of accurate dosing. Product contact surfaces shall be of stainless steel or other corrosion resistant material (LII, 2021a). Processed cheese is usually packed and wrapped in lacquered aluminum foil, tubes, cups, cans, cardboard or plastic cartons, and occasionally in glass jars (Carić and Kaláb, 1993). Immediately after packaging, labeling takes place, indicating the storage temperature, shelf life, and lot number of the product.

13.4.4.12 Cooling

As rapid cooling softens the product, cooling should be as fast as possible for processed cheese spreads and relatively slow for processed cheese blocks. However, slow cooling can intensify Maillard reactions and promote the growth of spore-forming bacteria (Carić and Kaláb, 1993).

13.4.4.13 Foreign material detection

All packaged processed cheese units are examined online for possible foreign bodies. A metal detector is intended for the detection of metal fragments and objects, such as ferrous and nonferrous metals, stainless steel and aluminum, while an X-ray inspection system can detect any hard foreign material (metal, glass, stone, dense plastic, etc.) (Anonymous, 2021).

13.4.4.14 Secondary packaging

The product cooled in the primary packaging undergoes secondary packaging (into packs, boxes, cartons, cases, etc.) using various packaging machines.

13.4.4.15 End product storage

The finished product should be stored at temperatures below 10°C, although such low temperatures may induce crystal formation (calcium monophosphate, calcium diphosphate, calcium pyrophosphate, etc.) (Carić and Kaláb, 1993).

13.4.5 On-site confirmation of flow diagram (Step 5)

The HACCP team should perform an on-site review of the operation to verify the accuracy and completeness of the flow diagram. Modifications should be made to the flow diagram as necessary and documented (NACMCF, 2017).

After these five preliminary steps have been completed, the seven principles of HACCP are applied.

13.4.6 Hazard analysis: Listing of relevant hazards, hazard evaluation, and control measures (Step 6/Principle 1)

A hazard is a biological, chemical or physical agent in, or condition of, food or feed with the potential to cause an adverse health effect (EU, 2016). The HACCP team should list all potential hazards that are likely to occur and associated with each step (including all inputs into that step) according to the scope of the food business operation. A hazard evaluation (risk assessment) should then be undertaken. Based on the severity of the hazards and their likelihood of occurrence, it consists of the evaluation of the potential hazards in order to identify which of them are significant for the specific food business operation. Subsequently, any measures to control the identified hazards should be considered (CAC, 2020a; LII, 2021c).

Hazards that are normally associated with the manufacture of processed cheese are (ASSIFONTE, 2018):

a) *Biological hazards*: Pathogenic bacteria from raw materials (*L. monocytogenes*, verotoxigenic *E. coli, S. aureus* and *Salmonella* spp.) and from ingredients (*Salmonella* spp., *C. perfringens, C. botulinum, B. cereus, L. monocytogenes*), bacterial toxins that may be present in raw materials (e.g., staphylococcal enterotoxin in certain raw milk cheese) and ingredients, additional hazards associated with flavoring foods (e.g., spices, herbs, ham, fruit, mushrooms, etc.), pathogens that may be transferred from staff, recontaminating molds and pathogenic bacteria from the processing and packaging environment.
b) *Chemical hazards*: Mycotoxins that may be present in cheese, pesticide residues in raw materials and ingredients (e.g., products rich in fat), environmental contaminants in raw materials and ingredients (e.g., dioxins and dioxin-like PCBs, heavy metals), contaminants from processing activities (coolants, lubricants, sanitizers), carry-over additives (with numerical acceptable daily intakes specified) from raw materials and ingredients, contaminants from food contact materials, including coating, waxes and soft plastic such as ripening films (e.g., mineral oils), additional hazards associated with flavoring foods (e.g., spices, herbs, ham, fruit, mushrooms, etc.), allergens in ingredients and from equipment (e.g., lubricants).
c) *Physical hazards*: Glass from unsecured windows, lamps, bulbs, neon tubes, thermometers, laboratory glass, spy holes on the tanks, insect killing tubes or other glass equipment, insect debris from raw materials, ingredients and packaging materials, wood from pallets or wooden equipment, metal from improperly constructed or maintained machines or equipment, plastic from buckets, boxes, brushes, stones from damaged floors or walls, utensils used in the production zone and lost by staff (pencils, knives, keys, etc.).

Regarding the control measures applied as PRPs, these are general conditions and activities for maintaining hygiene. In general, they are not specific to any product or hazard, but reduce the likelihood of hazards occurring. On the contrary, the control measures applied to HACCP are specific to a product and its production process steps, and are necessary to prevent, eliminate or reduce to an acceptable level a hazard considered significant (CAC, 2020a).

Table 13.4 gives a brief hazard analysis of the manufacture of processed cheese, listing all potential hazards by ingredient or production step, which can be controlled by the installed PRPs or which require measures through the implementation of the HACCP system. Significant hazards are addressed in the HACCP plan (Table 13.5). Such a document contains, at a minimum, the production steps/ingredients, CCPs, targeted hazards, critical limits, monitoring procedures as control parameters (what), control measures (how), frequency (when) and responsible (who), corrective actions (what, who), verification, and record keeping.

13.4.7 Determination of critical control points (Step 7/Principle 2)

The term "critical control point" refers to a point, step, or procedure in a food process at which a control measure can be applied and at which control is essential to reduce an identified food hazard to an acceptable level (LII, 2021c). The identification of a CCP requires a logical approach which can be facilitated by the use of a decision tree or other methods, according to the knowledge and experience of the HACCP team (EU, 2016). Examples of CCPs may include: thermal processing, chilling, testing ingredients for chemical residues, product formulation control, and testing product for metal contaminants (NACMCF, 2017).

The flow diagram (Fig. 13.2) shows the five CCPs identified in the manufacture of processed cheese, while the HACCP plan (Table 13.5) also presents the corresponding targeted hazards. Some details are given below.

Natural cheese (raw material) and other ingredients constitute CCP1, because biological hazards such as heat-resistant enterotoxins (Necidová et al., 2019) and pest (mites, vermin) cannot be eliminated during the manufacturing process. Likewise, chemical hazards such as heat-resistant mycotoxins (Raters and Matissek, 2008), and other contaminants (drug and pesticide residues, heavy metals, dioxins, PCBs), therefore respective control measures should be applied when selecting natural cheese and other ingredients.

Processing and storing in buffer tank-packaging-labeling steps constitute CCP 2 and CCP 4, respectively. They refer to the destruction of pathogenic microorganisms in natural cheese and other ingredients which takes place in two phases (1st and 2nd heat treatments), but also to their subsequent potential survival. In the case of hot filling, the pathogens are destroyed during processing (only one heat treatment). In addition, since airborne and packaging pathogens are destroyed in hot molten cheese, the finished product is microbiologically shelf stable during storage. Thus, the storing in buffer tank-packaging-labeling steps is no longer a CCP. The same is true for sterilized processed cheese, although, in these products, the destruction of the spores of *C. perfringens, C. botulinum,* and *B. cereus* is sought.

Filtering and foreign material detection constitute CCP 3 and CCP 5, respectively. Both are designed to eliminate the physical hazards of the product. Filters remove solid particles from the molten cheese, while metal detection and X-ray inspection systems

Table 13.4 Brief hazard analysis for processed cheese manufacture (adapted from ASSIFONTE, 2018).

Step/ingredient	Hazard	Control by
Raw materials		
Cheese suitable for direct consumption	Biological: *L. monocytogenes, E. coli, S. aureus, Salmonella* spp.	HACCP
Cheese material intended for further processing only	Biological: Pathogens, bacterial toxins, mites and vermin	HACCP
	Chemical: Mycotoxins, drug and pesticide residues, heavy metals	HACCP
	Physical: Fragments of metal, hard plastic and glass	HACCP
Other ingredients		
Dried dairy products (milk powder, whey powder, dried whey protein concentrates)	Biological: *Salmonella* spp.	HACCP
Liquid dairy products ($a_w > 0.92$)	Biological: *L. monocytogenes*	HACCP
Milk fat products and vegetable oils (for processed cheese analogues)	Chemical: Lead, certain pesticide residues, dioxins, PCBs	HACCP
Starches	Biological: *Salmonella* spp., thermoduric spores	HACCP
Spices (untreated) and dried herbs	Biological: *Salmonella* spp., spores of *C. perfringens, C. botulinum, B. cereus*	HACCP
Spices, herbs and similar ingredients (harvested)	Physical: Small stones	PRP
Additives	Chemical: Allergens	PRP
Shredding and mixing		
Shredding and mixing equipment	Chemical: Residues of sanitizers	PRP
	Physical: Metal fragments	HACCP
Shredding and mixing environment	Physical: Fragments (metal, plastic)	HACCP
Personnel	Biological: Infectious diseases	PRP
Processing		
Processing equipment	Biological: Thermoduric bacteria such as *B. cereus*	PRP
	Chemical: Residues of sanitizers	PRP
	Physical: Metal fragments	HACCP
Packaging		
Filling equipment	Biological: *S. aureus, L. monocytogenes*	PRP
	Chemical: Residues of sanitizers, allergenic lubricants	PRP
	Physical: Metal fragments	HACCP
Packaging environment	Biological: *Salmonella* spp., mycotoxin forming molds	PRP
	Physical: Wood fragments	PRP
Packaging material	Chemical: Acrylamides, print colors	PRP
Personnel	Biological: Infectious diseases	PRP
End products		
Storage of end products	Biological: *S. aureus, L. monocytogenes, Salmonella* spp., *C. perfringens, C. botulinum*	HACCP

Table 13.5 HACCP plan of processed cheese manufacture (adapted from ASSIFONTE, 2018; EDA/EUCOLAIT, 2018).

Monitoring procedure							Corrective action				
Production step/ingredient	**CCP No.**	**Targeted hazard**	**Critical limit**	**Control parameter (What)**	**Control measure (How)**	**Frequency (When)**	**Responsible (Who)**	**Action (What)**	**Responsible (Who)**	**Verification**	**Record Keeping**
Natural cheese and other ingredients	1	B: Enterotoxins	Not detected in 25 g	Staphylococcal enterotoxins	Detection os staphylococcal enteroroxins	Every lot	Quality control inspector	Remove contaminated material, and dispose of it and use it as an animal by-product_Return rejected lots to the supplier	Quality control manager	Testing for coagulase-positive staphylococci	Monitoring corrective action and verificatio records
		B: Pest	Absence	Mites, vermin	Macroscopic control					Microscopic control	
		C: Chemical contaminants	Values of Tables 13.9 and 13.10	Mycotoxins, drug and pesticide residues, heavy metals, dioxins, PCBs	Chemical analyses					Chemical examination of external laboratory	
Processing	2	B: Pathogens	90°C	Temperature	1st heat treatment (pasteurization)	Continuously	Operator	Rework Restore treatment conditions	Process manager	Thermometer calibration	Monitoring corrective action and verification methods
			1 min	Time						-	
Filtering	3	P: Physical hazards	0.2-0.8 mm	Filter size	Filtering	Continuously	Operator	Removal of hazards, filter change	Process manager	Inspection of filter integrity prior to eacg production run	Monitoing corrective action and verification records

(*continued on next page*)

Table 13.5 HACCP plan of processed cheese manufacture (adapted from ASSIFONTE, 2018; EDA/EUCOLAIT, 2018)—cont'd

Monitoring procedure							Corrective action				
Production step/ingredient	**CCP No.**	**Targeted hazard**	**Critical limit**	**Control parameter (What)**	**Control measure (How)**	**Frequency (When)**	**Responsible (Who)**	**Action (What)**	**Responsible (Who)**	**Verification**	**Record Keeping**
Storing in buffer tank Packaging-Labelling	4	B: *Salmonella* spp., Verotoxin producing *E. coli, L. monocytogens*	60°C	Temperature	2nd heart treatment	Continuously for every 10 min	Operator	Rework	Process manager	Thermometer calibration	Monitoring corrective action and verification records
			30 min	Time				Prolong holding time		[Instruction: Additional correction: Please move the two dashes in the "verification column further to the right"]	
		B: Any surviving pathogens	0-4 °C	Temperature	Labeling	Every packaging	Operator	Rejection of packaging	Packaging manager	Daily check of the labeling machine	Monitoring corrective action and verification records
			6 months	Time							
Foreign material detection	5	P: All physical hazards	5mm	Fragment size	X-ray inspection	Every package	Operator	Rejection of packaged product-identify the cause	Packaging manager	X-ray inspection system control and calibration	Monitoring corrective action and verification records
		P:Metal objects			Metal detection					Metal detection control and calibration	

B: biological, C: chemical and P: physical hazards [Instruction: This table is not properly shown in Proof. The last two columns and not shown].

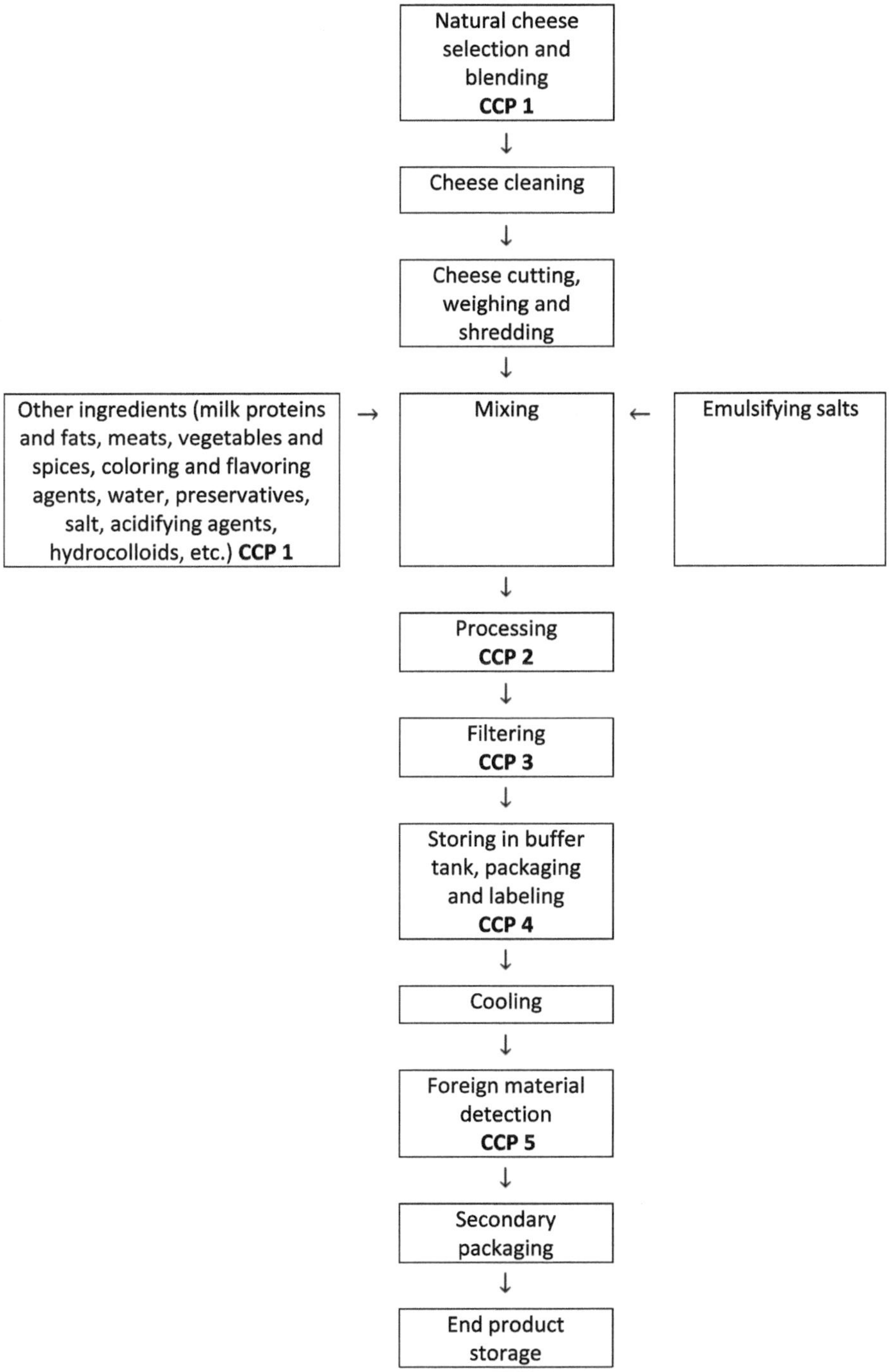

Figure 13.2 Typical flow diagram of processed cheese manufacture with CCPs (critical control points).

detect metal objects or any hard foreign material, respectively, in the packaged finished product. Filtering is very effective at removing foreign bodies and can often completely replace foreign material detection. The latter is however absolutely necessary when filtering cannot be used, for example in a product with pieces of meat or vegetables. It also provides additional safety in case the filter is damaged during the production process or the homogenizer loses a metal fragment. X-ray inspection has the advantage over metal detection that it can also be used on products in metal packaging (lacquered aluminum foil, cans).

13.4.8 Critical limits at CCPs (Step 8/Principle 3)

The term "critical limit" refers to the maximum or minimum value to which a biological, chemical or physical parameter must be controlled at a critical control point to prevent, eliminate, or reduce to an acceptable level the occurrence of the identified food hazard (LII, 2021c). Since a critical limit is used as a criterion to separate acceptable products from unacceptable ones, it should be measurable or observable and scientifically validated (CAC, 2020a).

The most obvious critical limits are specific temperature/time relationships for either processing or storage that are necessary to prevent, eliminate or reduce microbial hazards. Other processing factors such as available chlorine, contact time or flow rate can be critical limits. Food composition information such as moisture level, acidity, pH and a_w can also be used. Critical limits could be based on existing literature, regulations or guidance from competent authorities, or studies conducted by a third party (e.g., specifications on raw materials/ingredients and finished products) (Ross-Kung and Neves, 2003; CAC, 2020a).

In the manufacture of processed cheese, for the selection of raw materials and ingredients as well as for the quality control of the finished product, the biological hazards can have the following critical limits:

a) Microbiological criteria of European Union for processed cheese raw materials and ingredients (Table 13.6),
b) Microbiological criteria of European Union for processed cheese (Table 13.7),
c) Microbiological criteria of different countries for processed cheese (Table 13.8).

Regarding the chemical hazards in the manufacture of processed cheese, the following critical limits can be taken:

a) Maximum levels of European Union for chemical contaminants in processed cheese ingredients (Table 13.9),
b) Maximum levels of European Union for chemical contaminants in processed cheese (Table 13.10).

In addition, for the physical hazards of processed cheese, a ready-to-eat product may contain a hard or sharp foreign object that measures 7 to 25 mm, in length. If the product is intended for special risk groups such as infants, surgery patients, and the elderly, the foreign object shall be less than 7 mm (FDA, 2005).

Table 13.6 Microbiological criteria of European Union for processed cheese raw materials and ingredients (EU, 2005, 2007, 2010).

Raw material/ ingredient	Microorganisms/ toxins	Sampling plan		Limits	
		n	c	m	M
Pasteurized milk and other pasteurized liquid dairy products	Enterobacteriaceae	5	0	10 cfu/mL	
Cheeses made from milk or whey that has undergone heat treatment	*E. coli*	5	2	100 cfu/g	1000 cfu/g
Cheeses made from raw milk	Coagulase-positive staphylococci	5	2	1×10^4 cfu/g	1×10^5 cfu/g
Cheeses made from milk that has undergone a lower heat treatment than pasteurization and ripened cheeses made from milk or whey that has undergone pasteurization or a stronger heat treatment	Coagulase-positive staphylococci	5	2	100 cfu/g	1000 cfu/g
Unripened soft cheeses (fresh cheeses) made from milk or whey that has undergone pasteurization or a stronger heat treatment	Coagulase-positive staphylococci	5	2	10 cfu/g	100 cfu/g
Butter and cream made from raw milk or milk that has undergone a lower heat treatment than pasteurization	*E. coli*	5	2	10 cfu/g	100 cfu/g

(*continued on next page*)

Table 13.6 Microbiological criteria of European Union for processed cheese raw materials and ingredients (EU, 2005, 2007, 2010)—cont'd

Raw material/ ingredient	Microorganisms/ toxins	Sampling plan		Limits	
		n	c	m	M
Cheeses, butter and cream made from raw milk or milk that has undergone a lower heat treatment than pasteurization	*Salmonella* spp.	5	0	Absence in 25 g	
Milk powder and whey powder	Enterobacteriaceae	5	0	10 cfu/g	
	Coagulase-positive staphylococci	5	2	10 cfu/g	100 cfu/g
	Salmonella spp.	5	0	Absence in 25 g	
Milk products able to support the growth of *L. monocytogenes*	*L. monocytogenes*	5	0	100 cfu/g (Throughout the shelf life)	
		5	0	Absence in 25 g (Just after production)	
Milk products unable to support the growth of *L. monocytogenes*	*L. monocytogenes*	5	0	100 cfu/g (Throughout the shelf life)	
Cheeses, milk powder and whey powder	Staphylococcal enterotoxins	5	0	Not detected in 25 g (Throughout the shelf life)	

n = number of units comprising the sample; c = number of sample units giving values over m or between m and M (limits).

The HACCP plan (Table 13.5) presents the critical limits for each of the five CCPs identified in the manufacture of processed cheese, as follows (ASSIFONTE, 2018; EDA/EUCOLAIT, 2018).

The critical limits corresponding to natural cheese and other ingredients (CCP 1) are: (i) not detected in 25 g, for staphylococcal enterotoxins; (ii) absence, for mites and vermin; and (iii) values in Tables 13.9 and 13.10, for mycotoxins, drug and pesticide residues, heavy metals, dioxins, and PCBs.

For the destruction of pathogenic bacteria, heat treatments must reach a 6 log reduction. This can be achieved by heating the mixture to 90°C for at least 1 min during processing (CCP 2), and keeping the mass hot in the buffer tank, with a declining temperature during at least 30 min and with a filling temperature of at least 60°C

Table 13.7 Microbiological criteria of European Union for processed cheese (EU, 2005, 2007).

Processed cheese	Microorganisms/ toxins	Sampling plan		Limits	
		n	c	m	M
Processed cheese able to support the growth of *L. monocytogenes* (products with relatively high moisture content)	*L. monocytogenes*	5	0	100 cfu/g (Throughout the shelf life)	
		5	0	Absence in 25 g (Just after production)	
Processed cheese unable to support the growth of *L. monocytogenes* (products with relatively low moisture content)	*L. monocytogenes*	5	0	100 cfu/g (Throughout the shelf life)	
Processed cheese made with milk or whey as ingredient	*E. coli*	5	2	100 cfu/g	1000 cfu/g
Processed cheese	Staphylococcal enterotoxins	5	0	Not detected in 25 g (Throughout the shelf life)	

n = number of units comprising the sample; c = number of sample units giving values over m or between m and M (limits).

Table 13.8 Microbiological criteria of different countries for processed cheese.

Country	Microorganisms/ toxins	Sampling plan		Limits		References
		n	c	m	M	
Australia/New Zealand (All cheese)	*E. coli*	5	1	10 cfu/g	100 cfu/g	FSANZ (2017)
China	Total count plate, coliforms, *S. aureus*	5	2	100 cfu/g	1000 cfu/g	MHPRC (2010)
	Salmonella, *L. monocytogenes*	5	0	0 cfu/25g		
	Yeasts, molds			≤ 50 cfu/g		
United States (All dairy products)	Enterohemorrhagic *E. coli* (EHEC) O157:H7 or other EHEC			Absent		CFSAN (2010)
	Vegetative cells of *C. botulinum* or *C. botulinum* toxin			Absent		
	E. coli			≤ 100 MPN/g		
	S. aureus			$< 1 \times 10^4$ cfu/g		
	B. cereus			$< 1 \times 10^4$ cfu/g		

n = number of units comprising the sample; c = number of sample units giving values over m or between m and M (limits).

Table 13.9 Maximum levels of European Union for chemical contaminants in processed cheese ingredients.

Contaminant	Ingredient	Maximum level
Aflatoxin B1	A number of dry ingredients and spices	2–8 μg/kg, depending on the ingredient (EU, 2006)
Aflatoxins (sum of B1, B2, G1 and G2)	A number of dry ingredients and spices	4–15 μg/kg, depending on the ingredient (EU, 2006)
Aflatoxin M1	Milk	0.05 μg/kg (EU, 2006)
Ochratoxin A	Processed cereal products	3 μg/kg (EU, 2006)
Deoxynivalenol	Cereal flour	750 μg/kg (EU, 2006)
Zearalenone	Cereal flour	75 μg/kg (EU, 2006)
Fumonisins (sum of B1 and B2	Maize flour, refined maize oil	1000 μg/kg (EU, 2006)
Lead	Milk, meats, vegetables, cultivated fungi, berries and small fruit, fats and oils	0.02–0.3 mg/kg, depending on the ingredient (EU, 2006)
Cadmium	Meats, vegetables, fresh herbs, cultivated fungi	0.05–0.2 mg/kg, depending on the ingredient (EU, 2006)
Benzo(a)pyrene	Oils and fats, smoked meats	2–5 μg/kg, depending on the ingredient (EU, 2006)
Dioxins (WHO-PCDD/F-TEQ)	Meats, vegetable oils and fats, milk and dairy products with >2% fat	0.75–2.5 pg/g fat, depending on the ingredient (EU, 2011b)
Dioxins + dioxin-like PCBs (WHO-PCDD/F-PCB-TEQ)	Meats, vegetable oils and fats, milk and dairy products with >2% fat	1.25–5.5 pg/g fat, depending on the ingredient (EU, 2011b)
PCBs (PCB 28, 52, 101, 138, 153, 180) (ICES - 6)	Meats, vegetable oils and fats, milk and dairy products with >2% fat	40 ng/g fat (EU, 2011b)
Residues of pesticides	Milk and cream, not concentrated, nor containing added sugar or sweetening matter, butter and other fats derived from milk, cheese and curd	0.001–2 mg/kg (usually 0.01 mg/kg), depending on the pesticide (EU, 2021)

(CCP 4). The above combinations of temperatures and times can be the critical limit. For sterilized processed cheese, a heat treatment resulting in a 12 log reduction in *C. botulinum* spores is sought. For the control of microbial growth in cold-filled end products during storage, the critical limit is the correctly labeled storage temperature and shelf life (CCP 4). As for hot-filled pasteurized but also for sterilized processed cheese, the storing in buffer tank-packaging-labeling steps is not a CCP, the shelf life is not a critical point and is determined by organoleptic quality factors and not by microbiology. Regarding the removal of metal fragments and other physical hazards, the critical limit is the size of the filter (200–800 μm) which depends on the viscosity of the molten cheese blend and the pressure created by the feeding pump (CCP 3). For

Table 13.10 Maximum levels of European Union for chemical contaminants in processed cheese.

Substance	Product	Maximum level
Tin (inorganic)	Canned processed cheese	200 mg/kg (EU, 2006)
Dioxins (WHO-PCDD/F-TEQ)	Processed cheese with >2% fat	2.5 pg/g fat (EU, 2011b)
Dioxins + dioxin-like PCBs (WHO-PCDD/F-PCB-TEQ)	Processed cheese with >2% fat	5.5 pg/g fat (EU, 2011b)
PCBs (PCB 28, 52, 101, 138, 153, 180) (ICES - 6)	Processed cheese with >2% fat	40 ng/g fat (EU, 2011b)
Residues of pesticides	All processed cheese	0.001-2 mg/kg (usually 0.01 mg/kg), depending on the pesticide (EU, 2021)

the foreign material detection (CCP 5), the critical limit is typically a size of 5 mm. However, an advanced metal detector or X-ray inspection system on the production line can prevent contamination of the product by metal objects smaller than 1 mm. Dense nonmetallic contaminants smaller than 3 mm can also be detected by an X-ray inspection system (Anonymous, 2021).

13.4.9 Monitoring procedures at CCPs (Step 9/Principle 4)

Monitoring of CCPs is the scheduled measurement or observation at a CCP relative to its critical limits. The monitoring procedures should be able to detect a deviation (or a trend towards a deviation) at the CCP. Monitoring of CCPs should be continuous. In the event that it cannot be continuous, the frequency of monitoring should be sufficient to notice if the critical limit has been met (CAC, 2020a).

Physical and chemical measurements are usually preferred to microbiological testing because physical and chemical tests can be done rapidly and can often indicate the control of microbial hazards associated with the product and/or the process (CAC, 2020a).

The monitoring system shall consist of relevant procedures, instructions and records that cover the following (ASSIFONTE, 2018):

- Who is to perform monitoring and checking (responsibility and authority related to monitoring and evaluation of monitoring results),
- When monitoring and checking is performed (monitoring frequency),
- How monitoring and checking is performed (monitoring devices used),
- A record of requirements and methods.

The HACCP plan (Table 13.5) shows the monitoring procedures for the five CCPs identified in the manufacture of processed cheese. The method and frequency of monitoring are as follows (ASSIFONTE, 2018; EDA/EUCOLAIT, 2018).

On each lot receipt, the raw material and other ingredients (CCP 1) are examined for staphylococcal enterotoxins with immunoenzymatic or molecular biology methods, for mites and vermin macroscopically, and for mycotoxins, drug and pesticide residues, heavy metals, dioxins, and PCBs with the corresponding rapid or analytical methods.

During processing (CCP 2) and storing in buffer tank-packaging (CCP 4), the heating time and temperature should be monitored, preferably continuously. For labeling (CCP 4), the checking of each new packaging used is necessary. Filtering (CCP 3) monitoring consists of the automatic removal of foreign bodies, while foreign material detection systems (CCP 5) check each unit of product in the primary packaging, during production.

13.4.10 Corrective actions (Step 10/Principle 5)

Corrective actions should be developed for each CCP, in order to effectively respond to deviations when they occur at the critical limits. The corrective actions should ensure that the CCP has been brought under control and food that is potentially unsafe is handled appropriately and does not reach consumers (CAC, 2020a). Actions taken should include segregating the affected product and analyzing its safety to ensure proper disposition. It may be determined that the product can be reprocessed or diverted to another use. In other situations, the product may need to be destroyed (CAC, 2020a). A root-cause analysis should be conducted where possible to identify and correct the source of the deviation in order to minimize the potential for the deviation to reoccur (CAC, 2020a). The HACCP plan (Table 13.5) shows the planned corrective actions for each of the five CCPs identified in the manufacture of processed cheese, as follows (ASSIFONTE, 2018; EDA/EUCOLAIT, 2018).

When natural cheese and other ingredients (CCP 1) contain staphylococcal enterotoxins, mites and vermin or their chemical contaminants exceed defined limits, the contaminated material should be removed, discarded, and used as an animal by-product (EU, 2009). In the case of purchased materials, the rejected lots should be returned to the supplier. If a decrease in temperature or time is observed during processing (CCP 2) and storing in buffer tank-packaging (CCP 4), the product should be reworked and the treatment conditions should be restored (e.g., adjustment of flow, steam feed, holding time prolongation, etc.). If the storage temperature and shelf life are not correctly labeled (CCP 4), the packaging material in question should be rejected and the cause should be identified.

Concerning the filtering (CCP 3), at the end of each production run, it will be necessary to remove the foreign bodies retained by the filter and to identify their origin in order to put in place any additional previous control measures. In the event that the filter is damaged, it should be replaced, and the product should be discarded, if the foreign material detection system (CCP 5) is not sensitive enough. For the latter, if a contaminated package is detected, it is rejected and handled as a nonconforming product. The origin of the metal fragment or other foreign material should be identified. If origination comes from steps prior to the filter system, the reason why the filter system did not remove the hazard should be investigated. Actions to prevent or minimize further foreign bodies from various sources should be considered.

13.4.11 Verification (and validation) procedures (Step 11/Principle 6)

13.4.11.1 Validation of the HACCP plan

Before the HACCP plan can be implemented, its validation is needed; this consists of making sure that the following elements together are capable of ensuring control of the significant hazards relevant to the food business: identifying the hazards, critical control points, critical limits, control measures, frequency and type of monitoring of CCPs, corrective actions, frequency and type of verification and the type of information to be recorded. Validation could include a review of the scientific literature, the use of mathematical models, the performance of validation studies, and/or the use of guidance developed by authoritative sources (CAC, 2013a, 2020a).

13.4.11.2 Verification procedures

After the HACCP system has been implemented, procedures should be established to confirm that the HACCP system is working effectively. Verification activities include: reviewing monitoring records, reviewing corrective action records, calibrating or checking the accuracy of instruments used for monitoring, observing that control measures are being conducted in accordance with the HACCP plan, sampling and testing for biological, chemical and physical hazards, sampling and testing the environment for microbial contaminants, reviewing the HACCP system, etc. (CAC, 2020a).

Verification should be carried out by someone other than the person who is responsible for performing the monitoring and corrective actions. The frequency of verification activities should be sufficient to confirm the effectiveness of the HACCP system (CAC, 2020a). The HACCP plan (Table 13.5) shows the verification procedures for each of the five CCPs identified in the manufacture of processed cheese, as follows (ASSIFONTE, 2018; EDA/EUCOLAIT, 2018).

In order to verify the presence of staphylococcal enterotoxins, natural cheese and other ingredients (CCP 1) are tested for the development of coagulase-positive staphylococci. It is generally recognized that there is no risk of toxin formation as long as levels of *S. aureus* are kept below 1×10^5 cfu/g. A microscopic control for mites and vermin, and a set of chemical analyses performed by an external laboratory for chemical contaminants are necessary.

Verification of the effectiveness of heat treatment during processing (CCP 2) and storing in buffer tank-packaging (CCP 4) is achieved by microbiological examination of the finished processed cheese for the survival of pathogenic microorganisms. Another way is to have thermometers calibrated annually by an accredited calibration body. In addition, milk and fluid dairy products (such as the molten processed cheese) subjected to pasteurization should show a negative alkaline phosphatase (ALP) reaction immediately after the heat treatment (CAC, 2009). Regarding the verification of commercial sterilization of milk and milk products, sterilized products must be microbiologically stable at room temperature, either measured after storage until end of shelf life or incubated at 55°C for 7 days (or at 30°C for 15 days) in accordance with appropriate standards (CAC, 2009). For labeling (CCP 4) with the correct storage temperature and shelf life, a daily check of the marking setting of the labeling machine is necessary.

Filtering (CCP 3) is verified by inspecting the integrity of the filter prior to each production run, while the metal detector or X-ray inspection equipment (CCP 5) should be checked twice per shift with a seeded sample, and calibrated annually by an accredited calibration body.

13.4.12 Documentation and record keeping (Step 12/Principle 7)

Efficient and accurate documentation and record keeping are essential to the application of a HACCP system. Documents that should be developed are the composition of the HACCP team, hazard analysis, determination of CCPs, determination of critical limits, validation of control measures, and modifications of the HACCP plan. Records that should be kept are CCP monitoring activities, deviations and corrective actions, and verification (CAC, 2020a), as in the HACCP plan of processed cheese manufacture (Table 13.5).

13.4.13 Training

Training of personnel in food businesses, government and academia in HACCP principles and applications is an essential element for the effective implementation of HACCP. As an aid in developing specific training to support a HACCP plan, working instructions and procedures should be developed which define the tasks of the operating personnel in charge of each CCP (CAC, 2020a). For this purpose, Food and Agriculture Organization of the United Nations (FAO) has published a training manual on food hygiene and the HACCP system (FAO, 1998) and a guidance on the application of HACCP in small and/or less-developed food businesses (FAO/WHO, 2006), intended for food control personnel, including government authorities and food industry personnel carrying out food quality and safety assurance programs (CAC, 2020a).

Training programs should be designed to address the concepts at a level appropriate for the knowledge and skill level of the personnel being trained. Training programs should be reviewed periodically and updated where necessary (CAC, 2020a).

13.5 Quality assurance of processed cheese

In this section, quality assurance of processed cheese results from a review of the requirements of US processed cheese legislation (LII, 2021a,d). The Food and Drug Administration (FDA) defines three categories of processed cheese: pasteurized process cheese (PC), pasteurized process cheese food (PCF) and pasteurized process cheese spread (PCS). All three categories include a second variety with fruits, vegetables, or meats, while the first category also includes a third variety with pimento.

A generalized definition of processed cheese of these three categories is the food prepared by comminuting and mixing, with the aid of heat, one or more natural cheeses of the same or two or more varieties, with an emulsifying agent into a homogeneous plastic mass. One or more optional ingredients may be used. The three categories are distinguished from each other by different requirements. As a set of process-oriented activities, quality assurance refers to the processes related to the design and production of processed cheese products (PC, PCF, and PCS). In general, the processes described refer to the PC and also apply to the other two categories of products (PCF, PCS), except for the specific cases indicated in the text.

13.5.1 Design of processed cheese

The product design includes the quality specifications of the finished products and raw materials.

13.5.1.1 Quality specifications of finished products

They relate to the type and quantity of raw materials allowed, heat treatment, and labeling.

a) *Natural cheeses*: One or more cheeses of the same or two or more varieties are used for the manufacture, except cream cheese, Neufchatel cheese, cottage cheese, low fat cottage cheese, cottage cheese dry curd, cook cheese, hard grating cheese, semisoft part-skim cheese, part-skim spiced cheese, and skim milk cheese.

The weight of each variety of cheese in the finished product made from two varieties of cheese is $\geq$25% of the total weight of both, except for blue cheese, Nuworld cheese, Roquefort cheese or Gorgonzola cheese ($\geq$10%), and for Limburger cheese ($\geq$5%). The weight of each variety of cheese in the finished product made from three or more varieties of cheese is $\geq$15% of the total weight of all, except for blue cheese, Nuworld cheese, Roquefort cheese or Gorgonzola cheese ($\geq$5%) and Limburger cheese ($\geq$3%). These limits do not apply to the quantity of Cheddar cheese, washed curd cheese, Colby cheese and granular cheese in mixtures which are designated as "American cheese". Such mixtures are considered as one variety of cheese.

For PCF and PCS, a smaller amount of natural cheese is permitted, that is, the weight of the cheese ingredient is $\geq$51% of the weight of the finished product (LII, 2021d).

b) *Emulsifying agents*: The emulsifying agents added, either alone or in combination, at $\leq$3% w/w of the finished product, are monosodium phosphate, disodium phosphate, trisodium phosphate, dipotassium phosphate, sodium hexametaphosphate, sodium acid pyrophosphate, tetrasodium pyrophosphate, sodium aluminum phosphate, sodium citrate, potassium citrate, calcium citrate, sodium tartrate, and sodium potassium tartrate (LII, 2021d).

c) *Optional ingredients*: These are acidifying agents (vinegar, lactic acid, citric acid, acetic acid and phosphoric acid), milk fat products (cream, anhydrous milk fat and dehydrated cream giving fat <5% w/w of the product), water, salt, artificial coloring, spices or flavorings, mold-inhibiting ingredient (sorbic acid, potassium sorbate and sodium sorbate at $\leq$0.2% w/w, or sodium propionate and calcium propionate at $\leq$0.3% w/w) for slices and cuts), anti-sticking agent (lecithin at ≤0.03% w/w for slices and cuts), enzyme modified cheese, smoked cheese, or substances from wood smoke.

For PCF and PCS, the addition of optional dairy ingredients is allowed, such as milk, skim milk, buttermilk, cheese whey, albumin from cheese whey and skim milk cheese. For PCS, the addition of other optional ingredients is also allowed, such as hydrocolloids (carob bean gum, gum karaya, gum tragacanth, guar gum, gelatin, sodium carboxymethylcellulose, carrageenan, oat gum, sodium alginate, propylene glycol alginate and xanthan gum, at ≤0.8% w/w in the finished product), sweetening agents (sugar, dextrose, corn sugar, corn syrup, corn syrup solids, glucose syrup, glucose syrup solids, maltose, malt syrup, and hydrolyzed lactose, in *quantum satis*), and nisin (≤250 ppm in the product). When hydrocolloids are used, dioctyl sodium sulfosuccinate may be added in a quantity not in excess of 0.5% by weight of such ingredients (LII, 2021d).

d) *Pasteurization*: All processed cheese products (PC, PCF, PCS) are heated for ≥30 s at a temperature ≥65.6°C (150°F). In this case, the pasteurization conditions are a validation procedure (LII, 2021d).

e) *Labeling*: The names of the varieties of cheese used in the manufacture shall appear in the name of the finished product, except in the case of cheeses designated as "American cheese". The full name of the food shall appear on the principal display panel of the label in type of uniform size, style, and color. The name of the food shall include a declaration of any flavoring or spice that characterizes the product. Each of the ingredients used in the food shall be declared on the label, except cheeses designated as "American cheeses" (LII, 2021d).

13.5.1.2 Quality specifications of raw materials

The raw materials used in the manufacture of processed cheese shall meet the following specifications (LII, 2021a):

a) *Cheddar cheese, Colby cheese, washed or soaked curd, granular or stirred curd cheese*: They shall at least meet the requirements for Standard Grade (USDA, 1991). They shall possess a pleasing and desirable taste and odor consistent with the age of the cheese, and shall have body and texture characteristics which will impart the desired body and texture characteristics in the finished product. They shall also possess finish and appearance characteristics which will permit removal of all packaging material and surface defects.

b) *Swiss cheese*: It shall be Standard Grade or better (USDA, 2001b), except that the cheese may be blind or possess finish characteristics which do not impair the interior quality.

c) *Gruyere cheese*: It shall be of good, wholesome quality and, except for smaller eyes and sharper flavor, shall meet the same requirements as for Swiss cheese.

d) *Cream cheese, Neufchatel cheese*: Mixed with other foods, or used for spreads and dips, they shall possess a fresh, pleasing, and desirable flavor.

e) *Cream, plastic cream, and anhydrous milk fat*: They shall be pasteurized, sweet, have a pleasing and desirable flavor and be free from objectionable flavors.

f) *Nonfat dry milk*: It shall meet the requirements of Extra Grade (USDA, 2001a), except that the moisture content may be in excess of that specified for the particular grade.

g) *Whey*: Condensed or dry whey used shall meet the requirements for Extra Grade (USDA, 2000), except that the moisture requirement for dry whey may be waived.

h) *Flavor ingredients*: They shall be permitted for processed cheese, and in no way deleterious to the quality or flavor of the finished product. In the case of bulky flavoring ingredients such as pimento, the particles shall be, at least to a reasonable degree, uniform in size, shape, and consistency. The individual types of flavoring materials shall be uniform in color and shall impart the characteristic flavor desired in the finished product.

i) *Coloring*: It shall be annatto or any other permitted cheese or butter color.
j) *Acidifying agents*: If used, they shall be permitted for processed cheese.
k) *Salt*: It shall be free flowing, permitted white refined sodium chloride.
l) *Emulsifying salts*: They shall be permitted for processed cheese and shall be free from extraneous material.

13.5.2 Production of processed cheese

The production includes all the following steps of the production procedure of processed cheese.

13.5.2.1 Selection of cheese for processing

A representative sample shall have been examined to determine fat and moisture content. One sample unit from each vat of cheese shall have been examined to determine the suitability of the vat for use in processed cheese products in accordance with the flavor, body and texture characteristics permitted as applicable, and to determine the characteristics it will contribute to the finished product when blended with other cheese. The cheese included in each blend shall be selected on the basis of the desirable qualities which will result in the desired finished product. Rework from equivalent blends, and hot cheese from the filler may be added to the cooker in amounts which will not adversely affect the finished product (LII, 2021a).

13.5.2.2 Blending

Each vat of cheese should be divided and distributed throughout numerous cooker batches, the purpose being to minimize the preponderance and consequent influence of any one vat on the characteristics of the finished product, and to promote as much uniformity as is practical. In blending, the final composition requirements for fat and moisture shall also be considered. Quantities of salt, color, emulsifier and other allowable ingredients to be added shall be calculated and predetermined for each cooker batch (LII, 2021a).

13.5.2.3 Trimming and cleaning

The natural cheese shall be cleaned free of all nonedible portions. Paraffin and bandages as well as rind surface, mold or unclean areas or any other part which is unwholesome or unappetizing shall be removed (LII, 2021a).

13.5.2.4 Cutting and grinding

The trimmed and cleaned cheese should be cut into sections of convenient size to be handled by the grinder or shredder. The grinding and mixing of the blended lots of cheese should be done in such a manner as to insure a homogeneous mixture throughout the batch (LII, 2021a).

13.5.2.5 *Adding optional ingredients*

As each batch is added to the cooker, the predetermined amounts of salt, emulsifiers, color, or other allowable optional ingredients shall be added. However, a special blending vat may be used to mix the ground cheese and other ingredients before they enter the cooker to provide composition control (LII, 2021a).

13.5.2.6 *Cooking the batch*

Each batch of cheese within the cooker, including the optional ingredients, shall be thoroughly commingled and the contents pasteurized at a temperature of at least 70°C (158°F) and held at that temperature for not less than 30 s or any other equally effective combination of time and temperature. Care shall be taken to prevent the entrance of cheese particles or ingredients after the cooker batch of cheese has reached the final heating temperature. After holding for the required period of time, the hot cheese shall be emptied from the cooker as quickly as possible (LII, 2021a).

13.5.2.7 *Forming containers*

Containers either lined or unlined shall be assembled and stored in a sanitary manner to prevent contamination. The handling of containers by filler crews should be done with extreme care and observance of personal cleanliness. Preforming and assembling of pouch liners and containers shall be kept to a minimum and the supply rotated to limit the length of time exposed to possible contamination prior to filling (LII, 2021a).

13.5.2.8 *Filling containers*

Hot fluid cheese from the cookers may be held in hotwells or hoppers to assure a constant and even supply of processed cheese to the filler or slice former.

Filler valves shall effectively measure the desired amount of product into the pouch or container in a sanitary manner and shall cut off sharply without drip or drag of cheese across the opening. An effective system shall be used to maintain accurate and precise weight control. Damaged or unsatisfactory packages shall be removed from production, and the cheese may be salvaged into sanitary containers, and added back to cookers (LII, 2021a).

13.5.2.9 *Closing and sealing containers*

Pouches, liners, or containers having product contact surfaces, after filling shall be folded or closed and sealed in a sanitary manner, preferably by mechanical means, so as to assure against contamination. Each container in addition to other required labeling shall be coded in such a manner as to be easily identified as to date of manufacture by lot or sublot number (LII, 2021a).

13.5.2.10 Cooling the packaged cheese

After the containers are filled, they shall be stacked, or cased and stacked in such a manner as to prevent breaking of seals due to excessive bulging and to allow immediate progressive cooling of the individual containers of cheese. As a minimum, the cheese should be cooled to a temperature of 37.8°C (100°F) or lower within 24 h after filling (LII, 2021a). For sliceable processed cheese (block-type) cooling is usually performed slowly (10–12 h), whereas for processed cheese spreads it is performed at a fast rate (15–30 min), in order to facilitate the formation of the required consistencies (Gunasekaran and Ak, 2003). The temperature of the cheese should be reduced further, before being shipped or if storage is intended.

13.5.2.11 Storage

The pasteurization process yields cheeses with exceptionally long shelf lives. Recommended storage temperatures are from 0 to 4°C (32–39°F). Freezing is not recommended. When retained in the original factory sealed packaging, their shelf life is indeterminate and often measured in years. Sterilized products are shelf stable and require no refrigeration (Anonymous, 2016).

13.6 Sampling

In order to carry out quality control, sampling should be preceded in order to obtain representative samples of the product.

13.6.1 Sampling procedure

ISO provides guidance on methods of sampling milk and milk products for microbiological, chemical, physical and sensory analysis, except for (semi)automated sampling. The main instructions applicable to all natural and processed cheeses are as follows (ISO, 2008):

a) *Sampling personnel*: An authorized person, properly trained in the appropriate technique, for example, for microbiological purposes, and free from any infectious disease, shall perform sampling.
b) *Sampling equipment*: It consists of cheese triers of shape and size appropriate to the cheese to be sampled, knife with a pointed blade, spatula, and cutting wire of sufficient size and strength. Sampling equipment should be made of stainless steel, or other suitable material of adequate strength. All surfaces should be smooth and free from crevices. Sampling equipment should be clean and dry before use. For microbiological examination, it should also be sterilized.
c) *Sample containers*: Depending on the sample, containers should be the correct shape and size. Appropriate container materials include opaque glass, metallic materials (e.g., stainless steel) and plastics (e.g., polypropylene). Sample containers need to be airtight sealed to prevent contamination and air ingress. For microbiological examination, they should also be sterilized.

d) *Sampling*: It is carried out, depending upon the shape, mass and type, either by taking entire small cheeses or prepacked portions or by removing sectors, slices, or cores using the sampling equipment. A minimum recombined sample size of 100 g and a sampling temperature of 1–5°C are recommended. The heterogeneity of the product should be taken into account when collecting samples. Immediately after sampling, the samples should be placed in a sample container. The sample may be cut into pieces for insertion into the container, but it should not be compressed or ground. Storage of cheese samples tightly wrapped in aluminum foil, waxed paper (cheese paper) or closable plastic bags inside or even outside a sample container is an additional measurement to prevent molding of the cheese surface. Unless otherwise specified and whatever the method of sampling used, the sample should include any surface layer of the cheese (such as mold and rind). Samples should be taken at least in duplicate. For microbiological examination, samples should always be taken first using aseptic techniques. Whenever possible, they should be taken from the same cheese as those for chemical and physical analysis and for sensory examination.

e) *Sample labeling and sampling report*: Information on the product, sample, sampling personnel and conditions should be stated on the attached sample label. More details should be given in the sampling report.

f) *Storage and transportation*: The state of the sample at the time of sampling shall remain essentially unaltered until the time of initiation of the test procedure. The storage temperature of the cheese sample is 1–5°C. Samples should be dispatched immediately after sampling to the testing laboratory and within 24 h. After preparation of the test portion, analysis should be carried out immediately.

13.6.2 Sampling plans

For all milk products, both sampling for inspection by attributes and sampling for inspection by variables apply.

13.6.2.1 Sampling for inspection by attributes

The sampling for inspection by attributes is based on classifying a unit of a sample, taken at random, as "good" or "defective". A "good" unit is one which meets the requirements of a specification, while a "defective" unit is one which does not. In order to choose the size of a representative sample, adequate sampling plans are applied to the milk and milk products presented in separate lots, regardless of whether the lots are from the same production. The plans are indexed by batch or lot size and acceptance quality limit (AQL). AQL is the quality level (in percent nonconforming items or in nonconformities per hundred items) that is the worst tolerable process average, when a continuing series of lots is submitted for acceptance sampling. The acceptance or rejection of a lot is determined by the respective acceptance number (Ac) and rejection number (Re). The first is the maximum number of defective units permitted in the sample, while the second is the maximum number of defective units required in the sample to reject the lot (ISO, 2004).

Before selection of a sampling plan, all critical, major and minor defects should be defined.

Critical defect is a failure that renders the product unacceptable. It relates to the presence of contaminants at a critically high level. Examples include heavy metals and pesticide residues. The acceptance number for critical defects is always zero. When the inspection is nondestructive, it must be 100% inspection. If the inspection is destructive, a sample is needed, the size of which results from a formula relating a certain percentage of defectives to an acceptable risk of not detecting a defective (ISO, 2004).

Major defect is a failure liable to render the product unfit for use; that is, in the case of milk and milk products, unfit for sale to the consumer. A major defect would result in the product spoiling or becoming unfit for sale or processing. Examples include (a) composition defect, where this would affect keeping quality, (b) contamination with inhibitory substances, (c) integrity of packaging, and (d) visible contamination with dirt. Sampling plans for major defects shall be selected using an AQL of not more than 6.5% (ISO, 2004).

Minor defect is a failure to comply with a specification, but which does not render the unit unfit for use and sale, nor cause it to spoil. Examples include (a) a unit with a chemical composition or net content outside, but close to, a specification limit, and (b) small abnormalities in appearance. Sampling plans for minor defects shall be selected using an AQL of not more than 10% (ISO, 2004).

Next, the inspection level must be selected, namely the relation of the sample size compared to the lot size. The inspection level designates the relative amount of inspection. There are three general inspection levels (I, II, and III). Unless otherwise specified, level II shall be used. Level I may be used when less discrimination is needed or level III when greater discrimination is required. Four additional special levels (S-1, S-2, S-3, and S-4) may be used where relatively small sample sizes are necessary and larger sampling risks can be tolerated (ISO, 1999). The number of units per sample to be checked increases in the following order: S-1, S-2, S-3, S-4, I, II, and III.

It is then necessary to choose the inspection severity, namely the relation of the acceptance and rejection numbers compared to the sample size. In terms of inspection severity, normal inspection, tightened inspection and reduced inspection are distinguished (ISO, 1999):

a) *Normal inspection*: It is the use of a sampling plan with an acceptance criterion that has been devised to secure the producer a high probability of acceptance when the process average of the lot is better than the AQL. Normal inspection is used when there is no reason to suspect that the process average differs from an acceptable level.

b) *Tightened inspection*: It is the use of a sampling plan with an acceptance criterion that is tighter than that for the corresponding plan for normal inspection. Tightened inspection is invoked when the inspection results of a predetermined number of consecutive lots indicate that the process average might be poorer than the AQL.

c) *Reduced inspection*: It is the use of a sampling plan with a sample size that is smaller than that for the corresponding plan for normal inspection and with an acceptance criterion that is comparable to that for the corresponding plan for normal inspection.

The discriminatory ability under reduced inspection is less than under normal inspection. Reduced inspection may be invoked when the inspection results of a predetermined number of consecutive lots indicate that the process average is better than the AQL (ISO, 1999).

For milk and milk products of different lot sizes, Ac and Re are obtained from sampling plans based on AQL of 2.5, 4, 6.5 and 10% at inspection level I, S-4, S-3, S-2 and S-1, for normal, tightened, and reduced inspection. Inspection level I is preferred (ISO, 2004).

In the case of lots in isolation (isolated sequences of lots, an isolated lot, a unique lot or a short series of lots), an acceptance sampling system for inspection by attributes indexed by a series of specified values of limiting quality (LQ) is used. LQ is the quality level, when a lot is considered in isolation, which, for the purposes of acceptance sampling inspection, is limited to a low probability of acceptance. It is mostly equated to the consumer's risk quality, which corresponds to a consumer's risk (probability of acceptance at the LQ) usually below 0.10 (10%), except in some instances (ISO, 2020).

13.6.2.2 Sampling for inspection by variables

Inspection by variables is a method which consists in measuring a quantitative quality characteristic for each item of a sample taken from a population (lot). The acceptability of the population is established statistically from the number of items in the sample that meet specified requirements (ISO, 1988).

In the lot-by-lot inspection, for one or more independent quality characteristics, single sampling plans indexed by AQL have been designed for use under the following conditions: (i) the inspection procedure is applied to a continuing series of lots of discrete products all supplied by one producer using one production process; (ii) the quality characteristics of the items of product are measurable on a continuous scale; (iii) the production is stable (under statistical control) and the quality characteristics are distributed, at least to a close approximation, according to normal distributions; (iv) a contract or standard defines a lower specification limit, L, an upper specification limit, U, or both; an item is qualified as conforming if and only if its measured quality characteristic, x, satisfies the appropriate one of the following inequalities: $x \geq L$ (i.e., the lower specification limit is not violated); $x \leq U$ (i.e., the upper specification limit is not violated); $x \geq L$ and $x \leq U$ (i.e., neither the lower nor the upper specification limit is violated) (ISO, 2013b, c).

In an acceptance sampling plan by variables, acceptability constant, k, a constant depending on the specified value of the AQL and the sample size, is used in the criteria for accepting the lot (ISO, 2013b, c). For milk and milk products of different lot sizes, acceptability constant is obtained from sampling plans based on AQL of 1, 1.5, 2.5, 4, 6.5 and 10% at inspection level I, for normal, tightened, and reduced inspection (ISO, 1988).

13.7 Quality control of processed cheese

In this section, quality control of processed cheese results from a review of the requirements of US processed cheese legislation (LII, 2021a,d). As a reminder, the three categories of processed cheese defined by the FDA are: pasteurized process cheese (PC), pasteurized process cheese (PCF) and pasteurized process cheese spread (PCS). As a set of product-oriented activities, quality control mainly refers to the chemical analyses, examination of sensory characteristics, keeping quality estimation and weight control of the finished product.

13.7.1 Chemical analyses

For the natural cheese (raw material), a representative sample shall have been tested prior to usage to determine its moisture and fat content. For the processed cheese, as many samples shall be taken direct from the cooker, hopper, filler or other location as is necessary to assure compliance with composition requirements (moisture content, fat content, and pH). Efficiency of adequate pasteurization is also tested (LII, 2021d).

a) *Moisture content*: The moisture content of a PC made from a single variety of cheese is ≤43% w/w, except for a PC made from washed curd cheese or Colby cheese (≤40%), Swiss cheese or Gruyere cheese (≤44%) and Limburger cheese (≤51%).

The moisture content of a PC made from two or more varieties of cheese is ≤43%, except for a PC made from a mixture of Cheddar cheese, washed curd cheese, Colby cheese and granular cheese (≤40%), and from a mixture of Swiss cheese and Gruyere cheese (≤44%). The same applies to the other two categories of processed cheese, with the difference that the moisture content of a PCF is ≤44% w/w and that of a PCS is >44% but ≤60% w/w.

b) *Fat content*: The fat content of the solids of a PC made from a single variety of cheese is ≥43% w/w, except for a PC made from Gruyere cheese (≥45%). The fat content of the solids of a PC made from two or more varieties of cheese is ≥47% w/w, except for a PC made from a mixture of Swiss cheese and Gruyere cheese (≥45%). The same applies to the other two categories of processed cheese, with the difference that the fat content of a PCF is ≥23% w/w and that of a PCS is ≥20% w/w.

c) *pH*: The pH of the PC, PCF and PCS is ≥5.3, ≥5.0, and ≥4.0, respectively.

d) *Alkaline phosphatase activity*: Spot checks should be made on samples from the cooker as frequently as is necessary to indicate efficient pasteurization by means of the alkaline phosphatase test, as well as any other tests necessary to assure good quality control. The alkaline phosphatase test is used as a verification procedure, and shall give ≤3 μg of phenol equivalent per 0.25 g of processed cheese product (≤12 μg/g).

13.7.2 Examination of sensory characteristics

As many representative samples shall be taken from production as is necessary to assure meeting the required sensory characteristics of the products. The samples shall be examined at approximately 21°C (70°F) the first day of operation after the date of processing for the following characteristics: flavor, body and texture, color, and finish and appearance (LII, 2021a).

a) *Flavor*: Processed cheese shall have a pleasing and desirable mild cheese taste and odor characteristic of the variety or varieties of cheese ingredients used. If additional optional ingredients are used they shall be incorporated in accordance with good commercial practices and the flavor imparted shall be pleasing and desirable. Processed cheese may have a slight cooked or very slight acid or emulsifier flavor, and be free from any undesirable tastes and odors. Especially for the processed cheese spread, its cheese taste is not required to be mild, while it is allowed to have an acid flavor.

b) *Body and texture*: Processed cheese shall have a medium-firm, smooth and velvety body free from uncooked cheese particles. It is resilient and not tough, brittle, short, weak, or sticky. It shall be free from pin holes or openings except those caused by trapped steam. The cheese shall slice freely, and shall not stick to the knife or break when cut into approximately 1/8 inch slices. If in sliced form, the slices shall separate readily. Especially for the PCS, instead of having a medium-firm and velvety body, it is required to form a homogeneous plastic mass spreadable at around 21°C (70°F).

c) *Color*: Processed cheese may be colored or uncolored but shall be uniform throughout. If colored, it shall be bright and not be dull or faded. To promote uniformity and a common reference to describe color, the color designations as depicted by the National Cheese Institute standard color guide for cheese shall be used (or may use for processed cheese spread).

d) *Finish and appearance*: The wrapper may be slightly wrinkled but shall envelop the cheese, adhere closely to the surface, and be completely sealed and not broken or soiled. In addition for the processed cheese spread, other containers made of suitable materials shall be completely filled.

13.7.3 Keeping quality estimation

During processing or preferably from the cooled stock, sufficient samples shall be selected at random from the production run. The samples should be stored at approximately 10°C (50°F) for 3 months for evaluation of sensory characteristics. Additional samples may be selected and held at different temperatures or times (LII, 2021a).

13.7.4 Weight control

During the filling operation as many samples shall be randomly selected and weighed from each production run as is necessary to assure accuracy of the net weight established for the finished products (LII, 2021a).

13.8 Further quality control determinations

Processed cheese quality control involves the mandatory and basic quality control checks required to ensure the safety and quality of end products, which have been described previously. Manufacturers and R&D specialists may choose to assess some additional parameters regarding the quality of end products, in order to further strengthen the safety of processing, detect defects and carefully assess the quality specifications of processed cheese. Further quality control determinations could involve the following:

13.8.1 Intact casein

One of the main factors that can dictate body/texture consistency of processed cheese is intact casein, which is the amount of casein that has not been hydrolyzed during ripening of cheese. Intact casein is calculated from the level of nitrogen insoluble at pH 4.6 (casein isoelectric pH) (Guinee, 2009). It is generally expressed as relative casein content, a ratio of the amount of unhydrolyzed casein nitrogen to the total nitrogen in cheese. Determinations of intact casein can be significant for the selection of the ideal type/age of natural cheese which should be used for the manufacture of processed cheese of specific quality characteristics. The intact casein content of natural cheese is inversely related to its age. The level of intact casein decreases during maturation as a result of proteolysis (Garimella Purna et al., 2006). Other dairy ingredients can provide intact casein to processed cheese, such as nonfat dry milk or milk protein concentrate. It has been shown that the reduction of intact casein significantly reduces the firmness of processed cheese products and increases the extent of flow upon heating (meltability) (Kapoor et al., 2007). Garimella Purna et al. (2006) studied the effect of aging of Cheddar cheese on the properties of processed cheese foods of similar pH (5.65–5.75). The properties determined were apparent viscosity after manufacture, firmness of unmelted samples, hot apparent viscosity, and product thickening during cooling. Results showed that all parameters were significantly affected by natural cheese age. In general, block processed cheese requires high proportions of young cheese (75–90% relative casein content) whereas predominantly medium-ripe cheese (60–75% relative casein content) is required for processed cheese spreads (Fox et al., 2017).

Also, as the content of intact or relative casein decreases the amount of emulsifying salt to be added decreases. That is, if the total protein content is 30% and the relative casein content is 100%, the required amount of emulsifying salt (polyphosphate) will be 3.6%. The latter will be only 2.1% for the same total protein content but for a relative casein content of 50% (Meyer, 1973).

13.8.2 Free oil formation

Free oil formation is the tendency for liquid fat to separate from a sample of melted cheese and accumulate in the form of fat pockets. Extensive aggregation and rupture of fat globules contributes to the formation of free oil and the determination of this property may be an indicator of the stability of emulsions during storage (Kindstedt and Fox, 1991). In general, a certain amount of free oil is necessary to avoid local dehydration on the surface of processed cheese and the parallel appearance of surface browning during heating. In addition, fat globules that are partially broken down over time produce a small amount of free oil that is desirable for certain properties, such as increased spreadability, surface sheen or melting capacity (Everett and Auty, 2008). Low moisture in nonfat substances and low protein content have previously been shown to enhance free oil in pasteurized processed cheese samples, which has been attributed to the reduced ability of proteins to trap fat (Acharya and Mistry, 2005) due to their reduced hydration (Lee et al., 2004). Free oil in processed cheeses has been reported to range from 0% to 20% (Acharya and Mistry, 2005; Hassan et al., 2007; Chatziantoniou et al., 2015), as determined with the modified Gerber test (Kindstedt and Fox, 1991), which is an accelerated fat globule coalescence method, involving the use of a centrifugal force to recover free oil from heated samples.

13.8.3 Water activity

The shelf life of processed cheese depends significantly on its water activity (a_w). The value of a_w of processed cheese products ranges from $\approx$0.91 to 0.97 (Fox et al., 2017), and is usually between 0.93 and 0.96 (Kapoor and Metzger, 2008; Buňkova and Buňka, 2015). The environment of lower fat processed cheeses is usually marked by high a_w, which influences microbial growth. Products with higher a_w, require high temperature processing to eliminate microbial growth throughout the product's shelf life and are usually sterilized, especially if destined for storage at room temperatures. Different ingredients can modify the values of a_w. For example, added lactose (up to 5%, w/w) has been shown to result in lower a_w and lower spreadability of processed cheese products. However, these products show an increased propensity to nonenzymatic browning during processing (especially at high temperatures) and storage (Guinee, 2009). Spores that survive heating are prevented from growing by a_w of the product, among other factors. Table 13.11 details the effect of a_w on the inhibition of growth of spore-forming bacteria. It is important that a_w of processed cheese end products should be controlled and determined on a regular basis.

13.8.4 Color

The color of processed cheese can be evaluated using sensory tests, as well as through instrumental determinations. Colorimetry can be a useful tool for the inspection of color of processed cheese and can give information relative to the quality or progression of deterioration of finished products. Color of processed cheese has frequently been determined with the aid of colorimeters or spectrophotometers and expressed in the CIE 1976 L*a*b* (or CIELAB) color space.

Processed cheeses should have a glossy surface and uniform color. As fat decreases, processed cheese becomes more opaque. Also, the color is affected by the calcium-protein ratio, since as this ratio decreases, products becomes more opaque (Johnson et al., 2009).

According to Chatziantoniou et al. (2015), the measurement of the color coordinates (L*, a* and b*) of processed whey cheese samples, with the aid of a colorimeter, showed that the brightness (L*) of samples corresponded to off-white samples (90.1–91.7). The positive values of the b* coordinate (6.2–7.2) of all samples indicated the presence of a yellow hue, mainly due to the β-carotene of the fat of the added cream (Kneifel et al., 1992). Finally, the negative sign of the coordinate a* (-1.1 to -1.7) of all samples indicated the presence of a light green hue, organoleptically imperceptible. The green color has also been observed in cream cheese (Wendin et al., 2000). It is

Table 13.11 Effect of water activity (a_w) on the growth of *C. botulinum* and *C. perfringens* (EFSA, 2005).

Strains	Inhibition of growth at a_w values
C. botulinum Group I	< 0.94
C. botulinum Group II	< 0.97
C. perfringens	< 0.94–0.95

characteristic of whey color and is due to the riboflavin content (Paul-Sadhu, 2016). Similar results were reported by Cunha et al. (2010), where samples of spreadable processed cheese analogues with a fat content of 23–24% and a moisture content of 63% showed that the brightness L* was approximately 86, the coordinate a* from -0.2 to -0.7 and the coordinate b* from 9 to 11. Chen and Liu (2012) reported the determined color parameters of processed cheese prepared with different emulsifying salts, where values of L* ranged from 52.3 to 61.8, of a* from 3.2 to 3.8 and of b* from 10.8 to 11.3. The instrumental measurement of cheese color can also be used to determine the amount of cheese colorants to add in colored natural and/or processed cheeses (Sharma et al. 2019).

The determination of color changes can be used for the quality control of processed cheese and can identify the quality deterioration during storage. The color of pasteurized processed cheeses was determined at different storage temperatures for 12 months and it was found that a more intense browning appeared when samples were stored at 37°C, compared to samples stored at 4°C. Specifically, the brightness (L*) and green color (a*) decreased, while the yellow color (b*) increased significantly in the samples stored at 37°C (Kristensen et al., 2001). A more pronounced increase in parameter b* was found during the storage of sterilized processed cheeses (19% fat w/w) stored for 12 weeks at 23°C, than in samples stored at 8°C (Buňka et al., 2008). The instrumental determination of cheese color can also be used to quantify color defects such as pink discoloration (Quigley et al., 2016).

13.8.5 Hardness

Texture of processed cheese may be manipulated to range from soft and spreadable (in spreadable processed cheese) to firm enough to slice or cube (in block-type processed cheese). In general, hardness is the force required to attain a given deformation of the material tested. It depends on the shape and size of the sample or probe used as well as the rate of deformation. Hardness, usually expressed as hardness 1 (H_1), is the force at the end of the first compression or penetration cycle measured by Texture Profile Analysis (TPA) where the sample is subjected to two consecutive cycles of compression or penetration. Hardness can help the manufacturers to assess the texture specifications of end products or the textural-functional properties of novel formulations.

Cylindrical samples of 23 mm in diameter and 21.5 mm in height of block-type processed cheeses that were compressed to 80% deformation at a speed of 80 mm/min showed values of H_1 between 34.6 and 160.9 N (Dimitreli and Thomareis, 2007). In contrast, under similar conditions but with a deformation rate of 20 mm/min, samples of spreadable processed whey cheese with good spreadability showed lower values of hardness 1 (17.5–27.8 N) (Chatziantoniou et al., 2015). Kapoor and Metzger (2004) reported that processed cheeses with whey (up to 9.5%) had hardness ranging from 12 N for spreadable pasteurized processed cheese samples, up to 126 N for pasteurized processed cheese samples, under similar experimental conditions. Brummel and Lee (1990) reported that processed cheese spread firmness increased with hydrocolloid content and the texture of processed cheeses, with 40 and 50% reduced fat, was improved using various hydrocolloids at levels $<1.7\%$ w/w. A cylindrical probe with a diameter of 20 mm and a speed of 2 mm/s was used to assess the hardness of the processed cheese spreads by two sequential penetration events at a deformation of 25%. Hardness 1 was found to range between 1 and 32 N, depending on the concentration and type of emulsifying salts used (Buňka et al., 2014). Generally, there is a positive correlation between the protein content and firmness of processed cheese (Dimitreli and Thomareis, 2007; Guinee and O'Callaghan, 2010; Chatziantoniou et al., 2015).

13.8.6 Meltability

Melt properties of processed cheese end products should be evaluated taking into account the customer expectations. In this respect, product flow can be targeted to free-flowing or completely restricted from flow, depending on its performance requirements. A free-flowing melt is desirable for hamburgers, and in cooking applications. Extra-melt processed cheese is the product which melts easily on heating, does not thicken and can withstand high temperature hold for long periods. Partial melt-restriction, or controlled melt, is typically required for toasted sandwiches and cooked meals, where some flow is desired, but should be limited. Processed cheese food products usually melt quickly to give a smooth liquid. Full melt-restriction, where product maintains its shape at high temperature, otherwise known as slow melt, is useful when processed cheese is used in breaded cheese sticks or as inclusions within further processed foods, for example, sausages (Kapoor and Metzger, 2008; Kussy and Aylward, 2009).

Since heating is conducive to destabilization, during determinations of meltability, a visual evaluation of the appearance of the melted processed cheese can also be conducted, specifically the detection of uneven surface and browning. Excessive formation of bubbles during heating of end products could indicate loosely bound water in processed cheese. Browning during melting may suggest excessive lactose in the product. These indications could help manufacturers determine the type and level of emulsifying agents, lactose, moisture (Kussy and Aylward, 2009), or the functional properties of new formulations. Drake et al. (1999) evaluated the functional properties of reduced fat processed cheese with soy lecithin. They applied a modified Schreiber melt test (Muthukumarappan et al., 1999) and reported case-hardening, scorching, and inconsistent melt diameters. A study on the properties of reduced fat processed cheese spreads reported that texture and meltability was improved through the addition of hydrocolloids, at addition levels <1.7% w/w). Meltability was determined using a modified test of the method described by Olson and Price (1958). Values were 130 mm for the control sample, while reduced fat processed cheese samples had meltability ranging from 72 to 150 mm (Brummel and Lee, 1990). Savello et al. (1989) reported that the meltability of model processed cheese, which contained rennet and acid casein, was 5–80 mm, as determined by the method of Olson and Price (1958). Kapoor and Metzger (2004) studied the meltability of processed cheese with salt whey and reported values of 49–80 mm, as measured with the Schreiber melt test.

13.8.7 Sensory attributes

Sensory evaluation of the natural cheeses used to manufacture processed cheeses is critical in order to obtain finished products with desired organoleptic properties. It is necessary for the cheese selection and blending. Combinations of cheeses can be blended to produce intermediate levels of flavor, for example, flavorful cheese may be combined with mild cheese. Additionally, blending young cheese with older cheese provides a backbone of intact casein, which is useful in developing processed cheese texture.

For the sensory evaluation of the processed cheeses, a sensory lexicon has been described (Drake et al., 1999) and further developed (Drake et al., 2010). This lexicon incorporates mouth evaluated terms, such as firmness, cohesiveness, stickiness to teeth, elasticity/rubberiness, adhesiveness to teeth, slipperiness of the mass, smoothness, as well as hand evaluated terms, such as springiness, stickiness, firmness, brittleness/crumbliness, etc. The manipulation of a processed cheese sample with the fingers can reveal a lot of information regarding breakdown and spreadability characteristics. The relationship between sensory and instrumental texture measurements of processed cheeses has also been studied by Drake et al. (1999).

Sensory evaluation can help to detect various off-flavors in processed cheese. These off-flavors could be originating from milk (e.g., feed, rancid, oxidized), cheese, dry ingredients (e.g., acid, cooked, scorched, lacks freshness/stale) (Kussy and Aylward, 2009) or other ingredients. For example, the addition of starter culture distillate can add fresh cheese or buttery flavor to processed cheese, however artificial and/or popcorn-like flavor can arise in end products containing excessive amounts. Also, some yeast extracts can provide aged natural cheese flavor, but can also give an undesirable flavor or a specific peculiar aftertaste to processed cheese (Tamime et al., 2011).

In sensory evaluation of processed cheese, sensory scorecards can be useful tools. Scorecards could involve clear definitions of expected flavor and texture characteristics, derived from the manufacturer's developed lexicon or they could be more abstract, containing suggested attributes from panelists. An example of certain attributes of processed cheese which can be evaluated with scorecards is shown in Table 13.12.

Any tests based on customer specifications should be clearly identified as documentation of standard operating procedures (SOP) (Kussy and Aylward, 2009). Hanaei et al. (2015) attempted to identify the terms most frequently used to describe processed cheese by consumers. Commercial samples of processed with different types were studied and 550 different characteristic terms were gathered. Among those, the following nine terms were those most frequently elicited by the consumers: sticky, shiny, yellow, smooth, compact, hard, spreadable, creamy, and easy to spread.

Table 13.12 Attributes describing the appearance, texture and flavor during in-house sensory evaluation of processed cheese (Kussy and Aylward, 2009).

Evaluation	Attribute
Appearance	Shiny, matte, browning, pinking, yellow, orange, greasy
Texture	Soft, firm, pasty, brittle, chalky
Flavor	Aged cheese, astringent, barny, browned or toasted, burnt, diacetyl, cardboard, milk fat, sour aromatic, sweet aromatic, unclean, fruity, goaty, moldy, musty, old oil (rancid), soapy, acidic (sour), yeasty, acetaldehyde, brothy, cooked, cowy (phenolic), free fatty acid, nutty, sulfur, whey

The sensory characteristics of processed cheeses have been studied in various scientific works, mainly to evaluate their quality or modify their composition and possibly add other ingredients, such as whey proteins (Thapa and Gupta, 1992; Kapoor and Metzger, 2004; Pinto et al., 2007; Chatziantoniou et al., 2015), stabilizers (Brummel and Lee, 1990; Schäffer et al., 2001; Černíková et al., 2010; Chatziantoniou et al., 2019), or other additives (Tamime et al., 1999; Everard et al., 2007; Johnson et al., 2009).

13.9 Defects

The quality characteristics considered during the sensory evaluation of processed cheeses can be summarized as follows: Products must have a uniform and compact body, free from fermentation gas bubbles or crystals and have a homogeneous color. A high quality processed cheese can be sheared without crumbling or being sticky, while melting evenly (Kosikowski and Mistry, 1997). Processed cheeses should also have a smooth and glossy surface, a pleasant aroma and flavor, reminiscent of the cheeses used as raw material (Johnson et al., 2009). The aroma varies from mild to acid, depending on the natural cheese used or the combination of two or more cheeses (Kosikowski and Mistry, 1997). Depending on blend formulation, processing parameters and cooling rate, processed cheeses have a characteristic body, which can vary from firm and sliceable to soft and spreadable. Defects may be observed during manufacturing, after the cooling stage or even after days or weeks following manufacture. Defects in processed cheese may be carried over from the natural cheese used, may be due to the type of natural cheese, unsuitable ingredient blend, the use of unsuitable emulsifying salts or implementation of incorrect processing or storage conditions. Frequently, defects may arise due to a combination of the above mentioned factors.

13.9.1 Appearance defects

13.9.1.1 Color

Color of processed cheeses affects the acceptance of these products by consumers. Processed cheeses should have a uniform color (Kosikowski and Mistry, 1997).

a) *Browning*: Processed cheese is susceptible to possible alteration of color due to the relatively high temperatures involved in their production. A significant defect in the color of processed cheeses is nonenzymatic browning. Various factors have been found to influence the development of browning in processed cheese, which is mainly caused by the Maillard reaction. This reaction takes place at elevated temperatures between amino groups and the carbonyls of reducing sugars and can lead to objectionable color and flavor development (Thomas, 1969). Browning can also be caused by improper cleaning of heating equipment surfaces (Clark and Agarwar, 2007).

Pale to dark browning can be observed in products that have been processed at high temperatures or long processing times. The characteristics of the natural cheese blend used for manufacture will also influence the browning of the resulting processed cheese. Browning is initiated when ingredients with high lactose content are used during processed cheese manufacture. In particular, high levels of residual galactose and lactose in Cheddar cheese of high salt-in-moisture levels have resulted in increased levels of browning in the processed cheese (Thomas 1969; Bley et al. 1985). Addition of skim milk powder at a level of 3–5% of the blend has been found to increase the propensity to nonenzymatic browning on storage (Fox et al., 2017). Heat treatment of processed cheese in the presence of lactose is recommended not to be carried out at temperatures >85°C (Clark and Agarwar, 2007). Rapid cooling of processed cheeses can prevent the appearance of a brown hue (Bley et al., 1985; Kussy and Aylward, 2009). In this respect, a strong brown discoloration at a depth of 5 cm has been observed in processed cheeses that have been stored in large containers or drums, while the surface was normal. This defect arises due to the Maillard reaction which continues to take place during the slow cooling stage. Processed cheeses should be stored in small containers that is, shallow boxes. If tall containers are used for packaging, discoloration can be avoided by cooling the product successively in thin layers (Meyer, 1973).

In addition, products with a high pH, greater than 5.9, and a high storage temperature show accelerated browning (Thomas 1969; Kristensen et al., 2001). Thomas (1969) indicated that processed cheese should not be stored at temperatures greater than 35°C for more than 6 weeks. In a study on processed cheese samples which were stored at different temperatures for 12 months, it was found that a more intense browning was evident during the storage of samples at 37°C, compared to samples stored at 4°C. The brightness decreased, that is browning was observed over time, while the green color (-a*) and the yellow color (+b*) decreased significantly in the samples stored at 37°C. According to a study by Buňka et al. (2008), the color and flavor of sterilized processed cheese samples stored at 23°C deteriorated faster, in relation to samples stored at 8°C. Browning is often associated with changes in rheological characteristics and tends to coincide with an increase in product viscosity (Kristensen et al., 2001).

Therefore, careful selection of ingredients and optimum final pH during processed cheese manufacture, along with desirable processing and post manufacture storage conditions can help in preventing the browning defect in processed cheese. Processed cheese should be stored at low temperatures (≈5°C), so as to minimize nonenzymatic browning.

b) *Pink discoloration*: Pink discoloration can result in downgrading or rejection of processed cheese. In processed cheese, pink discoloration can manifest throughout the cheese block and not just at the surface (Shumaker and Wendorff, 1998). It has previously been associated with decomposition of colorant (Kosikowski and Mistry, 1997) in processed cheeses that were either artificially colored with annatto or products that had natural cheese which already contained annatto. An increase in the amount of colored natural cheese, as well as in the content of aged cheese (with extensive proteolysis) in the blend, has been shown to increase pink discoloration of processed cheese (Shumaker and Wendorff 1998; Daly et al., 2012). Alkaline extracts of annatto have shown a higher propensity to cause such a defect. Norbixin is the primary carotenoid in annatto responsible for the color formation (Kang et al., 2010). Processed cheese samples with emulsion-based annatto colorants in the formulation were more susceptible to pink discoloration compared to suspension-based annatto colorants (Zehren and Nusbaum 2000).

c) *Mottled surface*: Mottling could be observed in processed cheese blocks due to mechanical faults in filling or due to physicochemical action, in the case of the addition of a solution of citrates. In the former case, mottling can be observed when the product is filled in various steps and not with one stroke of the piston, when product has been left without agitation or when insufficient cleaning between batches has allowed hard or dehydrated pieces of product to be transferred to the next batch. In the latter case, micro-crystals of calcium citrate cause streaking in the body of processed cheese and manufacturers should select different emulsifying salts (Meyer, 1973).

13.9.1.2 Crystal development

Crystals have a considerable effect on the visual and textural appeal of cheese. Different crystals have been observed in natural cheeses. These include lactose, tyrosine, calcium lactate, various salts of calcium phosphate and calcium citrate. Crystals can also emerge in processed cheese, as crystalline deposits on the surface and interior of products, visible to the naked eye as spots or lines (Schär and Bosset, 2002). Crystals are often referred to as haze or bloom and can produce a grainy, uneven texture in processed cheeses, especially detrimental for products in slices, which would normally have a smooth and shiny surface. Crystal deposits are not aesthetically pleasing and processed cheese products are sometimes rejected by consumers or retailers because the white deposits are mistakenly regarded as mold contamination (Guinee et al., 2004; Guinee et al., 2007). The types of crystals identified in processed cheese are salts of calcium tartrate (Leather, 1949), calcium citrate (Morris et al., 1969), tertiary complexes of Na- and Ca-citrates (Klostermeyer et al., 1984), various salts of Na- and Ca-phosphates (Pommert et al., 1988; Guinee et al., 2004), and complexes of various materials such as calcium, fatty acids (Bester and Venter, 1986), protein and lactose (Guinee et al., 2004). Crystals can vary in shape and dimensions, e.g., calcium phosphate aggregates $\approx$30 μm in diameter, and calcium lactate crystals up to 80 μm in diameter (Guinee et al., 2007).

Major sources of crystal development in processed cheese are due to the use of emulsifying salts, which are added during product formulation. The most common inclusion seen in processed cheese is evidence of incomplete dissolution of emulsifying salts, which can be avoided by using solutions of salts instead of dry salts, sufficient mixing for complete dispersion, use of rapidly hydrated emulsifying salts and development of formulations which include added water for dissolution (Meyer, 1973).

The presence of emulsifying salts can also lead to the formation of various complexes/salts, which may have a lower solubility than the regular emulsifying salts. These include calcium phosphate, disodium phosphate dodecahydrate, monoclinic calcium pyrophosphate dihydrate, and calcium citrate (Kindstedt and Polowsky, 2020). Calcium phosphate crystals can be formed as a result of the interaction between the anion of the emulsifying salt and the Ca of paracasein. The formation of these crystals could be reduced by substitution of rennet casein or high-calcium hard cheeses in the original formulation with cheeses with a lower calcium level (Guinee, 2007). The solubility of these complexes/salts is further influenced by pH of the processed cheese or storage conditions. Crystal formation was found to accelerate when the processed cheese surface was directly exposed to cold air. A multiple regression model was reported by Scharpf and Kichline (1968) which described that the possibility for crystal formation increased with increasing pH and phosphorus content. The relationship between phosphate content and pH has been studied, where it was observed that the higher the phosphate content, the lower the pH required in order to prevent crystallization (Guinee, 2007). Manufacturing guidelines for processed cheeses suggest the control of pH to 5.2, not lower than 5.0, and the use of citrate/phosphate blends in an 85:15 ratio (Meyer, 1973).

On the contrary, tetrasodium pyrophosphate (TSPP) crystals form when TSPP is in contact with water, through an exothermic reaction that fuses TSPP and water. For the formation of crystals to be avoided, which can grow large in 1–2 months following manufacture, it is suggested to implement dry blending TSPP with sodium chloride in a fine suspension and when incorporating TSPP to avoid the simultaneous addition of water (Lucey et al., 2011).

Crystals in processed cheese may also originate as preformed crystals present in the ingredient cheeses, which are carried into the final product (Kindstedt and Polowsky, 2020). Any inherent crystals of amino acids in the aged cheese cannot be dissolved upon cooking. Tyrosine crystals are insoluble white particles that form and are distributed throughout the products' mass. They could be observed as a thin layer resembling a mold mycelium on the cheese block's surface. This may be addressed by either filtering out the crystals or limiting the addition of aged cheese or both (Meyer, 1973).

Regarding the formation of lactose crystals in processed cheese, it has been attributed to the lactose present in dairy ingredients (Kindstedt and Polowsky, 2020), such as nonfat dry milk or whey powder (Templeton and Sommer, 1932; Johnson et al., 2009). The crystallization of lactose in processed cheese can be prevented by avoiding excessive use of skim milk or whey powder and by maintaining the lactose level below its maximum solubility level in the water phase of cheese. The maximum concentration of lactose that is soluble in water is 17% (w/w) at 20°C (Harper, 1992). The final lactose content in processed cheese has been suggested not to exceed 7.48% (w/w) for processed cheese foods (44% moisture w/w) and 10.20% (w/w) for processed cheese spreads (60% moisture w/w) (Kapoor and Metzger, 2008). Crystal development in processed cheese could also be promoted by manufacturing equipment, such as casting lines and slice cooling belts. When not thoroughly cleaned, equipment can act as nucleation sites for crystallization (Berger et al., 2002). Treatments that lead to dehydration of processed cheese, such as smoking, can also lead to crystal growth on the surface of the product (Guinee, 2007).

13.9.2 Functional defects

13.9.2.1 Formation of holes or fissures

Holes or fissures in the body of processed could be a result of various factors, such as:

a) *Physical*: Air bubbles introduced during mixing can be trapped in the body of processed cheese during processing (Kapoor and Metzger, 2008). Air could be trapped in a tough body of processed cheese or by inadequate vacuum applied during processing. To avoid trapping of air within the cheese mass, a fairly high vacuum should be used during processing, the molten cheese mass should be thin and fluid and the heating temperature should be 90–95°C. Air could be also trapped during the filing process (Meyer, 1973). Any air bubbles which are present in processed cheese are typically very small and concentrated on the bottom of the packaging (Kosikowski and Mistry, 1997). The emergence of holes in processed cheese due to air bubbles can be controlled with appropriate control of processing parameters, such as duration of heat treatment and mixing speed (Kapoor and Metzger, 2008).

Holes in processed cheese could be also due to the production of carbon dioxide, in the case that citrates are used as emulsifying salts. When citrates are used, the solution should be previously heated so as to expel carbon dioxide. Holes, often filled with liquid, could also occur due to insufficient heating time, resulting in a poor dissolution of emulsifying salts. This can be avoided by increasing heating time and using solutions of emulsifying salts (Meyer, 1973).

b) *Chemical*: The formation of holes in the upper surface of processed cheese is caused by the production of hydrogen through the contact of cheese with the aluminum foil, in cases where the foil is porous. Manufacturers should carefully select the packaging materials and replace them if necessary (Meyer, 1973).

c) *Microbial*: Holes created in processed cheese during storage could be due to microbial growth, for example, *E. coli* or bacterial spores (Clark and Agarwal, 2007) in the case of insufficient heating temperature and/or time during manufacture or cross-contamination. This factor is discussed in detail in 13.9.5.

13.9.2.2 Fat/water separation

Separation of liquid is considered a major defect in processed cheeses. It involves the separation of fat and water in the final processed cheese products, due to improper emulsion formation. This may include (Guinee, 2007):

- Incomplete emulsification of free fat droplets which are released during heating and shearing of the blend of natural cheeses and other materials
- De-emulsification of fat after over-emulsification and protein dehydration, known as over-creaming, which is further discussed in detail in 13.9.3.

Causes of improper emulsification of processed cheese are (Guinee, 2007; Kapoor and Metzger, 2008):

- Low protein-fat ratio
- Addition of excessive fat/oil to the blend, for example, butter oil
- Low final pH of product (<5.6), either due to selected formulation or due to microbial causes
- Inadequate levels of emulsifying salts, low level of intact casein, due to use of highly aged natural cheese in the cheese blend, and short processing time and/or low processing temperatures during manufacture, or
- Excessive levels of emulsifying salts, high level of intact casein, due to use of very young natural cheese in the cheese blend, and excessive shear and/or long processing time and/or high processing temperatures during manufacture.

An example of a processed cheese spread with an uneven consistency, which leaks liquid during compression is shown in Fig. 13.3.

Fat and/or water separation can be minimized by the appropriate control of various formulation and processing parameters during processed cheese manufacture, such as optimization of emulsifier usage, pH and processing conditions (Kapoor and Metzger, 2008).

Exudation of water or leakage could be also observed in processed cheeses due to bacteriological action, which result in changes in the product's structure. Unsuitable

Figure 13.3 Processed cheese spread with uneven consistency, leaking liquid during compression (Chatziantoniou, 2020).

storage temperatures and pressure applied onto products' packages could also lead to water or even fat leakage in processed cheese (Meyer, 1973).

13.9.2.3 Meltability, flowability

Very high or very low meltability may be considered defects when arising in processed cheeses during cooking, depending on their intended end use (Kapoor and Metzger, 2008). Block-type processed cheese production (e.g., processed American Cheddar cheese) requires high levels of calcium and intact casein to impart good sliceability and moderate meltability to final products (Guinee and O'Callaghan, 2010). Resistance to flow on cooking has been associated with low pH (Gupta et al., 1984; Lee and Klostermeyer, 2001; Lee et al., 2006), fat reduction (Swenson et al., 2000; Schäffer et al., 2001) and the addition of whey proteins, even at relatively low levels (Mleko and Foegeding, 2001).

Flowability has been found to continuously decrease as the level of substitution of caseins or cheese proteins by whey proteins increased (Savello et al., 1989; Gupta and Reuter, 1993; Mleko and Foegeding, 2001; Mounsey et al., 2007). This has been attributed to the inability to degrade proteins during heating, due to the formation of high molecular weight whey protein aggregates (Lee et al., 2006; Lee et al., 2013). On the contrary, there are some reports of improvement of spreadability and meltability of processed cheese spreads by the addition of whey proteins up to 4.5% (w/w) (Pinto et al., 2007). Also, Kapoor and Metzger (2004) reported that the incorporation of whey (up to 9.5%) as an ingredient in processed cheeses did not adversely affect product quality. This defect can be overcome by preheating whey proteins and subsequently adding to the blend. Predenatured whey proteins have been found to increase meltability and decrease hardness of processed cheese. Higher levels of denaturation of whey proteins produced softer model processed cheeses, with lower elastic modulus which were meltable, due to the fact that denatured proteins were incorporated as large aggregates and dispersed within the cheese matrix. These aggregates did not contribute to, and possibly disrupted the structure of the processed cheese (Lee et al., 2013).

Low meltability can also be observed in processed cheeses that have been processed with high mixing rates and/or processing temperatures. Overcreamed processed cheese exhibits a higher viscosity and lower degree of fluidity and flow/spreadability on heating (Guinee, 2007). Cooling rate also affects the firmness and meltability of processed cheese, since it determines the interactions between emulsified proteins. Specifically, a slow cooling rate produces less meltable products, whereas a rapid cooling rate leads to more meltable processed cheeses (Kapoor and Metzger, 2008).

13.9.2.4 Adherence to packaging foil

When processed cheese adheres to the packaging foil it could be due to high moisture of product, the use of cheese which was too young, insufficient creaming of the mixture, excessive time of product's holding without agitation or high pH (>6.2). In order to avoid this defect, moisture of processed cheese could be reduced or water could be added in two portions, more mature cheese should be used, processing should involve continuous agitation and pH of product should be kept between 5.7 and 6.0. If this defect is due to defective lacquering of foil, another type of foil could be selected (Meyer, 1973).

13.9.3 Texture defects

Texture is a key feature of the sensory evaluation of processed cheese. Finished product may also be prone to defects in texture, such as oily body, very high to very low firmness, presence of stickiness, brittleness, crumbliness or graininess.

13.9.3.1 Oily body

Defects such as oily or fatty processed cheese could be due to the use of excessively mature cheese in the blend, low pH, inappropriate amount of emulsifying salts added or even inadequate quantities of added water. To overcome such problems manufacturers should optimize formulation using natural young cheeses, adjust pH and level of emulsifying agents, and possibly add water to the blend (Meyer, 1973; Gupta et al., 1984).

13.9.3.2 Soft, thin body

A soft, spreadable, adhesive texture is undesirable in processed cheese slices and blocks. Processed cheese could exhibit defects such as soft, thin or sticky body, resulting in adherence of product to the knife blades during manufacture or handling. Products with these defects may be produced by excessively mature cheese, excessive water or fat in the blend, weak emulsifying salts or low added quantities, pH higher than 5.8, inadequate processing times, low agitation speed, as well as processing at temperatures >110°C and rapid cooling. Processed cheese samples with pH>6.0 have been shown to be very soft and flow excessively on heating (Gupta et al., 1984, Guinee, 2007).

To overcome such problems manufacturers should optimize formulations using younger, protein stable cheeses, appropriate emulsifying agents and possibly add the water in two stages during processing or add a creamed typed precooked cheese. If processed cheese appears thin in body, moisture and/or pH could be reduced and heating time increased. In the case that processed cheese has too thin consistency when destined for production of block-type cheese, manufacturers could mix very young cheese to the blend, avoid the use of precooked cheese, decrease the processing temperature to 80-85°C, reduce the holding time during heating and allow the finished packaged blocks to cool down more slowly (Meyer, 1973).

13.9.3.3 Sticky, gummy consistency

Stickiness and gumminess are observed when processed cheese has been produced with large quantities of young cheese, unsuitable emulsifying salts which do not result in creaming, as well as short processing times and low agitation speed (Meyer, 1973). The above result in (Guinee, 2007):

- Insufficient fat dispersion during processing
- Incomplete emulsification of free fat by the cheese protein (undercreaming).

Commercial processed cheese samples with decreased fat content have been found to have increased stickiness (Muir et al., 1997). When encountering defects such as stickiness or gumminess, manufacturers should mix more mature cheese to the blend, use creaming types of emulsifying salts, add precooked cheese, and add water in small portions as well as increase processing time and agitation speed (Meyer, 1973).

13.9.3.4 Grainy texture

When the texture of processed cheese is rough and grainy instead of being smooth and homogeneous, it can manifest as an inability to flow smoothly right after the processing stage. Causes of such a defect could be insufficient levels of emulsifying salts, low pH of processed cheese, as well as low temperature, low agitation speed, or short heating times during processing. The grainy texture of processed cheese could be prevented by increasing mixing time during heating, raising the pH by adding more emulsifying salts, using more alkaline emulsifying salts, or adding more mature cheese to the blend

(Meyer, 1973). Grainy texture of processed cheese could also be a result of fat reduction (Muir et al., 1997) or the presence of crystals. When the size of crystals increases enough, processed cheese acquires a sandy or grainy texture. The prevention of crystals is further discussed in 13.9.1.

13.9.3.5 Firm body

The defect of excessively firm body of processed cheese is manifested as products which do not slice smoothly. It could be observed in processed cheeses with low moisture levels or low pH. Excessively firm body could also be observed due to the addition of high levels of emulsifying salts or excess polyphosphates (Meyer, 1973). Minimization of polyphosphate usage level can help eliminate the emergence of this defect (Kapoor and Metzger, 2008).

Generally, an increase in mixing rate or time and heating temperature has been found to increase firmness of processed cheese. Increasing the mixing rate and time causes the amplification of interactions between casein and emulsified fat, while increasing the heating temperature results in strengthening of hydrophobic interactions between constituents. The rate of cooling will also influence the characteristics of the gel under formation, since it affects the protein-protein interactions. Porous gels with weak cross-links are formed via rapid cooling of processed cheese, whereas less porous gels with strong cross-links are produced with slow cooling of products. As a result, firm processed cheeses are produced by slow cooling rates, whereas softer products can be formed via rapid cooling rates (Guinee, 2009).

Another common problem encountered in processed cheese manufacture is known as the rework defect. This includes the formation of a stiff, heavy, dense, gummy or hard body, even while the formulation is still being processed. It is manifested due to the addition of previously produced processed cheese to a new batch (Meyer, 1973). During further heating, addition of the precooked processed cheese possibly leads to thermally induced dehydration and aggregation of paracasein, which increases the end product's elasticity (Guinee et al., 2004).

Similar defects can arise when processed cheese is held at a high temperature for too long (Meyer, 1973). Prolonged holding of a processed cheese at a high temperature (70–90°C), for example due to delays or disruptions in the packaging line, can lead to the defect of overcreaming, which is a colloidal chemical change in the body of cheese (Guinee et al., 2004). Overcreaming could also be the result of high mixing rates during processing or excessive levels of emulsifying salts (Meyer, 1973). This defect is highly undesirable, since it may lead to problems during manufacture of the product (e.g., pumping or filling lines) and deteriorate the end products' quality. End products may be not spreadable, less meltable, greasy, and have an uneven appearance with reduced surface sheen (Guinee et al., 2004). Overcreamed processed cheese can have an over-firm and heavy, porous, pudding-like consistency and dull appearance, especially in high moisture processed cheese spreads (Guinee, 2007). In block-type processed cheese, the product's surface may resemble an "orange-peel", where water or free oil accumulates in small areas, evident postmanufacture, after the product has been cooled (Guinee et al., 2004). In extreme cases, processed cheese may leak moisture, especially on cooling (Guinee, 2007). The above mentioned defects arise due to:

- De-emulsification of milk fat globules and coalescence of the liquefied oil droplets, and
- Partial dehydration, aggregation and shrinkage of the paracasein matrix, through various interactions between proteins.

Overcreaming is accompanied by a decrease in nitrogen solubility, which reveals that it is possibly associated with hydrophobic-induced aggregation of paracasein taking place during prolonged holding at high temperature. Thus, as the paracasein matrix becomes dehydrated, it loses the ability to stabilize emulsified fat globules, resulting in fat coalescence. Young cheese, rennet casein or sodium caseinate, which have a high level of intact casein, can lead to overcreaming of processed cheese. On the contrary, hydrolysis of proteins during cheese ripening renders the proteins less likely to confer a firm, elastic texture on the end product. Therefore, processed cheeses prepared from aged natural cheeses are in general softer or more spreadable than those made with young cheese (Guinee, 2007).

Manufacturers encountering the defect of overcreaming should consider increasing the amount of water in the blend, adding the water at the beginning of mixing, reducing the amount of emulsifying salts, as well as raising pH to 5.7–5.9 and cooling the melted mass at a faster rate (Meyer, 1973). Raising the pH to 5.7–6.2 has been found to lead to an increase in casein hydration, a lower degree of paracasein aggregation, a finer gel network and a more uniform structure (Guinee, 2009).

As can be deduced from the above, the variation of manufacturing conditions, such as heating time/temperature, mixing time/rate and cooling rate can largely affect the properties of processed cheese. The manipulation of processing conditions can lead to the production of various processed cheeses with completely different functionalities using the same formulation.

13.9.3.6 Brittle, crumbly texture

A dry, short, brittle or crumbly processed cheese falls apart when slicing. These types of defects arise when the final pH of the processed cheese is too low (<5.4) (Gupta et al., 1984), desired pH being 5.6–6.1 (Guinee, 2007). This is due to its closeness to the isoelectric point of proteins, which results in a decreased net negative charge (Gupta et al., 1984). Brittle and crumbly texture could also arise due to addition of high levels of skim milk in the blend (Guinee, 2009) and very low levels of moisture or fat. An example of a processed cheese sample with a dry, crumbly consistency is shown in Fig. 13.4, where these defects were clearly demonstrated during compression testing. Thus, such textural determinations can help detect similar defects in processed cheese. Major causes of brittleness and crumbliness are the reduced hydration of paracasein and excessive creaming with a high degree of fat emulsification with little, or no, free oil. Reduced hydration of the protein matrix, thus, a heterogeneous structure, can occur in processed cheese due to inadequate dissolution, mixing and interaction between materials, as a result of low mixing rates, heating temperature or short processing times. On the contrary, factors which contribute to a high degree of fat emulsification

Figure 13.4 Processed cheese spread which is dry and crumbly, evident when subjected to compression (Chatziantoniou, 2020).

which result to overcreaming (e.g., high mixing rates, processing times and intact casein content) also cause shortness in the end product (Guinee, 2007). Such defects can be overcome with optimization of moisture, pH, emulsifier levels and processing conditions (Meyer, 1973; Guinee, 2007).

This defect could be also observed postmanufacturing, during storage time of processed cheese. Hardening and reduction of spreadability of processed cheese during storage is mainly attributed to overcreaming or microbiological factors which cause the reduction of pH. Specifically, pH is reduced by acids produced by the lactic and butyric acid bacteria which have survived the heating process in lactose containing products. In order to avoid this postmanufacturing defect, processed cheese should be produced by careful selection of natural cheese, use of processing temperatures from 95°C to 98°C or 120°C (ultrahigh temperature/short time method), ensuring that a final temperature of 85–90°C is reached (Meyer, 1973).

13.9.3.7 Rubbery, less spreadable texture

Defects in low-fat natural cheeses eventually carry over into processed cheeses. In this respect, low-fat natural and processed cheeses are generally characterized as firm and rubbery, with reduced spreadability (Muir et al., 1997; Gwartney et al., 2002). When fat is reduced, the protein's influence on the structure of processed cheese is increased. A study on commercial samples of low-fat processed cheeses reported that fat globule aggregation was less extensive, as studied by scanning electron microscopy (SEM), which contributes to their firmness. In fat-free commercial processed cheese samples, the protein matrix also appeared rough and dense, with large cracks, whereas fat was dispersed as very few, small fat globules (Mistry and Anderson, 1993). Results of a study on fat-free processed cheese showed that as the pH of fat-free processed cheeses increased, hardness also increased, with a significant increase in pH above 6.0, whereas the melting capacity increased and the spreadability decreased (Swenson et al., 2000).

Low-fat processed cheeses with acceptable body and texture can be produced, by the successful mimicking of the structural role of fat. Apart from using a natural lower-fat cheese in the formulation, research has also involved the incorporation of various fat replacers in a processed cheese formulation for the manufacture of products with acceptable flavor, texture, and functional properties. The characteristics and regulatory status of various commercially available fat replacers has been described by Akoh (2002). Brummel and Lee (1990) reported that the texture of processed cheeses, with 40 and 50% reduced fat, was improved using various hydrocolloids. Samples with a fat content of 15% (w/w) and added hydrocolloids, such as λ-carrageenan (2.2% w/w), pectin (1.7–2.2% w/w) or low viscosity guar gum (1.7% w/w), had a texture similar to the control sample with a fat content of 25% w/w. The samples of processed cheese spreads with 15% fat (w/w), 62% moisture (w/w), and 1.7% pectin (w/w) were the most similar to the control sample.

13.9.4 Flavor defects

All ingredients used for the production of processed cheeses should be of high quality, so as to ensure the manufacture of, in every respect, satisfactory end products. The addition of inferior quality milk fat or low-quality cheese could lead to the production of processed cheese with acid, sharp, old, oxidized, soapy, or even rancid flavor. Sweet flavor could also develop in processed cheese when blown cheese is used in the formulation, which is cheese with undesirable openness of the interior as a result of microbiological spoilage. Milk protein of inferior quality could yield products having a bitter taste (Clark and Agarwal, 2007).

Acids may be useful to adjust processed cheese pH, but may lead to sharp, acidic or even bitter flavor defects. Emulsifying salts can impart off-flavors to processed cheeses, for example, phosphates can give rise to chemical or metallic, acid or bitter off-flavors, especially when added in high levels, while citrates have relatively little influence on flavor (Karahadian and Lindsay, 1984). In a study regarding commercial processed cheese samples, it was shown that as the fat content of the samples decreased, the acid and bitter attributes increased, whereas creaminess decreased (Muir et al., 1997). Chemical/metallic/bitter flavors have been found to be increased in processed cheese samples with reduced NaCl (Karahadian and Lindsay, 1984). Better flavor and color has been attributed to processed cheese containing whey products with reduced mineral and/or reduced lactose content (Guinee et al., 2004).

The most common off-flavors found in processed cheese are presented in Table 13.13, along with their causes and proposed corrective actions in order to eliminate or reduce the off-flavors. Generally, manufacturers should consider minimizing the usage of phosphates and carefully select cheese base and other raw materials for formulations. In addition to high quality ingredients in the blend, appropriate formulation and processing conditions are also important for the production of high quality processed cheese (Clark and Agarwal, 2007).

13.9.5 Postmanufacturing defects

13.9.5.1 Physical and chemical changes

Processed cheese is expected to be a much more stable product than natural cheese (Guinee et al., 2004). Shelf life of processed cheeses depends on the ingredients used, the additives, the heating process (temperature and time), the packaging, etc. It does not normally exceed 6–12 months. In order to exceed this period and have a shelf life of at least 24 months, processed cheeses must be sterilized (Buňka et al., 2008).

As in most foods, during storage, structure and flavor of processed cheese will progressively change, through physical and chemical aging processes. These changes can be influenced by composition and processing conditions, however, the deterioration of processed cheese is largely affected by handling processes during postmanufacture storage (Schär and Bosset, 2002). Specific factors that contribute to stress during storage (in production plants and in distribution/retail establishments or during transportation) that can greatly reduce the shelf life of processed cheese are light exposure and variable storage temperatures (Christensen et al., 2003).

A series of changes occur to processed cheese during storage due to the following (Schär and Bosset, 2002; Guinee et al., 2004):

1. Loss of water vapor, where depending on storage temperature, the weight of product can be reduced through packaging materials if the packaging is not completely airtight and lead to an increase in firmness.
2. Changes in ionic equilibrium, where pH gradually changes and eventually stabilizes and the interactions between proteins, emulsifying salts, sodium and calcium influence water binding and texture.
3. Hydrolysis of polyphosphates, where a part of the used polyphosphates has already undergone hydrolysis during the heating process, whereas the rest is completely hydrolyzed after 7–10 weeks of storage. This reaction lowers the pH-value and consequently increases the hardness-firmness of the product, due to the resulting protein aggregation (Thomas, 1970; Awad et al., 2002, 2004; Buňka et al., 2008).
4. Crystal formation, which is accentuated by high pH (especially where sodium orthophosphates are used) and long storage at low temperatures. Crystals may cause a gritty or sandy texture, white specks throughout the mass of processed cheese or a white mold-like appearance on the surfaces of slices or blocks (Guinee, 2007). This is further discussed in 13.9.1.
5. Lipid oxidation, which leads to oxidation flavors, attributable to methional, carbonyl compounds, cholesterol oxides, etc. The resulting off-flavors have been described as fatty, fried, plastic, tallowy, fishy, metallic or cardboard-like (Meyer, 1973). Oxidation is caused by

Table 13.13 Off-flavors in processed cheese, causes and corrective actions (Meyer, 1973; Clark and Agarwal, 2007; Guinee, 2007; Tamime et al., 2011).

Off-flavor	Cause	Corrective action
Metallic-oily	Presence of metals which oxidize fats, e.g., copper, iron, manganese etc. Additives, e.g., ammonium chloride, glycinamide hydrochloride Excessive lactic acid addition	Eliminate containers made from copper, brass, zinc Eliminate the specific additives Lactic acid content should be kept <0.4%
Tallowy	Presence of oxidants e.g., bromate, hydrogen peroxide	Eliminate oxidants in formulation
Chemical	Raw cheese with salty taste Impure emulsifying salts Phosphates Preservatives Steam used as an ingredient	Select raw cheese of good quality Verify purity of emulsifying salts Minimize the use of phosphates Alter preservatives Verify purity of steam
Soapy, detergent	High pH (>6.2) Formation of soaps (sodium palmitate or potassium oleate) as a result of the interaction of the cations of emulsifying salts (e.g., Na^+, K^+) and medium to long chain fatty acids Carried over from natural cheese High levels of fat and/or moisture High processing temperatures	Reduce pH by adding more young cheese or a more acid emulsifier Reduce the use of cheese or other materials with high levels of free fatty acids Eliminate the use of natural cheese with soapy off-flavor Reduce fat or moisture levels Reduce processing temperatures
Flat, empty	Large quantities of very young cheese	Add more mature cheese or flavored additives, seasonings
Sharp, old	Excessively mature cheese	Add young cheese
Bitter, soapy or rancid	Overripe or mold ripened cheese Low quality cream or butter	Add young cheese Select high quality raw materials
Bitter	Sour raw cheese Low pH Production of lactic or butyric acid	Add more mature cheese with a high pH Ensure adequate heat processing
Burnt	Overheating product in the presence of lactose	Keep heating temperature <90°C, when lactose is present Raise the pH by 0.2–0.3 units, when sterilization is appropriate

(*continued on next page*)

Table 13.13 Off-flavors in processed cheese, causes and corrective actions (Meyer, 1973; Clark and Agarwal, 2007; Guinee, 2007; Tamime et al., 2011)—cont'd

Off-flavor	Cause	Corrective action
Sour, astringent	Sulfuric acid by the use of sodium sulfite as preservative	Select alternative preservative
Salty	High levels of emulsifying salts Cheese or other ingredients which were salty	Reduce level of emulsifying salts Add young unsalted cheese, fresh rennet curd or quark
Moldy or musty	Moldy cheese Mold growth caused by air contamination in improperly sealed packages	Select raw cheese of good quality Carefully select packaging materials and verify sealing efficiency
Putrid	Very overripe cheese e.g., overripe Cantal Growth of *C. sporogenes* in end product	Reject cheese with sweet flavor Ensure adequate heat processing
Sweet	Blown cheese Development of blowing in processed cheese	Reject cheese with sweet flavor Ensure adequate heat processing
Sweet and salty	Excessive whey powder	Reduce the quantity of whey ingredients
Artificial	Excessive starter culture distillate	Reduce the quantity

reactions induced by light and oxygen, due to traces of oxygen in the packaged product and oxygen permeation through the packaging material. It is a process that consists of several stages and can be monitored by methods such as HPLC and dynamic headspace GC-MS (Emmons et al., 1986; Sunesen et al., 2002). The selected packaging materials should have low oxygen and vapor permeability and sufficient light barrier properties in order to minimize oxidation reactions (Alves et al., 2007). Another option could be the incorporation of active packaging technology, where oxygen scavengers can be included in multilayer packaging films (Oyugi and Buys, 2007).

6. Nonenzymatic browning, where processed cheese is deteriorated through the Maillard reaction. This involves the carbonyl-amine addition between reducing sugars and amines and numerous further reactions. These reactions lead to discoloration and off-flavor development in processed cheese during storage, especially at elevated temperatures. The Maillard reaction and its prevention have been reviewed (Pizzoferrato et al., 1998). Processed cheese which contains high levels of lactose (10–80 g/kg), is processed in high temperatures or is stored in elevated temperatures could develop a brownish discoloration, as well as off-flavors during storage (Buňka et al., 2008). In order to minimize nonenzymatic browning, processed cheese should be stored at low temperatures ($\approx$5°C) (Fox et al., 2017).
7. Enzymatic activity, where a minority of enzymes are not deactivated during processing. Denatured proteases, such as plasmin, can exhibit some proteolytic activity during storage (Mulvihill and McCarthy, 1993). Heat stable lipases have been found to degrade the quality of processed cheese during storage.
8. Interactions with packaging materials, which involves interactions with lacquers and coating materials used with metal containers in order to prevent interaction of food acids with steel, tin or aluminum, as well as the migration of plasticizers, pigments, metal ions and

other components of packaging into the processed cheese body. Interactions with packaging materials have been reviewed by Gallmann (1995) and Bhunia et al., (2013). Since processed cheese can achieve sufficient shelf life when packaged in low O_2 and high CO_2 levels, modified atmosphere packaging, active packaging and/or good barrier materials should be preferred. The migration of substances, such as butylhydroxytoluene (BHT) and Irganox 1010, into processed cheese slices has been determined, where it was shown that migration decreased with increasing molecular weight of the migrating substance (Schwope et al., 1987). Packaging material selection of optimal quality and produced in accordance to the relevant legislation is of utmost importance.

13.9.5.2 Microbial growth

Contaminations constitute a major cause of microbial-induced spoilage to processed cheeses and adverse health effects to consumers, resulting from the ingredients used, as well as from the postprocessing steps, due to the packaging and storage conditions (Meyer, 1973). The most important microorganisms in the processed cheese industry include pathogenic and nonpathogenic spore-formers such as *Clostridium* spp. and *Bacillus* spp. and postprocessing nonspore-forming pathogenic bacteria such as *L. monocytogenes, Salmonella* spp., *S. aureus*, and *E. coli* O157:H7. Other nonpathogenic bacteria such as lactobacilli, as well as yeasts and molds can also cause spoilage (Glass and Doyle, 2013). Microorganisms which contaminate the product after processing are called secondary contaminants (Lazárková et al., 2010).

The genus *Clostridium* has been identified as the most common microorganisms associated with spoilage of processed cheese (Sinha and Sinha, 1988). These spore-forming bacteria that survive pasteurization can grow under anaerobic conditions, if low temperatures are not maintained throughout the product shelf life, and can decompose proteins and alter the flavor of products (Meyer, 1973).

During quality control, processed cheeses should not present any bulging, which is of microbial cause. Gas production is caused by bacteria of the genus *Clostridium.* Lycken and Borch (2006) studied 42 spoiled samples of processed cheese spreads and reported that the most harmful microorganism present in the products was *C. sporogenes*, but *C. cochlearium* and *C. tyrobutyricum* were also isolated. As these clostridia are not pathogenic, the production of gas in processed cheeses is not accompanied by a safety risk for the consumers. Bulging can also be caused by contamination with propionibacteria (Guinee et al., 2004).

In canned processed cheeses, which are low-acid canned foods, major concerns regarding quality and safety are related to the survival of *Clostridium* spp. Thus, commercial sterilization is adamant for these products. The microbiological safety of canned processed cheeses can be achieved by:

- Heat treatment at 121°C for 2.5–3 min, which is the heating regime required to inactivate the pathogenic *C. botulinum*, resulting in a 12 log reduction in numbers of spores, or
- Manipulation of formulation, pH and/or aw so as to prevent outgrowth of spores.

However, processing at high temperatures may have undesirable effects on the microstructure and functional properties of canned processed cheese (Glass and Doyle, 2013).

Furthermore, pH, salt content and a_w are major factors that influence the development of microorganisms. Apart from the quantities of NaCl added to the blend, emulsifying salts, such as monophosphates, polyphosphates and citrates, also tend to reduce the a_w of the product. Citrate-based emulsifying salts are less effective at inhibiting the growth of *C. botulinum* compared to phosphate-based emulsifiers (ter Steeg et al., 1995). The inhibition of microbial growth and the antibotulinal effects in processed cheese of phosphate-based emulsifying salts have been extensively studied (Tanaka et al., 1986). These authors reported that toxins were not detected when a_w of processed cheese spreads was lower than 0.944, whereas, formulations of the same product supported toxin production when a_w was higher than 0.957. The final pH of processed cheese is also important for controlling microbial growth and toxin production (Tanaka et al., 1986; ter Steeg et al., 1995; Glass and Doyle, 2013). All the processed cheeses associated with outbreaks were found to have high water activity ($\approx$0.96–0.97) and high pH ($\approx$5.7–5.8), which have possibly led to the production of toxins by *C. botulinum* during storage (Briozzo et al., 1983; Glass and Doyle, 2013). Low pH in processed cheese has been reported to enhance the inhibitory activity of sorbic acid (Glass and Doyle, 2013).

In order to reduce microbial growth, basic raw materials such as natural cheeses with a large number of spores (e.g., excessively mature cheeses) should be avoided, the pH of the final processed cheese products should be between 5.6 and 5.8 and the NaCl concentration maintained at 2.5–3.5% (w/w). Specifically, for spreadable processed cheese, pH should be lower than 5.2 (Kosikowski and Mistry, 1997). Manufacturers should also select the type and level of emulsifying salts and should also consider prior treatment of additives and use approved antimicrobial agents such a nisin and/or sorbate (Clark and Agarwal, 2007; Guinee and O'Callaghan, 2010).

Based on the study by Tanaka et al. (1986), the Canadian Food Inspection Agency (CFIA, 2021) suggests an adapted chart (Fig. 13.5) providing recommended combinations of pH (5.05–6.2) and percentage of sodium chloride plus phosphate emulsifying salt solids (2–6% w/w) for nonrefrigerated processed cheese spreads of moisture content from 51 to 60% (w/w), to prevent the growth of spore-forming bacteria after heat treatment.

The influence of fat content of processed cheeses on the growth of anaerobic bacteria is also significant. Fat protects bacteria by reducing the ability of many antimicrobials (Glass et al., 1998; McLay et al., 2002). Reduced-fat processed cheeses are considered safer than high-fat products (ter Steeg et al., 1995). *C. botulinum* toxin production was found to be delayed in low-fat cheese substitutes compared to regular-fat cheese substitutes with the same moisture, NaCl, emulsifier and pH levels (Glass et al., 1998).

There are sufficient studies involving the hygienic quality of commercial processed cheese which reveal certain microbiological safety concerns. Warburton et al. (1986) studied the microbiological quality of 78 lots of domestic (Canadian) and imported processed cheese products and concluded that all samples had counts of *E. coli* and coliforms <1.8 cfu/g, and had no staphylococcal thermonuclease activity. However, counts >500 spores/g regarding aerobic and anaerobic spore-formers were observed in 24% and 15% of the tested samples, respectively. The authors believe that the high level of spore-formers found must be controlled due to an increased risk

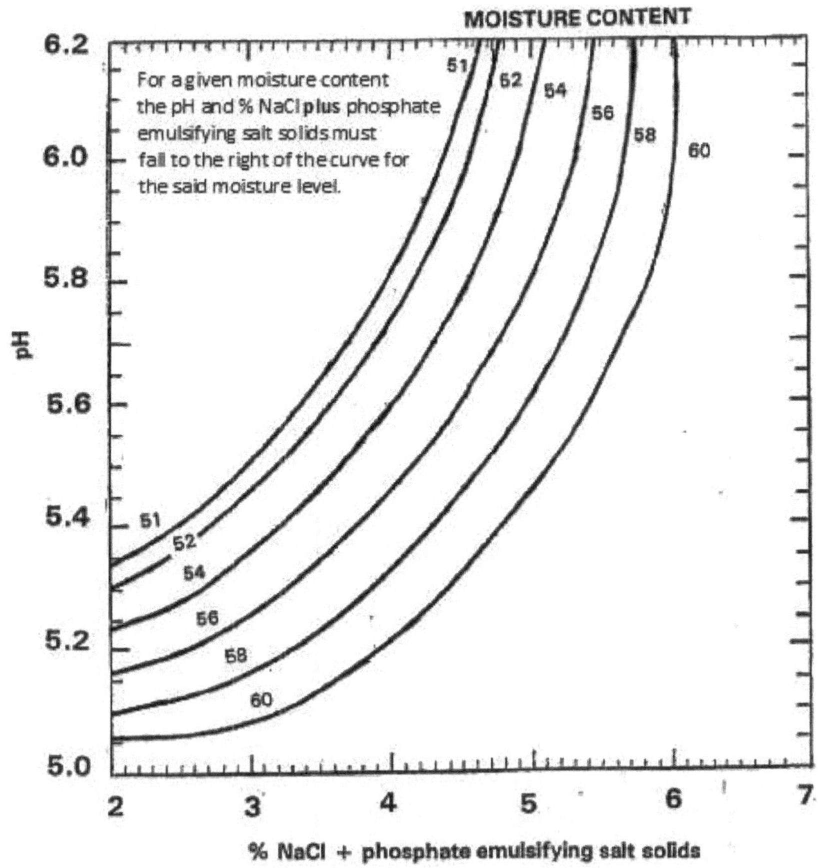

Figure 13.5 Microbiological stability chart for nonrefrigerated processed cheese spreads (CFIA, 2021).

of foodborne disease. Benamara et al. (2016) isolated *B. cereus* spores from sixty samples of processed cheeses manufactured and marketed (four brands) in Algeria and reported that all samples were contaminated by *B. cereus sensu lato* at increased levels (1.4×10^3–3.4×10^4 cfu/g). These levels could be dangerous to consumers, in the case that they improperly handled and allowed to increase further. Thirty samples of processed cheeses marketed in Egypt were studied by Abdel-Salam and Soliman (2019).

Where it was found that the majority of samples were highly contaminated with coliforms, proteolytic bacteria (*Bacillus* spp.), lipolytic bacteria (*Pseudomonas* spp., *Enterobacter* spp., *Alcaligenes* spp.), enterococci, yeasts and molds, with the following maximum values (cfu/g): 1.3×10^{10}, 3.4×10^7, 5.7×10^4, 6.7×10^9, 3.1×10^4 and 1×10^7, respectively. Domestic and imported processed cheeses marketed in Sudan were studied by Hussein et al. (2011), where it was found that samples had high counts of total bacteria and coliforms with maximum values of 4.8×10^4 cfu/g and 1.3×10^4 cfu /g, respectively. A study on commercial samples of processed cheese in Hungary (Varga, 2007) revealed that, at the end of shelf life, 10% of the samples tested had bacterial load which exceeded the regulatory standards (total plate count $>1 \times 10^5$ cfu/g), and *S. aureus* was present in two samples, one of which had staphylococcal counts as high as 3×10^5 cfu/g. Laslo and György (2018) studied commercially available dairy products in Romania and concluded that certain processed cheese samples were contaminated and, thus, potentially hazardous. Particularly, processed cheese samples were found to have increased levels of coliforms, *E. coli, Salmonella* spp., *S. aureus, B. cereus*, yeasts and molds, with the following maximum values (cfu/g): 8×10^3, 7×10, 6×10^2, $>3 \times 10^2$, $>3 \times 10^2$, 1×10^4 and 1×10^3, respectively. Two of the five varieties of processed cheese tested were also positive for the foodborne pathogen *C. perfringens.* Kim et al. (2018) conducted microbiological analysis of 800 commercial processed cheese products from different countries (Republic of Korea, Australia/New Zealand, EU, USA, and UK). Bacteria, such as *C. perfringens*, coliforms, enterohemorrhagic *E. coli, L. monocytogenes, Salmonella* spp., and *S. aureus* were not detected in the samples studied. However, 79.9% of the samples contained aerobic plate count ranging up to 6.3×10^7 cfu/g, including *Bacillus* spp., *Enterococcus* spp., *Staphylococcus* spp., and other microorganisms.

From the above it appears that processed cheeses available on the international market contain microbial load capable of causing product spoilage and/or endangering the health of consumers. Therefore, the need to implement quality assurance systems such as GMPs, GHPs, and HACCP is absolutely imperative. Moreover, most of the authors of the above studies have pointed out the necessity to take more measures for the prevention and control of undesirable microorganisms during manufacture, storage and distribution. This is also confirmed by Ammar et al. (2017), who by implementing the HACCP system in processed cheese production line achieved a dramatic reduction in the total microbial load from 1700 to 8 cfu/g and yeasts and molds from 4300 to 100 cfu/g.

13.10 Concluding remarks and further perspectives

Quality assurance and quality control of processed cheese involves the knowledge and compliance with customer needs and expectations, as well as statutory and regulatory requirements throughout the food chain. The importance of adhering to GMPs, GHPs, and HACCP is reflected in the quality and safety of processed cheese products, as these systems are able to prevent or minimize biological, chemical or foreign material contamination of raw materials or finished products. For even greater efficiency, the dairy establishments must implement a combination of a quality management system (QMS) and a food safety management system (FSMS). These systems further include a quality and food safety policy and objectives and seek feedback on the satisfaction of customers and other interested parties, as well as on continual improvement. Of course, it is also necessary to complete the scheme with an environmental management system (e.g., ISO 14001) to protect the environment.

Research on processed cheeses has identified the factors affecting the quality characteristics of such products and has revealed the appropriate methods for their determination. Many of the disadvantages of processed cheeses are due to the quality of the cheeses used to make them, while others are directly related to their processing conditions. Because some defects of natural cheeses eventually carry over into processed cheeses, many defects could be avoided by the careful selection of natural cheese. Disadvantages related to processing conditions are due to poor mixing of the raw material, insufficient or excessive heating and incorrect choice of the type and amount of emulsifier. In order to avoid further defects in processed cheese, there should be a careful selection of added ingredients, emulsifying salts and their levels, final pH, processing conditions, packaging, as well as hygiene monitoring during manufacture and implementation of appropriate transportation and storage conditions of the finished products.

However, there is limited understanding of the relationship between different formulations or manufacturing conditions and the functionality of processed cheese end products. Further research in this field should focus on the effects of formulation manipulation and change of processing conditions on the properties of processed cheese, assisted by instrumental and sensory quality control. Also, more real-time online testing protocols have to be developed and applied in the industry, so as to prevent unsafe and defective finished products, which result in adverse health effects for the consumers and economic losses for the dairy organizations.

References

Abdel-Salam, A.B., Soliman, N.S.M., 2019. Open J. Appl. Sci. 9, 620–630.

Acharya, M.R., Mistry, V.V., 2005. J. Dairy Sci. 88, 3037–3043.

ACS, 2017. Best Practices Guide for Cheesemakers. The American Cheese Society, Englewood, USA.

Akoh, C.C., 2002. Lipid-based synthetic fat substitutes. In: Akoh, C.C., Min, D.B. (Eds.), Food Lipids, 2nd Edition. Marcel Dekker, Inc., New York, NY, pp. 695–727.

Alves, R.M.V., van Dender, A.G.F., Jaime, S.B.M., Moreno, I., Pereira, B.C., 2007. Int. Dairy J. 17, 365–373.

Ammar, E.M.A., Reyad, M.A., Abdel-Kader, Y.I., Farag, A.M.K., 2017. J. Food Dairy Sci. 8 (2), 121–125.

Anonymous, 2009. Request for Proposal 10-X-21123, Foods: Perishable, Cheeses & Oleomargarine – DSS. The Department of the Treasury, New Jersey. Available at: https://www.nj.gov/treasury/purchase/noa/attachments/a1097-rfp.pdf.

Anonymous, 2016. Reference Manual for U.S. Cheese. U.S. Dairy Export Council, Arlington, USA, p. 135.

Anonymous, 2021. What food processors should know: metal detection or X-ray inspection. Thermo Fisher Scientific Inc., Waltham, USA, p. 4.

Awad, R.A., Abdel-Hamid, L.B., El-Shabrawy, S.A., Singh, R.K., 2002. LWT - Food Sci. Technol. 35 (1), 54–61.

Awad, R.A., Abdel-Hamid, L.B., El-Shabrawy, S.A., Singh, R.K., 2004. Int. J. Food Prop. 7 (3), 429–448.

Benamara, R.N., Ziane, M., Medjahdi, K., Moussa-Boudjemâa, B., 2016. Afr. J. Microbiol. Res. 10 (30), 1173–1181.
Berger, W., Klostermeyer, H., Merkenich, K., Ulhmann, C., 2002. Processed Cheese Manufacture - A JHA Guide. B.K. Giulini Chemie GmbH & Co. OHG, Ladenburg, Germany, p. 238.
Bester, D.J.W., Venter, B.G., 1986. South African J. Dairy Sci. 18, 51–53.
Bhunia, K., Sablani, S.S., Tang, J., Rasco, B., 2013. Compr. Rev. Food Sci. Food Saf. 12, 523–545.
Bley, M.E., Johnson, M.E., Olson, N.F., 1985. J. Dairy Sci. 68, 555–561.
Briozzo, J., Lagarde, E.A., Chirife, J., Parada, J.L., 1983. Appl. Environ. Microbiol. 45, 1150–1152.
Brummel, S.E., Lee, K., 1990. J. Food Sci. 55 (5), 1290–1292.
Bubelová, Z., Tremlová, B., Buňková, L., Pospiech, M., Vítová, E., Buňka, F., 2015. J. Food Sci. Technol. 52 (8), 4985–4993.
Buňka, F., Doudová, L., Weiserová, E., Cerníková, M., Kuchař, D., Slavíková, S., Nagyová, G., Pońížil, P., Grůber, T., Michálek, J., 2014. LWT - Food Sci. Technol. 58, 247–255.
Buňka, F., Štětina, J., Hrabě, J., 2008. Eur. Food Res. Technol. 228, 223–229.
Buňkova, L., Buňka, F., 2015. Crit. Rev. Food Sci. Nutr. 228, 223–229.
CAC/Codex Alimentarius Commission, 2001a. Code of Hygienic Practice for the Transport of Food in Bulk and Semi-Packed Food (CAC/RCP 47-2001). Food and Agriculture Organization of the United Nations – World Health Organization, Rome.
CAC/Codex Alimentarius Commission, 2001b. Code of Practice Concerning Source Directed Measures to Reduce Contamination of Food with Chemicals (CAC/RCP 49-2001). Food and Agriculture Organization of the United Nations – World Health Organization, Rome.
CAC/Codex Alimentarius Commission, 2006. Principles for Traceability/Product Tracing as a Tool within a Food Inspection and Certification System (CAC/GL 60-2006). Food and Agriculture Organization of the United Nations – World Health Organization, Rome.
CAC/Codex Alimentarius Commission, 2008a. Codex General Standard for Named Variety Process(ed) Cheese and Spreadable Process(ed) Cheese (CODEX STAN 285-1978). Food and Agriculture Organization of the United Nations – World Health Organization, Rome.
CAC/Codex Alimentarius Commission, 2008b. Codex General Standard for Process(ed) Cheese and Spreadable Process(ed) Cheese (CODEX STAN 286-1978). Food and Agriculture Organization of the United Nations – World Health Organization, Rome.
CAC/Codex Alimentarius Commission, 2008c. Codex General Standard for Process(ed) Cheese Preparations (Process(ed) Cheese Food and Process(ed) Cheese Spread (CODEX STAN 287-1978). Food and Agriculture Organization of the United Nations – World Health Organization, Rome.
CAC/Codex Alimentarius Commission, 2009. Code of Hygienic Practice for Milk and Milk Products (CAC/RCP 57-2004). Food and Agriculture Organization of the United Nations – World Health Organization, Rome.
CAC/Codex Alimentarius Commission, 2013a. Guidelines for the Validation of Food Safety Control Measures (CAC/GL 69-2008). Food and Agriculture Organization of the United Nations – World Health Organization, Rome.
CAC/Codex Alimentarius Commission, 2013b. Principles and Guidelines for the Establishment and Application of Microbiological Criteria Related to Foods (CAC/GL 21-1997). Food and Agriculture Organization of the United Nations – World Health Organization, Rome.
CAC/Codex Alimentarius Commission, 2018. General Standard for the Labelling of Prepackaged Foods (CXS 1-1985). Food and Agriculture Organization of the United Nations – World Health Organization, Rome.

CAC/Codex Alimentarius Commission, 2019. General Standard for Contaminants and Toxins in Food and Feed (CXS 193-1995). Food and Agriculture Organization of the United Nations – World Health Organization, Rome.
CAC/Codex Alimentarius Commission, 2020a. General Principles of Food Hygiene (CXC 1-1969). Food and Agriculture Organization of the United Nations – World Health Organization, Rome.
CAC/Codex Alimentarius Commission, 2020b. Code of Practice on Food Allergen Management for Food Business Operators (CXC 80-2020). Food and Agriculture Organization of the United Nations – World Health Organization, Rome.
Carić, M., Kaláb, M., 1993. Processed cheese products. In: Fox, P.F. (Ed.), Cheese – Chemistry, Physics and Microbiology. Springer, Boston, pp. 467–505.
Černíková, M., Buňka, F., Pospiech, M., Tremlová, B., Hladkác, K., Pavlínek, V., Březina, P., 2010. Int. Dairy J. 20 (5), 336–343.
CFIA/Canadian Food Inspection Agency, 2021. SFCR/Safe Food for Canadians Regulations, Appendix F: Processed cheese microbiological stability chart for non-refrigerated products. Available at: https://inspection.canada.ca/preventive-controls/dairy-products/appendix-f/eng/1543384250985/1543384251251.
CFSAN, 2010. Guidance for FDA staff: Compliance policy guide Sec. 527.300 Dairy Products - Microbial contaminants and alkaline phosphatase activity. United States Department of Health and Human Services, Food and Drug Administration, Center for Food Safety and Applied Nutrition, Office of Regulatory Affairs. College Park, Rockville, USA.
Chatziantoniou, S.E., 2020. Development and Study of Processed Whey Cheeses. Ph.D. Thesis. University of Ioannina, Ioannina, Greece, p. 297.
Chatziantoniou, S.E., Thomareis, A.S., Kontominas, M.G., 2015. Eur. Food Res. Technol. 241, 737–748.
Chatziantoniou, S.E., Thomareis, A.S., Kontominas, M.G., 2019. Eur. Food Res. Technol. 245, 2401–2412.
Chen, L., Liu, H., 2012. J. Dairy Sci. 95 (9), 4823–4830.
Christensen, J., Povlsen, V.T., Sørensen, J., 2003. J. Dairy Sci. 86, 1101–1107.
Clark, S., Agarwal, S., 2007. Pasteurized process and related cheeses. Ch. 25. In: Hui, Y.H. (Ed.), Handbook of Food Products Manufacturing. John Wiley & Sons, Inc., Hoboken, NJ, USA, pp. 596–614.
Cunha, C.R., Dias, A.I., Viotto, W.H., 2010. Food Res. Int. 43, 723–729.
Daly, D., McSweeney, P., Sheehan, J., 2012. Dairy Sci. Technol. 92 (5), 439–453.
Dimitreli, G., Thomareis, A.S., 2007. J. Food Eng. 79, 1364–1373.
Drake, M.A., Truong, V.D., Daubert, C.R., 1999. J. Food Sci. 64, 744–747.
Drake, S.L., Yates, M.D., Drake, M., 2010. J. Sens. Stud. 25 (5), 720–739.
EDA/EUCOLAIT, 2018. Guidance on Cheese as Raw Material in the Manufacture of Food Products. European Dairy Association (EDA) and the European Association of the Dairy Trade (EUCOLAIT), Brussels, p. 52.
EFSA, 2005. Opinion of the Scientific Panel on Biological Hazards on a request from the Commission related to *Clostridium* spp. in foodstuffs. EFSA J. 199, 1–65.
Emmons, D.B., Paquette, G.J., Froehlich, D.A., Beckett, D.C., Modler, H.W., Butler, G., Brackenbridge, P., Daniels, G., 1986. J. Dairy Sci. 69, 2437–2450.
EU, 2001. Directive 2001/95/EC of the European Parliament and of the Council of 3 December 2001 on general product safety. Off. J. Eur. Commun., L 11, 4–17.
EU, 2002. Regulation (EC) No 178/2002 of the European Parliament and of the Council of 28 January 2002 laying down the general principles and requirements of food law, establishing

the European Food Safety Authority and laying down procedures in matters of food safety. Off. J. Eur. Commun., L 31, 1–24.
EU, 2004a. Regulation (EC) No 852/2004 of the European Parliament and of the Council of 29 April 2004 on the hygiene of foodstuffs. Off. J. Eur. Union, L 139, 1–54.
EU, 2004b. Corrigendum to Regulation (EC) No 852/2004 of the European Parliament and of the Council of 29 April 2004 on the hygiene of foodstuffs. Off. J. Eur. Union, L 226, 3–21.
EU, 2004c. Corrigendum to Regulation (EC) No 853/2004 of the European Parliament and of the Council of 29 April 2004 laying down specific hygiene rules for food of animal origin. Off. J. Eur. Union, L 226, 22–82.
EU, 2004d. Regulation (EC) No 853/2004 of the European Parliament and of the Council of 29 April 2004 laying down specific hygiene rules for on the hygiene of foodstuffs. Off. J. Eur. Union, L 139, 55–205.
EU, 2004e. Regulation (EC) No 1935/2004 of the European Parliament and of the Council of 27 October 2004 on materials and articles intended to come into contact with food and repealing Directives 80/590/EEC and 89/109/EEC. Off. J. Eur. Union, L 338, 4–17.
EU, 2005. Commission Regulation (EC) No 2073/2005 of 15 November 2005 on microbiological criteria for foodstuffs. Off. J. Eur. Union, L 338, 1–26.
EU, 2006. Commission Regulation (EC) No 1881/2006 of 19 December 2006 setting maximum levels for certain contaminants in foodstuffs. Off. J. Eur. Union, L 364, 5–24.
EU, 2007. Commission Regulation (EC) No 1441/2007 of 5 December 2007 amending Regulation (EC) No 2073/2005 on microbiological criteria for foodstuffs. Off. J. Eur. Union, L 322, 12–29.
EU, 2009. Regulation (EC) No 1069/2009 of the European Parliament and of the Council of 21 October 2009 laying down health rules as regards animal by-products and derived products not intended for human consumption and repealing Regulation (EC) No 1774/2002 (Animal by-products Regulation). Off. J. Eur. Union, L 300, 1–33.
EU, 2010. Commission Regulation (EU) No 365/2010 of 28 April 2010 amending Regulation (EC) No 2073/2005 on microbiological criteria for foodstuffs as regards Enterobacteriaceae in pasteurised milk and other pasteurised liquid dairy products and Listeria monocytogenes in food grade salt. Off. J. Eur. Union, L 107, 9–11.
EU, 2011a. Commission Implementing Regulation (EU) No 931/2011 of 19 September 2011 on the traceability requirements set by Regulation (EC) No 178/2002 of the European Parliament and of the Council for food of animal origin. Off. J. Eur. Union, L 242, 2–3.
EU, 2011b. Commission Regulation (EU) No 1259/2011 of 2 December 2011 amending Regulation (EC) No 1881/2006 as regards maximum levels for dioxins, dioxin-like PCBs and non dioxin-like PCBs in foodstuffs. Off. J. Eur. Union, L 320, 18–23.
EU, 2011c. Regulation (EU) No 1169/2011 of the European Parliament and of the Council of 25 October 2011 on the provision of food information to consumers, amending Regulations (EC) No 1924/2006 and (EC) No 1925/2006 of the European Parliament and of the Council, and repealing Commission Directive 87/250/EEC, Council Directive 90/496/EEC, Commission Directive 1999/10/EC, Directive 2000/13/EC of the European Parliament and of the Council, Commission Directives 2002/67/EC and 2008/5/EC and Commission Regulation (EC) No 608/2004. Off. J. Eur. Union, L 304, 18–63.
EU, 2016. Commission Notice on the implementation of food safety management systems covering prerequisite programs (PRPs) and procedures based on the HACCP principles, including the facilitation/flexibility of the implementation in certain food businesses. Off. J. Eur. Union, C 278, 1–32.

EU, 2020. Directive (EU) 2020/2184 of the European Parliament and of the Council of 16 December 2020 on the quality of water intended for human consumption. Off. J. Eur. Union, L 435, 1–62.
EU, 2021. EU Pesticides Database. Available at: https://ec.europa.eu/food/plant/pesticides/eu-pesticides-database/mrls/?event=search.pr.
Everard, C.D., O'Donnell, C.P., O'Callaghan, D.J., Sheehan, E.M., Delahunty, C.M., O'Kennedy, B.T., Howard, V., 2007. J. Sci. Food Agric. 87, 641–650.
Everett, D.W., Auty, M.A.E., 2008. Int. Dairy J. 18, 759–773.
FAO, 1998. Food Quality and Safety Systems - A Training Manual on Food Hygiene and the Hazard Analysis and Critical Control Point (HACCP) System. Food and Agriculture Organization of the United Nations, Rome, p. 178.
FAO, 2003. Development of a Framework for Good Agricultural Practices. Committee on Agriculture, 17th Session. Food and Agriculture Organization of the United Nations, Rome.
FAO/IDF, 2011. Guide to Good Dairy Farming Practice. Animal Production and Health Guidelines. Food and Agriculture Organization of the United Nations, Rome, p. 36.
FAO/WHO, 2006. Guidance to governments on the application of HACCP in small and/or less-developed food businesses. FAO Food and Nutrition Paper 86. Food and Agriculture Organization of the United Nations–World Health Organization, Rome, p. 86.
FAO/WHO, 2019. Safety and Quality of Water Used in Food Production and Processing. Microbiological Risk Assessment Series 33, Meeting Report. Food and Agriculture Organization of the United Nations–World Health Organization, Rome, p. 83.
FDA, 2005. CPG Sec. 555.425 Foods, Adulteration Involving Hard or Sharp Foreign Objects. Available at: https://www.fda.gov/media/71953/download.
Fox, P.F., Guinee, T.P., Cogan, T.M., McSweeney, P.L.H., 2017. Processed cheese and substitute/imitation cheese products. Fundamentals of Cheese Science, 2nd Edition. Springer Science+Business Media LLC, New York, pp. 589–626.
FSANZ, 2017. Australia New Zealand Food Standards Code – Schedule 27 – Microbiological limits in food. Food Standards Australia New Zealand, Kingston, Australia.
Gallmann, P., 1995. Migration. Technical Guide for the Packaging of Milk and Milk Products, Document No. 300, 3rd Edition. International Dairy Federation, Brussels, pp. 8–11.
Garimella Purna, S.K., Pollard, A., Metzger, L.E., 2006. J. Dairy Sci. 89, 2386–2396.
Glass, K.A., Doyle, M.E., 2013. Safety of processed cheese, FRI Briefings. Food Research Institute, University of Wisconsin, Madison, USA.
Glass, K.A., Kaufman, K.M., Johnson, E.A., 1998. J. Food Prot. 61, 290–294.
Guinee, T.P., 2007. Processed cheese. In: McSweeney, P.L.H. (Ed.), Cheese Problems Solved. Woodhead Publishing Ltd., Oxford, UK, pp. 365–383.
Guinee, T.P., 2009. The role of dairy ingredients in processed cheese products, Ch. 20. In: Corredig, M. (Ed.), Dairy Derived Ingredients. Woodhead Publishing Ltd., Oxford, UK, pp. 507–538.
Guinee, T.P., Carić, M., Kaláb, M, 2004. Pasteurized processed cheese and substitute/imitation cheese products. In: Fox, P.F., McSweeney, P.L.H., Cogan, T.M., Guinee, T.P. (Eds.), Cheese–Chemistry, Physics and Microbiology, 3rd Edition, Vol. 2. Elsevier Academic Press, Amsterdam, pp. 349–394.
Guinee, T.P., O'Callaghan, D.J., 2010. Control and prediction of quality characteristics in the manufacture and ripening of cheese, Ch. 8. In: Law, B.A., Tamime, A.Y. (Eds.), Technology of Cheesemaking, 2nd Edition. John Wiley & Sons Ltd., West Sussex, UK, pp. 260–314.
Gunasekaran, S., Ak, M.M., 2003. Cheese Rheology and Texture. CRC Press, Boca Raton, USA, p. 435.

Gupta, S.K., Karahadian, C., Lindsay, R.C., 1984. J. Dairy Sci. 67, 764–778.
Gupta, V.K., Reuter, H., 1993. Lait 73, 381–388.
Gwartney, E.A., Foegeding, E.A., Larick, D.K., 2002. J. Food Sci. 69 (9), S333–S339.
Hanaei, F., Cuvelier, G., Sieffermann, J.M., 2015. Food Qual. Prefer. 40, 316–325.
Harper, W.V., 1992. Lactose and lactose derivatives. In: Zadow, J.G. (Ed.), Whey and Lactose Processing. Elsevier Science Publishers Ltd., New York, pp. 326–328.
Hassan, A.N., Awad, S., Mistry, V.V., 2007. J. Dairy Sci. 90, 3604–3612.
Health Canada, 2020. Guidelines for Canadian Drinking Water Quality – Summary Table. Water and Air Quality Bureau. Healthy Environments and Consumer Safety Branch, Health Canada, Ottawa.
Hussein, F.S.Z., El Zubeir, I.M., Fadlelmoula, A.A., 2011. Jordan J. Biol. Sci. 4 (4), 231–236.
IFST, 2018. Food & Drink – Good Manufacturing Practice: A Guide to its Responsible Management. The Institute of Food Science & Technology Trust Fund, 7th Edition. John Wiley & Sons Ltd, Hoboken, USA, p. 361.
ISO, 1988. Milk and milk products−Sampling−Inspection by variables. International Standard ISO 8197, 1st Edition International Organization for Standardization, Geneva, Switzerland.
ISO, 1999. Sampling procedures for inspection by attributes–Part 1: Sampling schemes indexed by acceptance quality limit (AQL) for lot-by-lot inspection. International Standard ISO 2859-1, 2nd Edition International Organization for Standardization, Geneva, Switzerland.
ISO, 2004. Milk and milk products−Sampling−Inspection by attributes. International Standard ISO 5538|IDF 113, 2nd Edition International Organization for Standardization, Geneva, Switzerland.
ISO, 2007. Traceability in the feed and food chain−General principles and basic requirements for system design and implementation. International Standard ISO 22005:2007, 1st Edition International Organization for Standardization, Geneva, Switzerland.
ISO, 2008. Milk and milk products–Guidance on sampling. International Standard ISO 707|IDF 50, 3rd Edition International Organization for Standardization, Geneva, Switzerland.
ISO, 2009. International Standard ISO 22935-3|IDF 99-3, 1st Edition International Organization for Standardization, Geneva, Switzerland.
ISO, 2009. Prerequisite programmes on food safety–Part 1: Food manufacturing. Technical Specification ISO/TS 22002-1:2009, 1st Edition International Organization for Standardization, Geneva, Switzerland.
ISO, 2011. Prerequisite programmes on food safety−Part 3: Farming. Technical Specification ISO/TS 22002-3:2011, 1st Edition International Organization for Standardization, Geneva, Switzerland.
ISO, 2013a. Prerequisite programmes on food safety−Part 4: Food packaging manufacturing. Technical Specification ISO/TS 22002-4:2013, 1st Edition International Organization for Standardization, Geneva, Switzerland.
ISO, 2013b. Sampling procedures for inspection by variables−Part 1: Specification for single sampling plans indexed by acceptance quality limit (AQL) for lot-by-lot inspection for a single quality characteristic and a single AQL. International Standard ISO 3951-1, 2nd Edition, International Organization for Standardization, Geneva, Switzerland Corr. Version 2016.
ISO, 2013c. Sampling procedures for inspection by variables−Part 2: General specification for single sampling plans indexed by acceptance quality limit (AQL) for lot-by-lot inspection of independent quality characteristics. International Standard ISO 3951-2, 2nd Edition International Organization for Standardization, Geneva, Switzerland Corr. Version 2016.

ISO, 2015a. Quality management systems–Fundamentals and vocabulary. International Standard ISO 9000:2015, 4th Edition International Organization for Standardization, Geneva, Switzerland.

ISO, 2015b. Quality management systems – Requirements. International Standard ISO 9001:2015, 5th Edition International Organization for Standardization, Geneva, Switzerland.

ISO, 2016. Prerequisite programmes on food safety–Part 6: Feed and animal food production. Technical Specification ISO/TS 22002-6:2016, 1st Edition International Organization for Standardization, Geneva, Switzerland.

ISO, 2018. Food safety management systems–Requirements for any organization in the food chain. International Standard ISO 22000:2018, 2nd Edition International Organization for Standardization, Geneva, Switzerland.

ISO, 2019. Prerequisite programmes on food safety–Part 5: Transport and storage. Technical Specification ISO/TS 22002-5:2019, 1st Edition International Organization for Standardization, Geneva, Switzerland.

ISO, 2020. Sampling procedures for inspection by attributes–Part 2: Sampling plans indexed by limiting quality (LQ) for isolated inspection. International Standard ISO 2859-2, 2nd Edition International Organization for Standardization, Geneva, Switzerland.

Johnson, M.E., Kapoor, R., McMahon, D.J., McCoy, D.R., Narasimmon, R.G., 2009. Compr. Rev. Food Sci. Food Saf. 8, 252–268.

Kang, E.J., Campbell, R.E., Bastian, E., Drake, M.A., 2010. J. Dairy Sci. 93, 3891–3901.

Kapoor, R., Metzger, L.E., 2004. J. Dairy Sci. 87, 1143–1150.

Kapoor, R., Metzger, L.E., 2008. Compr. Rev. Food Sci. Food Saf. 7, 194–214.

Kapoor, R., Metzger, L.E., Biswas, A.C., Muthukummarappan, K., 2007. J. Dairy Sci. 90, 1625–1634.

Karahadian, C., Lindsay, R., 1984. J. Dairy Sci. 67, 1892–1904.

Kim, N.H., Lee, N.Y., Kim, M.G., Kim, H.W., Cho, T.J., Joo, I.S., Heo, E.J., Rhee, M.S., 2018. Food Res. Int. 106, 468–474.

Kindstedt, P.S., Fox, P.F., 1991. J. Food Sci. 56 (4), 1115–1116.

Kindstedt, P.S., Polowsky, P.J., 2020. J. Dairy Sci. 104 (2), 1233–1250.

Klostermeyer, H., Uhlmann, G., Merkenich, K., 1984. Milchwissenschaft 39, 195–197.

Kneifel, W., Ulberth, F., Schaffer, E., 1992. Lait 72, 383–391.

Kosikowski, F.V., Mistry, V.V., 1997. Process cheese and related products. In: Kosikowski LLC, F.V. (Ed.), Cheese and Fermented Milk Foods: Origins and Principles, 3rd Edition, 1. Westport, CT, USA, pp. 204–225.

Kristensen, D., Hansen, E., Arndal, A., Appelgren, T., Skibsted, L.H., 2001. Int. Dairy J. 11, 837–843.

Kussy, D., Aylward, E., 2009. Pasteurized process cheese. In: Clark, S., Costello, I.M., Drake, I.M., Bodyfelt, I.F. (Eds.), The Sensory Evaluation of Dairy Products. Springer Science & Business Media, New York, USA, pp. 387–401.

Laslo, E., György, E., 2018. Acta Universitatis Sapientiae. Alimentaria 11, 27–44.

Lazárková, Z., Buňka, F., Buňková, L., Valášek, P., Kráčmar, S., Hrabě, J., 2010. Czech J. Food Sci. 28, 168–176.

Leather, A.N., 1949. Analyst 74, 51.

Lee, S.K., Anema, S., Klostermeyer, H., 2004. Int. J. Food Sci. Technol. 39, 763–771.

Lee, S.K., Huss, M., Klostermeyer, H., Anema, S.G., 2013. Int. Dairy J. 32, 79–88.

Lee, S.K., Klostermeyer, H., 2001. LWT – Food Sci. Technol. 34 (5), 288–292.

Lee, W., Clark, S., Swanson, B.G., 2006. J. Food Process. Preserv. 30, 164–179.

LII/Legal Information Institute, 2021a. 7 CFR Subpart B–General Specifications for Dairy Plants Approved for USDA Inspection and Grading Service 1. Available at: https://www.law.cornell.edu/cfr/text/7/part-58/subpart-B.

LII/Legal Information Institute, 2021b. 21 CFR Part 117–Current Good Manufacturing Practice, Hazard Analysis, and Risk-Based Preventive Controls for Human Food. Available at: https://www.law.cornell.edu/cfr/text/21/part-117.

LII/Legal Information Institute, 2021c. 21 CFR Part 120–Hazard Analysis and Critical Control Point (HACCP) Systems. Available at: https://www.law.cornell.edu/cfr/text/21/part-120.

LII/Legal Information Institute, 2021d. 21 CFR Part 133–Cheeses and Related Cheese Products. Available at: https://www.law.cornell.edu/cfr/text/21/part-133.

Lucey, J.A., Maurer-Rothmann, A., Kaliappan, S., 2011. Functionality of ingredients: Emulsifying salts. In: Tamime, A.Y. (Ed.), Processed Cheese and Analogues. Blackwell Publishing Inc., West Sussex, UK, pp. 110–132.

Lycken, L., Borch, E., 2006. J. Food Prot. 69, 1887–1891.

McLay, J.C., Kennedy, M.J., O'Rourke, A.L., Elliot, R.M., Simmonds, R.S., 2002. Int. J. Food Microbiol. 73, 1–9.

Meyer, A., 1973. Processed Cheese Manufacture. Food Trade Press Ltd, London, p. 330.

MHPRC, 2010. National Standard of the People's Republic of China. GB 25192-2010–National food safety standard Process(ed) cheese. Ministry of Health of People's Republic of China, Beijing, China, p. 5.

Mistry, V.V., Anderson, D.L., 1993. Food Structure 12 (2), 259–266.

Mleko, S., Foegeding, E.A., 2001. Milchwissenschaft 56 (11), 612–615.

Morris, H.A., Manning, P.B., Jenness, R., 1969. J. Dairy Sci. 52, 900.

Mounsey, J.S., O'Kennedy, B.T., Kelly, P.M., 2007. Milchwissenschaft 62, 44–47.

Muir, D.D., Williams, S.A.R., Tamime, A.Y., Shenana, M.E., 1997. Int. J. Food Sci. Technol. 32, 279–287.

Mulvihill, D.M., McCarthy, A., 1993. J. Dairy Res. 60, 431–438.

Muthukumarappan, K., Wang, Y.C., Gunasekaran, S., 1999. J. Dairy Sci. 82, 1068–1071.

NACMCF, 2017. HACCP Principles & Application Guidelines. Available at: https://www.fda.gov/food/hazard-analysis-critical-control-point-haccp/haccp-principles-application-guidelines.

Necidová, L., Bursová, Š., Haruštiaková, D., Bogdanovičová, K., Lačanin, I., 2019. J. Dairy Sci. 102, 3924–3932.

Olson, N.F., Price, W.V., 1958. J. Dairy Sci. 41 (7), 999–1000.

Oyugi, E., Buys, E.M., 2007. Int. J. Dairy Technol. 60, 89–95.

Paul-Sadhu, S., 2016. Acta Alimentaria 45 (3), 390–397.

Pinto, S., Rathour, A.K., Prajapati, J.P., Jana, A.H., Solanky, M.J., 2007. Nat. Prod. Radiance 6 (7), 398–401.

Pizzoferrato, L., Manzi, P., Vivanti, V., Nicoletti, I., Corradini, C., Cogliandro, E., 1998. J. Food Prot. 61, 235–239.

Pommert, J.F., Klaebe, A., Perie, J., Lebugle, A., Puech, J., 1988. J. Food Sci. 53, 1367–1369.

Quigley, L., O'Sullivan, D.J., Daly, D., O'Sullivan, O., Burdikova, Z., Vana, R., Beresford, T.P., Ross, R.P., Fitzgerald, G.F., McSweeney, P.L.H., Giblin, L., Sheehan, J.J., Cotter, P.D., 2016. Appl. Environ. Sci. 1 (3), 1–12.

Raters, M., Matissek, R., 2008. Mycotoxin Res. 24 (3), 130–134.

Ross-Kung, P., Neves, P., 2003. Validation and Verification of HACCP Plans in Retail Food Establishments. Massachusetts Department of Public Health, Jamaica Plain, Massachusetts, USA, p. 98.

Savello, P.A., Ernstrom, C.A., Kaláb, M., 1989. J. Dairy Sci. 72, 1–11.

Schäffer, B., Szakály, S., Lõrinczy, D., Schäffer, B., 2001. J. Therm. Anal. Calorim. 64, 671–679.

Schär, W., Bosset, J.O., 2002. LWT - Food Sci. Technol. 35, 15–20.

Scharpf, L.G., Kichline, T.P., 1968. J. Dairy Sci. 51, 853–857.

Schwope, A.D., Till, D.E., Ehntholt, D.J., Sidman, K.R., Whelan, R.H., Schwartz, P.S., Reid, R.C., 1987. Food Chem. Toxicol. 25, 317–326.

Sharma, P., Segat, A., Kelly, A.L., Sheehan, J.J., 2019. Compr. Rev. Food Sci. Food Saf. 1–23.

Shumaker, E.K., Wendorff, W.L., 1998. J. Food Sci. 63, 828–831.

Sinha, R.N., Sinha, P.R., 1988. J. Food Sci. Technol. 25, 101–102.

Sunesen, L.O., Lund, P., Sørensen, J., Hølmer, G., 2002. LWT – Food Sci. Technol. 35, 128–134.

Swenson, B.J., Wendorff, W.L., Lindsay, R.C., 2000. J. Food Sci. 65 (5), 822–825.

Tamime, A.Y., Muir, D.D., Shenana, M.E., Kaláb, M., Dawood, A.H., 1999. LWT - Food Sci. Technol. 35 (1), 50–59.

Tamime, A.Y., Muir, D.D., Wszolek, M., Domagala, J., Metzger, L., Kneifel, W., Dürrschmid, K., Domig, K.J., Hill, A., Smith, A., Guinee, T.P., Auty, M.A.E, 2011. Quality control in processed cheese manufacture, Ch. 10. In: Tamime, A.Y. (Ed.), Processed Cheese and Analogues. Blackwell Publishing Inc., West Sussex, UK, pp. 246–339.

Tanaka, N., Traisman, E., Plantinga, P., Finn, L., Flom, W., Meske, L., Guggisberg, J., 1986. J. Food Prot. 49, 526–531.

Templeton, H.L., Sommer, H.H., 1932. J. Dairy Sci. 15, 155–162.

ter Steeg, P.F., Cuppers, H.G.A.M., Hellemons, J.C., Rijke, G., 1995. J. Food Prot. 58, 1091–1099.

Thapa, T.B., Gupta, V.K., 1992. Indian J. Dairy Sci. 45 (2), 88–92.

Thomareis, A.S., 1986. Propriétés Thermophysiques des Produits Laitiers. Cas des Fromages Fondus. Ph.D. Thesis. Institut National Polytechnique de Lorraine, France, p. 251.

Thomas, M.A., 1969. Austr. J. Dairy Technol. 24, 185–189.

Thomas, M.A., 1970. Austr. J Dairy Technol. 25, 23–26.

USDA, 1991. United States Standards for Grades of Bulk American Cheese. Agricultural Marketing Service. United States Department of Agriculture, Washington, p. 9.

USDA, 2000. United States Standards for Dry Whey. Agricultural Marketing Service. United States Department of Agriculture, Washington, p. 5.

USDA, 2001a. United States Standards for Grades of Nonfat Dry Milk (Spray Process). Agricultural Marketing Service, United States Department of Agriculture, Washington, p. 6.

USDA, 2001b. United States Standards for Grades of Swiss Cheese, Emmentaler Cheese. Agricultural Marketing Service, United States Department of Agriculture Washington, p. 14.

Varga, L., 2007. Microbiological quality of commercial dairy products. In: Méndez-Vilas, A. (Ed.), Communicating Current Research and Educational Topics and Trends in Applied Microbiology. Formatex Microbiology Series (1), Formatex, Badajoz, Spain, pp. 487–494.

Warburton, D.W., Peterkin, P.I., Weiss, K.F., 1986. J. Food Prot. 49 (3), 229–230.

Wendin, K., Langton, M., Caous, L., Hall, G., 2000. Food Chem. 71, 363–378.

WHO, 2006. Five Keys to Safer Food Manual. World Health Organization, Geneva, Switzerland, p. 28.

WHO, 2017. Guidelines for Drinking-water Quality, 4th Edition. World Health Organization, Geneva, Switzerland, p. 541.

Zehren, V.L., Nusbaum, D.D., 2000. Process Cheese, 2nd Edition. Cheese Reporter, Madison, USA, p. 362.

Microbiology of processed cheese

14

Gustavo Luís de Paiva Anciens Ramos[a,b], *Adriana Torres Silva e Alves*[c], *Leila Maria Spadoti*[c], *Patrícia Blumer Zacarchenco*[c] *and Adriano Gomes da Cruz*[b]
[a] Universidade Federal Fluminense (UFF), Faculdade de Veterinária, Niterói, Rio de Janeiro, Brazil, [b] Instituto Federal de Educação, Ciência e Tecnologia do Rio de Janeiro (IFRJ), Departamento de Alimentos, Rio de Janeiro, Brazil, [c] Instituto de Tecnologia de Alimentos (ITAL), Centro de Tecnologia de Laticínios (Tecnolat), Campinas, São Paulo, Brazil

14.1 Introduction

The production of processed cheese began from a requirement to increase the shelf-life of cheeses or to create a new type of cheese more stable. The main steps in the manufacture of processed cheese is melting and heating cheeses, adding the emulsifying or melting salts, agitating to produce a homogeneous mixture, packaging and/or cooling (Tamime, 2011).

The microbiological stability and safety of processed cheeses depends on the characteristics of the ingredients used, the intensity of the heat treatment, the rate of cooling after this step, the material and packaging technology, and the storage conditions. If the treatment is as mild as pasteurization, the processed cheese will need refrigeration after production. The mild thermal treatment inactivates the vegetative cells of pathogenic or deteriorating agents, however any bacterial spores present are not inactivated, which represents a great challenge to guarantee the quality and safety of these products. Consequently, ensuring the microbiological safety and stability of processed cheese are important considerations for the cheese industry to meet the growing market demand for this product (Cruz et al., 2019).

In general, processed cheeses have good microbiological stability, with aerobic and anaerobic mesophilic counts below 10 CFU/g, in most cases. There are samples, however, that showed high counts of aerobic mesophilic counts, especially when they are prepared from cheese or curd produced using lactic starter. In other hand, the microbiological quality of industrially processed cheeses is better than the product obtained by hand, where there is less processing control (Van Dender et al., 2012).

The presence of molds and yeasts in processed cheeses is rarely reported, as many of these products are packaged in partial anaerobiosis. However, there are reports of counts of up to 4 log CFU/g of molds and yeasts, where there was no packaging in an anaerobic atmosphere, also considering the permeability of package materials (plastic or glass, for example) (Alves et al., 2007). At Fig. 14.1 different types of packages commonly used for a typical Brazilian spreadable processed cheese (called requeijão cremoso). The more common plastic packages are polypropylene plastic cups with

Processed Cheese Science and Technology: Ingredients, Manufacture, Functionality, Quality, and Regulations.
DOI: https://doi.org/10.1016/B978-0-12-821445-9.00016-9

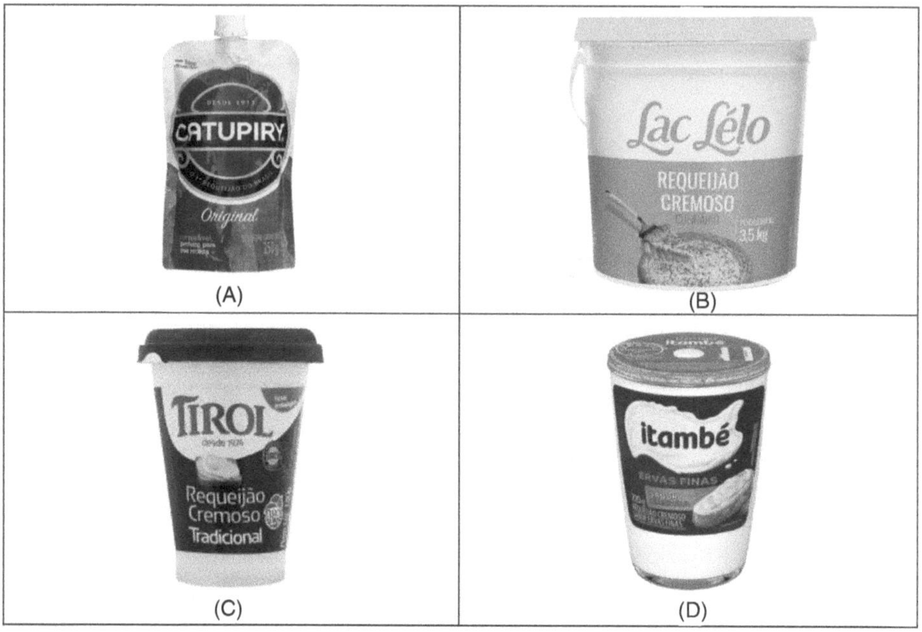

Figure 14.1 Different types of packages and packages materials used for spreadable processed cheese. **(A)** stand up pouche multilayer package; **(B)** plastic bucket; **(C)** plastic cups with aluminum seal and a plastic cover and **(D)** glass cup with "easy-open" closure cap (tinplate).

aluminum seal and a plastic cover and plastic bucket (polypropylene). One of the packages at Fig. 14.1 contain 3.5 kg and represents a risk for germination of spores during the cooling step because of the long time until refrigeration temperatures be achieved.

Spore-forming bacteria are contaminants in the dairy industry and these microorganisms can be responsible for serious economic losses due to deterioration and outbreaks of foodborne diseases. The spore-forming bacteria most commonly found in dairy products belong to the genera *Bacillus* and *Clostridium*, which contain several species associated with food spoilage and public health prospects. Thus, it is necessary to have knowledge about the main routes of contamination of processed cheese by spore-forming bacteria, as well as control measures. The spores survive in environments with minimal nutritional availability, and when favorable conditions are available, they germinate and become normal vegetative cells (Zeinab et al., 2015; Cruz et al., 2019).

In addition, postpasteurization contamination, which has been identified especially in dairy products, cannot be disregarded. The main contaminants in this case are *Listeria monocytogenes, Salmonella* spp., *Escherichia coli, Bacillus cereus,* and *Staphylococcus aureus* (Glass and Doyle, 2005). Other genera such as *Pseudomonas* and *Aeromonas* are also isolated from processed cheese. If the processing is not well controlled and not follows good manufacturing practices, there is a high risk of food outbreaks due to these pathogens. Experimental studies on processed cheeses showed that *S. aureus* may be viable after 180 days of storage at 5°C and after 270 days of storage at 22°C, while *L. monocytogenes* can multiply from 1°C (Buňková and Buňka, 2015, Cruz et al., 2019).

The emulsifying salts are essential ingredients for the manufacturing process of processed cheeses. They are responsible for the physical-chemical process of melting, acting on the texture and flavor of the product. In addition, they play an important role in the microbiological safety of processed cheeses, with several properties that inhibit or reduce microbial development (Kapoor and Metzger, 2008).

Considering the above, in this chapter the characteristics of the microbiota of processed cheeses will be addressed with an emphasis on the main aspects of processing, sources and routes of contamination and the action of melting or emulsifying salts as inhibitors of deteriorating microorganisms and pathogens. The characteristics of the spore-forming bacteria presented in processed cheese will be emphasized. The action points to carry out the microbiological control of processed cheeses, such as conservative agents, biopreservatives, control of extrinsic factors, and predictive microbiology will also presented. Finally, the recent findings regarding the use of probiotics in processed cheeses will be reviewed.

14.2 Processing of processed cheeses and sources of contamination

Although the processed cheeses are essentially characterized by their composition (fat and moisture content), pH and consistency, the water activity defines the appropriate heat treatment applied in the processing. The products with high water activity (>0.95) require more intense thermal processing, such as sterilization, to reduce the risk of germination of bacterial spores, and multiplication of vegetative cells during the shelf life. On the other hand, pasteurized products must be stored under refrigeration (Oliveira et al., 2016).

Various types of processed cheese are found on the market according to their different characteristics, including the presentation in slices, pieces, blocks and tubes, which can be spreadable or creamy. The recent consumer demands for healthier products, with reduced fat and salt content, can lead to relevant changes in processing technologies and produce major challenges in ensuring stable and safe products during storage. It is important to highlight that the unique composition of milk fat, with a favorable fatty acid profile and increased by the presence of biologically active lipids, contributes to the differentiated nutritional properties of milk and derived products (Cruz et al., 2019).

The detailed steps involved in the processing of the processed cheeses are slicing of the cheeses, followed by mixing it with the other ingredients, adding melting salts, melting of the mixture, heat treatment, hot filling of the liquid mixture using various types of packaging, cooling and storage at an appropriate temperature. Each ingredient used in the production of processed cheese can affect the physical-chemical, sensory and functional properties, the microbiological characteristics and, consequently, product safety. The cheeses used in the manufacture of processed cheese is essentially produced by the microbial fermentation of milk by selected acid bacteria, however, it can be contaminated by undesirable organisms. The cheese curds can also be produced using heat and lactic acid addition, but if good manufacture practices were not applied, they will be contaminated (Kapoor and Metzger, 2008; Cruz et al., 2019).

Cheeses of different types and degrees of ripeness are used in the production of processed cheese. These ingredients are mixed and heated to produce a homogeneous mixture, which is packaged and cooled. The temperature of the melting process normally varies between 70°C and 85°C, but this heating must be rapid and involve stirring (mechanical treatment) to achieve the desired texture. After the melting process, additional heat treatment, such as pasteurization, can be applied, where the product is heated to 72–95°C by indirect heating or direct injection of steam under vacuum. This heating generally occurs in heating pans and can be carried out with high or low mechanical agitation/shear. Another heat treatment that can be applied is sterilization, which requires a temperature above 100°C (usually heating to 140°C for 10 s). The time/temperature combinations that are widely used vary between 128°C and 145°C for 3–15 s (Kapoor and Metzger, 2008; Cruz et al., 2019).

The cooling step after heat treatment is also of great importance for the properties of processed cheese, such as firmness and adhesiveness. However, it also has an important influence on the microbiological safety of the product, since very long times can be used in order to obtain a slow cooling. With a slower cooling, a firmer processed cheese is obtained, however this can allow microbial development during the applied time. There are treatments of up to 50 h of cooling at 20°C, showing a great potential for microbial development during the process (Piska and Štětina, 2004).

The microbial contamination of processed cheese can originate from several sources. Microorganisms can contaminate products at all stages of production, from the farm (soil, feces, silage, and raw milk) to the industry (ingredients, environment, equipment, and packaging material). In addition, the problem with germination and multiplication of bacterial spores can arise during product distribution and storage, leading to deterioration and outbreaks of foodborne diseases (Oliveira et al., 2016).

In the farm environment, there is a clear association between spore-forming bacteria in the soil and milk contamination, which can reach the place where processed cheese is made. Common spores include strictly anaerobic (*Clostridium* spp.) and aerobic species (*Bacillus* spp. and related genera), and their concentrations can reach 5–6 log spores/g of soil (Doyle et al., 2015). Another important source of contamination is bacteria that colonize the intestines of dairy animals, with spores spreading through the feces. In fact, the dispersion of spores by the fecal material of warm-blooded animals is an important cause of contamination of dairy foods. The silage used as dairy cattle feed is also another source of spore contamination, mainly by *Clostridium tyrobutyricum*, which is an important deteriorating bacterium. All sources of contaminants present in the first stage of the process can be transferred to the milk used for processed cheese (Oliveira et al., 2016; Cruz et al., 2019).

During the manufacture of processed cheeses, the final amount of bacterial spores is the result of the contribution of each ingredient used in the process. The type and quantity of cheese, as well as the proportions used to produce the product, comprise another important source of contamination by the spores of the final product.

Furthermore, the cheese curd can be considered a major source of clostridia for processed cheese. Other dairy ingredients (fluid and powdered milk and whey, milk, and whey protein concentrates) and nondairy ingredients (spices, starches, gums and others) can also be relevant sources, since the concentration of spores in some ingredients, such as powdered milk and spices, can be quite high (greater than 3 log spores/g). The amount of bacterial spores depends on factors such as the local microbial ecology, the type of food supplied to the animals, the agricultural practices adopted, and the hygiene practices during processing (Cruz et al., 2019).

The microbiological contamination in the dairy industry environment is a major challenge for food processors. Spore-forming bacteria are established in this environment based on their ability to form biofilms and survive to hygiene processes. Spores can be found on the floor, equipment and utensils used during the production of processed cheese. The main issue with the introduction of spores in the dairy processing environment is its high resistance. Although the spore count can be reduced by heat treatment, the effectiveness of the treatment is influenced by several factors, such as the combination of time and temperature used, the physico-chemical properties of the formulation and the thermal resistance of the spores. Due to their longer decimal reduction time, spores can survive to food processing and persist for long periods in processed cheese (Oliveira et al., 2016; Cruz et al., 2019).

The packaging material is also a potential source of contamination and when not properly handled, stored and sanitized before being filled with processed cheese, it can serve as a source of spore recontamination after heat treatment. Once present in the final product, the spores can slowly germinate and multiply if the pH of the product and water activity is favorable, even at low storage temperature, leading to product deterioration or food risk. The time required for pathogenic deteriorating agents to reach infectious levels depends on the microbiological stability of the formulation, the storage temperature conditions, and the characteristics of the spore-forming bacteria present in the final product (Cruz et al., 2019).

14.3 Action of emulsifying salts in the microbiology of processed cheeses

The melting or emulsifying salts play an important role in the microbiological control of processed cheeses. Phosphates are the most used type and have bacteriostatic properties, that is, they inhibit microbial multiplication. In addition, they can exert a bactericidal effect on certain microorganisms, increasing the shelf life of the product. Within the phosphate group, polyphosphates are reported to have a more significant action on microorganisms than orthophosphates (Berger et al., 1997; Nagyová et al., 2014).

The inhibitory effect of phosphates can be affected by internal factors such as the number of phosphorus atoms (size of the chain, the larger ones have more affinity with the calcium ion) and acid-base properties of the phosphate. External factors also influence the action of salts, such as pH of the food matrix and different cell membrane structures. Synergistic factors positively affect the inhibitory effect, such as pH change, regulated atmosphere, and addition of sodium chloride and lysozyme (Lorencová et al., 2012).

The mechanism of action of polyphosphates is based on the fact that the cell walls and membranes of microorganisms require calcium for their proper functioning. Polyphosphates are unable to cross cell membranes and bind to the calcium ions in the medium, thus making them unavailable for stabilizing the cell lining of microorganisms and causing change in the structure and disturbance of selective permeability of the cell membrane, causing damage to osmotic regulation and decreased metabolic activities of the microorganism. Also, the interaction of phosphates with the cell membrane can affect the pH of the medium and increase ionic strength, resulting in the inhibition of some cell transport reactions (Lorencová et al., 2012; Berger, 1997).

The action of polyphosphates is more effective in Gram-positive microorganisms, as is the case of the main genera of sporulated microorganisms associated with processed cheeses, *Bacillus* and *Clostridium*. This is due to differences in the surface structure of the cell. Gram-positives have teichoic acid in their cell wall, which plays an important role in calcium balance and allows an ion reserve close to the cell surface that may be required for the activity of some enzymes. This mechanism can be considered as similar to that exerted by lipopolysaccharides and lipoproteins in Gram-negative microorganisms. In addition, phosphates also have an inhibitory action on some molds and yeasts (Lorencová et al., 2012; Nagyová et al., 2014).

A concentration of 0.1% (w/v) of polyphosphate is capable of inhibiting important species of the genus *Bacillus*, such as *B. cereus* and *B. subtilis*, while a concentration of 0.5% (w/v) has the ability to inhibit Gram-negative food pathogens, such as *Escherichia coli* and *Salmonella* spp. The evaluation of a greater range of Gram-positive indicated that a concentration of 0.3% (w/v), in general, has inhibitory capacity on vegetative cells of these, while for Gram-negative a concentration from 0.5% (w/v) is required, depending on the microorganism. Also, polyphosphates are able to inhibit the germination of *Bacillus cereus* spores, showing their bacteriostatic and bactericidal effect for this pathogen (Novaková, 2008).

Species of the genus *Clostridium* are also inhibited by polyphosphates, where a concentration of 0.1% (w/v) is able to inhibit the vegetative form of *C. tyrobutyricum* in spreadable processed cheese. Regarding the germination of spores of the same species, the concentration of 0.5% (w/v) of polyphosphate is able to provide an efficient control when the spore count is low and the product is kept under refrigeration. With a concentration of 1% (w/v), the spores and vegetative form are completely inhibited (Loessner et al., 1997).

The action of polyphosphates on Gram-negative in general is extremely weak, even when using long chain salts. Among Gram-positives, *Enterococcus faecalis* exhibits greater resistance to the inhibitory effects of phosphates, requiring greater concentration for its control (Lorencová et al., 2012).

In the processing of processed cheeses, it is also common to use combinations of different melting salts, such as polyphosphate/phosphate and orthophosphate/phosphate. Another type of melting salt used is citrates, although they are less inhibitory than phosphates. It is also possible to use different types of salts in combination, such as citrate and polyphosphate. The action of melting salts in combinations proves to be very efficient, causing a significant increase in the shelf life of processed cheeses, which may inhibit the presence of botulinum toxin in products manufactured with 3 log CFU/g of *C. botulinum* spores and preserved at 30°C for about a year (Salek et al., 2015; Cruz et al., 2019).

The melting salts are generally used in their sodium forms, however with the constant initiative to reduce sodium in food, the possibility arose of using potassium salts. However, its effects of antimicrobial action seem to correspond to 75% of the effect obtained when using sodium salts to control *C. botulinum* in processed cheeses (Glass et al., 2017; Buňková and Buňka, 2015).

The presence of monoacylglycerols, called real emulsifiers, also affects the microbiology of processed cheeses. Different monoacylglycerols in concentrations of 0.15% (w/v) are capable of inhibiting important species of the genus *Bacillus*, such as *B. cereus* and *B. subtilis* in cheeses processed for 5 months under refrigeration. This effect is little observed when it comes to the genus *Clostridium*. The mechanism of effect of monoglycerides is not clear, and its use is limited due to the probable changes in the organoleptic properties of processed cheeses, especially short chain monoglycerides (Hauerlandová et al., 2014).

14.4 Sporulated bacteria of importance for processed cheeses

The two main spore-forming bacterial families in the processed cheese industry are *Bacillaceae* and *Clostridiaceae*. These microorganisms are a serious concern for the international dairy industry because their spores can easily contaminate milk throughout the production chain and survive the heat treatment process. If they are present in enough amount in the processed cheese, these bacteria can cause spoilage or foodborne illness. Table 14.1 shows some of the main sporulated Gram-positive microorganisms of interest for the microbiology of processed cheeses, classified by optimal growth temperature and atmosphere and pH conditions (Buňková and Buňka, 2015).

14.4.1 Bacillus *spp. in processed cheeses*

Under conditions of environmental stress such as nutrient depletion, members of the genus *Bacillus* are able to produce spores. Sporulation can occur in the soil, in the gut of animals and in processing lines, as well as in food. It is well known that spores are able to withstand chemical and physical stresses such as air drying, high temperature, high pressure, UV light, and acidity. Under suitable conditions, the spores germinate and become vegetative cells that produce toxins. This is an important aspect in processed cheeses, because the ability to form spores favors the long-term survival and persistence of *Bacillus* spp. throughout the production chai (Fernández-No et al., 2011; Lücking et al., 2013).

The heat treatments such as pasteurization are sufficient to kill most vegetative cells, but insufficient to kill spores. Thus, pasteurized processed cheeses are susceptible

Table 14.1 Examples of the main Gram-positive sporulated microorganisms of interest in the microbiology of processed processes.

Psychrotrophic and psychrophiles	Facultative anaerobic	*Bacillus circulans* *Bacillus mycoides*
	Anaerobic	*Clostridium gasigenes* *Clostridium putrefaciens*
Thermophilic	Facultative anaerobic	*Bacillus coagulans* *Geobacillus stearotermophilus*
	Anaerobic	*Bacillus subtilis* *Thermoanaerobacterium thermosaccharolyticum*
Mesophilic	Anaerobic proteolytic	*Clostridium sporogenes* *Clostridium putrefaciens*
	Facultative anaerobic	*Bacillus circulans* *Paenibacillus macerans*
	Facultative anaerobic and acidophilic	*Brevibacillus laterosporus* *Alyciclobacillus acidoterrestris*
	Anaerobic	*Clostridium tyrobutyricum* *Clostridium butyricum*

Adapted from Buňková and Buňka, 2015.

to contamination by *Bacillus* due to their ability to survive to moderate heating and to grow at refrigerated temperatures. In addition, surviving bacteria have limited competition and can grow faster than vegetative cells destroyed by pasteurization (Lücking et al., 2013).

With the advance of molecular biology and consequent changes in the taxonomy of microorganisms, some species of the genus *Bacillus* were dismembered and originated new bacterial genera, such as *Paenibacillus, Brevibacillus, Viridibacillus, Solibacillus, Psychrobacillus, Anoxybacillus, Geobacillus*, and *Alicyclobacillus* (Buňková and Buňka, 2015).

Species of the genus *Bacillus* can cause problems associated with product quality, whereby deterioration promotes changes that make a product unacceptable for human consumption, resulting in economic losses. However, there are no exact figures for the degree to which this occurs, but considering the great potential of the processed cheese industry, the problems related to the deterioration of *Bacillus* can reach considerable economic proportions (Panel and Hazards, 2016).

Bacillus species can be classified as psychrotolerant, mesophilic, and thermophilic types. *B. cereus, B. circulans,* and *B. mycoides* are considered to be predominantly psychrotolerant species, while *B. licheniformis, B. pumilus* and *B. subtilis* are mesophilic. *B. stearothermophilus,* and some thermotolerant isolates of *B. licheniformis* are the most common thermophilic species. As psychrotolerant species grow best at lower temperatures, pasteurized processed cheeses stored under refrigeration are susceptible to their growth. Defects caused by psychrotolerant species are quite similar to those caused by other deteriorating bacteria, such as unpleasant taste and structural defects caused by proteolytic, lipolytic, and/or phospholipase enzymes. Changes in flavor can be manifested by bitter, putrid, rancid, and fruity flavors. A bitter taste is caused by the activity of the proteases, while the rancid and fruit aromas are caused by lipases. Sterilized processed cheese can also be changed as a result of recontamination during filling, especially when it is contaminated by heat-resistant thermophilic or mesophilic proteolytic species. *Bacillus sporothermodurans* can break down dairy casein, resulting in the release of small peptides, although only minor changes occur in the product (Panel and Hazards, 2016; Cruz et al., 2019).

From the point of view of deterioration, it is notable that many *Bacillus* spp. produce thermotolerant lipolytic enzymes, which lead to the deterioration of dairy products. These enzymes exhibit optimal activity at temperatures between 60°C and 75°C, similar to those used for pasteurization. If the cooling step is slow, these enzymes can also keep their activities during a long time. In addition, they produce a wide variety of extracellular enzymes, including amylases, hemolysins, lecithinases, phospholipases, proteases, beta-lactamases, and sphingomyelinases. Proteolytic and lipolytic enzymes resistant to heat negatively affect the quality of dairy products during storage. Thus, aroma quality and texture problems are attributed to heat-stable protease activity. Lipolytic deterioration of milk due to enzymes produces taste defects that are associated with fat breakdown. Free fatty acids are released and generate unpleasant flavors that can be described as rancid, butyric, bitter, impure, soapy, and astringent. Deteriorating microorganisms are responsible for several sensory defects (off-odor, discoloration), but these sensory descriptors are not easily associated with enzymatic functions or metabolic pathways (Cruz et al., 2019).

Various events of deterioration of processed cheese can occur as a result of the growth of species of the genus *Bacillus* such as *B. licheniformis, B. subtilis, B. mycoides, B. megaterium, B. stearothermophilus, B. circulans, B. amyloliquefaciens,* and *B. pumilus*. Examples include the coagulation of casein by chymosin-like proteases produced by some species, the fermentation of lactose by *B. circulans* and the acid produced by *B. stearothermophilus*, due to its ability to produce acid but not gas. However, the safety of processed cheese is affected by the growth of *B. cereus*, which can cause two types of food poisoning syndromes: emetic and diarrheal. Other species of *Bacillus* such as *B. licheniformis, B. amyloliquefaciens,* and *B. pumilus* have also been reported to produce toxic components that may play a role in food poisoning (Panel and Hazards, 2016; Cruz et al., 2019).

A recent concern with the *Bacillus* genus is the increasing tolerance or resistance that its spores exhibit in relation to conditions supposedly controlled by growth or inactivation treatments such as ultrahigh temperature (UHT). Consequently, both pasteurized and commercially sterilized processed cheeses are, to some extent, susceptible to risks of deterioration or food security. Some *Bacillus* spores have an adhesive characteristic that facilitates their attachment to the surface of the processing equipment with the formation of biofilms. Such a scenario may result in the selection of *Bacillus* adapted with attributes of resistance to the hygienic procedures of the processed cheese industry, which will further result in the deterioration of the product (Cruz et al., 2019).

B. cereus is the most representative pathogen of this genus, as it can promote human diseases such as diarrheal and emetic intoxications. The emetic syndrome results from toxins produced in the ingested food, and the diarrheal syndrome results from the ingestion of food contaminated by *B. cereus* spores, whose cells pass through the digestive tract and produce the diarrheal toxin in the small intestine. The *Bacillus cereus* group consists of eight formally recognized species: *Bacillus cereus* sensu stricto, *B. anthracis, B. thuringiensis, B. weihenstephanensis, B. mycoides, B. pseudomycoides, B. citotoxicus,* and *B. toyonensis*. Most cases of food-borne outbreaks caused by the *B. cereus* group have been associated with concentrations above 5 log CFU/g, however, cases of emetic and diarrheal diseases involving lower levels of cell concentration have been reported (Arslan et al., 2014; Abbas et al., 2014).

Bacillus licheniformis proves to be one of the most present in raw milk and in the processing of dairy products. Despite not being characterized as a pathogen, its sporulation causes technological defects in the products and its high prevalence is attributed to its wide environmental distribution. On the other hand, *B. thuringiensis* is part of the *B. cereus* sensu lato group, being biochemically identical to *B. cereus* sensu stricto, and producing similar enterotoxins. As the standard method for counting has no sensitivity to differentiate species, there is an indication that contamination by *B. thuringiensis* may be underestimated, as well as other potentially pathogenic *Bacillus*, such as *B. pumilus, B. subtilis,* and *B. weihenstephanensis*. Also, species like *B. subtilis, B. megaterium, B. licheniformis,* and *B. amyloliquefaciens* can produce biogenic amines from the decarboxylation of free postproteolysis amino acids, causing toxic reactions to human health (Gopal et al., 2015).

14.4.2 Clostridium *spp. in processed cheese*

The genus *Clostridium* is classified as Gram-positive, rod, and strict anaerobic. However, *C. botulinum* tolerates traces of oxygen due to the enzyme superoxide dismutase, an important antioxidant defense in almost all cells exposed to oxygen. Cases of food outbreaks have already been reported due to the occurrence of *C. botulinum* in processed cheeses, with isolation of the botulinum toxin and causing consumer deaths. The spores of *C. botulinum* may be in the ingredients and survive the entire processing chain, although pH 4.7 and a temperature of at least 5°C are necessary for their growth (Briozzo et al., 1983; Chaturvedi and Shukla, 2015; Gómez-Torres et al., 2015).

With the development of molecular biology, new bacterial genera were originated from the genus *Clostridium*, such as *Moorella* and *Thermoanaerobacterium* (Buňková and Buňka, 2015). *Clostridium* species can also lead to quality and safety problems in processed cheese. In fact, extensive proteolysis can occur during the aging of ripening cheeses, which if used in the manufacture of processed cheese lead to an increase in pH due to the release of amino acids, which may favor the growth of *Clostridium* and result in the production of gas and acid butyric. As spores of this genus are not inactivated by mild heat treatments, they can germinate and produce gases, butyric and acetic acids, carbon dioxide and hydrogen being the common deterioration products. The stability of processed cheese in the presence of these bacteria is only achieved when the number of spores is low, the pH is not higher than 5.8 and the salt concentration is at least 6%. These levels of salt are not achieved in many formulations of processed cheese, especially in that with sodium reduction (Olguín-Araneda et al., 2015; Ter Steeg et al., 1995).

Clostridium species are the main cause of late cheese blowing and, consequently, represent a great potential impact on the stability and safety of processed cheese. Although *C. tyrobutyricum* is the main relevant microorganism, other species such as *C. sporogenes, C. beijerinckii,* and *C. butyricum* have also been shown to contribute to this technological defect. The occurrence of such defects depends on different factors, including the number and species of *Clostridium* that originally contaminated the milk, the manufacturing process and the characteristics of the final product (Tanaka et al., 1979; Cruz et al., 2019).

Butyric acid fermentation in cheese is caused by the growth of *Clostridium* spp. capable of fermenting lactic acid, producing butyric acid, acetic acid, carbon dioxide, and hydrogen. Its growth is critically influenced by the concentration of lactic acid, concentration of salt, pH, moisture, and fat content. In addition, storage time and temperature and the presence of other microorganisms influence the growth of species of the genus *Clostridium* (Brändle et al., 2016).

The presence of *Clostridium* in processed cheeses is reported to be around 25%, with the most frequent species being *C. perfringens, C. cochlearius, C. butyricum, C. tyrobutyricum,* and *C. sporogenes. C. perfringens* is apparently more isolated in processed cheeses than in raw milk and other types of cheese. These differences may be due to the type of raw materials and ingredients, their quality and the quantities used in processed cheese formulations (Cruz et al., 2019.

In processed cheeses inoculated with *Clostridium* species and stored for 2 weeks, *C. tyrobutyricum* shows no signs of sporulation, while *C. cochlearius* and *C. sporogenes* exhibit visible sporulation, indicated by cavitation due to gas production, syneresis with the separation of casein and color change in cheese (Buňková and Buňka, 2015).

C. botulinum spores can maintain viability in processed cheese stored for up to 5 years at 2–4°C. During the first year of storage, a toxin production of up to five times the initial occurs, remaining stable until the third year. Thereafter, the concentration of toxin decreases and reaches the end of the storage period with half the toxin of the first year. The production of botulinum toxin can be inhibited by the presence of lactic acid. *C. butyricum* spores can develop and produce toxins from 12°C and pH 4.8. In 5 days of storage at 25°C the presence of toxin is observed. It is known that about one spore of *C. tyrobutyricum* per milliliter is sufficient to cause a late blowing in hard cheeses. However, the minimum spore concentration needed to cause this type of deterioration in processed cheese remains unknown. Avoiding contamination of raw milk and removing bacterial spores from contaminated milk before processing processed cheese (using bacterofugation) are potential methods to prevent spoilage and should be applied in the industry (Buňková and Buňka, 2015; Olguín-Araneda et al., 2015).

Pathogenic species are present in the environment (soil, sewage, and feces), as well as in the intestines of healthy humans. *C. perfringens* can cause histotoxic infections, including gastric gangrene, anaerobic cellulitis, infection of simple wounds, and various human or animal diseases originating in the intestines. In addition, *C. sporogenes* is an important cause of food spoilage from vacuum-packed meat and dairy products, similar to some types of processed cheese, but is considered a nontoxic substitute for *C. botulinum* with regard to safety (Gómez-Torres et al., 2015).

Table 14.2 Some species of molds and yeasts already isolated from processed cheeses.

Yeasts	*Debaryomyces hansenii*
	Geotrichum candidum
	Candida parapsilosis
	Kluyveromyces marxianus
Molds	*Penicillium chrysogenum*
	Penicillium frequentans
	Penicillium notatum
	Aspergillus flavus
	Aspergillus niger
	Penicillium viridicatum
	Byssochlamys fulva

Adapted from Buňková and Buňka, 2015.

Some pathogenic species of *Clostridium* can also affect dairy products, such as *C. botulinum* and *C. butyricum*, both associated with botulism. *C. perfringens* can cause enterocolitis and *C. botulinum* and *C. perfringens* can result in sudden infant death syndrome. In addition, some strains of *C. butyricum* may be related to neonatal necrotizing enterocolitis, and several species of *Clostridium* are considered to be potential producers of harmful biogenic amines in cheese and its derivatives (Gómez-Torres et al., 2015).

14.5 Molds and yeasts in processed cheeses

Although little reported, the presence of molds and yeasts in processed cheeses can lead to lipolytic and proteolytic action, leading to the formation of substances that cause organoleptic and color changes, in addition to the fact that molds are precursors of mycotoxins. The deterioration by these microorganisms in processed cheeses is attributed in most cases to the genus *Penicillium*, mainly *P. commune* and *P. roqueforti*. Table 14.2 shows some species of molds and yeasts already isolated from processed cheeses (Buňková and Buňka, 2015).

The isolation of *Byssochlamys fulva*, which is resistant to higher temperatures, is noteworthy. Also, the presence of molds associated with the production of mycotoxins B_1 and B_2, such as *Aspergillus flavus*, has also been reported. *P. viridicatum* is associated with the production of citrinin, another mycotoxin. The genera *Aspergillus, Penicillium, Cladosporium,* and *Geotrichum* are associated with the production of mycotoxins, which are resistant to industrial processing and reach the final product. There are reports of filamentous fungi isolated from processed cheeses capable of producing mycotoxins B_1 and B_2, citrinine, cyclopiazonic acid, mycophenolic acid, and Ergot alkaloids (Buňková and Buňka, 2015).

Molds of the genus *Penicillium* are shown to be viable for more than 270 days in processed cheese stored at 5°C, while *Cladosporium* is inhibited after 180 days of storage. At 22°C, the number of cells continues to increase until the end of storage (Buňková and Buňka, 2015).

In some countries, the maximum limit the counts of yeasts and molds or their absence is established. For example, the Syrian processed cheese legislation states that yeasts and molds must be absent. In Turkey, legal microbiological criteria are set at maximum yeasts and molds count less than 100 per g. In Brazil, the legal standards allow maximum yeasts and molds count less than 5000 CFU/g for the category ground processed cheese. For the others types of processed cheese like blocks, slice or spreadable, is not established parameters for yeasts and molds count. Also, there are no specific microbiological requirements laid down for processed cheese and related products in European Commission Regulation (EC) No. 2073/2005 (Brasil, 1997; EU, 2005).

14.6 Microbiological control in processed cheeses

The reduction in the number of spores in processed cheese can be achieved through the implementation of control measures throughout the production chain. Various measures can be applied in agricultural and industrial environments, such as soil, garbage and feed management, good hygiene and sanitation practices, and careful handling. An important measure is to guarantee the quality of the silage, as it represents the main source of contamination of milk. Likewise, contamination of raw milk can be avoided through proper management of animals and waste, since these processes are essential to prevent the accumulation, survival and transmission of spores. In addition, ensuring the health of dairy animals, as well as hygienic milking, handling and transportation practices are ways to maintain the quality of milk on the farm (Oliveira et al., 2016; Kousta et al., 2010).

The reduction in the load of milk spores during cheese production is achieved successively by the use of high speed centrifugation, which is called bacterofugation. This procedure is widely used in the dairy industry for the production of semihard and hard cheeses and is an interesting alternative for the production of pasteurized processed cheese. Bacterofugation removes 15–33% of the bacteria from milk, a percentage that is increased to 95% when the double treatment is performed (Faccia et al., 2013; Cruz et al., 2019).

Microfiltration can also be applied to reduce microbial contamination, but this approach is restricted to skimmed milk products. The high effectiveness of microfiltration in reducing bacteria to levels of 99.99% of the initial count has been reported for whole milk and a 99.98% reduction in milk spores is because the bacteria are about the same size as fat globules in whole milk. Therefore, this technology could be applied to increase the shelf life of skim processed cheese products (Cruz et al., 2019).

An appropriate heat treatment design is a crucial aspect to extend the life of dairy products because gentle heating during milk processing may not be sufficient to eliminate heat resistant spores. The appropriate temperature/time binomial for the elimination of heat-resistant spores is an important research topic since spore concentrations close to 5 log CFU/g or mL^{-1} are sufficient to promote the deterioration of dairy products. Still in relation to processing, the important inhibitory action of emulsifying salts, already discussed in a previous topic, stands out (Cogan, 2011; Cruz et al., 2019).

The higher lipid content in processed cheeses can reduce the antimicrobial action of emulsifying salts and preservatives such as potassium sorbate and sorbic acid, due to the interaction of their lipophilic parts with the lipid molecules, thus making them unavailable. By decreasing the fat content, the growth of anaerobic bacteria, as well as their sporulation and toxin production, is hampered. Comparatively, a fat content of 1–5% does not produce botulinum toxin for 1 year in storage at 27°C, while at a 20% fat content, the toxin is produced after 4 weeks of storage at the same temperature. In the other hand, low lipid content in processed cheese can increase the shelf life of these products because the sensorial defects related to lipid oxidation are reduced (Buňková and Buňka, 2015).

Depending on the technology employed and the duration of storage, spore inhibition can be achieved through the use of chemical additives, biopreservation, modification of water activity, reduction of pH, or a combination of inhibitory factors. The temperature of storage and pH control of processed cheese are crucial to avoid the possible germination of spores present. In processed cheeses manufactured with concentrations of 1.5–2.0% (w/v) of melting salts (polyphosphate/citrate), germination of spores of *C. botulinum* does not occur if the pH is maintained close to 5.5, up to a temperature of 25°C. If the pH is maintained at 5.9, spores germinate in 3 months under storage at 18°C, while under storage at 25°C, germination occurs in 1 week (Cruz et al., 2019; Cogan, 2011; Oliveira et al., 2016).

The addition of nitrate (maximum 150 mg/kg) to milk is the most common protection measure against late blowing, presenting low cost and being safe and simple. In addition, during maturation, nitrate-reducing bacteria (coliforms, lactobacilli, and superficial microflora) and the xanthine oxidase enzyme, which occurs naturally in the membrane of fat globules in milk, reduce nitrate, which is considered a germination inhibitor of spores. However, as nitrate decomposition occurs very quickly and a portion of the nitrogen is washed off during brine and evaporates as a gas, germination of the spores is not completely inhibited, but is delayed until the saline concentration of the cheese becomes homogeneous after the brine. The presence of nitrate in cheese occurs naturally in the range of 1–8 mg/kg, although the amount decreases during ripening and with storage (Amer, 2013).

Hexamethylenetetramine (HMT) is a precursor to formaldehyde and decomposes under acidic conditions into ammonia and formaldehyde. Formaldehyde acts as an inhibitor of spore development in cheese, reducing the availability of certain macromolecules when combined with them. However, due to its high reactivity with nucleic acids, the addition of HMT to milk cannot exceed approximately 15–25 ppm. However, this substance cannot be used because of legal restrictions in some countries (Cruz et al., 2019).

Propionic acid and calcium propionate are able to inhibit *B. subtilis* at pH 6.0, while at pH 4.5, it can inhibit molds and yeasts. Sorbic acid can be used in processed cheeses with a high fat content, due to its low water solubility, and does not cause organoleptic changes. Potassium sorbate is able to inhibit *C. botulinum* and its toxin in concentrations of 0.13–0.26% (Buňková and Buňka, 2015).

Biopreservatives, such as lysozyme (at a level between 250 and 400 mg/kg) and bacteriocins, especially nisin (at a maximum level of 12.5 mg/kg), are also effective in preventing spores and represent alternatives to additives chemicals in the production of processed cheese. Lysozyme promotes cell wall lysis of the vegetative form of *C. tyrobutyricum* through enzymatic cleavage and, consequently, controls clostridial growth and butyric acid fermentation during cheese storage. However, there are some limitations regarding its application against a large number of spores or certain resistant strains, in addition to having low effectiveness against Gram-negative microorganisms (Buňková and Buňka, 2015; Ávila et al., 2014).

Nisin is a bacteriocin produced by *Lactococcus lactis* that is widely used to prevent spore germination. Nisin has the ability to inhibit the proliferation of *Clostridium* and *Bacillus* spores during the production of processed cheeses, canned vegetables, and other products. For processed cheese, which has high moisture content and a reduced salt concentration, nisin can be added directly during cheese making at levels as high as 250 ppm, depending on the water activity and pH of the product formulation. When applied in combination with other processing technologies, such as thermal processing, other bacteriocins or high pressure, nisin can promote a long shelf life and greater food safety (Ávila et al., 2014; Aouadhi et al., 2014).

Traditionally, decreases in water activity and pH have been applied to inhibit the growth of any microorganism in food. However, the addition of salts ($>2\%$) and various acidulants that reach inhibitory values (pH < 5.7) for most spore-forming bacteria is difficult because these levels can affect organoleptic attributes. Thus, an interesting alternative is a combination of techniques that lead to the establishment of multiple preventive measures that are more restrictive than those achieved with individual ones. In fact, low temperature storage, reduced pH and reduced water activity, heating and other approaches, including preservatives, often work in synergy to control spores in food (Kapoor and Matzger, 2008; Cruz et al., 2019).

An important ally to the microbiological control of foods that has been gaining prominence is predictive microbiology, which allows the behavior of microorganisms of interest in the food matrices to be predicted under certain conditions. This prediction supports the most appropriate decision making in each situation and a better understanding of the problem in the face of processing steps. A predictive analysis results in a mathematical model that offers important parameters related to the kinetics of microbial development in the product, such as the decimal reduction time, which corresponds to the time required for the reduction of 1 log CFU of the microorganism under the applied conditions (Oliveira et al., 2018).

For example, in a study with two formulations of processed cheese (normal and light), the growths of *Paenibacillus sanguinis, Clostridium sporogenes,* and *Bacillus thuringiensis* were modeled and were able to withstand a 140°C treatment for 1 min. From this, important information is obtained, such as the fact that the time of decimal reduction is longer in cheese with reduced fat formulation, that is, the formulation considerably affects the kinetics of microorganisms. Furthermore, *C. sporogenes* was able to resist for 10 min at 105°C, for 3 min at 110°C and for 1.3 min at 115°C, being the most resistant species among those tested. Thus, it is interesting to consider it for the determination of process parameters in the manufacture of processed cheese (Oliveira et al., 2018).

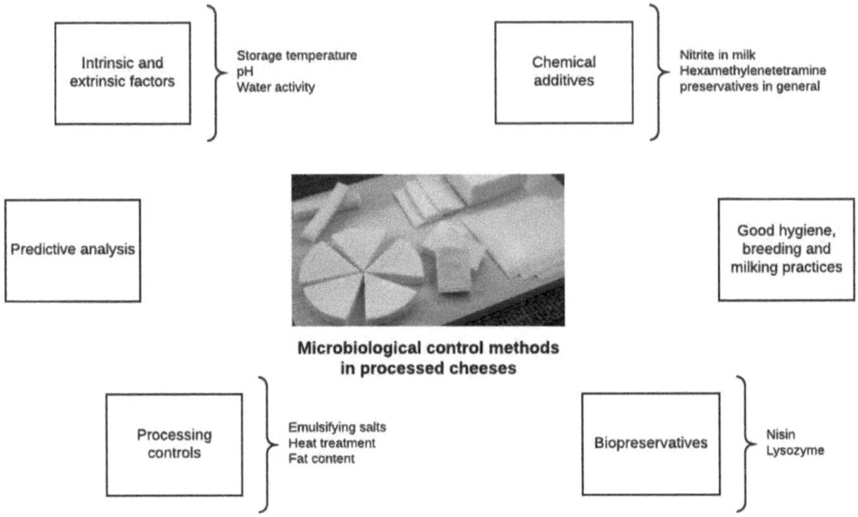

Figure 14.2 Microbiological control methods in processed cheeses.

Another example is the evaluation of technological factors in processed cheese, such as determining whether the substitution of sodium salt for potassium salt in the formulation would cause any change in the microbiological safety of the product. It appears that if the substitution is calculated on a molar basis, there is security from a microbiological point of view. Considering the emulsifying salts, it was mentioned above that the effects of antimicrobial action of potassium salts seem to correspond to 75% of the effect obtained when using sodium salts to control *C. botulinum* in processed cheeses (Glass et al., 2017; Buňková and Buňka, 2015).

Quantitative risk analysis is also shown to assist in the microbiological control of food, indicating the numerical probability of a foodborne disease occurring in a given population that consumes a certain product processed in a certain way. Based on the identification and characterization of the hazard (using dose-response models) and assessing exposure to individuals (consumption and storage habits), it is possible to estimate a low risk of disease acquired by consuming natural and processed cheese, per person per day, for *Clostridium perfringens*, both in the order of 10^{-14} (Lee et al., 2016).

Fig. 14.2 shows a summary of the microbiological control methods in processed cheeses.

14.7 Probiotics in processed cheese

Probiotics are defined as living microorganisms that confer benefits to human health, when consumed in an adequate amount, and their use, especially in dairy products, has been extensively studied. In the case of processed cheeses, the use of probiotics is an alternative to inhibit the growth of undesirable microorganisms, especially spore-forming bacteria (Sadek et al., 2017).

The genera *Lactobacillus* and *Bifidobacteria* are the most used as probiotics, and a range of 6 log CFU/100 g of product is usually required for the product to be considered as a probiotic. Cheeses in general have interesting properties for the application of probiotics, such as high fat and total solids and buffering properties. There are few reports of the application of probiotics in processed cheeses due to the processing steps applied in the manufacture of these products. Processing requires application of temperatures up to 90°C, which makes beneficial microorganisms unfeasible. In order to get around this, there is the possibility of adding probiotics after the cooling step of the processed cheeses; however, in this case there is a risk of recontamination, since the product will not undergo any further heat treatment. Other obstacles faced by probiotics in processed cheeses are long storage times and high salt concentrations, knowing that for the product to be considered a probiotic, it must present viability of microorganisms throughout its shelf life, which is extensive in the case of cheeses processed. A content of 4% of salt is sufficient for the total reduction of the cell viability of probiotic microorganisms usually employed (Talbot-Walsh et al., 2018).

Thus, the use of probiotics in processed cheeses is restricted to *Bacillus* strains recognized as safe and with probiotic properties, having high resistance to more severe processing. The most used strains are of the species *B. coagulans*, although there are also reports of the use of *B. subtilis* and *B. flexus* (Soares et al., 2019).

Some reported benefits related to *Bacillus* strains are improved macronutrient digestion, reduced muscle damage and better recovery after exercise, reduced abdominal pain and bloating, modulation of the elderly microbiota and obesity prevention, in addition to having antimicrobial properties against pathogens in the gastrointestinal tract (Soares et al., 2019).

The pasteurization step causes a decrease of only 1 CFU log of *B. coagulans* in processed cheese, while in subsequent processing steps, the count remains stable. Therefore, the melting step is shown to be the best time for adding the microorganism, as it will provide better homogenization and a subsequent thermal treatment will occur, avoiding recontamination (Soares et al., 2019).

The viability of probiotics can reach 60 days in cheeses processed with 6 log CFU counts at the end of storage; however, it is necessary to individually assess the sensory impacts, since antimicrobial compounds produced by *Bacillus* can generate unpleasant odors and flavors. The addition of *Bacillus* as a probiotic reduces the contaminating microbiota due to the production of inhibitory substances, thus increasing the shelf life of processed cheese. However, it is necessary to control the pH of the product, since there is a large formation of lactic acid from cell respiration and causing a decrease in pH. *Bacillus* is sensitive to pH less than 4.6, so the pH must be monitored during the product's shelf life so that the probiotic property is not lost. (Talbot-Walsh et al., 2018).

In processed cheese added with about 8 log CFU of *B. coagulans* as a probiotic, after 60 days of storage there is viability in the necessary quantity, but sensory attributes are slightly affected, with a decrease in pH due to the conversion of lactose into lactic acid and consequent increased acidity. To circumvent the various factors that may restrict the growth of the probiotic microorganism, such as low water activity, excessive acidity, temperature and high concentration of melting salts, a solution presented would be the microencapsulation of the vegetative cell or spores (Ehsannia and Sanjabi, 2016).

In processed cheese with reduced salt content inoculated with a probiotic strain of *B. coagulans*, a marked proteolytic activity is observed, as this microorganism is capable of releasing intracellular peptidases. Compared to a control cheese, the count of indicators such as coliforms and molds and yeasts are lower in probiotic cheese, indicating a biopreservative function. At the end of the 60 day storage, there are off flavors, however acceptable in the sensory analysis. The control proved to be unacceptable at 45 days, emphasizing the biopreservative property of *B. coagulans* (Ehsannia and Sanjabi, 2015).

It is possible to add conventional probiotic microorganisms such as *Lactobacillus* after the thermal step and their effects are satisfactory, including antimicrobial activity. However, as already mentioned, the risk of recontamination is very high. *Lactobacillus curvatus* and *L. acidophilus* have inhibitory properties against *Bacillus* and *Leuconostoc mesenteroides* is more effective against *Clostridium* (Sadek et al., 2017).

The probiotics can lose the viability at the processing and this loss enhances at the storage when added to processed cheeses. The postbiotics and paraprobiotics are more stable and safer for industrial products. Its applications in food products may represent many technological advantages for the manufacturers when compared to the viable microorganisms. Paraprobiotics, also called "inactivated probiotics" or "ghost probiotics," are defined as nonviable microbial cells (intact or ruptured) or raw cellular extracts (with complex chemical composition), which when administered (orally or topically) in adequate amounts, confer a benefit to the human or animal consumer (Taverniti, Guglielmetti, 2011;. Barros et al, 2020).

Part of the benefits associated with the consumption of probiotic products may occur possibly due to the metabolic by-products/postbiotics or from inactivated microorganisms/paraprobiotics. Considering the industrial handling and commercialization, the dairy products containing paraprobiotics and postbiotics are more simple and convenient when correlated to the probiotic products, because are few or no interaction with compounds of the components of the food. This aspect may increase shelf life. They remain stable in a wide range of pH and temperature, which allow the addition to matrix with high acidity and before thermal processing, reducing the chances of microbial contamination after processing. Another advantage is the enlarged shelf life without altering the sensory properties of the product (Shigwedha et al, 2014; Aguilar-Toala et al, 2018; de Almada et, 2016, Barros et al, 2020).

14.8 Conclusion

The processed cheeses are very microbiologically safe foods when compared to other cheeses with equal moisture content. However, they require strict control during the processing steps, especially in the cooling of those products filled in large packages. In addition, processed cheeses kept at room temperature must have their water activity and pH fixed within safe limits. In both of these cases, the germination of spores from deteriorating or pathogenic microorganisms can occur and cause many defects and risks. The high temperatures used in the processing of processed cheeses and the hot filling represent limitations to the addition of some species of probiotic to these products. The processed cheeses can be favored by the presence of paraprobiotics from probiotics present in the cheeses, cream or butter used in the manufacture of these products.

References

Abbas, A.A., Planchon, S., Jobin, M., Schmitt, P., 2014. Absence of oxygen affects the capacity to sporulate and the spore properties of *Bacillus cereus*. Food Microbiol. 42, 122–131.

Aguilar-Toala, JE, Garcia-Varela, R, Garcia, HS, Mata-Haro, V, Gonzalez, Cordova AF, Vallejo-Cordoba, B, Hernandez Mendoza, A, 2018. Postbiotics: an evolving term within the functional foods field. Trends Food Sci. Technol. 75, 105–114.

Alves, R.M.V., Van Dender, A.G.F., Jaime, S.B.M., Moreno, I., Pereira, B.C., 2007. Effect of light and packages on stability of spreadable processed cheese. Int. Dairy J. 17 (4), 365–373.

Amer, Z.M., 2013. Determination of nitrate and nitrite contents of Syrian white cheese. Pharmacol. Pharm. 4, 171–175.

Aouadhi, C., Rouissi, Z., Mejri, S., Maaroufi, A., 2014. Inactivation of *Bacillus sporothermodurans* spores by nisin and temperature studied by design of experiments in water and milk. Food Microbiol. 38, 270–275.

Arslan, S., Eyi, A., Küçüksari, R., 2014. Toxigenic genes, spoilage potential, and antimicrobial resistance of *Bacillus cereus* group strains from ice cream. Anaerobe 25, 42–46.

Ávila, M., Gómez-Torres, N., Hernández, M., Garde, S., 2014. Inhibitory activity of reuterin, nisin, lysozyme and nitrite against vegetative cells and spores of dairy-related Clostridium species. Int. J. Food Microbiol. 172, 70–75.

Barros, C.P., Guimarães, J.T., Esmerino, E.A., Duarte, M.C.K., Silva, M.C., Silva, R., Cruz, A.G., 2020. Paraprobiotics and postbiotics: concepts and potential applications in dairy products. Curr. Opin. Food Sci. 32, 1–8.

Berger, W., Klostermeyer, H., Merkenich, K, Uhlmann, G., 1997. Elaboración de Queso Fundido. Una Guía Joha, Ladenburg.

Brasil, 1997. Portaria n° 356, de 04 de setembro de 1997. Regulamento técnico para fixação de identidade e qualidade de queijo processado ou fundido, processado pasteurizado e processado ou fundido u.h.t (uat).

Brändle, J., Domig, K.J., Kneifel, W., 2016. Relevance and analysis of butyric acid producing clostridia in milk and cheese. Food Control 67, 96–113.

Briozzo, J., Lagarde, E.M., Chirife, J., Parada, J.L., 1983. *Clostridium botulinum* type A growth and toxin production in media and process cheese spread. Appl. Environ. Microbiol. 45, 1150.

Buňková, L., Buňka, F., 2015. Microflora of processed cheese and the factors affecting it. Crit. Rev. Food Sci. Nutr. 57 (11), 2392–2403.

Chaturvedi, A., Shukla, S., 2015. Occurrence of *Clostridium* species in different dairy products and its associated health risk. Int. J. Recent Sci. 6, 2827–2829.

Cogan, T.M., 2011. Microbiology of Cheese. Fundamentals of Cheese Science. Verlag, Springer, New York, p. 803.

Cruz, A.G., Zacarchenco, P.B., Oliveira, C.A.F., Corassin, C.H., 2019. Processamento de Produtos Lácteos: Queijos, Leites Fermentados, Bebidas Lácteas, Sorvete, Manteiga. Creme de Leite, Doce de Leite, Soro em Pó e Lácteos Funcionais. Ed. Elsevier: Rio de Janeiro, p. 360.

de Almada, C.N., Almada, C.N., Martinez, R.C.R., Sant'Ana, A.S., 2016. Paraprobiotics: evidences on their ability to modify biological responses, inactivation methods and perspectives on their application in foods. Trends Food Sci. Technol. 58, 96–114.

Doyle, C.J., Gleeson, D., Jordan, K., Beresford, T.P., Ross, R.P., Fitzgerald, G.F., Cotter, P.D., 2015. Anaerobic spore formers and their significance with respect to milk and dairy products. Int. J. Food Microbiol. 197, 77–87.

Ehsannia, S., Sanjabi, M.R., 2015. Quality characterization of processed cheese inoculated by *Bacillus coagulans* during cold storage: compositional and sensorial attributes and probiotic microorganism viability. J. Food Process. Pres. 40 (4), 667–674.

Ehsannia, S., Sanjabi, M.R., 2016. Physicochemical, microbiological and spoilage analysis of probiotic processed cheese analogues with reduced emulsifying salts during refrigerated storage. J. Food Sci. Technol. 53 (2), 996–1003.

EU, 2005. Commission Regulation (EC) No. 2073/2005 of 15 November 2005 on microbiological criteria for foodstuffs. Off. J. Eur. Comm. L338, 1–26.

Faccia, M., Mastromatteo, M., Conte, A., Del Nobile, M.A., 2013. Influence of the milk bactofugation and natural whey culture on the microbiological and physico-chemical characteristics of Mozzarella cheese. J. Food Process. Technol. 4 (4), 1–7.

Fernández-No, I.C., Guarddon, M., Böhme, K., Cepeda, A., Calo-Mata, P., Barros-Velázquez, J., 2011. Detection and quantification of spoilage and pathogenic *Bacillus cereus, Bacillus subtilis* and *Bacillus licheniformis* by real-time PCR. Food Microbiol. 28 (3), 605–610.

Glass, K., Doyle, M.E., 2005. Fri Food Safety Review: Safety of Processed Cheese. Fri Briefings, Madison http://fri.wisc.edu/docs/pdf/ProcCheese_may2005_v2.pdf.

Glass, K.A., Mu, M., Levine, B., Rossi, F., 2017. Inhibition of *Clostridium botulinum* in model reduced-sodium pasteurized prepared cheese products. J. Food Protect. 80 (9), 1478–1488.

Gómez-Torres, N., Garde, S., Peirotén, Á., Ávila, M., 2015. Impact of *Clostridium* spp. on cheese characteristics: microbiology, color, formation of volatile compounds and off-flavors. Food Control 56, 186–194.

Gopal, N., Hill, C., Ross, P.R., Beresford, T.P., Fanelom, M.A., Cotter, P.D., 2015. The Prevalence and control of *Bacillus* and related spore-forming bacteria in the dairy industry. Front. Microbiol. 6, 1418.

Hauerlandová, I., Lorencová, E., Buňka, F., Navrátil, .J., Janečková, K., Buňková, L., 2014. The influence of fat and monoacylglycerols on growth of spore-forming bacteria in processed cheese. Int. J. Food Microbiol. 182-183, 37–43.

Kapoor, R., Metzger, L., 2008. Process cheese: Scientific and technological aspects - a review. Compr. Rev. Food Sci. Food Saf. 7 (2), 194–214.

Kousta, M., Mataragas, M., Skandamis, P., Drosinos, E.H., 2010. Prevalence and sources of cheese contamination with pathogens at farm and processing levels. Food Control 21 (6), 805–815.

Lee, H., Lee, S., Kim, S., Lee, J., Ha, J., Yoon, Y., 2016. Quantitative microbial risk assessment for *Clostridium perfringens* in natural and processed cheeses. Asian-Australas. J. Anim. Sci. 29 (8), 1188–1196.

Loessner, M.J., Maier, S.K., Schiwek, P., Scherer, S., 1997. Long-chain polyphosphates inhibit growth of *Clostridium tyrobutyricum* in processed cheese spreads. J. Food Protect. 60 (5), 493–498.

Lorencová, E., Vltavská, P., Budinský, P., Koutný, M., 2012. Antibacterial effect of phosphates and polyphosphates with different chain length. J. Environ. Sci. Health. 47 (14), 2241–2245.

Lücking, G., Stoeckel, M., Atamer, Z., Hinrichs, J., Ehling-Schulz, M., 2013. Characterization of aerobic spore-forming bacteria associated with industrial dairy processing environments and product spoilage. Int. J. Food Microbiol. 166 (2), 270–279.

Nagyová, G., Buňka, F., Salek, R.N., Černíková, M., Mančík, P., Grůber, T., Kuchař, D., 2014. Use of sodium polyphosphates with different linear lengths in the production of spreadable processed cheese. J. Dairy Sci. 97 (1), 111–122.

Nováková, A., 2008. Antimicrobial Effects of Linear Polyphosphates for Chosen Microorganisms. Tomas Bata University in Zlín. Faculty of Technology (2012). J. Environ. Sci. Health, Part A, 47, 2241–2245. Available at http://dspace.k.utb.cz.

Olguín-Araneda, V., Banawas, S., Sarker, M.R., Paredes-Sabja, D., 2015. Recent advances in germination of *Clostridium* spores. Res.Microbiol 166 (4), 236–243.

Oliveira, R.B., Margalho, R.P., Nascimento, J.S., Costa, L.E.O., Portela, J.B., Cruz, A.G., Sant'Ana, A.S., 2016. Processed cheese contamination by spore-forming bacteria: a review of sources, routes, fate during processing and control. Trends Food Sci. Technol. 57, 11–19.

Oliveira, R.B., Baptista, R.C., Chincha, A.A.I.A., Conceição, D.A., Nascimento, J.S., Costa, L.E.O., Cruz, A.G., Sant'Ana, A.S., 2018. Thermal inactivation kinetics of Paenibacillus sanguinis 2301083PRC and *Clostridium sporogenes* JCM1416MGA in full and low fat "requeijão cremoso. Food Control 84, 395–402.

Panel, E., Hazards, B., 2016. Risks for public health related to the presence of *Bacillus cereus* and other *Bacillus* spp. including *Bacillus thuringiensis* in foodstuffs. EFSA Panel on Biol. Hazards. 14 (7), 4254.

Piska, I., Štětina, J., 2004. Influence of cheese ripening and rate of cooling of the processed cheese mixture on rheological properties of processed cheese. J. Food Eng. 61 (4), 551–555.

Sadek, Z.I., Refaat, B.M., El-Shakour, E.H., Mehanna, N., Hassan, M.S., 2017. Biocontrol of processed cheese by incorporation of probiotic bacteria and its metabolites. Int. J. Dairy Sci. 12 (2), 93–104.

Salek, R.N., Černíková, M., Nagyová, G., Kuchař, D., Bačová, H., Minarčíková, L., Buňka, F., 2015. The effect of composition of ternary mixtures containing phosphate and citrate emulsifying salts on selected textural properties of spreadable processed cheese. Intl. Dairy J. 44, 37–43.

Shigwedha, N, Sichel, L, Jia, L, Zhang, L, 2014. Probiotical cell fragments (PCFs) as "Novel Nutraceutical Ingredients". J Biosci. Med. 2, 43–55.

Soares, M.B., Almada, C.N., Almada, C.N., Martinez, R.C.R., Pereira, E.P.R., Balthazar, C.F., Cruz, A.G., Ranadheera, C.S, Sant'Ana, A.S., 2019. Behavior of different Bacillus strains with claimed probiotic properties throughout processed cheese ("requeijão cremoso") manufacturing and storage. Int. J. Food Microbiol. 307, 108288.

Talbot-Walsh, G., Kannar, D., Selomulya, C., 2018. A review on technological parameters and recent advances in the fortification of processed cheese. Trends Food Sci. Technol. 81, 193–202.

Tamime, A., 2011. Processed Cheese and Analogues. Wiley-Blackwell, England, p. 368.

Tanaka, N., Goepfert, J.M., Traisman, E., Hoffbeck, W.M., 1979. A challenge of pasteurized process cheese spread with *Clostridium botulinum* spores. J. Food. Prot. 42, 787.

Taverniti, V., Guglielmetti, S., 2011. The immunomodulatory properties of probiotic microorganisms beyond their viability (ghost probiotics: proposal of paraprobiotic concept). Genes Nutr. *6* (3), 261–274.

Ter Steeg, P.F., Cuppers, H.G.A.M., Hellemons, J.C., Rijke, G., 1995. Growth of proteolytic *Clostridium botulinum* in process cheese products: I. data acquisition for modeling the influence of pH, sodium chloride, emulsifying salts, fat dry basis, and temperature. J. Food Protect. 58 (10), 1091–1099.

Van Dender, A.G.F., Spadoti, L.M., Zacarchenco, P.B., Trento, F.K.H.S., Ormenese, R.C.S.C., Morgano, M.A., 2012. Efeito dos sais fundentes nas características do requeijão cremoso sem adição de gordura e com teor reduzido de sódio. Rev. Inst. Latic. Cândido Tostes. 387 (67), 38–47.

Zeinab, S., Refaat, B., Shakour, E.A., Sh, M.N., Ms, H., 2015. Potential sources of aerobic and anaerobic spore former bacteria in processed cheese. Res. J. Pharm., Biol. . Chem. Sci. 6 (1830), 1830–1837.

Regulations and legislations on processed cheese

15

Nurcan Koca[a], *Zafer Erbay*[b] *and Müge Urgu Öztürk*[a]
[a] Ege University, Engineering Faculty, Food Engineering Department, Bornova, Izmir, Turkey, [b] Adana Alparslan Turkes Science and Technology University, Engineering Faculty, Food Engineering Department, Saricam, Adana, Turkey

15.1 Introduction

Processed cheese has been produced since the mid-1890s to prolong the shelf-life of natural cheese and develop a new cheese type. The primarily manufacturing step of processed cheese is melting of blends of different types of cheeses with the aid of heat and with an addition of emulsifying salts. Other ingredients such as cream, butter, butter oil, additives, flavorings, colors, etc. can be added to the formulation in permitted levels specified in the countries' legislation. In addition, nondairy proteins and vegetable oils can be used in the formulations to produce imitation processed cheese if the use of these ingredients is governed by the legal regulations in the countries. The naming and the compositional and labeling requirements for processed cheese and related products in the legislations are also varied from country to country. They can be designated as processed cheese, processed cheese food, processed cheese spread, processed cheese product, etc. depending on the ingredients used. Because of this diversity, the fundamentals including the classifications, definitions and requirements specified in national and/or international regulations and legislation should be revealed for ensuring the safety and quality of processed cheese and regulating the import and/or export facilities in different countries.

Since ages, food legislations have been developed and issued in many countries for many purposes. The main aims of legislations for food commodities are regulating the market and ensuring fair trade practices. Moreover, preventing the consumer from being misled and protecting their health are other important aims in preparing legislation. The international food standards, guidelines and codes of practices by Codex Alimentarius are important in this sense and contribute to the safety, quality, and fairness of the international food trade. It is known as the global reference regulation for food producers and processors, consumers, and the international food trade. Although international standards shall be considered to achieve food safety and quality and have the function of regulating the international market, the import approval of any food is subject to the importing country's rules and regulations. Many countries and some regions have developed their own individual national and regional legislation and standards for food safety, retail sales and imports over the years. Therefore, the general terms, descriptions, and/or legislation related to processed cheese should be discussed from the Codex Alimentarius as a starting point to national regulations in the individual countries. This led us to the consideration of the legislations on processed cheese in

Processed Cheese Science and Technology: Ingredients, Manufacture, Functionality, Quality, and Regulations.
DOI: https://doi.org/10.1016/B978-0-12-821445-9.00004-2

three main topics: Codex Alimentarius as an international standard, legislations in particular regions and different countries. In addition, due to the use of cheese as a raw material for the manufacturing of processed cheese, deciding on the suitability of the cheeses used is challenging for the hygienic manufacture of processed cheese and for the reduction of cheese waste. Therefore, the guidance on cheese as a raw material in the manufacture of food products developed by the European Dairy Association (EDA) and the European Association of the Dairy Trade (EUCOLAIT) is discussed in detail.

The processed cheese legislations are usually included in food and cheese legislations, but there are specific standards for processed cheese in some countries. These standards in various countries are available electronically on their official website with the updated latest version of texts. However, some of the national legislations are unavailable or published in the national language of the countries concerned which could result in many difficulties in application and introduction of the products into the markets of the relevant countries. Considering these regulations and legislations in a general view, the naming and classification of processed cheese; definitions, compositional aspects of the processed cheese, that is, minimum cheese content, the percentages of dry matter and fat content; acceptability of the use of ingredients such as gelatin, starches, stabilizers and permitted levels of emulsifying salts, salts, additives, flavoring, etc. in the manufacture of processed cheese may differ due to the differences in needs and demands related to their native countries or regions and jurisdictions.

15.2 Codex Alimentarius

The Codex Alimentarius is a collection of internationally adopted food standards and related texts presented in a uniform manner. They aim to protect the health of consumers and to ensure fair trade practices in the food industry. The publication of the Codex Alimentarius is intended to guide and promote the elaboration and establishment of definitions and requirements for foods to assist in their harmonization and in doing so to facilitate international trade (FAO/WHO, 2016a). The World Trade Organization mentioned the Codex Alimentarius in their Agreement on Sanitary and Phytosanitary measures and as a result, Codex has become an important reference for resolving trade disputes. To apply stricter precaution in food trade and/or safety than those set by the Codex Alimentarius, the countries should scientifically justify their reasons.

Additionally, the Codex Alimentarius does not only have the function of regulating the international market. It also sets an example in the preparation of national standards. Considering that countries representing almost half of the world's population are currently rewriting their food safety laws and standards, the importance of the Codex Alimentarius would be revealed. Therefore, if the general terms, descriptions and/or legislation related to any food product are to be discussed or processes are evaluated in terms of international trade, the starting point should be the place in Codex Alimentarius. Moreover, having an impact on what is written in the Codex Alimentarius means influencing individual national regulations and getting an advantage in the trade with those countries. Because of this, the processes of drafting the Codex Alimentarius standards, guidelines, and codes of practice witness both various scientific and political discussions. In this regard, the discussions about processed cheese at the Codex Alimentarius Commission meetings give illustrative examples.

15.2.1 Dairy products in the Codex Alimentarius

The need for international regulations was dated back to the beginning of the 20th century. It should be mentioned that the International Dairy Federation (IDF) was one of the most important organizations which made attempts in preparing international regulations for dairy products. Therefore, the Codex Alimentarius Commission itself emphasized that IDF was an important catalyst in the conception of the Codex Alimentarius Commission (FAO/WHO, 2016a). After the World War II, the need for international regulations became vital and the Food and Agriculture Organizations of the United Nations (FAO) and the World Health Organization (WHO) took the lead in unraveling the skein of food regulations that were impeding trade and providing mostly inadequate protection for consumers. In 1953, the World Health Assembly stated that a new public health problem was presented by the increase in the use of chemicals in food production. Afterward, these two organizations started working together on the use of food additives and convened the first joint FAO/WHO Conference on Food Additives in 1955. Similarly, the Joint FAO/WHO Committee of Government Experts on the Code of Principles Concerning Milk and Milk Products (CGECPMMP) was established to discuss the international regulations for dairy products and held its first meeting in 1958. Finally, the establishment of the Codex Alimentarius Commission was carried out and in 1963, the first meeting was held in Rome. The working groups such as CGECPMMP continued their studies under the Codex Alimentarius Commission. In 1994, the name of the CGECPMMP changed as the Codex Committee on Milk and Milk Products (CCMMP) (FAO/WHO, 1997). CCMMP has prepared more than 40 standards about cheese, one of which is about cheese in general (Codex Alimentarius, CXS 283-1978) and the rest about original cheeses since 1963. However, (apart from General Standard for Cheese) only 20 cheese standards in force recently. In other words, more than half of the prepared cheese standards were revoked, and the standards related to process cheeses are also among those that were revoked. Nevertheless, more than half of the standards related to dairy products focused on cheeses, and in this sense, the cheese standards still maintain their importance in the Codex Alimentarius.

15.2.2 Processed cheese in the Codex Alimentarius

As a specific group of cheese or cheese products, drafting the international standard of processed cheese dated back to the beginning of the 70s. As a result, three standards about processed cheeses were prepared in 1978, came into force in 1979 and revoked in 2010. These standards are:

- Named variety of process(ed) cheese and spreadable process(ed) cheese (FAO/WHO, 2008a).
- Process(ed) cheese and spreadable process(ed) cheese (FAO/WHO, 2008b).
- Cheese preparations as process(ed) cheese food and process(ed) cheese spread (FAO/WHO, 2008c).

While there was a consensus about these three standards in 1978, it was noticed by many countries that these standards had become outdated because of the developments in the processing technologies and product varieties of processed cheese. Therefore, the revision of these three existing standards was initiated by the Codex Committee on Milk and Milk Products (CCMMP) before its 1st session (FAO/WHO, 1994).

After the first review, it was recommended to combine the existing three standards into one (FAO/WHO, 1994). However, the discussions in the second review showed that there were important disagreements about the minimum cheese content between the member countries (FAO/WHO, 1996). Afterward, the CCMMP decided to establish a Working Group on Cheese, and the group was tasked to focus on whether the use of minimum cheese content was adequate to characterize the products covered, and if so, identification of the relevant minimum contents, and if not adequate, provide guidance on an alternative approach (FAO/WHO, 1998; FAO/WHO, 2000). The CCMMP agreed that only one standard (excluding Processed Cheese Products and Preparations) for Processed Cheese would be enough to meet the needs. On the other hand, the replies to the questionnaire of the Working Group from the member counties showed that specifications of minimum cheese content in processed cheese and their preparations were not generally practiced (FAO/WHO, 2001, 2002a,b). Therefore, the CCMMP agreed not to specify minimum cheese content in the standard (FAO/WHO, 2004a,b,c). However, the key problem in the further development of the standard was still to get consensus on the general approach, especially on the description and the source of raw materials to be used in processed cheese production. While there was a suggestion to discontinue the work on this agenda item, other delegations pointed out that the standard was important and that the document was a good starting point for further elaboration (FAO/WHO, 2006a,b,c). Finally, the CCMMP decided to redraft the standard (FAO/WHO, 2006d). For redrafting, a physical Working Group, under the coleadership of France and New Zealand, was established (FAO/WHO, 2008d). After 2 years of this Working Group's efforts, the cochairs suggested not to continue the drafting standard process (FAO/WHO, 2010a,b). Although progress has been made toward an agreement in some issues such as regarding conditions for the use of starches and preponderance of opinion in favor of the exclusion of gelatin, the issues about the use of stabilizers could not be finalized because of the absence of a stable consensus on the issue of composition, including cheese content. In the report of the physical Working Group, it was mentioned that it would not be possible to reach an agreement on the issue of composition and cheese content and it would also not be possible to reach an agreement on other related issues, including labeling, stabilizers, and starches. The fundamental difficulty with attempting to develop the standard arises from the requirement for the standard to address the very large variety of products marketed as processed cheese while retaining scope for innovation. The report clearly emphasized that the unwavering choice of some delegations to insist on a standard reflecting very closely their own national product specifications, as distinct from a more global approach, suggests that it may not be possible to resolve this problem (FAO/WHO, 2010c). Finally, the Commission agreed to revoke the existing standards, postponed the work on drafting standards, and decided to ask for the opinions of

Regional Coordinating Committees about the discontinuation of the draft (FAO/WHO, 2010c,d).

The countries against the drafting process claimed that processed cheese and related products as currently produced and traded internationally represent a diverse range of products with varied compositional and functional characteristics and are not readily amenable to standardization. Moreover, they did not identify any significant trade associated with processed cheese products. Furthermore, existing Codex standards (both generic and general commodity) provide adequate guidance on essential health and safety aspects of these products, and drafting a more detailed one would limit the product variety and international trade competition. Finally, despite all efforts, it is obvious that it was not possible to reach a consensus in core topics. Oppositely, the justifications for the countries claiming that an international standard for processed cheese was required were: (i) processed cheese production and trade were considerable and were increasing, there was diversity of legislation, and that standard(s) were needed for protection of consumers in terms of health and preventing fraudulent practice and for fair trade; (ii) differing standards in importing countries, leading to increased production costs to produce various formulations, and increased marketing costs; (iii) lack of standards in importing countries, leading to difficulty obtaining a permit to market the products; (iv) there were few national standards and no international standard that countries could reference to ensure safety and quality and protect interests of consumers and producers, and for use by enforcement and antifraud officials; (v) there were differences in requirements, even among countries that subscribe to Codex standards, that prevented marketing; (vi) products differed in composition; some products did not meet minimum requirements; replacement, or partial replacement, of milk fat by vegetable fat; addition of small quantities of starch as an ingredient; (vii) lack of reference documentation for sanitary certificates and sales permits; (viii) lack of resources in countries to develop scientific and technically-based standards; (ix) guidelines would encourage innovation and new technologies of production, and foster trade and technology sharing; (x) the standards would establish the required functional properties, that is, thermal stability; (xi) the products were also a source of milk protein and a standard would be helpful to meet consumer expectations.

According to the discussions of six regional Coordinating Committees, only two of them (Coordinating Committee for Europe and Coordinating Committee for North America and the South West Pacific) were fully supported the discontinuation decision and others mentioned different views from the Commission (FAO/WHO, 2011). As a consequence, the Codex Alimentarius Commission stood behind the decision not to start a new draft work as no new or pragmatic basis for continuing work on the development of standards for processed cheese was provided whereas they also rejected the proposal to end discussions on a new standard (FAO/WHO, 2012). Then, the six FAO/WHO Coordinating Committees, which met between September 2012 and February 2013, discussed again the need for a standard for processed cheese (FAO/WHO, 2013). The Commission confirmed that several countries (in particular in Latin America and the Caribbean and Near East region) still consider feasible and necessary the development of a standard to respond to their needs to ensure that processed cheese products traded internationally to respond to certain compositional aspects and to be used as the basis of their national legislation. Therefore, the Commission concluded a decision to establish an electronic Working Group which will prepare a project document for new work on a standard for processed cheese that would clearly define: (i) the scope of the standard and the products to be covered; (ii) the compositional aspects of these products, i.e. minimum cheese content; and (iii) acceptability of the use of gelatin, starches and stabilizers in the manufacture of processed cheese (FAO/WHO, 2013). The Commission considered the analysis and conclusion of the Electronic Working Group (EWG) on the development of an international standard for processed cheese and the proposal to develop a more general standard, which would provide a framework for these types of products (FAO/WHO, 2014a). Therefore, a Physical Working Group led by New Zealand and co-chaired by Uruguay was established to prepare a proposed draft standard (FAO/WHO, 2014b). After the work of this group, the Commission adopted the proposed draft standard at Step 5 with a target of finalizing work in 2016 (FAO/WHO, 2015). Despite all efforts made to find consensus, some outstanding issues remained to be resolved and the Commission suggested discussing the discontinuation of the work (FAO/WHO, 2016b). Finally, the Commission approved the discontinuation of the work on the development of the general standard for processed cheese and noted the reservations of Algeria, Argentina, Belize, Bolivia, Brazil, Chile, Costa Rica, Cuba, Dominican Republic, Ecuador, Egypt, El Salvador, Honduras, Jamaica, Jordan, Lebanon, Morocco, Qatar, Saudi Arabia, Saint Vincent and Grenadines, Sudan, Suriname, Trinidad and Tobago, Tunisia, and Uruguay (FAO/WHO, 2017).

15.2.3 Discussions during drafting the standard

Although there was a consensus on many topics in the debates over 23 years, a few core topics remained unsolved. These topics can be summarized as (i) description of the processed cheese, (ii) essential characteristics of processed cheese, (iii) the manufacturing process as an integral part of the standard, (iv) minimum cheese content, and (v) use of gelatin, starch and stabilizers.

It was agreed to use the term "processed cheese" without looking at an alternative terminology. An essential characteristic of processed cheese was that it was produced through a process of grinding, melting by heat, and emulsifying the cheese mix and milkfat for standardization with or without added dairy ingredients or food flavorings/flavoring additives, usually within specified limits. A general agreement was obtained about the manufacturing processes as an integral part of the standard that it should be brief and only showing its essential steps, as the existence of detailed manufacturing processes in the standard could stifle product innovation as underlined by the wide range of processed cheese and processed cheese products traded internationally.

Some of the essential characteristics listed include cheese content, spreadability or sliceability, smooth blended texture and product stability. Additionally, some countries mentioned organoleptic characteristics. Briefly, processed cheese is generally regarded as a dairy-based food, in which there should be no substitution of dairy ingredients. There was however a range of views on matters related to product composition and how the cheese content should be controlled and declared (FAO/WHO, 2014a). Some countries emphasized that the term processed cheese should be limited to products containing cheese and milkfat (for standardization) and if other dairy ingredients were used in the production, the name of the product should be differentiated. Other countries that allowed the use of dairy ingredients suggested they should be justified according to technological need or should be limited by imposing maximum lactose content. Moreover, some countries (esp. USA) insisted that the minimum cheese content in a quantified form should be a criterion. An important problem for this criterion was the control method. Although the use of cheese was recognized as essential in the production of processed cheese products all over the world, there is no analysis method making it possible to check the quantity of cheese used in the end product. A realistic approach to limit the addition of dairy ingredients other than cheese and milkfat (for standardization) was to specify the maximum lactose content of 5% which was used in some countries' national legislation.

Another discussion is about the use of gelatin, starch and stabilizers. In processed cheese, the primary stabilization is made by casein and/or paracasein. Casein and/or paracasein become acting as active stabilizers by the addition of emulsifying salts. As a result of the addition of emulsifying salts, which cause calcium sequestration and increase pH of the cheese slurry, casein and/or paracasein mostly become soluble (60–80% of the total casein) and act as an active stabilizer by binding water. Moreover, the free fat released during the processing of cheese (heating and shearing) was emulsified by these proteins. However, if the products contain a greater degree of hydrolyzed protein (contains less intact casein), the emulsifying and stabilizing properties of the soluble protein phase decrease and their contribution to consistency, texture and viscosity will be less especially in spreadable processed cheese. In such cases, the addition of stabilizers (and/or starch and gelatin) can take an important role in texture without altering the organoleptic properties. High-temperature processed products, low-pH products, low-fat products (<40% FDM), and/or products with specific properties such as sliceability, melting and stringiness are some particular examples for the use of stabilizers in the production of processed cheese products. Some countries allow the use of stabilizers (and gelatin and starch as ingredients with stabilizing properties) in the production of processed cheese whereas others do not. In some countries where the use is permitted the resulting products are defined as processed cheese whereas in some others they are defined as processed cheese preparations. The use of food additives is justified only when such use has an advantage, does not present an appreciable health risk to consumers, does not mislead the consumer, and serves one or more of the technological functions set out by Codex. The discussion was focused specifically on the necessity of the use of these stabilizers. Some countries emphasized that stabilizers needed to be applied to the manufacture of stringy/sliceable and/or low fat processed cheese whereas others against their inclusion in the list of permitted ingredients because they claimed that these ingredients were not necessary for these products (FAO/WHO, 2010c). Although progress towards agreement regarding conditions for the use of starches and preponderance of opinion in favor of the exclusion of gelatin was obtained, a clear consensus did not appear as a result of the lack of an agreement on processed cheese composition (esp. in minimum cheese content).

15.3 Processed cheese legislations for a particular region

15.3.1 The legislations in the European Union

The European Union legislations (regulations, directives and decisions) regarding food products can be available using the EUR-Lex website (https://eur-lex.europa.eu). Although specific European legislation for processed cheese is not available in their list, many legislations consisting of general measures and specific detailed requirements can be useful for facilitating the production. First, processed cheese manufacturers shall ensure that the production, processing and distribution stages should comply with the relevant hygiene and safety requirements laid down in EU Regulation 178/2002 (EU, 2002), 852/2004 (EU, 2004a), 853/2004 (EU, 2004b), 2073/2005 (EU, 2005), 1881/2006 (EU, 2006a) and with the implementation of the good hygienic practices and the hazard analysis and critical control points (HACCP) system. The guidelines on cheese as a raw material in the manufacture of food products and for the hygienic manufacture of processed cheese developed in compliance with European Parliament and Council regulations, Codex recommendations and ISO 22000:2018 (ISO, 2018) and 22002-1:2009 (ISO, 2009) also provide practical information for producers to assess the cheese used as raw material and to design the food safety control systems for processed cheese. In addition, the European legislation on additives, labeling and packaging materials includes general principles and requirements for producers.

The general information and requirements in the guidance on cheese as raw material in the manufacture of food products and the European guide for the hygienic manufacture of processed cheese covering the related EU Regulations and the legislation on additives, labeling and packaging materials are given in the subtitles.

15.3.1.1 Guidance on cheese as raw material in the manufacture of food products

The guidance on cheese as a raw material in the manufacture of food products developed by the European Dairy Association (EDA) and the European Association of the Dairy Trade (EUCOLAIT) provides information whether the cheese is appropriate for further processing and treatment or must be disposed of and used in accordance with Regulation 1069/2009 (EDA/EUCOLAIT, 2018). The selection, handling, treatment and use of cheese as a raw material in the production of processed cheese, recommendations for control measures and procedures that are capable of controlling contamination and/or restoring control of material are given in the context. The guide describes the "*out-of-specification cheese*" that does not meet food safety and/or quality criteria and "*recovered cheese*" which represents an excellent raw material used in the manufacture of various foods, possibly after some form of treatment. The manufacturing and handling of cheese intended for direct consumption and the use of other types of raw materials and ingredients are not included in the guidance.

15.3.1.1.1 The suitability of raw materials used in the processed cheese production

The cheese is not accepted for processing if contaminated or decomposed or foreign substances to such an extent that, even after the food business operator had hygienically applied normal sorting and/or preparatory or processing procedures, the final product would be unfit for human consumption (EU, 2004a). According to the guidance developed by EDA/EUCOLAIT (2018), the suitability of the different types of cheeses for futher processing is given in Table 15.1. The cheese samples which are not complying with these aspects are not suitable for the processed cheese production. The measures for the preparation, handling, treatment and use of cheese for processing are detailed in the guide context.

15.3.1.2 European guide for the hygienic manufacture of processed cheese

The European Commission issued three main food hygiene regulations which are:

- *Regulation (EC) 852/2004* on the hygiene of foodstuffs,
- *Regulation (EC) 853/2004* laying down specific hygiene rules for food of animal origin,
- *Regulation (EC) 854/2004* laying down specific rules for the organization of official controls on products of animal origin intended for human consumption.

In addition to these regulations, the microbiological criteria for foodstuffs are laid down in *Regulation 2073/2005* (EU, 2005) to give guidance on the acceptability of the product and their manufacturing, handling and distribution processes. Also, the *Commission Regulation 1881/2006* (EU, 2006a) setting maximum levels for certain contaminants in foodstuffs provides information to keep contaminants at toxicologically acceptable levels.

Association of the Cheese Processing Industry (ASSIFONTE), the partner of the European Commission in discussions, has a special working group from different countries of Europe. It provides innovations and best practices and co-operates with the European Dairy Association (EDA) to represent the specific points of the processed cheese industry. The guideline for the hygienic manufacture of processed cheese developed by ASSIFONTE provides advice and guidance for proper implementation of the requirements in EU hygiene regulations used in the production of processed cheese and takes account of the principles contained in the Codex Alimentarius (ASSIFONTE, 2018). The guidelines should be applied with EDA/EUCOLAIT guidance on cheese as a raw material in the manufacture of food products. The information about the resource management, prerequisite programs, HACCP system, handling of nonconforming products and verification are included throughout the guide context.

15.3.1.2.1 Resource management

The guide emphasizes that all the facilities shall be designed according to relevant hygienic and safety standards (EU, 2004a). The plant shall be designed to have adequate spaces for the hygienic flow of materials, products and personnel. The design of the ceilings, walls, doors, windows, floors, layout of processes, outdoor areas are given in

Table 15.1 The assessment of the use of cheese as a raw material for further processing and specific measures*.

Raw material	Assessment of the usage	Measures prior to use
Cheese (both ripened and unripened) intended for direct consumption	Suitable for processing without any restriction.	Remove packaging material, waxes, unclean spots and coating material by peeling, brushing or scrubbing, or apply a heat-based removal process
Pre-packed cheese and cheese cuts returned from the market or retailers	Suitable for processing without any restriction if the cheese returned from wholesale enterprises and retailers approved in accordance with Regulation 853/2004.	Remove packaging material, waxes, unclean spots and coating material by peeling, brushing or scrubbing, or apply a heat-based removal process
	The returned cheese from other retailers (outlets) is not permitted.	-
Samples intended for testing and analysis	Unopened shelf-life reference samples and laboratory samples held in controlled chilled storage can be suitable for processing.	Remove packaging, unclean spots and coating material. Keep at temperature and specified on the label.
	Unopened accelerated reference shelf-life samples under elevated controlled storage conditions and leftovers of samples used for professional sensory testing are suitable when supported by a full food safety evaluation.	
Cheese not complying with quality specifications such as cheese with eyes, white layers or spots on the surfaces	Suitable for processing. If the microorganism causing the deviation is identified as potentially hazardous, cheese must be disposed of.	Remove packaging, unclean spots and coating material. Keep at temperature and specified on the label.
Physically contaminated cheeses	The cheese contaminated with unspecified foreign material and glass or hard plastic is not suitable for processing.	-
	The cheese contaminated with metal is suitable for processing after the fragments are efficiently removed.	Remove with efficient methods (magnetic fields, filters, etc.) accepted by competent authority. Metal detectors should be applied to the products.

(*continued on next page*)

Table 15.1 The assessment of the use of cheese as a raw material for further processing and specific measures*—cont'd

Raw material	Assessment of the usage	Measures prior to use
Chemically contaminated cheeses	The cheese contaminated with chemicals exceeding the established maximum levels or maximum residue levels according to Regulation 1881/2006 is not suitable for processing.	-
Contaminated with yeast	Suitable for processing without any restriction.	-
Cheese exceeding microbiological process hygiene criteria	Suitable for processing if not exceeding 10^5 cfu/g of *S. aureus* or exceeding the levels without any staphylococcal enterotoxins. If staphylococcal enterotoxins are detected, cheese must be disposed of.	Heat treatment (or equivalent treatments) can be carried out to achieve at least 8 log reductions of coagulase positive staphylococci.
	Cheese exceeding other food safety criteria for pathogens that do not produce toxins can be used for processing.	Heat treatment (or equivalent treatments) that delivers at least 8 log reductions of the pathogens.
Cheese contaminated with undesired mold colonies	Cheese contaminated with visible mold colonies that are most likely to be species typically used in the manufacture of mold-ripened cheese is suitable for processing.	Excessive mold on the surface shall be removed. Moldy material must not exceed 10% of ingoing raw materials used in the recipe. Heat treatment at least 75°C for at least 1 min.
	Visible mold where it cannot be substantiated that they are likely to be species typically used in the manufacture of mold-ripened cheese.	If no more than 10% of the surface of hard cheese (moisture on fat free basis < 56%) is covered with mold, cheese can be released and must be heat treated at least 75°C for at least 1 min. If more than 10% of the hard cheese surface or less than 10% of other cheese surface is covered, visible molds must be removed by removing 2-3 cm in depth. If other cheese with only smaller mold spots on the surface (<2–3 cm in diameter), the mold spots should be scraped off.

(*continued on next page*)

Table 15.1 The assessment of the use of cheese as a raw material for further processing and specific measures*—cont'd

Raw material	Assessment of the usage	Measures prior to use
	Cheese combined with flavouring foods (herbs, spices, etc.) is suitable if the specific assessment determines that any additional mold species can be controlled by the measures already in place to ensure that the possibility of mycotoxin formation is minimised.	-
Line recoveries	Cut offs of excessive cheese and cheese curd are suitable for processing. Cheese recovered from floors are not used for further processings.	-
Cheese exceeding age specifications	Cheese exceeding pre-established durability (best before) date is suitable for processing whereas cheese exceeding pre-established use-by date is not suitable for processing.	-
Cheese contaminated with cheese mites	Non infested parts of cheese is suitable for processing. Infested parts of cheese shall be disposed of.	The contaminated material should be cleaned. Heat treatment at least 75°C for at least 1 min should be applied.
Deteriorated cheese	Cheese with protein or fat decompositions is suitable for processing. Cheese contaminated with pests and with nauseous taste and smell is not used for processing.	-

* Modified from the guide issued by EDA/EUCOLAIT (2018).

detail to achieve prevention of the contamination and effective cleaning. The supply lines, drains and equipment shall be located in a way allowing adequate maintenance, designed to be easily cleaned and protect the product from external contamination. All the surfaces in contact or close to the product shall be made with corrosion-resistant, smooth, impermeable and nontoxic material. The suitability and design of processing equipment which shall be constructed according to EU regulations on materials in contact with food (EU, 2004c) are described. The general requirements for the equipment used in the manufacture of processed cheese such as containing vessels for mixing or processing, pumps, pipelines, valves, cleaning-in-place, heat treatment, homogenizers, filling and cooling equipment, waste containers, and drains and sewage system are given in details. All equipment used in the processing shall be designed, installed and equipped to prevent contamination and need to be cleaned efficiently.

The guide also involves some requirements for staff facilities and laboratories which need to be designed to reduce the risk of product contamination from people and plants. Storage areas shall also be arranged to allow the segregation of raw materials (cheese and other dairy ingredients), ingredients, packages, chemicals, waste and processed cheese and designed to prevent contamination and minimize deterioration with temperature and humidity controls.

The importance of the traceability systems for processed cheese production is emphasized in the guide. Processors must be able to trace and follow their food, feed, food-producing animal or substance intended to be, or expected to be, incorporated into a food or feed, through all stages of production, processing and distribution in accordance with EU Regulation 178/2002 (EU, 2002). The manufacturers shall be able to identify suppliers and businesses to which their products have been supplied. Also, relevant data shall be recorded as a minimum fulfills EU Regulation 931/2011 (EU, 2011b) to allow effective traceability of any materials throughout the whole purchasing, production, storage, and releasing stages.

15.3.1.2.2 The prerequisite programs

ASSIFONTE (2018) is referred to the basic conditions and activities of prerequisite programs which are applied to maintain hygienic conditions on which specific hazard controls can be based upon. The prerequisites listed in the guide are briefly given below.

- Maintenance operations shall be recorded and their correct conduct must be verified.
- Pest control program and correction actions shall be applied to prevent pests (insects, spiders, birds, etc.) according to procedures.
- Supply utilities shall be sufficient to meet the needs of the production process and suitable quality and purity.
- Foreign materials shall be prevented and controlled.
- Allergen management program shall include the assessment of the possible sources and nature of potential allergens listed in Annex II to EU Regulation 1169/2011 (EU, 2011a).
- Personnel hygiene and effective training procedures on food safety and hygiene shall be achieved according to the specifications detailed in the guide.
- All materials (raw materials and ingredients) shall be purchased from approved and registered suppliers. The cheese material for processing should be checked and stored according to EDA/EUCOLAIT. All packages, either product contact or outer protective wrapping and crates, and labeling requirements shall be complied with EU Regulation 1935/2004 (EU, 2004c) and 1169/2011 (EU, 2011a), respectively.

15.3.1.2.3 HACCP system

The guide emphasizes the HACCP system for manufacturers and gives information about developing a model HACCP plan for training purposes in the annexes

(ASSIFONTE, 2018). HACCP is a monitoring system that can include detailed analysis of potential hazards of the manufacturing process, storage, transport, and distribution. Each producer must have a functioning HACCP system according to the seven principles established by Codex Alimentarius. Data and information needed to conduct the hazard analysis shall be described in line with the information briefly explained in the following items.

- Each raw material, food ingredients, food additives, processing aids, packaging materials shall be described with their microbiological status and specifications.
- The main steps of the process (melting and emulsification, heat processing, filling process) shall be described and maintained with flow charts or diagrams of the sequence and interaction of the processing steps. The flow diagram shall indicate where raw materials, ingredients, intermediate products enter the flow, where reworking and recycling take place, where end products, intermediate products, by-products and waste are released.
- The adequate composition, organoleptic characteristics, packaging, intended shelf life with storage conditions, the expected distribution methods of the product shall be described and specified.
- Several biological, physical and chemical hazards associated with the manufacture of processed cheese are listed and detailed in the guide. All manufacturers and processors must be aware of these possible sources of hazards and be prepared to take measures to prevent hazards from occurring. The acceptable level of each hazard shall be determined by considering the regulatory requirements and not exceeded in the manufacture of processed cheese and end product. The limits of contaminants established for ingredients, raw materials and processed cheese are listed in Table 15.2. The other significant hazards which should be assessed and controlled by control measures are also documented in the guidelines.

The guide also assists the producers in determining whether a processing step in the flow diagram is a critical control point and provides control strategies. The information required to conduct the hazard assessment and a model HACCP plan for processed cheese is provided for training purposes throughout the context.

15.3.1.3 The European Union legislations on food additives

Many directives about the use of additives, colors, sweeteners, flavorings used in the foods have been established and amended by the EU over the years. Currently, Regulation 1333/2008 (EU, 2008a) in forced simplifies the repealed legislations on colors (EU, 1994a), sweeteners (EU, 1994b) and other food additives (EU, 1995) and harmonizes the use of food additives and food enzymes thus ensuring the safety and quality and facilitating the storage and use.

Food additives are defined as any substances that are not normally consumed as the food itself but are added to food for technological purposes such as the preservation of food, as stated by Regulation 1333/2008 (EU, 2008a). They should be safe when used and there should be a technological need for their use. On the other hand, substances used to impart flavor and/or taste or for nutritional purposes, such as salt replacers, vitamins and minerals should not be considered as food additives. The authorized additives and relevant conditions of their use laid down in EU Regulation 1129/2011 (EU, 2011d) of 11 November 2011 amending Annex II to Regulation 1333/2008 are

Table 15.2 Limits of contaminants established for ingredients, raw materials, processed cheese.*

	Contaminants	**Foods to which the criterion applies**	**Maximum levels**
Ingredients and raw materials used in the processed cheese manufacture	Aflatoxin B1, B2, G1 and G2 (sum)	Dry ingredients and spices	See EU Regulation 1881/2006 (EU, 2006a)
	Aflatoxin M1	Raw milk	
	Other mycotoxins	Flours etc.	
	Lead	Raw milk, meats, vegetables, berries, fruits, oils and milk fats	
	Cadmium	Meats, vegetables, fresh herbs, cultivated fungi	
	Dioxin	Vegetable oils and fats, meat, raw milk and dairy products	
	Benzo(a)pyrene	Oils and fats	
Processed cheese	Sum of dioxins (WHO PCDD/F-TEQ)	Processed cheese with >2% fat	2.5 pg/g fat (EU, 2011c)
	Sum of dioxins and dioxin-like PCBs (WHO PCDD/F-PCB-TEQ)	Processed cheese with >2% fat	5.5 pg/g fat (EU, 2011c)
	Tin (inorganic)	Canned processed cheese	200 mg/kg (EU, 2006a)

* Modified from the guide issued by ASSIFONTE (2018).

stated with the division of foods into categories and specific subcategories. The list of foods in which the presence of additives and colors may not be permitted by virtue of the carry-over principle set out in Article 18 (Part A), all additives including colors, sweeteners, additives other than colors and sweeteners (Part B), the definitions of groups of additives (Part C), the food categories (Part D), and the authorized food additives and conditions of use in food categories (Part E) are provided in the annexes.

Processed cheese is included in the category 01.7.5 (EU, 2011d). It is allowed to use carotenes (E160a) and paprika extract, capsanthin, capsorubin (E160c) at *quantum satis* level, and annatto, bixin and norbixin (E160b) at 15 mg kg^{-1} in processed cheese. The sorbic acid-sorbates (E 200-203) which may be added individually or in combination, nisin (E 234), phosphoric acid-phosphates-di, tri, and polyphosphates (E 338-452) are permitted to use at maximum levels of 2,000, 12.5 and 20,000 mg kg^{-1} in processed cheese, respectively. The other permitted additives for processed cheese are cassia gum (E 427) and silicates (E 551-559) which are allowed at maximum levels of 2500 and 10,000 mg kg^{-1}, respectively. The list contains some additives permitted for only flavored processed cheese. In addition, the permitted additives for cheese products is included in Regulation 1129/2011. All food additives used in the manufacture of processed cheese should be approved and used only when they fulfill the criteria laid down in this Regulation. Moreover, the specifications for food additives including colors, sweeteners and other than colors and sweeteners listed in Annexes II and III to Regulation 1333/2008 are laid down in Regulation 231/2012 (EU, 2012).

Flavorings, food ingredients with flavoring properties, source materials for flavorings, food containing flavorings should only be used to improve or modify the odor and/or taste of processed cheese when they fulfill the criteria laid down in the Regulation 1334/2008 (EU, 2008b), which repealed Directive 88/388 (EU, 1988) and 91/71 (EU, 1991). Annex III of this Regulation includes maximum levels of substances (such as 1-Allyl-4-methoxybenzene and 4-Allyl-1,2-dimethoxy-benzene) for dairy products. All substances that should comply with EU requirements can be used in the manufacture of processed cheese.

15.3.1.4 The European Union legislations on food labeling

European Commission issued the general labeling requirements for all food products laid down in the EU Regulation 1169/2011 (EU, 2011a). The regulation serves the interests of the internal market by simplifying the law, ensuring legal certainty and reducing administrative burden, and benefits citizens by requiring clear, comprehensible and legible labeling of foods and provides greater clarity especially for nutritional and potential allergen information to consumers. It includes the definitions, general principles, responsibilities and requirements of labeling. According to the Regulation, the name of the food, list of ingredients, any ingredient or processing aid causing allergies or intolerances used in the manufacture and still present in the end product, the quantity of certain ingredients or categories of ingredients, net quantity of the food, date of minimum durability or use-by date, special storage conditions and/or conditions of use, nutrition declaration, the origin of the product, the name and address of manufacturer or seller shall be included on the label of the processed cheese. In addition, an identification mark shall be indicated according to Regulation 853/2004 (EU, 2004b).

It should be noted that Regulation 1169/2011 (EU, 2011a) requires an indication of the quantity of an ingredient or category of ingredients used in the manufacture or preparation of all prepacked foods. Therefore, QUID, stands for Quantitative Ingredient Declaration, shall be required in the label and located in or next to the name of the food or in the list of ingredients to indicate the percentage of the certain ingredients. QUID is not compulsory for all ingredients. It is necessary when the ingredient appears in the name of the food, be usually associated with that name by the consumer, be emphasized on the labeling in words, pictures or graphics, and be essential to characterize a food and to distinguish it from products with a similar name or appearance. When considering the processed cheese, Hickey (2011) stated that cheese content would be subject to QUID labeling if the variations in the cheese content are governed by the choice of consumers in the countries. The other relevant requirements for labeling of processed cheese shall comply with the Regulation 1169/2011 (EU, 2011a), and the detailed information about the obligations to indicate QUID and exemptions is given in Commission Notice 2017/7605 (EU, 2017).

15.3.1.5 The European Union legislations on food packaging materials

Processed cheese is packed with packaging materials which should be suitable for contact with food. The material shall not be a hazard for health and not cause unacceptable changes in the composition and/or deterioration of organoleptic characteristics of the product. The EU Regulation 1935/2004 (EU, 2004c) provides general requirements and principles of all food contact materials, taking into account Regulation 178/2002 (EU, 2002) and 882/2004 (EU, 2004d), and any other materials and articles (containers, food machinery, etc.). In addition, good manufacturing practices for food contact materials are laid down in Commission Regulation 2023/2006 (EU, 2006b). All packaging material including product contact materials such as plastic containers, tubs, form-fill-seal materials or outer protective wrapping and crates shall comply with the requirements of EU Regulation 1935/2004 (EU, 2004c), and also the requirements of Commission Regulation 10/2011 (EU, 2011e) which establish the specific rules for plastic materials. The guide developed by ASSIFONTE (2018) also emphasized that the packaging materials shall be purchased from approved and registered suppliers, and the documentation and required test results such as specific migrations tests for constituents shall be carried out and available to demonstrate the compliances.

15.3.2 Processed cheese regulations in Gulf countries

Gulf Cooperation Council Standardization Organization (GSO) established by the resolution of the Supreme Council of the Gulf Cooperation Council (22nd Session, Muscat, Oman, 30-31 December 2001) is a Regional Standardization Organization. Its operation has been assumed with the membership of the governments of The State of the United Arab Emirates, The Kingdom of Bahrain, The Kingdom of Saudi Arabia, The Sultanate of Oman, The State of Qatar and The State of Kuwait since May 2004. The main purpose of GSO is to prepare and publish the Gulf standards or technical regulations through technical committees. Each member has responsibilities for adopting the developed and defined GSO standards as national standards, providing them to customers in their country, and withdrawing any existing standards if a conflict with the amended ones. GSO has a wide variety of standard lists to regulate the market and ensure fair trade practices for the food industry. At least 80% of the standards are adopted as identical international standards (GCO, 2020).

GSO has the following regulations related to processed cheese:

- GSO 160:2006, "Processed cheese and its products" and its amendment Amd 1:2016
- GSO 182:1994, "Processed cheese with vegetable oils" and its amendment Amd 1:2016

Although GSO has prepared and issued a final draft regulation named as "Processed cheese, spreadable processed cheese, processed cheese with vegetable oils and spreadable processed cheese with vegetable oils" (GSO 05 FDS 160:2018) which should repeal the above mentioned regulations, the regulation has not been officially approved and published on the official website. Saudi Food and Drug Authority (SFDA) approved the technical regulation "Processed cheese and spreadable processed cheese and processed cheese with vegetable oils and spreadable processed cheese with vegetable oils" (SFDA FD 58:2018) on 28 December 2018 and replaces and supersedes GSO160:2006 and GSO182:1994 in Saudi Arabia.

GSO 160:2006 defined three product varieties: (a) spreadable processed cheese, (b) processed cheese with vegetable oils (c) processed cheese with vegetable oils and natural flavor. According to the regulation, spreadable processed cheese is defined as a spreadable product with semi-solid texture obtained by chopping, mixing, melting and emulsifying one or more types of cheese with heat and emulsifying salts. The definition of processed cheese with vegetable oils is also given as the processed edible product made by grinding, mincing one or more varieties of cheese and adding vegetable oils (hydrogenated partially or wholly or nonhydrogenated), after the addition of suitable emulsifiers and other additives permitted to be used. Processed cheese with vegetable oils and natural flavors complies with this definition with the addition of natural flavors. According to the regulation, minimum heat treatment at 70°C for 30 s or any equivalent heat treatment is specified for spreadable processed cheese production. In addition to cream, butter and milk fat (in appropriate amounts), the use of starter culture and enzymes used in cheese production, salt, vinegar or acetic acid, spices, natural flavorings, water and processed food ingredients which shall not exceed 17% of the total dry weight is also permitted.

In the GSO 182:1994, the maximum levels of sodium chloride (4%) and emulsifying salts (3.5%) are specified. The other ingredients such as fruits, vegetables, etc. are restricted with a maximum level of 15% of the total weight of the final product. As regards the compositional specifications, the moisture and ash contents shall not exceed 60% and 8%, respectively. Also, the minimum fat to dry matter and milk fat shall not be less than 40% and 5%, respectively and milk fat shall not exceed 10% of total fat content.

GSO 05 FDS 160:2018 (draft) and SFDA FD 58:2018 regulations define processed cheese and spreadable processed cheese as products obtained by chopping, mixing, melting and emulsifying one or more types of cheese with heat and emulsifying salts with or without other milk products. The vegetable oil is also included in this definition for defining the processed cheese with vegetable oils and spreadable processed cheese with vegetable oils. All the products can be spreadable, sliceable, or in other forms. As per SFDA FD 58:2018 and GSO draft regulation (GSO 05 FDS 160:2018), general requirements for spreadable processed cheese and spreadable processed cheese with vegetable oils are almost similar to GSO 160:2006 and GSO 182:1994. However, some general requirements are added and stated in SFDA FD 58:2018. For example, the amount of salt is restricted with a maximum of 1.8 g/100 g (720 mg sodium/100 g) and other optional ingredients such as fruits, vegetables, etc. shall not exceed 16.5% of the total weight of the final product. Also, the moisture and fat requirements for spreadable processed cheese and spreadable processed cheese with vegetable oils are similar to those amounts specified in GSO 182:1994. Although the requirements in the GSO draft regulation (GSO 05 FDS 160:2018) are identical to SFDA FD 58:2018, the ash content is specified at a maximum level of 8% in spreadable processed cheese with vegetable oils in the draft version.

The additives permitted for use in processed cheese products are listed in GSO 2500:2015 under the categories of 01.6.4 processed cheese, 0.1.6.4.1 plain processed cheese, 01.6.4.2 flavored processed cheese and 0.1.6.5 cheese analogues. However, current GSO 2500:2015 for additives is being updated by GSO and it is in the final draft stage. The microbiological limits and shelf life of processed cheese are laid down in GSO1016:2015 and GSO150-2:2013, respectively. Packing, storage, transportation and labeling requirements are contained in the current and draft GSO regulations.

15.4 Processed cheese legislations, regulations and standards in different countries

Countries take into account international and regional regulations to achieve food safety and quality while many countries have developed their own national regulations. It is important to discuss these regulations because the import approval is subject to the rules and regulations in importing countries. In some countries, processed cheese is included in the general food standards which are separate whereas some countries have separate standards for processed cheese. The countries are given in reverse alphabetical order. In order to understand the specific standards for processed cheese, the regulations in the United States are given in more detail due to the fact that their standards are detailed.

15.4.1 The United States

In the United States, the U.S. Department of Agriculture (USDA) and the Food and Drug Administration (FDA) are involved in regulating the nation's food supply. The majority of the food supply (80% or more) including dairy is regulated by the FDA. In addition, The FDA Food Safety Modernization Act (FSMA) signed in 2011 strengthens the food safety system to protect public health, including a collection of preventive food safety laws designed to protect the U.S. food supply from production to consumption for imported foods. It is mandatory to comply with the rules of good manufacturing practice (GMP), sanitation performance standard (SPS), and sanitation standard operating procedures (SSOPs) for cheese making (Hui, 2015). Additionally, the HACCP is used in the cheese industry on a voluntary basis (Sommer, 2016).

USDA has established U.S. Grade Standards, U.S. Quality Specifications, and Commercial-Item-Descriptions (CID) for dairy products (USDA, 2020). U.S. Grade Standards for the most popular cheeses are voluntary. The cheese manufacturer must use the official requirements and processing specifications issued by USDA to get the USDA seal of approval for the products. U.S. Federal Standards of Identity are established by the FDA with a broader variety of cheeses including process cheese compared with the U.S. Grade Standards.

The Code of Federal Regulations (CFR) is a codification of the general and permanent rules published in the Federal Register by the departments and agencies of the Federal Government. CFR contains 50 titles representing broad areas that are divided into chapters. Each chapter is further subdivided into parts covering specific regulatory areas and large parts may be subdivided into subparts. These 50 titles are updated once each calendar year as follows: titles 1–16: January 1; titles 17–27: April 1; titles 28–41: July 1; and titles 42–50: October 1 (CFR, 2020a; e-CFR, 2020).

U.S. Federal Dairy Plant Specifications are provided in CFR Title 7 - Agriculture Part 58. Subpart A covers "Provisions Governing the Inspection and Grading Services of Manufactured or Processed Dairy Products" while subpart B covers "General Specifications for Dairy Plants Approved for USDA Inspection and Grading Service". General Specifications for Approved Plants and Standards for Grades of Cheese Manufacturing are given in 7 CFR 58.405-58.530. Supplemental Specifications for Plants Manufacturing, Processing and Packaging Pasteurized Process Cheese and Related Products are given in 7 CFR 58.705-58.738 under this subpart (CFR, 2020b). The requirements for equipment and utensils (7 CFR 58.706-710), the quality of cheeses used in process cheese making and of other ingredients such as cream, nonfat dry milk, whey, salt, flavor ingredients, acidifying and emulsifying agents (7 CFR 58.711-722) are specified in detail for process cheese and related products. In addition, the operations and operating procedures for process cheese are given from selecting cheese to quality control tests for the packaged cheese (7 CFR 58.723-733). Quality specifications for finished process cheese and related products are specified in terms of flavor, color, body, texture and appearance in 7 CFR 58:736-738.

The FDA standards of identities for cheeses are provided in Title 21 Food and Drugs, Chapter 1, Subchapter B Food for Human Consumption, Part 133, Cheese and Related Cheese Products. They describe the major varieties of cheese and identify the manufacturing procedures, the permitted ingredients and the moisture and milk fat contents. In CFR standards (21 CFR 133.169- 133-180), the process cheeses with emulsifying salt are classified into three categories including pasteurized process cheese (PPC), pasteurized process cheese food (PPCF) and pasteurized process cheese spread (PPCS) (CFR, 2019). The types, names, pH values, moisture and fat contents of process cheeses in CFR standards are given in Table 15.3. In addition to these categories, the pasteurized cheese blend is also defined in 21 CFR 133:167-133:168. Despite the similarity of description and requirements of pasteurized blend cheese to pasteurized process cheese, the emulsifying salt is not used in its process. Similarly, pasteurized cheese spread which is defined in 21 CFR 133:15-133:176 is subject to the requirements for PPCS except that the emulsifying salt is not used.

15.4.1.1 Pasteurized process cheese

Pasteurized process cheese is described as "the food prepared by comminuting and mixing, with the aid of heat, one or more cheeses of the same or two or more varieties, except cream cheese, Neufchatel cheese, cottage cheese, low-fat cottage cheese, cottage cheese dry curd, cook cheese, hard grating cheese, semisoft part-skim cheese, part-skim spiced cheese and skim milk cheese for manufacturing with an emulsifying agent" according to 21 CFR 133.169

The cheese must be heated not less than 65.6 C for 30 s which is verified by phosphatase test. The emulsifying agent can be used as one or any mixture of two or more of the monosodium phosphate, disodium phosphate, dipotassium phosphate, trisodium phosphate, sodium metaphosphate (sodium hexametaphosphate), sodium

Table 15.3 FDA standards for process cheese and related products (created using CFR, 2019).

	Name	Moisture content (%)	Fat content in dry matter (%)	pH
21 CFR 133.169 Pasteurized process cheese (PPC)	*From a single variety of cheese* "Pasteurized process ___ cheese " *From the varieties of cheeses* "Pasteurized process ___ and ___ cheese", or "Pasteurized process ___ blended with ___ cheese", or "Pasteurized process blend of ___ and ___ cheese, in order of predominance by weight "Pasteurized process gruyere cheese"[c] "Pasteurized process American cheese"[d] "American cheese"[e]	*From a single variety of cheese* PPC <43[a] PPC (washed curd or colby) <40 PPC (swiss or gruyere) <44 PPC (limburger) <51 *From the varieties of cheeses* PPC <43[a] PPC (cheddar, washed curd, colby, granular) <40 PPC (swiss or gruyere) <44	*From a single variety of cheese* PPC >47[b] PPC (swiss) >43 PPC (gruyere) >45 *From the varieties of cheeses* PPC >47[b] PPC (swiss and gruyere) >45	>5.3
21 CFR 133.170 PPC with fruits, vegetables, or meats	Naming as 21 CFR 133.169 "with ___", in order of predominance by weight.	1% more than PPC	1% less than PPC	
21 CFR 133.171 Pasteurized process pimento cheese	"Pasteurized process pimento cheese"	<41	>49	

(*continued on next page*)

Table 15.3 FDA standards for process cheese and related products (created using CFR, 2019)—cont'd

	Name	Moisture content (%)	Fat content in dry matter (%)	pH
21 CFR 173 Pasteurized process cheese food (PPCF)	"Pasteurized process cheese food", including a declaration of any flavoring and any spice	<44	>23 (milk fat content)	>5.0
21 CFR 133.174 PPCF with fruits, vegetables, or meats	"PPCF with ___", in order of predominance by weight "Pimento PPCF" or "Pasteurized process pimento cheese food"[f]	>22		
21 CFR 133.179 Pasteurized process cheese spread (PPCS)	"Pasteurized process cheese spread", including a declaration of any flavoring and any spice	<40–<60	>20 (milk fat content)	>4.0
21 CFR 133.180 PPCS with fruits, vegetables, or meats	"PPCS with ___", in order of predominance by weight			

[a] The moisture content of a pasteurized process cheese made from a single variety of cheese is not more than 1% greater than the maximum moisture content prescribed by the definition. The moisture content of a pasteurized process cheese made from two or more varieties of cheese is not more than 1% greater than the arithmetical average of the maximum moisture contents prescribed by the definitions and standards of identity.

[b] The fat content of the solids of a pasteurized process cheese made from a single variety of cheese is not less than the minimum prescribed by the definition and standard of identity. The fat content of the solids of a pasteurized process cheese made from two or more varieties of cheese is not less than the arithmetical average of the minimum fat contents prescribed by the definitions and standards of identity.

[c] In case it is made from gruyere cheese and swiss cheese, and the weight of gruyere cheese is not less than 25% of the weight of both.

[d] In case it is made of cheddar cheese, washed curd cheese, colby cheese, or granular cheese or any mixture of two or more of these.

[e] When cheddar cheese, washed curd cheese, colby cheese, granular cheese, or any mixture of two or more of these is combined with other varieties of cheese in the cheese ingredient.

[f] If the only vegetable ingredient is pimento, and no meat or fruit ingredient is used.

acid pyrophosphate, tetrasodium pyrophosphate, sodium aluminum phosphate, sodium citrate, potassium citrate, calcium citrate, sodium tartrate and sodium potassium tartrate. The weight of the solids of emulsifying agent is not more than 3% of the weight of the pasteurized process cheese. The acidifying agents such as vinegar, lactic acid, citric acid, acetic acid and phosphoric acid can be used. Cream, anhydrous milkfat and dehydrated cream are the other optional ingredients in such quantity that the weight of the fat-derived therefrom is less than 5% of the weight of the pasteurized process cheese. Safe and suitable enzyme modified cheese, harmless artificial coloring and spices or flavorings are also permitted to use. Specific mold-inhibiting ingredients including sorbates and propionates and an anticaking agent such as lecithin can be used in permitted amounts for PPC in the form of slices or cuts in consumer-sized packages. PPC or the cheese used for production may be smoked or may contain substances prepared by condensing or precipitating wood smoke.

PPC with fruits, vegetables, or meats is subject to the requirements for label statement of ingredients, prescribed for pasteurized process cheese by 21 CFR 133.169 according to 21 CFR133.170. Any properly cooked, canned and dried fruits and vegetables and cooked and canned meat can be used alone or in combinations. For the pasteurized process pimento cheese (21 CFR 133.171), the only fruit, vegetable, or meat ingredient is pimentos. The cheese ingredient is cheddar cheese, washed curd cheese, Colby cheese, granular cheese while the smoking and spices and flavorings are not used.

15.4.1.2 Pasteurized process cheese food

The regulations of PPCF and PPCF with fruits, vegetables and meats (21 CFR 133.173-133:174) are similar to those of PPC and PPC with fruits, vegetables and meats except permitted optional dairy ingredients, which are cream, milk, skim milk, buttermilk, cheese whey, their partly or totally dried forms, anhydrous milkfat, dehydrated cream, albumin from cheese whey, and skim milk cheese for manufacturing according to 21 CFR 133.171. The permitted amount of emulsifying salts and acidifying agents are the same as PPC and the additional exceptions for optional cheese ingredients are creamed cottage cheese and skim-milk cheese for manufacturing.

15.4.1.3 Pasteurized process cheese spread

The regulations of PPCS and PPCS with fruits, vegetables and meats (21 CFR 133.179) are similar to those of PPCFs, but some other ingredients including carob bean gum, gum karaya, gum tragacanth, guar gum, gelatin, sodium carboxymethylcellulose (cellulose gum), carrageenan, oat gum, algin (sodium alginate), propylene glycol alginate, or xanthan gum which are not more than 0.8% of the weight of the finished food are permitted to use for PPCS. In addition, the dioctyl sodium may be used in a maximum amount of 0.5% by weight of such ingredients. A sweetening agent consisting of sugar, dextrose, corn sugar, corn syrup, corn syrup solids, glucose syrup, glucose syrup solids, maltose, malt syrup and hydrolyzed lactose, in a quantity necessary for seasoning, can be used alone or in combination. In addition to mold-inhibiting ingredient and antisticking agent, Nisin is also permitted at a level of 250 mg kg^{-1}. The cheese must be spreadable at 21.1 C.

15.4.2 The United Kingdom

The first legislation on cheese in the UK was "The Cheese Regulations 1965" and the definitions about processed cheese and cheese spread were included in this regulation. This regulation was renewed in 1970 (HMSO, 1970) and the amendments were made to the renewed Cheese Regulations in 1974 and 1984 (HMSO, 1974,1984). In 1995, it was replaced by "The Cheese and Cream Regulations 1995" (HMSO, 1995) which was a short-lived one and repealed in 1996. After 1996, there was no specific regulation on cheese and UK cheese legislations were included in "The Food Labeling Regulations 1996" (HMSO, 1996). However, the definitions and regulations regarding processed cheese were not mentioned in this regulation and/or in any other legislation after this time.

According to "The Cheese Regulations 1970," processed cheese was defined as the dairy product which has been produced by a melting process of cheese and mixing with or without the addition of emulsifying salts, while milk products other than cheese can be mixed in the production of cheese spread (HMSO, 1970). This regulation classified the processed cheeses according to their fat-in-dry matter (FDM) and moisture contents and requested to specify the cheese variety if a unique type used in production. Moreover, the use of starch and gelatin is allowed in this regulation. While the definition and types of processed cheese were clear in UK legislations until 1996, there is no legislation regarding processed cheeses today and acted as a customary product. Processed cheese is accepted as a group of product in the "cheese and cheese products" group by the Food Standard Agency (FSA) and it is defined as "cheese product obtained from cheese and other ingredients, among which emulsifier salts, like phosphates or citrates" (FSA, 2017).

15.4.3 Turkey

Turkey is in the process of harmonizing its laws with those of the European Union. Turkish Food Codex Cheese Standard was published in 2015 by the Ministry of Agriculture and Forest, which is compulsory (TFC, 2015). In this regulation processed cheese is defined as a product which is produced using curd, one or more cheese types, if needed, the dairy ingredients such as milk powder, whey powder, butter, cream with the aid of emulsifying salt, and heat at pasteurization norm or higher. It has characteristic features of its varieties such as sliceable or spreadable. The cheese used in the manufacture of processed cheese should be suitable for the same regulation. In addition, the cheese can be named according to the shape and purpose of use. The moisture should be less than 60%. Another processed cheese standard, which is obligatory, is published by the Turkish Standard Institute (TSE, 2011) and amended in 2015 and 2016. In this standard, the processed cheese is classified in terms of its fat content and whether it is plain or with fruits, vegetables, etc. and the methods of analysis are referred.

15.4.4 Sweden

Almost all food rules are drawn up within the EU and apply to the whole of the EU. In Sweden, national rules on food also apply in the form of laws, ordinances and the National Food Administration's regulations. The processed cheese is described in the regulations on milk and cheese (Livsmedelsverkets föreskrifter om mjölk och ost, Livsmedelsverket, 2003:39) in Sweden, which is a product produced by melting and emulsifying the cheese. Milk ingredients, milk products and water may be added. Processed cheese with the designation of light or similar expressions shall have a maximum 15% of fat in dry matter.

15.4.5 Spain

All cheeses made in Spain must comply with the rules in Royal Decree 1113/2006 in annexes 1 for cheese and II for processed cheese, which its latest revision is in 2015 (Real Decreto, 2006). Processed cheese is described as "the product obtained by grinding, mixing, melting and emulsification of one or more varieties of cheese, with or without the addition of milk, dairy products and other foodstuffs. The dry matter content must be at least 35% except spreadable processed cheese having at least 30%. It can be named with a variety of cheese when one variety of cheese is higher than 50% of the raw material, and at least 75% of the mixture of cheeses used in the production of the product, and the remaining 25% must belong to a variety or similar varieties.

15.4.6 South Africa

In South Africa, the general information and requirements of processed cheese are laid down in the Agricultural Product Standards Act (Act No. 119 of 1990), amended in 2019 (Anonymous, 2019a). Two types of process cheeses are given: process cheese and cheese spread. Each product shall be manufactured by the grinding, mixing, melting and emulsifying with the addition of emulsifying salts and heat-treated for 30 s at a temperature of 70°C, or any other equivalent or greater time/temperature combination. It is indicated in the regulation that if the designation is used in connection with the sale of process cheese or cheese spread, the specified cheese must be included at least 50% of the formulation except for blue mold cheese which must be included at least 42% of the formulation. In addition, the specified cheese shall comply with the requirements on QUID, as specified in the regulations of the Foodstuffs, Cosmetics and Disinfectants Act (Act No. 54 of 1972). The product shall comply with the minimum dry matter content ranged between 30-52% for process cheese and 30–44% for cheese spread which is depended on the milk fat in dry matter content of the product.

15.4.7 Singapore

Singapore has a Food Regulation (SSO, 1988) which includes definitions, general provisions, additives, incidental constituents, labeling requirements for several food categories. Processed or emulsified cheese and cheese spread or cheese paste are given as two cheese types included in milk and milk product categories. Processed or emulsified cheese is defined as a cheese that has been comminuted, emulsified and pasteurized. It shall contain not more than 45% (w/w) moisture, not less than 45% (w/w) milk fat on a dry basis. The permitted level of emulsifying salt is 3%. Cheese spread or cheese pasta is a pasteurized spreadable cheese and the same requirements described for processed cheese are applied, except the moisture content shall not be more than 60% (w/w). The use of sorbic acid is permitted at a maximum level of 3000 mg kg^{-1} for processed cheese.

15.4.8 Russia

In Russia, the technical regulation for milk and dairy products (Anonymous, 2015) was adopted with respect to milk and dairy products for establishing safety requirements and processes of their production, storage, transportation, disposal, labeling and packaging. According to the regulation 126-ϕ3, processed cheese is defined as a dairy product or a dairy composite product made from cheese and (or) curd using dairy products and (or) by-products of milk processing, emulsifying salts or structure forming agents by grinding, mixing, melting and emulsifying the mixture for processing whether or not with the addition of nondairy components for the purpose of replacing of the milk constituents. Moisture, fat in dry matter and salt content shall comply with the requirements laid down in the regulation. Mass fraction of fat in solids (in percent) shall be given in the package of the processed cheese.

15.4.9 The Netherlands

The 1981 legislation, as amended, was withdrawn by the Warenwetbesluit Zuivel in 1994 (Lexius, 1994). This legislation included processed cheese, processed cheese product, cheese powder and processed cheese powder. After that, it was partly replaced by the Keuringsreglement COKZ Kaas that was adopted in 1998, and the latest text (to April 2008). Processed cheese and related products were not mentioned in this form. Therefore, there is no specific legislation in the Netherlands that defines or regulates processed cheese (Hickey, 2011).

15.4.10 Malaysia

Malaysia has a Food Regulations which contain processed cheese definition and requirements (Ministry of Health Malaysia, 1985). Processed cheese shall be the product obtained by grinding, mixing, melting and emulsifying one or more varieties of cheese and may include cultures of harmless bacteria, cream, butter and other milk products, salt, sugar, vinegar and any other food. The product shall contain not less than 51% of the dry matter derived from cheese. Cheese paste, cheese spread or cheese mixture is another product which is obtained by mixing cheese with condiments and other cheese food and shall contain not less than 75% of cheese and not contain more than 3% permitted emulsifier and 50% of water.

15.4.11 Latvia

In Latvia, the definition and classification of processed cheese are given in Regulation No. 97 which is the Regulation regarding requirements for the classification, quality and labeling of milk products and composite milk products (Anonymous, 2011). In this Regulation, processed cheese is defined as the product produced from a mixture of one or several cheese varieties and emulsifying salts with or without or without adding cream, butter or butter oil and other food products or components. When the certain cheese variety is used and the content thereof in the cheese mixture is not less than 75%, the product is named as processed "X" cheese where "X" is the name of the cheese variety. Moreover, the rest part of the cheese mixture is formed by cheese of similar variety. The product is also called as processed "X" and "Y" cheese or "X" and "Y" processed cheese where the names of cheese varieties in descending order or weight if the several cheese varieties are used in the production.

15.4.12 Kenya

Kenya has a standard for processed cheese coded as DKS 2509:2019 which specifies the requirements and sampling and analysis methods for processed cheese (Kenya Bureau of Standards (KEBS), 2019). Processed cheese is defined as a product obtained by the grinding, mixing, melting and emulsifying of one or more varieties of cheeses with the aid of heat and emulsifying salts. The product may be either sliceable or spreadable processed cheese depending on the type of emulsifying agent and the pH. The product shall contain suitable food additives and optional ingredients such as cream, butter, butter oil, salt (sodium chloride), vinegar, citric acid, lactic acid, phosphoric acid, nisin, sodium citrate, permitted emulsifiers/stabilizers, acidifying agents, spices and other seasonings, cultures of harmless bacteria and enzymes. The product shall comply with the fat (minimum 25% fat in dry matter), dry matter (30–75%) and salt (sodium, calcium or potassium; maximum 3% in dry matter) requirements. Also, the product shall comply with the requirements for microbiological criteria, additives, contaminants, packaging, and labeling established in the standard.

15.4.13 Japan

Ministerial Ordinance on Milk and Milk Products Concerning Compositional Standard regulates the cheese in Japan (Anonymous, 1951). In this Ministerial Ordinance, "cheese" means natural cheese and processed cheese. Processed cheese means the product which is made from natural cheese by grinding, heating to melt and emulsifying. The minimum milk solids content of processed cheese is 40.0%.

15.4.14 Italy

The Italian legislation (the Presidential Decree of November 1953) on cheese includes processed cheese which is described as "a product obtained by melting one cheese or a mixture of cheeses, with the possible addition of other milk products, including milk powder, casein and whey concentrates, with or without the addition of mineral salts, spices and aromas or when it is authorized by internal legislation, with the possible addition of vitamins; in addition, dissolving or emulsifying salts may be added in an amount not to exceed 3 g 100 g^{-1}" (Anonymous, 1954). The use of an "appellation d'origine" is authorized when one cheese with "appellation d'origine" is used. The use of a denomination is also authorized for processed cheese containing 75% of the cheese specified in the legislation. Additionally, the form and external characteristics of processed cheese cannot be the same as those of the specified cheese in the legislation. When the processed cheese with ham is stated on the product, the addition of lean ham is permitted.

15.4.15 Iran

The Iranian standard for processed cheese is contained in Iranian Standard (2003). The standard describes three types of processed cheese: processed cheese blocks, processed cheese spread and processed cheese food. The product contains different types of cheese, emulsifying salts with a maximum level of 4 g 100 g^{-1}. Other optional ingredients including butter, cream, vinegar, permitted spices, flavoring agents, sweeteners (sugar, glucose, natural sweeteners), whey powder, milk powder, microbial cultures, enzymes, acidity regulators (citric, phosphoric, acetic and lactic acids) can also be added to the formulation. The permitted level of salt is 1 g 100 g^{-1}, and anatto, β-carotene, and sorbates are 1000 mg kg^{-1}. The product shall comply with the minimum dry matter content which is specified according to the fat content and the minimum protein content of 12 g 100 g^{-1}.

15.4.16 India

India has Food Safety and Standards (Food Products Standards and Food Additives) Regulations (FSSAI, 2011) which contain the definition of food products, fat and moisture requirements, additives including colors, sweeteners, flavors. Processed cheese is defined as the product obtained by grinding, mixing, melting and emulsifying one or more cheese varieties with the aid of emulsifying salts and heat. Cream, butter, butter oil and other milk products can be added without exceeding 5% lactose content in the final product. Also, edible common salt, vinegar/acetic acid, spices and other vegetable seasoning and foods other than sugars properly cooked or prepared for flavoring and characterization of the product provided these additions can be added to the formulation without exceeding one-sixth of the weight of the total solids of the final product on dry matter basis. The moisture content of the product shall not exceed 47% and milk fat on a dry basis shall not be less than 40%. Processed cheese spread is another term given in their text. Unlike processed cheese, natural carbohydrate sweetening agents namely sucrose, dextrose, corn syrup, corn syrup solids, honey, maltose, malt syrup, and hydrolyzed lactose can also be added to the processed cheese spreads without exceeding one-sixth of the weight of total solids of the final product on a dry basis. The maximum moisture content and the minimum milk fat on the dry matter are 60% and 40%, respectively. According to their regulation, sorbic acid including its sodium, potassium, and calcium salt and nisin can be added to the processed cheese with maximum levels of 3,000 mg kg^{-1} and 12.5 mg kg^{-1}, respectively. Polyphosphate containing less than 6 phosphate moieties as sequestering and buffering agents can be included in the formulation at maximum levels of 40,000 mg kg^{-1}. Other detailed information about the additives permitted in the product and microbiological requirements of processed cheese are given in Appendix A and B of the regulation (FSSAI, 2011).

15.4.17 Hungary

The foodstuffs are regulated in Hungary according to the Food Codex, Decree No. 152 of 2009 (XI. 12) (Codex Alimentarius Hungaricus, 2009). These rules must be applied to foodstuffs produced and marketed within the territory of Hungary. Processed cheese is given under the Regulation 1-3/51-1 of the Hungarian Codex Alimentarius on dairy products (USDA, 2018). In this Codex, dairy products obtained from one or more types of cheese, with or without the addition of milk ingredients and/or other foodstuffs, by comminution, mixing, heat treatment (melting) and emulsification are defined in three categories such as named variety processed cheese, processed cheese, and processed cheese preparations. If a product is obtained exclusively from cheese (s) by the addition of milk fat (cream, butter, butterfat, butteroil) for the adjustment fat, it is called as named variety processed cheese. The amount of the named cheese (s) in the product must be at least 75% of cheeses. The remaining cheese should consist of a similar type of cheese. Bulk cheeses made from cut cheeses may have a dry matter content of not more than 4% (m/m) less than the named cheese used in their manufacture or as the arithmetic mean of the dry matter of the cheeses used. Processed cheese is a product obtained from cheese (s) by the addition of water and milk fat and other milk products for the adjustment of fat. The amount of the latter may not exceed 5% in the finished product. Processed cheese preparation is a product obtained from cheese(s) by the addition of milk fat for the adjustment of water and fat, as well as other dairy products. The amount of cheese (s) in the dry matter of the final product is required to be at least 51%. The milk sugar content is not limited and the use of sugar is permitted for this category.

15.4.18 Germany

The cheese legislation in Germany was delivered in 1965 and the last amendment was in 2017, which is Käseverordnung (cheese order) (Käseverordnung, 2017). In this legislation, the products made from cheese include processed cheese, processed cheese preparations, cheese preparations, and cheese compositions. In processed cheese, at least 50% of dry matter should come from cheese and it is produced using cheese and other milk products by the aid of heat and emulsifying salt. When the only lactic quark or quark with enzyme with the addition of the sliceable cheese without rind or skin is used for the production up to 8% in the product, the processed cheese can be described as cooking cheese. Processed Cheese Preparations are products produced by the aid of heat and emulsifying salt using other milk products, or added foods, cheese, or processed cheese, or mixtures of cheese and processed cheese. For the cheese preparations, the melting process by heat is not used. The cheese compositions are the products composed of two or more varieties of cheese, processed cheese, cheese preparations, or processed cheese preparations. The preparations of cheese and processed cheese can contain starch, edible gelatine, lactase, and also inulin if technologically needed. The rules on labeling and the needs in dry matter contents of processed cheeses are defined in the legislation.

15.4.19 France

The latest French legislation relating to cheese and cheese specialties, Decree No 2007-628 of April 27, 2007, was amended in 2013 and 2015 (Anonymous, 2007). Processed cheese (fromage fondu) and processed cheese specialty (spécialité fromagère fondue) are described in the amended version Decree No 2013:1010 of this legislation. Processed cheese is described as the product obtained by melting and emulsification, using heat (at a temperature of at least 70°C for 30 s or any other equivalent combination), cheese, or a mixture of cheeses, possibly supplemented with other dairy products. A minimum dry matter content of processed cheese is 40 g per 100 g of the finished product. When its fat is reduced, the dry matter content can be at least 30 g per 100 g of the finished product. Processed cheese specialty is the product which is produced by similar production techniques to processed cheese using cheese and other dairy products. A minimum dry matter content of which is 25%. Its fat-reduced version has at least a minimum dry matter content of 20%. Unlike processed cheese, the use of starch and gelatine is permitted up to 20 g kg^{-1} processed cheese specialty.

15.4.20 Egypt

The Egyptian Organization for Standards and Quality (EOS) is the only national institution which develops and issues many standards in Egypt. There have been two national standards related to the processed cheese in their official website (EOS, 2013): Processed cheese and spreadable processed cheese (999/2013) and processed cheese and spreadable processed cheese with vegetable oils and fats (1132/2013). The latter is mainly focused on processed cheese analogue. It is noted that the main reference of these standards is Codex standards (Codex Standard 286/1978 and 285/1978). Processed cheese is defined in the Standard 999/2013 as the product made by heating a blend of shredded natural cheeses (different varieties and degrees of maturity) and an emulsifying agent, with slow constant stirring, until a homogeneous emulsified mass is obtained. The dry matter content which can be changed by the fat content of the product must be at a minimum level to fulfill the standard specification.

15.4.21 Denmark

The legislation on cheese in Denmark is given in the Mælkeproduktbekendtgørelsen (Executive Order on Milk Products). The latest change to this regulation was in 2016 (Anonymous, 2016). In this order, cheese and cheese products are given in two different sections, including descriptions, manufacturing, and labeling. Processed cheese is categorized under cheese products and described as "a milk product made from cheese, with or without the addition of other milk products, milk ingredients, and water, which are mixed, melted and emulsified using emulsifying salts and possibly emulsifiers." Additionally, cheese powder is also described in this chapter as "a milk product made by drying of processed cheese, optionally with the addition of other milk products or ingredients." In processes cheese production, the largest ingredient must be cheese other than water. The processed cheese can be termed processed cream cheese when its fat content in dry matter is at least 60%, and spreadable cheese when it is spreadable, or can be designated with a characteristic name for a cheese variety when this variety is 75% of the total used.

15.4.22 The Czech Republic

The national regulation on dairy products including processed cheese is the Decree No: 397/2016 (milk, dairy products, frozen creams, edible fats and oils), amended in 2019, in the Czech Republic (Anonymous, 2019b). This decree provides the definition, minimum dry matters and fat contents, ingredients for processed cheese.

15.4.23 China

China has a national food safety standard for processed cheese (China National Standards, 2010). Process(ed) cheese is defined as the product made with cheese (the usage of cheese should be higher than 15%) as the main raw material and processed by heating, mixing and emulsifying with the aid of emulsifying salt, with or without the addition of other ingredients. The product shall comply with the requirements of dry matter and fat in dry matter. The minimum dry matter level of the final product is required to be in the range of 25 and 44% depending on the fat in dry matter contents. The cheese used as a raw material shall comply with requirements regulated in GB 5420. Contaminants, mycotoxins, microbiological limits, food additives, and nutrition enhancers shall conform to requirements established in GB 2762, GB 2761, GB 29921, GB 2760, GB 14880, respectively.

15.4.24 Canada

Federal food legislation implemented in Canada has been consolidated into a single standard under the name of Food and Drug Regulations in the Consolidated Regulations of Canada (CRC) (CRC, 2020). In part B of this regulation (entitled "Foods"), various types of foods were listed and defined with characteristics and in section 8 of this part (entitled "Dairy Products") includes dairy products. One of the seven main dairy products (milk, cheese, butter, ice cream, sherbet, ice milk, cream) mentioned in this section was Cheese which was evaluated with 26 subsections. Processed cheese was defined as a diverged product group in cheese. According to the legislation in Canada, processed cheese was classified in 3 main groups with two subgroups of each: Processed (naming the variety) cheese (B.08.040), Processed (naming the variety) cheese with (naming the added ingredients) (B.08.041), Processed cheese food (B.08.041.1), Processed cheese food with (naming the added ingredients) (B.08.041.2), Processed cheese spread (B.08.041.3) and Processed cheese spread with (naming the added ingredients) (B.08.041.4) (CRC, 2020). With the definitions of types of processed cheeses and the compositional properties, legislations in Canada and the USA are very similar. There are minor differences in the details of the composition of these products and in the additives that are allowed to be used.

15.4.25 Austria

In Austrian Food Codex (Codex Alimentarius Austriacus), processed cheese, spreadable processed cheese and processed cheese preparations are made from cheese, with or without ingredients, by heating and melting with the addition of emulsifying agents (Anonymous, 2019c). The dry matter must be at least 51% for processed cheese and preparations. Spreadable cheese and preparations contain at least 75% unripened soft cheese in the mixture and have a dry matter of at least 25%. As ingredients, aside from permitted additives, only cream, butter, butterfat, table salt and drinking water are used. It can be named with a variety of cheese when the mentioned variety is at least 75% of the mixture of cheeses, and the remaining must come from a similar type of cheeses.

15.4.26 Australia and New Zealand

Although Australia and New Zealand had had separate legislation, they were decided in 1991 to develop a unique standard. For this purpose, they established the Food Standards Australia New Zealand (FSANZ) and this organization produced a standard named the Food Standard Code. The FSANZ aims to review and develop the food-related standards in Australia and New Zealand in cooperation with the food industry. While the FSANZ in New Zealand is responsible for standards relevant to labeling, composition, and contaminants, it operates in a broader framework in Australia (such as standards connected to primary production and processing, coordination of recall systems, food safety, maximum residue limits and policies about imported food, etc.) (MPI, 2020).

15.5 Conclusion

A wide range of processed cheese and related products are produced and consumed all over the world. Therefore, current legislations on processed cheese including international, regional and national concern all countries. The regulations and legislations issued in the countries are also important for the import approval of the product. However, the classifications, definitions, ingredients added to formulation, compositional requirements and naming of the product stated in the legislations are varied among countries. After the Codex standards on processed cheese were rewoked in 2010, much effort has been made to establish an international standard. However, it was not possible to reach a consensus in core topics. In recent years, it is noteworthy that attention has been paid to the development of guidance on the suitability of cheeses as raw materials. This effort will contribute to both food safety and reduction of waste.

References

Anonymous, 1951. Ministerial Ordinance on Milk and Milk products Concerning Compositional Standards, etc. Ministry of Health and Welfare Ordinance No. 52, December 27, 1951. Available at https://www.mhlw.go.jp/english/topics/foodsafety/dl/t-1.pdf.

Anonymous, 1954. Decreto del Presidente della Ruppubblica 18 November 1953 n. 1099. Gazzetta Ufficiale della Repubblica Italiana 47, 618–629. Available at https://www.gazzettaufficiale.it/eli/id/1954/02/26/053U1099/sg.

Anonymous, 2007. Decree n ° 2007-628 of April 27, 2007 relating to cheeses and cheese specialties. Last modified September 1, 2015. Available at https://www.legifrance.gouv.fr/affichTexte.do?cidTexte=JORFTEXT000000644875&dateTexte=20200704 (Accessed on 4 July 2020).

Anonymous, 2011. Regulations regarding requirements for the classification, quality and labelling of milk products and composite milk products, Regulation No 97, Riga, Latvia, Available at https://likumi.lv/ta/en/en/id/225278.

Anonymous, 2015. Technical regulations for milk and dairy products, Federal Law No. 126. Available at http://www.consultant.ru/.

Anonymous, 2016. Executive order on dairy products. No. 1360 of 24/11/2016. Ministry of the Environment and Food Available at https://www.retsinformation.dk/eli/lta/2016/1360.

Anonymous, 2019a. Agricultural Product Standards Act, Act No. 119 of 1990. Department of Agricultural, Foresty and Fisheries. Government Gazette, 22 November 2019. Available at https://www.gov.za/sites/default/files/gcis_document/201911/42850rg11007gon1510.pdf.

Anonymous, 2019b. Decree No. 397/2016 Coll. on requirements for milk and dairy products, frozen creams and edible fats and oils. The Ministry of Agriculture. Available at https://www.zakonyprolidi.cz/cs/2019-274/zneni-20200101 (Accessed on 15 July 2020).

Anonymous, 2019c. Chapter B 32 of the Codex - Milk and milk products. Codex Alimentarus Austria. Last amendment on 5.7.2019. Available at http://www.lebensmittelbuch.at/milch-und-milchprodukte/kaese/schmelzkaese/.

Assifonte, 2018. European guide for the hygienic manufacture of prosessed cheese. Available at https://ec.europa.eu/food/sites/food/files/safety/docs/biosafety_fh_guidance_manuf-process-cheese_en.pdf (Accessed on 14 July 2020).

CFR, 2019. Code of Federal Regulations, U.S. Goverment Publishing Office. Tile 21. Food and Drugs. Revised as of April 1, 2019. Available at https://www.govinfo.gov/content/pkg/CFR-2019-title21-vol2/pdf/CFR-2019-title21-vol2.pdf. (Accessed on 15 May 2020).

CFR, 2020a. Code of Federal Regulations, U.S. Goverment Publishing Office. Available at https://www.govinfo.gov/app/collection/cfr.

CFR, 2020b. Code of Federal Regulations, U.S. Goverment Publishing Office. Tile 7, Agriculture. Revised as of January 1, 2020, Available at https://www.govinfo.gov/content/pkg/CFR-2020-title7-vol3/pdf/CFR-2020-title7-vol3.pdf. (Accessed on 15 May 2020).

China National Standards, 2010. National food safety standard process(ed)cheese, GB 25192 2010. Available at https://www.gbstandards.org/GB_standard_english.asp?code=GB%2025192-2010&word=National%20food%20safety%20standard.

Codex Alimentarius Hungaricus, 2009. Decree No. 152 of 2009 (XI. 12). Ministry of Agriculture and Rural Development of Hungary Available at https://net.jogtar.hu/jogszabaly?dbnum=1&docid=A0900152.FVM&mahu=1.

CRC, 2020. Food and Drug Regulations. Government of Canada, Department of Justice. Consolidated Regulations of Canada (CRC), c. 870. Current to June 17, 2020. Available at https://laws-lois.justice.gc.ca/eng/regulations/C.R.C.%2C_c_870/index.html#docCont. (Accessed on 12 July 2020).

e-CFR, 2020. Electronic Code of Food Regulations. Available at https://www.ecfr.gov.

EDA/EUCOLAIT, 2018. Guidance on cheese as raw material in the manufacture of food products. Available at https://ec.europa.eu/food/sites/food/files/safety/docs/biosafety_fh_guidance_cheese-raw-material_en.pdf (Accessed on 14 July 2020).

EU, 1988. Council Directive 88/388/EEC of 22 June 1988 on the approximation of the laws of the Member States relating to flavourings for use in foodstuffs and to source materials for their production. Off. J. Eur. Comm. L184, 61–66. Available at https://eur-lex.europa.eu/legal-content/EN/ALL/?uri=CELEX%3A31988L0388.

EU, 1991. Commission Directive 91/71/EEC of 16 January 1991 completing Council Directive 88/388/EEC on the approximation of the laws of the Member States relating to flavourings for use in foodstuffs and to source materials for their production. Off. J. Eur. Comm. L42, 25–26. Available at https://eur-lex.europa.eu/legal-content/en/ALL/?uri=CELEX%3A31991L0071.

EU, 1994a. European Parliament and Council Directive 94/35/EC of 30 June 1994 on sweeteners for use in foodstuffs. Off. J. Eur. Comm. L237, 3–12. Available at https://eur-lex.europa.eu/legal-content/EN/TXT/?uri=celex%3A31994L0035.

EU, 1994b. European Parliament and Council Directive No. 94/36/EC of 30 June 1994 on colours for use in foodstuffs. Off. J. Eur. Comm. L237, 13–29. Available at https://eur-lex.europa.eu/legal-content/EN/ALL/?uri=celex:31994L0036.

EU, 1995. European Parliament and Council Directive No. 95/2/EC of 20 February 1995 on food additives other than colours and sweeteners. Off. J. Eur. Comm. L61, 1–40. Available at https://eur-lex.europa.eu/legal-content/EN/ALL/?uri=CELEX%3A31995L0002.

EU, 2002. Regulation (EC) No. 178/2002 of the European Parliament and of the Council of 28 January 2002 laying down the general principles and requirements of food law, establishing the European Food Safety Authority and laying down procedures in matters of food safety. Off. J. Eur. Comm. L31, 1–24. Available at https://eur-lex.europa.eu/legal-content/EN/ALL/?uri=CELEX%3A32002R0178.

EU, 2004a. Regulation (EC) No. 852/2004 of the European Parliament and of the Council of 29 April 2004 on the hygiene of foodstuffs. Off. J. Eur. Comm. L139, 1–54. Available at https://eur-lex.europa.eu/legal-content/EN/TXT/?uri=CELEX%3A32004R0852.

EU, 2004b. Regulation (EC) No 853/2004 of the European Parliament and of the Council of 29 April 2004 laying down specific hygiene rules on the hygiene of foodstuffs. Off. J. Eur. Comm. L139, 55–205. Available at https://eur-lex.europa.eu/legal-content/EN/ALL/?uri=CELEX%3A32004R0853.

EU, 2004c. Regulation (EC) 1935/2004 of the European Parliament and of the Council of 27 October 2004 on materials and articles intended to come into contact with food and repealing Directives 80/590/EEC and 89/109/EEC. Off. J. Eur. Comm. L338, 4–17. Available at https://eur-lex.europa.eu/legal-content/EN/ALL/?uri=CELEX%3A32004R1935.

EU, 2004d. Regulation (EC) No 882/2004 of the European Parliament and of the Council of 29 April 2004 on official controls performed to ensure the verification of compliance with feed and food law, animal health and animal welfare rules. Off. J. Eur. Comm. L165, 1–141. Available at https://eur-lex.europa.eu/legal-content/EN/ALL/?uri=CELEX%3A32004R0882.

EU, 2005. Commission Regulation (EC) No. 2073/2005 of 15 November 2005 on microbiological criteria for foodstuffs. Off. J. Eur. Comm. L338, 1–26. Available at https://eur-lex.europa.eu/legal-content/EN/ALL/?uri=CELEX%3A32005R2073.

EU, 2006a. Commission Regulation (EC) No. 1881/2006 of 19 December 2006 setting maximum levels for certain contaminants in foodstuffs. Off. J. Eur. Comm. L364, 1–29. Available at https://eur-lex.europa.eu/legal-content/EN/TXT/?uri=CELEX:02006R1881-20200401.

EU, 2006b. Commission Regulation (EC) No 2023/2006 of 22 December 2006 on good manufacturing practice for materials and articles intended to come into contact with food. Off. J. Eur. Comm. L384, 75–78. Available at https://eur-lex.europa.eu/legal-content/EN/TXT/?uri=celex%3A32006R2023.

EU, 2008a. Regulation (EC) No 1333/2008 of the European Parliament and of the Council of 16 December 2008 on food additives. Off. J. Eur. Comm. L354, 16–33. Available at https://eur-lex.europa.eu/legal-content/EN/TXT/?uri=celex%3A32008R1333.

EU, 2008b. Regulation (EC) No. 1334/2008 of the European Parliament and of the Council of 16 December 2008 on flavourings and certain food ingredients with flavouring properties for use in and on foods and amending Council Regulation (EEC) No. 1601/91, Regulations (EC) No. 2232/96 and (EC) No. 110/2008 and Directive 2000/13/EC. Off. J. Eur. Comm. L354, 34–50. Available at https://eur-lex.europa.eu/legal-content/EN/ALL/?uri=CELEX%3A32008R1334.

EU, 2011a. Regulation (EC) No 1169/2011 of the European Parliament and of the Council of 25 October 2011on the provision of food information to consumers, amending Regulations (EC) No 1924/2006 and (EC) No 1925/2006 of the European Parliament and of the Council, and repealing Commission Directive 87/250/EEC, Council Directive 90/496/EEC, Commission Directive 1999/10/EC, Directive 2000/13/EC of the European Parliament and of the Council, Commission Directives 2002/67/EC and 2008/5/EC and Commission Regulation (EC) No 608/2004. Off. J. Eur. Comm. L304, 18–63. Available at https://eur-lex.europa.eu/legal-content/EN/ALL/?uri=CELEX%3A32011R1169.

EU, 2011b. Commission Regulation (EC) No 931/2011 of 19 September 2011 on the traceability requirements set by Regulation (EC) No 178/2002 of the European Parliament and of the Council for food of animal origin. Off. J. Eur. Comm. L242, 2–3. Available at https://eur-lex.europa.eu/legal-content/EN/TXT/?uri=CELEX%3A32011R0931.

EU, 2011c. Commission Regulation (EU) No 1259/2011 of 2 December 2011 amending Regulation (EC) No 1881/2006 as regards maximum levels for dioxins, dioxin-like PCBs and non dioxin-like PCBs in foodstuffs. Off. J. Eur. Comm. L320, 18–23. Available at https://eur-lex.europa.eu/legal-content/EN/TXT/?uri=celex:32011R1259.

EU, 2011d. Commission Regulation (EU) No 1129/2011 of 11 November 2011 amending Annex II to Regulation (EC) No 1333/2008 of the European Parliament and of the Council by establishing a Union list of food additives. Off. J. Eur. Comm. L295, 131–307. Available at https://eur-lex.europa.eu/legal-content/EN/ALL/?uri=CELEX%3A32011R1129.

EU, 2011e. Commission Regulation (EU) No 10/2011 of 14 January 2011 on plastic materials and articles intended to come into contact with food. Off. J. Eur. Comm. L12, 1–89. Available at https://eur-lex.europa.eu/legal-content/EN/ALL/?uri=CELEX%3A32011R0010.

EU, 2012. Commission Regulation (EU) No 231/2012 of 9 March 2012 laying down specifications for food additives listed in Annexes II and III to Regulation (EC) No 1333/2008 of the European Parliament and of the Council. Off. J. Eur. Comm. L83, 1–295. Available at https://eur-lex.europa.eu/legal-content/EN/TXT/?uri=celex:32012R0231.

EU, 2017. Commission Notice on the application of the principle of quantitative ingredients declaration (QUID). Off. J. Eur. Comm. C393, 5–12. Available at https://eur-lex.europa.eu/legal-content/EN/TXT/?uri=CELEX:52017XC1121(01)%0D.

FAO/WHO, 1994. Report of the First Session of the Codex Committee on Milk and Milk Products. Joint FAO/WHO Standard Programme, Codex Committee on Milk and Milk Products,

First Session, Rome, Italy 28 November–2 December 1994. Available at http://www.fao.org/input/download/report/400/al95_11e.pdf.

FAO/WHO, 1996. Consideration of Proposed Draft and Proposed Draft Revised Standards at Step 4, Agenda Item 5. Joint FAO/WHO Standard Programme, Codex Committee on Milk and Milk Products, Second Session, Rome, Italy 27–31 May 1996. Available at http://www.fao.org/3/w2198e/w2198e00.htm#Contents.

FAO/WHO, 1997. Codex Alimentarius Commission: Procedural Manual, 1997, Tenth Edition Joint FAO/WHO Standard Programme, Rome.

FAO/WHO, 1998. Report of the Third Session of the Codex Committee on Milk and Milk Products, Third Session Joint FAO/WHO Standard Programme, Codex Committee on Milk and Milk Products, Montevideo, Uruguay 18–22 May 1998. Available at http://www.fao.org/tempref/codex/Reports/Alinorm99/al99_11e.pdf.

FAO/WHO, 2000. Proposed Draft and Proposed Draft Revised Standards at Step 4, Agenda Item 4, Fourth Session Joint FAO/WHO Standard Programme, Codex Committee on Milk and Milk Products, Wellington, New Zealand 28 February –3 March 2000. Available at http://www.fao.org/fao-who-codexalimentarius/meetings/detail?meeting=CCMMP&session=4.

FAO/WHO, 2001. Circulation Letter: Request for Comments on the Minimum Cheese Contents in Processed Cheese. Joint FAO/WHO Standard Programme, Codex Committee on Milk and Milk Products, Fifth Session, Wellington, New Zealand, pp. 8–12 April 2002. Available at http://www.fao.org/fao-who-codexalimentarius/sh-proxy/en/?lnk=1&url=https%253A%252F%252Fworkspace.fao.org%252Fsites%252Fcodex%252FShared%2BDocuments%252FArchive%252FMeetings%252FCCMMP%252Fccmmp5%252Fcl01_20e.pdf.

FAO/WHO, 2002a. Background Information to Working Document CX/MMP 02/9, Agenda Item 4 (d). Joint FAO/WHO Standard Programme, Codex Committee on Milk and Milk Products, Fifth Session, Wellington, New Zealand, pp. 8–12 April 2002. Available at http://www.fao.org/fao-who-codexalimentarius/sh-proxy/en/?lnk=1&url=https%253A%252F%252Fworkspace.fao.org%252Fsites%252Fcodex%252FShared%2BDocuments%252FArchive%252FMeetings%252FCCMMP%252Fccmmp5%252FMM0209ae.pdf.

FAO/WHO, 2002b. Processed Cheese (Minimum Cheese Content) Comments Submitted in Response to CL 2001/20– MMP, Agenda Item 4 (d). Joint FAO/WHO Standard Programme, Codex Committee on Milk and Milk Products, Fifth Session, Wellington, New Zealand, pp. 8–12 April 2002. Available at http://www.fao.org/fao-who-codexalimentarius/sh-proxy/en/?lnk=1&url=https%253A%252F%252Fworkspace.fao.org%252Fsites%252Fcodex%252FShared%2BDocuments%252FArchive%252FMeetings%252FCCMMP%252Fccmmp5%252FMM02_09e.pdf.

FAO/WHO, 2004a. Proposed Draft Standard for Processed Cheese at Step 3, Agenda Item 4 (d). Joint FAO/WHO Standard Programme, Codex Committee on Milk and Milk Products, Sixth Session, Auckland, New Zealand, pp. 26–30 April 2004. Available at http://www.fao.org/fao-who-codexalimentarius/sh-proxy/en/?lnk=1&url=https%253A%252F%252Fworkspace.fao.org%252Fsites%252Fcodex%252FShared%2BDocuments%252FArchive%252FMeetings%252FCCMMP%252Fccmmp6%252FMM04_07e.pdf.

FAO/WHO, 2004b. Proposed Draft Standard for Processed Cheese Comments at Step 3, Agenda Item 4 (d). Comments from: Australia, Canada, Colombia, France, Germany, New Zealand, Sweden, United Kingdom, United States, and Uruguay. Joint FAO/WHO Standard Programme, Codex Committee on Milk and Milk Products, Sixth Session, Auckland, New Zealand, pp. 26–30 April 2004. Available at http://www.fao.org/fao-who-codexalimentarius/sh-proxy/en/?lnk=1&url=https%253A%252F%252Fworkspace.fao.org%252Fsites%252Fcodex%252FShared%2BDocuments%252FArchive%252FMeetings%252FCCMMP%252Fccmmp6%252FMM0407ae.pdf.

FAO/WHO, 2004c. Proposed Draft Standard for Processed Cheese Comments at Step 3, Agenda Item 4 (d). Comments from: Malaysia, Mexico, Japan and Thailand. Joint FAO/WHO Standard Programme, Codex Committee on Milk and Milk Products, Sixth Session, Auckland, New Zealand, pp. 26–30 April 2004. Available at http://www.fao.org/fao-who-codexalimentarius/sh-proxy/en/?lnk=1&url=https%253A%252F%252Fworkspace.fao.org%252Fsites%252Fcodex%252FShared%2BDocuments%252FArchive%252FMeetings%252FCCMMP%252Fccmmp6%252FMM0407be.pdf.

FAO/WHO, 2006a. Proposed Draft Standard for Processed Cheese Comments at Step 3, Agenda Item 4 (q). Comments from: IDF. Joint FAO/WHO Standard Programme, Codex Committee on Milk and Milk Products, Seventh Session, Queenstown, New Zealand 27 March –1 April 2006. Available at http://www.fao.org/fao-who-codexalimentarius/sh-proxy/en/?lnk=1&url=https%253A%252F%252Fworkspace.fao.org%252Fsites%252Fcodex%252FShared%2BDocuments%252FArchive%252FMeetings%252FCCMMP%252Fccmmp7%252Fmm0708be.pdf.

FAO/WHO, 2006b. Proposed Draft Standard for Processed Cheese Comments at Step 3, Agenda Item 4 (q). Comments from: Canada, France, India, Thailand and United States of America. Joint FAO/WHO Standard Programme, Codex Committee on Milk and Milk Products, Seventh Session, Queenstown, New Zealand 27 March–1 April 2006. Available at http://www.fao.org/fao-who-codexalimentarius/sh-proxy/en/?lnk=1&url=https%253A%252F%252Fworkspace.fao.org%252Fsites%252Fcodex%252FShared%2BDocuments%252FArchive%252FMeetings%252FCCMMP%252Fccmmp7%252Fmm0708ce.pdf.

FAO/WHO, 2006c. Proposed Draft Standard for Processed Cheese Comments at Step 3, Agenda Item 4 (q). Comments from: Argentina, Australia, Japan, New Zealand and Venezuela. Joint FAO/WHO Standard Programme, Codex Committee on Milk and Milk Products, Seventh Session, Queenstown, New Zealand 27 March –1 April 2006. Available at http://www.fao.org/fao-who-codexalimentarius/sh-proxy/en/?lnk=1&url=https%253A%252F%252Fworkspace.fao.org%252Fsites%252Fcodex%252FShared%2BDocuments%252FArchive%252FMeetings%252FCCMMP%252Fccmmp7%252Fmm0708ae.pdf.

FAO/WHO, 2006d. Proposed Draft Standard for Processed Cheese at Step 3, Agenda Item 4 (q). Joint FAO/WHO Standard Programme, Codex Committee on Milk and Milk Products, Seventh Session, Queenstown, New Zealand 27 March –1 April 2006. Available at http://www.fao.org/fao-who-codexalimentarius/sh-proxy/en/?lnk=1&url=https%253A%252F%252Fworkspace.fao.org%252Fsites%252Fcodex%252FShared%2BDocuments%252FArchive%252FMeetings%252FCCMMP%252Fccmmp7%252Fmm07_08e.pdf.

EOS, 2013. Egyptian Organization for Standards and Quality. Available http://www.eos.org.eg/.

FAO/WHO, 2008a. Codex General Standard for Named Variety of Process(ed) Cheese and Spreadable Process(ed) Cheese (CODEX STAN 285-1978). Codex Alimentarius International Standards. Formerly CODEX-STAN A-8(a) –1978, adopted in 1978, amendment in 2008, revoked in 2010.

FAO/WHO, 2008b. Codex General Standard for Process(ed) Cheese and Spreadable Process(ed) Cheese (CODEX STAN 286-1978). Codex Alimentarius International Standards. Formerly CODEX-STAN A-8(b) –1978, adopted in 1978, amendment in 2008, revoked in 2010.

FAO/WHO, 2008c. Codex General Standard for Cheese Preparations as Process(ed) Cheese Food and Process(ed) Cheese Spread (CODEX STAN 287-1978). Codex Alimentarius International Standards. Formerly CODEX-STAN A-8(c) –1978, adopted in 1978, amendment in 2008, revoked in 2010.

FAO/WHO, 2008d. Proposed draft standard for processed cheese (Step 3), Agenda Item 4 (b). Joint FAO/WHO Standard Programme, Codex Committee on Milk and Milk Products,

Eighth Session, Queenstown, New Zealand, pp. 4–8 February 2008. Available at http://www.fao.org/fao-who-codexalimentarius/meetings/detail?meeting=CCMMP&session=8.

FAO/WHO, 2010a. Provisional Agenda. Joint FAO/WHO Standard Programme, Codex Committee on Milk and Milk Products, Ninth Session, Auckland, New Zealand, pp. 1–5 February 2010. Available at http://www.fao.org/fao-who-codexalimentarius/sh-proxy/en/?lnk=1&url=https%253A%252F%252Fworkspace.fao.org%252Fsites%252Fcodex%252FShared%2BDocuments%252FArchive%252FMeetings%252FCCMMP%252Fccmmp9%252Fmm09_01e.pdf.

FAO/WHO, 2010b. Matters Referred by the Codex Alimentarius Commission and other Codex Committees and Task Forces, Agenda Item 2. Joint FAO/WHO Standard Programme, Codex Committee on Milk and Milk Products, Ninth Session, Auckland, New Zealand, pp. 1–5 February 2010. Available at http://www.fao.org/fao-who-codexalimentarius/sh-proxy/en/?lnk=1&url=https%253A%252F%252Fworkspace.fao.org%252Fsites%252Fcodex%252FShared%2BDocuments%252FArchive%252FMeetings%252FCCMMP%252Fccmmp9%252Fmm09_02e.pdf.

FAO/WHO, 2010c. Report of the Physical Working Group on the Proposed Draft Standard for Processed Cheese, Agenda Item 4. Joint FAO/WHO Standard Programme, Codex Committee on Milk and Milk Products, Ninth Session, Auckland, New Zealand, pp. 1–5 February 2010. Available at http://www.fao.org/fao-who-codexalimentarius/sh-proxy/en/?lnk=1&url=https%253A%252F%252Fworkspace.fao.org%252Fsites%252Fcodex%252FShared%2BDocuments%252FArchive%252FMeetings%252FCCMMP%252Fccmmp9%252Fmm09_04e.pdf.

FAO/WHO, 2010d. Report of the Physical Working Group on the Proposed Draft Standard for Processed Cheese (Submitted by the European Union), Agenda Item 4. Joint FAO/WHO Standard Programme, Codex Committee on Milk and Milk Products, Ninth Session, Auckland, New Zealand, pp. 1–5 February 2010. Available at http://www.fao.org/fao-who-codexalimentarius/sh-proxy/en/?lnk=1&url=https%253A%252F%252Fworkspace.fao.org%252Fsites%252Fcodex%252FShared%2BDocuments%252FArchive%252FMeetings%252FCCMMP%252Fccmmp9%252Fmm09_04ae.pdf.

FAO/WHO, 2011. Proposals for the Elaboration of New Standards and Related Texts for the Discontinuation of Work, Agenda Item 9. Joint FAO/WHO Standard Programme, Codex Alimentarius Commission, 34th Session, Geneva, Switzerland 4–9 July 2011. Available at http://www.fao.org/fao-who-codexalimentarius/meetings/detail?meeting=CAC&session=34.

FAO/WHO, 2012. Matters Referred to the Commission by Codex Committees and Task Forces, Agenda Item 9. Consideration of the Possibility of Further Work to Develop Standard(s) Covering Processed Cheese (Prepared by the CCMMP and Codex Secretariats). Joint FAO/WHO Standard Programme, Codex Alimentarius Commission, 35th Session, Rome, Italy 2–7 July 2012. Available at http://www.fao.org/fao-who-codexalimentarius/meetings/detail?meeting=CAC&session=35.

FAO/WHO, 2013. Matters referred to the Commission by Codex Committees and task forces, Agenda Item 10. Matters Referred From the 35th Session of the Commission New Work on Processed Cheese (Prepared by the Codex Secretariat). Joint FAO/WHO Standard Programme, Codex Alimentarius Commission, 36th Session, Rome, Italy 1–5 July 2013. Available at http://www.fao.org/fao-who-codexalimentarius/meetings/detail?meeting=CAC&session=36.

FAO/WHO, 2014a. Matters referred to the Commission by Codex Committees, Agenda Item 9. Matters referred from the 36th Session of the Commission (Processed Cheese: Report of the Electronic Working Group). Joint FAO/WHO Standard Programme, Codex Alimentarius

Commission, 37th Session, Geneva, Switzerland, pp. 14–18 July 2014. Available at http://www.fao.org/fao-who-codexalimentarius/sh-proxy/en/?lnk=1&url=https%253A%252F%252Fworkspace.fao.org%252Fsites%252Fcodex%252FShared%2BDocuments%252FArchive%252FMeetings%252FCAC%252Fcac37%252Fcac37_10_add1e.pdf.

FAO/WHO, 2014b. (REP14/CAC). Joint FAO/WHO Standard Programme, Codex Alimentarius Commission, 37th Session, Geneva, Switzerland, pp. 14–18 July 2014. Available at http://www.fao.org/fao-who-codexalimentarius/sh-proxy/en/?lnk=1&url=https%253A%252F%252Fworkspace.fao.org%252Fsites%252Fcodex%252FShared%2BDocuments%252FArchive%252FMeetings%252FCAC%252Fcac37%252FREP14_CACe.pdf.

FAO/WHO, 2015. Adoption at Step 5, Agenda Item 5 (b). Joint FAO/WHO Standard Programme, Codex Alimentarius Commission, 38th Session, Geneva, Switzerland, pp. 6–11 July 2015. Available at http://www.fao.org/fao-who-codexalimentarius/sh-proxy/en/?lnk=1&url=https%253A%252F%252Fworkspace.fao.org%252Fsites%252Fcodex%252FShared%2BDocuments%252FArchive%252FMeetings%252FCAC%252Fcac38%252Fcac38_04e.pdf.

FAO/WHO, 2016a. Codex Alimentarius: Understanding Codex, 2016. Food and Agriculture Organization of the United Nations and World Health Organization, Rome.

FAO/WHO, 2016b. Matters referred to the Commission by Codex Committees, Agenda Item 12. Matters for action: Committee on Milk and Milk Products (CCMMP). Joint FAO/WHO Standard Programme, Codex Alimentarius Commission, 39th Session, Rome, Italy 27 June –1 July 2016. Available at http://www.fao.org/fao-who-codexalimentarius/sh-proxy/en/?lnk=1&url=https%253A%252F%252Fworkspace.fao.org%252Fsites%252Fcodex%252FMeetings%252FCX-701-39%252FWD%252Fcac39_13Add1e.pdf.

FAO/WHO, 2016c. Report of the seventy-first session of the Executive Committee of the Codex Alimentarius Commission. Joint FAO/WHO Standard Programme, Codex Alimentarius Commission, 39th Session, Rome, Italy. Available at http://www.fao.org/fao-who-codexalimentarius/sh-proxy/en/?lnk=1&url=https%253A%252F%252Fworkspace.fao.org%252Fsites%252Fcodex%252FMeetings%252FCX-701-39%252FWD%252FREP16_ExeE.pdf.

FAO/WHO, 2017. Report (REP17/CAC). Joint FAO/WHO Standard Programme, Codex Alimentarius Commission, 40th Session, Geneva, Switzerland 17–22 July 2017. Available at http://www.fao.org/fao-who-codexalimentarius/sh-proxy/en/?lnk=1&url=https%253A%252F%252Fworkspace.fao.org%252Fsites%252Fcodex%252FMeetings%252FCX-701-40%252FREPORT%252FREP17_CACe.pdf.

FSA, 2017. Processed Cheese (as part-nature). A06BP. Available at https://data.food.gov.uk/codes/foodtype/id/A06BP. (Accessed on 12 July 2020).

FSSAI, 2011. Food Safety and Standards Authority of India. Available at https://www.fssai.gov.in/cms/product-standards.php.

GCO, 2020. Gulf Council Organisation. Available at https://www.gso.org.sa/en/standards/.

Hickey, M., 2011. Current legislation on processed cheese and related products. In: Tamime, A.Y. (Ed.), Processed Cheese and Analogues. John Wiley & Sons, Chichester, UK, pp. 25–80.

HMSO, 1970. The Cheese Regulations 1970. Food and Drugs: Composition and Labelling, No.94. Available at http://www.legislation.gov.uk/uksi/1970/94/contents/made. (Accessed on 12.07.2020).

HMSO, 1974. The Cheese (Amendments) Regulations 1974. Food and Drugs: Composition and Labelling, No.1122. Available at http://www.legislation.gov.uk/uksi/1974/1122/contents/made. (Accessed on 12 July 2020).

HMSO, 1984. The Cheese (Amendments) Regulations 1984. Food: Composition and Labelling, No.649. Available at http://www.legislation.gov.uk/uksi/1984/649/contents/made. (Accessed on 12 July 2020).
HMSO, 1995. The Cheese and Cream Regulations 1995. No.3240. Available at http://www.legislation.gov.uk/uksi/1995/3240/contents/made. (Accessed on 12 July 2020).
HMSO, 1996. The Food Labelling Regulations 1996. No. 1499. Available at http://www.legislation.gov.uk/uksi/1996/1499/contents/made. (Accessed on 12 July 2020).
Hui, Y.H., 2015. Plant Sanitation for Food Processing Food Service, 2nd Ed. CRC Press, Taylor & Francis Group, Boca Raton, p. 1308p.
Iranian Standard, 2003. Milk and Milk Products – Processed cheese – Specifications, ISIRI No. 4659. Institute of Standards and Industrial Research of Iran, Pounak, Tehran West Bahar Street, Sardar Jangal Boulivard.
ISO, 2009. ISO/TS 22002-1:2009. Prerequisite programmes on food safety. Available at https://www.iso.org/standard/44001.html.
Käseverordnung, 2017. Bundesministeriumm der Justiz un für Verbraucherschutz. Revised by V. April 14, 1986/412 and last amended by 5.7.2017/2272. Available at http://www.gesetze-im-internet.de/k_sev/BJNR511800965.html#BJNR511800965BJNG000100328. (Accessed on 1 July 2020).
ISO, 2018. ISO 22000:2018. Food Safety Management Systems - Requirements for Any Organization in the Food Chain. International Standardization Organization, Geneva, Switzerland Available at https://www.iso.org/standard/65464.html.
Kenya Bureau of Standards (KEBS), 2019. Processed Cheese-Specification. Kenya Standard, Nairobi, Kenya 2509:2019Available at http://onlinecatalogue.kebs.org/webquery.dll.
Livsmedelsverket, 2003. Livsmedelsverkets föreskrifter om mjölk och ost, LIVSFS 2003:39, Available at https://www.livsmedelsverket.se/om-oss/lagstiftning1/gallande-lagstiftning/livsfs-200339.
Ministry of Health Malaysia, 1985. Food Regulations. Available at http://fsis2.moh.gov.my/fosimtestsite/HOM/frmHOMFARSec.aspx?id=21.
MPI, 2020. Food Standards Australia New Zealand (FSANZ). New Zealand Food Safety, Ministry for Primary Industries. Available at https://www.mpi.govt.nz/law-and-policy/legal-overviews/food-safety/australia-new-zealand-co-operation/food-standards-australia-new-zealand-fsanz/. (Accessed on 12 July 2020).
Real Decreto 1113/2006. Ministry of the Presidency. Available at https://agriculturaganaderia.jcyl.es/web/jcyl/AgriculturaGanaderia/es/Plantilla100Detalle/1143210597398/_/12844 30547895/Redaccion 2006
Sommer, D., 2016. Safety of U.S. Cheese and Dairy Products. Reference Manual for U.S. Cheese. U.S. Dairy Export Council. Available at https://www.thinkusadairy.org/resources-and-insights/resources-and-insights/product-resources/reference-manual-for-us-cheese. (Accessed on 15 May 2020).
SSO, 1988. Food Regulation No. 264/1988. Available at https://sso.agc.gov.sg/SL/283-RG1?DocDate=20180327.
TFC, 2015. Turkish Food Codex, Cheese Standard, No:2015/6. Available at https://www.mevzuat.gov.tr/mevzuat?MevzuatNo=20523&MevzuatTur=9&MevzuatTertip=5. (Acessed on 20 June 2020)
TSE, 2011. Processed Cheese. Turkish Standard Enstitute No: 2176, Amended in 2015 and 2016. Ankara.
USDA, 2018. Hungary, Food and Agricultural Import Regulations and Standards Report. United States Department of Agricultural, Foreign Agricultural Service. Available at

https://agriexchange.apeda.gov.in/IR_Standards/Import_Regulation/FoodandAgricultural ImportRegulationsandStandardsReportBudapestHungary10242018.pdf.

USDA, 2020. Grades and Standards for Dairy Products, U.S. Department of Agriculture, Agriculturel Marketing Service. Available at https://www.ams.usda.gov/grades-standards/dairy-products. (Accessed on 15 May 2020).

Processed cheese: applications, challenges, and market trends

16

Yasser Bouhlal[a] *and Oral Capps Jr.*[b]
[a]Corteva Agriscience, Indianapolis, Indiana, United States, [b]Department of Agricultural Economics, Texas A&M University, College Station, Texas, United States

16.1 Introduction

In this chapter, we initially discuss market trends in per capita consumption of natural cheese as well as processed cheese in the United States. On the basis of this discussion, we focus exclusively on the challenge of expanding the demand for processed cheese. In particular, to address this challenge we center attention on the economic feasibility of fortifying processed cheese with omega-3 fatty acids.

16.2 Market trends in U.S. per capita consumption of cheese

Per capita consumption of cheese in the United States has grown steadily over the past 20 years. Key determinants of this trend may be attributed to advances in technology that have helped propel a variety of cheeses into prepared foods, snacks, and frozen entrees (Shoup, 2019), generic advertising and promotion activities conducted by the dairy industry largely by Dairy Management, Inc. (DMI) as well as increases in U.S. inflation-adjusted per capita income (Capps et. al, 2020), and perceived health and nutrition benefits (Robbins, 2019). Further, higher stocks of cheese together with lower milk prices in recent times have contributed to weaker domestic cheese prices. In accord with the law of demand in economics, this situation in regard to prices also has been a factor associated with the rise in per capita cheese consumption. As exhibited in Fig. 16.1, according to the United States Department of Agriculture (USDA), Economic Research Service (ERS), the rise of per capita U.S. cheese consumption over the period between 1995 and 2018, the latest data available, has been primarily for natural cheese as opposed to processed cheese (e.g., American singles, Velveeta, nacho cheese, pimento cheese, and provel cheese).

Natural cheeses consist of three classes—American types, Italian types, and other types of cheeses typically from cows. Cheddar, Colby, Monterey and other Jack cheeses comprise the American type category, while Mozzarella, Provolone, and Asiago among others make up the Italian category. The other category of natural category consists of cheeses that are neither of the American-type nor of the Italian-type, including but not limited to Swiss, Blue, Brick, Muenster, cream cheese, and Hispanic.

As exhibited in Fig. 16.2, from 1995 to 2018, the per capita consumption of cheddar cheese ranged from 9 pounds per capita to slightly more than 11 pounds

Processed Cheese Science and Technology: Ingredients, Manufacture, Functionality, Quality, and Regulations.
DOI: https://doi.org/10.1016/B978-0-12-821445-9.00003-0

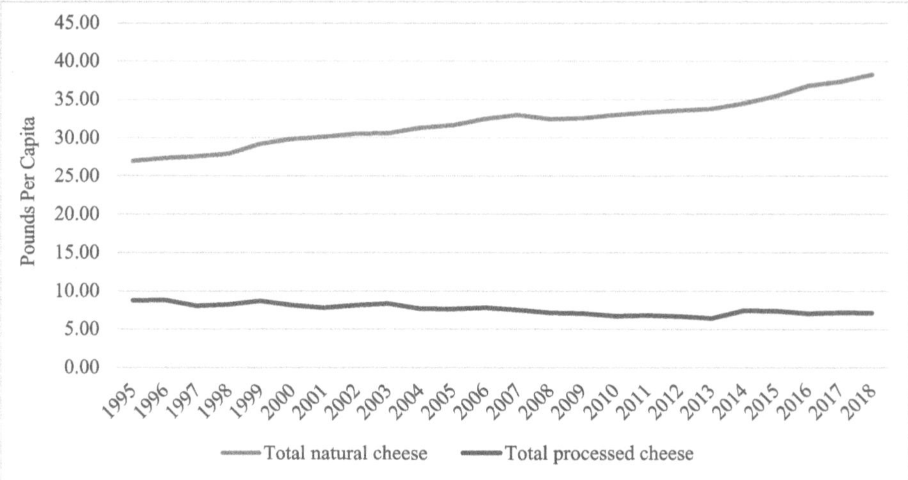

Figure 16.1 Per capita consumption of natural cheese and processed cheese in the United States, 1995 to 2018, pounds per capita.
Source: USDA, National Agricultural Statistics Service; USDA, Farm Service Agency; USDA, Foreign Agricultural Service; U.S. Department of Commerce, Bureau of the Census; and USDA, Economic Research Service calculations.

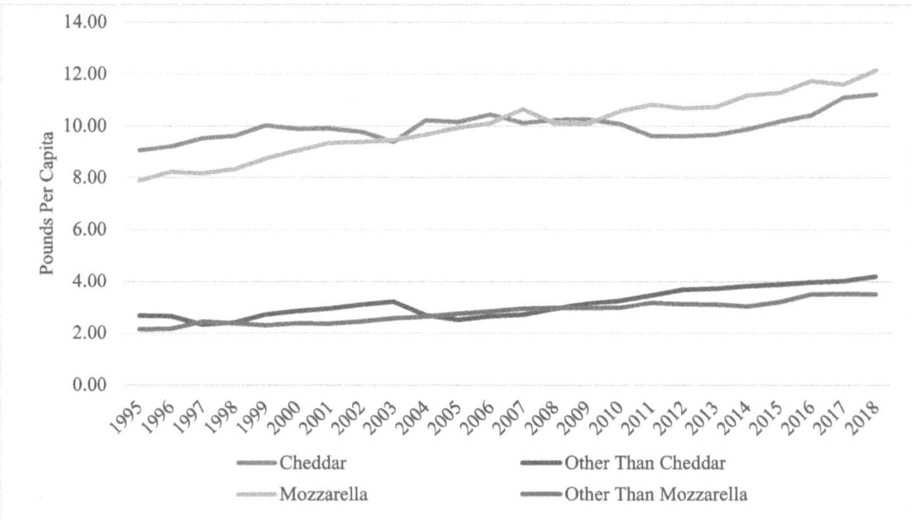

Figure 16.2 Per capita consumption of cheddar cheese, other than cheddar cheese, mozzarella cheese, and other than mozzarella cheese in the United States, 1995 to 2018, pounds per capita.
Source: USDA, National Agricultural Statistics Service; USDA, Farm Service Agency; USDA, Foreign Agricultural Service; U.S. Department of Commerce, Bureau of the Census; and USDA, Economic Research Service calculations.

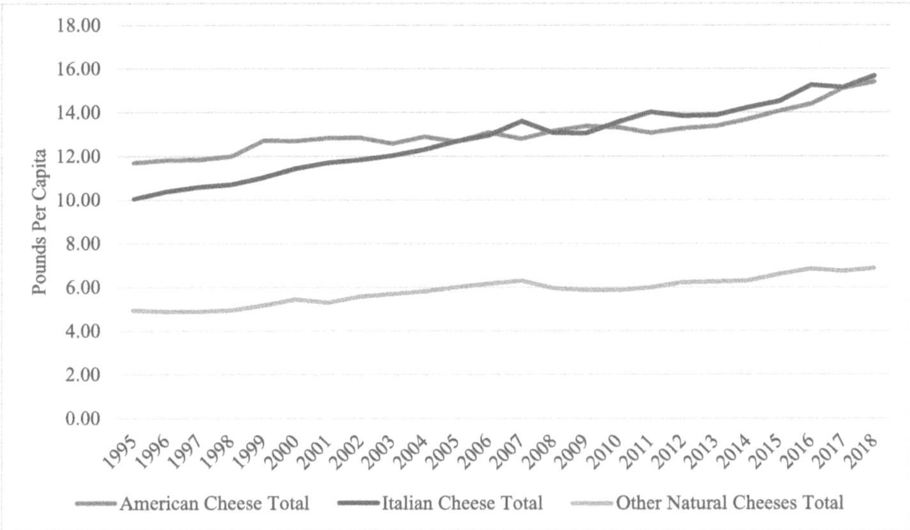

Figure 16.3 Per capita consumption of American cheese, Italian cheese, and other natural cheese in the United States, 1995 to 2018, pounds per capita.
Source: USDA, National Agricultural Statistics Service; USDA, Farm Service Agency; USDA, Foreign Agricultural Service; U.S. Department of Commerce, Bureau of the Census; and USDA, Economic Research Service calculations.

per capita, averaging 10 pounds per capita. The per capita consumption of Colby, Monterey and other Jack cheeses ranged from 2.3 pounds per capita to 4.2 pounds per capita, averaging slightly more than 3 pounds per capita. From 1995 to 2018, the per capita consumption of mozzarella cheese ranged from 8 pounds per capita to slightly more than 12 pounds per capita, averaging 10 pounds per capita. The per capita consumption of other Italian cheeses ranged from 2.1 pounds per capita to 3.5 pounds per capita, averaging slightly less than 3 pounds per capita. Without question, in the United States cheddar cheese is the kingpin of the American-types natural cheese category, while mozzarella cheese is the leader of the Italian-types natural cheese category.

In regard to per capita cheese consumption not associated with the American or Italian categories, as of 2018, cream cheese was the next largest category at 2.6 pounds per person, followed by Swiss cheese at 1.1 pounds per person, Hispanic cheese at 0.9 pounds per person, Muenster cheese at 0.5 pounds per person, and Blue cheese at 0.3 pounds per person. As shown in Fig. 16.3, growth in per capita consumption is evident for all categories of natural cheeses (American, Italian, and other natural cheeses) over the period from 1995 to 2018.

The share of per capita consumption of natural cheese has been on the rise from 74% to 84% over the period between 1995 and 2018, while the share of per capita consumption has been on the decline from 24% to 16% over this same period. Despite the decline in the share of the processed cheese category, Americans continue to consume American singles and Velveeta (Shoup, 2019). As well, the food service sector has increased the amount of processed cheese used in meal preparation in recent times (Shoup, 2019). Importantly, per capita consumption of processed cheese has rebounded modestly after hitting a historic low of 6.37 pounds per person in 2013. From 2014 to 2018, U.S. per capita consumption of processed cheese has varied from 6.98 pounds per person to 7.36 pounds per person. As of 2018, per capita consumption of processed cheese was close to 7.1 pounds per person, still well behind per capita consumption of cheddar cheese and mozzarella cheese at 11.2 pounds per person and 12.15 pounds per person, respectively.

16.3 The challenge of how to expand the demand for processed cheese

Given the market trends previously described, we center attention on the challenge of expanding the demand for processed cheese. To address this challenge, we focus on the economic feasibility of fortifying processed cheese with omega-3 fatty acids. Studies have been conducted in the extant literature dealing with the fortification of reduced-fat cheddar cheese with n-3 fatty acids (Martini et al., 2009), processed cheese fortified with fish oil emulsion (Ye et al., 2009), selected cheeses fortified with vegetable and animal sources of omega-3 fatty acids (Bermudez-Aguirre and Barbosa-Canovas, 2011), and the fortification of processed cheese spread with omega-3 fatty acids (Renuka et al., 2016). But no studies currently exist that deal with the economic feasibility of fortification of processed cheese. In this light, we wish to determine the effects of fortifying processed cheese with omega-3 fatty acids on the profits of manufacturers.

Fortification implies additional costs of production for cheese processors. This ex-ante analysis takes into account simulated market conditions, via demand and supply curves, and evaluates the percentage increase in the demand for processed cheese needed to offset the incremental costs of fortification in order to maintain producer profitability. This work then makes a contribution to the extant literature in regard to the feasibility of fortifying processed cheese with omega-3 fatty acids. Without question, this analysis provides valuable information to all stakeholders in the dairy industry, including dairy farmers, manufacturers, retailers, and consumers.

16.4 Fortification

Consumers purchase fortified foods and beverages, including dairy-based beverages, cheese, yogurt, and ice cream, to prevent illness or to manage specific health concerns (Beery, October 25, 2018). Fortification started as a public health intervention about 100 years ago in the United States and proved to be an effective approach to eradicating nutrient-deficiency illnesses, such as goiter, addressed by the iodization of salt in 1924; rickets, addressed by the addition of vitamin D to milk in 1933; and neural tube birth defects, addressed by a mandate from the U.S. Food and Drug Administration to enrich cereal grains to be fortified with folic acid starting in 1998 (Institute of Medicine Committee on Use of Dietary Reference Intakes in Nutrition Labeling, 2003).

The World Health Organization (WHO) and the Food and Agriculture Organization of the United Nations (FAO) define fortification as "the practice of deliberately increasing the content of an essential micronutrient in a food irrespective of whether the nutrients were originally in the food before processing or not, so as to improve the nutritional quality of the food supply and to provide a public health benefit with minimal risk to health" (Allen, de Benoist, Dary, and Hurell, 2006). Ingredients for example may be added as concentrated, purified compounds, such as fiber, minerals, omega-3 fatty acids, plant extracts, probiotics, proteins, and vitamins.

Over the past two decades, there has been a surge in interest among public and health professionals of the health effects associated with omega-3 fatty acids, a group of polyunsaturated fatty acids which include α-linolenic acid as well as its long-chain metabolites eicosapentaenoic acid (EPA) and docosahexaenoic acid (DHA) (Simopoulos, 2002). DHA and EPA are considered to have beneficial effects in the prevention and management of cardiovascular disease, hypertension, diabetes, arthritis, inflammatory diseases, cancers and autoimmune disorders (Kris-Etherton et al., 2002; Gebauer et al., 2006; MacLean et al., 2006; Goldberg and Katz, 2007; Lombardi and Terranova, 2007; Yokoyama et al., 2007; Gogus and Smith, 2010; and Renuka et al., 2016). Products rich in omega-3 fatty acids represent one of the fastest developing trends in the food industry (Ganesan et al., 2014).

According to the Report of the Joint WHO / FAO Expert Consultation on the Risks and Benefits of Fish Consumption (2010), the daily recommended intake of omega-3 fatty acids should be at least 250 mg/day for healthy life. But, the intake level of omega-3 fatty acids typically is far below the recommended level (Kolanowski, 2005). Typically, consumers do not get enough intake of omega-3 fatty acids through their diet alone since the most widely available source is cold water oily fish such as salmon which are not always consumed on a regular basis. Fortification of commonly consumed food products with omega-3 fatty acids is considered as a viable way of providing the health benefits to people without major alteration in their dietary habits (Garg et al., 2006). The increased recognition of the importance of omega-3 fatty acids in the diet, coupled with its limited availability in natural food sources, makes fortification a noteworthy solution in closing the nutrition gap for better health and disease prevention/management.

Most dairy products provide the minimal fat basis needed to more easily incorporate omega-3 oils (Bouhlal, 2012). Additionally, dairy products share the image of "being good-for-you" given that they deliver essential nutrients to consumers. Interventions in the dairy industry include yogurt carrying added health benefits in the way of probiotics. This exposure is very important when looking for new opportunities to proliferate healthy functional food ingredients such as omega-3 fatty acids. Dairy products, particularly cheese and butter, have been shown to possess good delivery systems for elevated levels of omega-3 fatty acids (Kolanowski and Weißbrodt, 2007). As such, dairy products represent potential for the use of fortification with omega-3 fatty acids. Overall, omega-3 fatty acids can be easily added to cheese with only minor changes (Bouckley, 2017).

16.5 Methodology

The methodology involves the use of a partial equilibrium framework, often referred to as Equilibrium Displacement Modeling (EDM) (Davis and Espinoza, 1998). With this approach, the industry is represented by demand and supply relationships for processed cheese. The impact of the fortification of processed cheese with omega-3 fatty acids is modeled as shifts in demand and supply from an initial equilibrium. Changes in prices and quantities that arise when the system equilibrium is displaced are estimated, as are the consequent changes in producer surplus reflecting welfare changes to manufacturers of processed cheese.

To invoke this methodology, we initially obtain estimates of the own-price elasticity of demand for processed cheese products from the economic literature. Subsequently, we determine producer surplus (synonymous with profit) considering the case of linear demand and supply functions. Finally, we establish by how much the demand for the new product (fortified processed cheese) would have to shift to the right so that producer surplus remains the same after the fortification. That is, we determine the minimum demand increase on a percentage basis required so that manufacturers would at least cover the marginal costs in producing processed cheese fortified with omega-3 fatty acids.

16.5.1 Estimates of the own-price elasticity of demand for processed cheese

From the extant economic literature, the own-price elasticities for processed cheese (as a category) have been estimated to be −0.36 (Schmit et al., 2003); −0.68 (Bouhlal, 2012); −0.70 (Gould and Lin, 1994); and −0.99 (Davis et al., 2010). This review of previous work related to the demand for processed cheese demonstrates that there is no unique value for the own-price elasticity of demand. Based on past work, the own-price elasticity for processed cheese has ranged from −0.36 to −0.99. The variation in these own-price elasticities is attributed to the difference in the nature of data used, the time period considered, and in the type of economic model estimated.

16.5.2 Profitability of fortification with omega-3s

Upon taking into account the own-price elasticities of demand for the category processed cheese, we subsequently calculate the producer surplus for the industry and determine by how much the demand would have to increase on a percentage basis so that manufacturers would at least cover their marginal costs in producing processed cheese fortified with omega-3 fatty acids.

In the ensuing calculations, we assume both the demand and the supply functions are linear and that the shift in supply due to the change in marginal costs is a parallel shift. We also assume that the shift in demand is not only parallel but also to the right

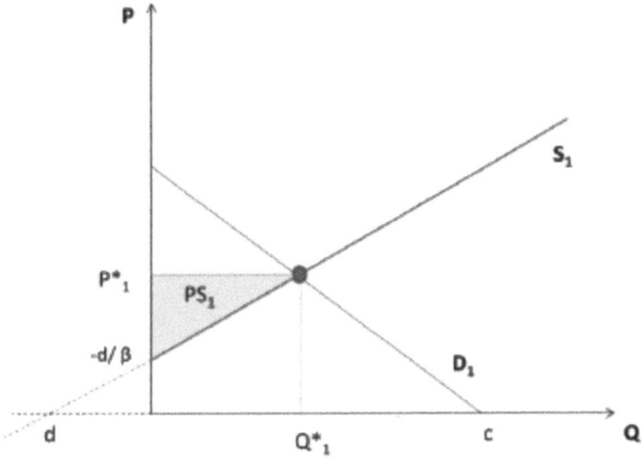

Figure 16.4 Demand and supply relationship before fortification.

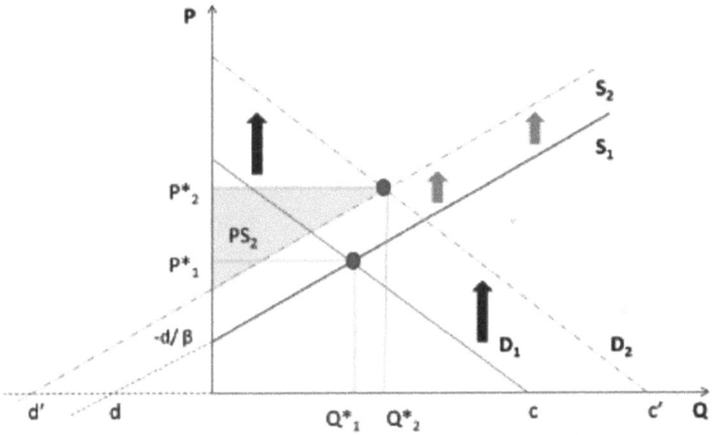

Figure 16.5 Demand and supply relationship after fortification.

due the health benefits associated with omega-3 fatty acids. Further, we assume that the elasticity of demand is the same for processed cheese with and without fortification.

$Q_d = -\alpha P + c$: Demand

$Q_s = \beta P + d$: Supply

$Q_d = Q_s$: Equilibrium condition

In Fig. 16.4, we graphically portray the demand and supply of processed cheese before fortification. The demand curve reflects the inelastic nature of the demand for processed cheese.

After fortification, as illustrated in Fig. 16.5, we have

$Q_d' = -\alpha P + c'$: Demand

$Q_s' = \beta P + d'$: Supply

Subsequently, we determine the producer surplus before and after the fortification, labeled respectively as PS_1 and PS_2.

At equilibrium, $PS_1 = \frac{1}{2}\left(P_1^* + {}^{d}/_{\beta}\right) * Q_1{}^*$ and $Q_d = Q_s$

Then $- \alpha P^* + c = \beta P^* + d$ and $P_1^* = \frac{c-d}{\alpha+\beta}$, $Q_1^* = -\alpha * \left(\frac{c-d}{\alpha+\beta}\right) + c$

Thus $PS_1 = \frac{1}{2}\left(\frac{c-d}{\alpha+\beta} + \frac{d}{\beta}\right) * \left(-\alpha * \left(\frac{c-d}{\alpha+\beta}\right) + c\right)$

$$PS_1 = \frac{-\alpha}{2} * \left(\frac{c-d}{\alpha+\beta}\right)^2 + \frac{c(c-d)}{2(\alpha+\beta)} - \frac{d\alpha}{2\beta}\left(\frac{c-d}{\alpha+\beta}\right) + \frac{dc}{2\beta} \tag{16.1}$$

Similarly, we calculate PS_2 as

$$PS_2 = \frac{-\alpha}{2} * \left(\frac{c'-d'}{\alpha+\beta}\right)^2 + \frac{c'(c'-d')}{2(\alpha+\beta)} - \frac{d'\alpha}{2\beta}\left(\frac{c'-d'}{\alpha+\beta}\right) + \frac{d'c'}{2\beta} \tag{16.2}$$

To determine by how much the demand for processed cheese would have to increase so that the manufacturers would at least cover their marginal costs in producing omega-3 fortified cheese, we set the change in producer surplus, ΔPS, equal to 0 and we solve for c':

The change in producer surplus is given by

$$\Delta PS = \frac{\beta}{(\alpha+\beta)^2}c'^2 + \frac{\alpha d'}{(\alpha+\beta)^2}c' + \left[\frac{-\alpha\left(d'^2 - (c-d)^2\right)}{2(\alpha+\beta)^2} - \frac{c(c-d)}{2(\alpha+\beta)} + \frac{\alpha\left(d'^2 + d(c-d)\right)}{2\beta(\alpha+\beta)} - \frac{dc}{2\beta}\right]$$

Solving for c' yields

$$c' = \frac{\frac{-\alpha d'}{(\alpha+\beta)^2} \pm \left[\left(\frac{\alpha d'}{(\alpha+\beta)^2}\right)^2 - \frac{4\beta}{(\alpha+\beta)^2} * \left(\frac{-\alpha\left(d'^2-(c-d)^2\right)}{2(\alpha+\beta)^2} - \frac{c(c-d)}{2(\alpha+\beta)} + \frac{\alpha\left(d'^2+d(c-d)\right)}{2\beta(\alpha+\beta)} - \frac{dc}{2\beta}\right)\right]^{\frac{1}{2}}}{\frac{2\beta}{(\alpha+\beta)^2}} \tag{16.3}$$

All parameters in (Eq. 16.3) are known. α (the slope) and c (the intercept) are the demand equation parameters, β and d represent the slope and the intercept of the supply equation, and d' is related to the shift in the supply function due to the increase in the marginal cost.

From Fig. 16.6, we determine d' as $d - d' = \frac{\Delta MC}{\sin\delta}$, where $\tan\delta = \beta$ and ΔMC is the change in the marginal cost of production due to the fortification.

Then

$$d' = d - \frac{\Delta MC}{\sin(\operatorname{atan}\beta)} \tag{16.4}$$

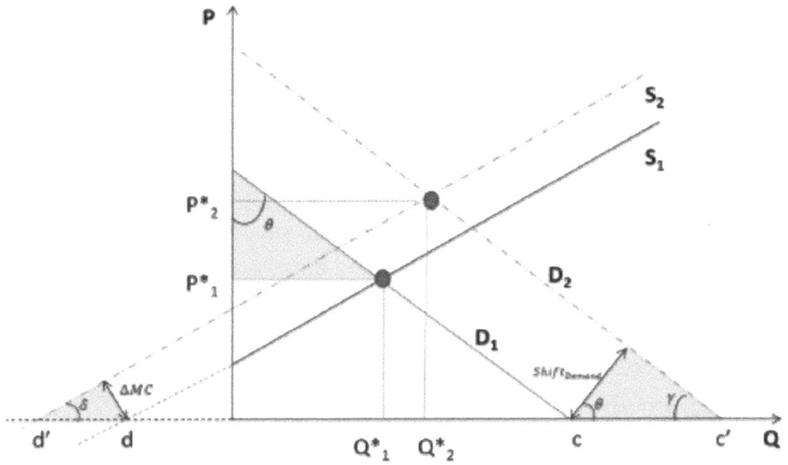

Figure 16.6 A Detailed graphical analysis of demand and supply relationship after fortification.

Graphically we can see that the shift in demand is determined as

$$Shift_{Demand} = \left(c' - c\right) * \sin\gamma \text{ where } \gamma = \frac{\pi}{2} - \theta \text{ and } \tan\theta = -\alpha$$

Then $Shift_{Demand} = (c' - c)^{*}\cos(\text{atan}(-\alpha))$.

Therefore, the percentage change in demand necessary to at least offset the marginal costs of producing processed cheese fortified with omega-3 fatty acids is given by

$$\%\ Shift_{Demand} = \frac{\left(c' - c\right)\cos[\text{atan}(-\alpha)]}{c} \times 100 \tag{16.5}$$

16.5.3 Demand parameters

Initially, we adopt the estimated own-price elasticity of demand for processed cheese from the work of Bouhlal (2012). This estimated value of −00.68 is the most recent from the economic literature, and coincidentally is in the mid-range of published own-price elasticities of demand for processed cheese. From Bouhlal (2012), the parameters α and c are equal to −046.69 and 22.98, respectively. Hence $Q_d = -\alpha P + c$ the demand for processed cheese is expressed as $Q_d = -46.69P + 22.98$.

16.5.4 Supply parameters

To the best of our knowledge, no information concerning the supply function of processed cheese products is available in the extant literature. Recall that the linear supply function is expressed as $Q_s = \beta P + d$. In this study we use different supply parameters β and d, relating to different values of the own-price elasticity of supply ε_s.

The values chosen for ε_s range from 0.001 (extremely inelastic) to 10 (very elastic), accounting for the short-run and the long-run impacts of the change in marginal cost of production due to the fortification process.

The own-price elasticity of supply ε_s is equal to $\beta P/Q_S$. Hence, only the parameter β can be derived from ε_s and as such we need to impose values for the parameter d. To investigate the impact of different values of the intercept d, we perform a sensitivity test measuring the effects in the percentage change in demand needed at least to cover the marginal costs in producing processed cheese fortified with omega-3 fatty acids. If we keep the change in marginal cost (ΔMC) and the elasticity of supply ε_s constant, it turns out that imposing different values of the intercept d does not impact the empirical results (for details, see Bouhlal (2012)).

16.5.5 Change in marginal costs of production with fortification

Incremental or marginal costs of production with fortification include but are not necessarily limited to costs associated with retro-fitting of processing equipment, labor costs, and the ingredient costs for omega-3 fatty acids. The Global Organization for EPA and DHA Omega-3 (GOED), a trade association founded in 2006 which represents the worldwide EPA and DHA omega-3 industry (GOED, 2020), suggests that the marginal cost of production associated with fortification should not exceed 2 cents per serving (Bouhlal, 2012). As one example, a serving of a slice of American processed cheese roughly is equal to 1.5 ounces. Then according to the GOED, the change in marginal cost of production of processed cheese would be 3 cents per ounce. To check on the sensitivity of the change in marginal cost, we choose different values of ΔMC, ranging from \$0.01/ounce to \$0.20/ounce. Hence, we perform a sensitivity analysis in evaluating the impact of the various changes in marginal cost on the profitability of fortifying processed cheese with omega-3 fatty acids.

16.5.6 Changes in demand needed to offset incremental fortification costs

According to the results exhibited in Table 16.1, in most cases the percentage change in demand needed to offset the extra cost due to the fortification is negligible. Apart from cases of very inelastic supply or very high change in the marginal cost, the fortification of processed cheese with omega-3 fatty acids turns out to be economically feasible. For the situation where ΔMC = \$0.01/ounce, in the very short run, when the own-price elasticity of supply is very inelastic (0.001), the shift needed in demand is 9.37%. However, in moving toward larger values of the elasticity of supply, the shift in demand needed diminishes. To illustrate, when the own-price elasticity of supply is equal to 0.1, this percentage change in demand drops to 0.006%. When the own-price elasticity of supply is equal to 0.6, the shift in demand needed barely exceeds 0.001%. According to our findings, the more elastic the supply, the less the demand for processed cheese products needs to shift to the right in order to cancel out the incremental fortification costs. Fortification is not likely a viable option only in extreme cases when the supply is very inelastic.

Table 16.1 Percentage change in demand associated with various measures of marginal cost and own-price elasticities of supply.

ΔMC	Es	%change demand	ΔMC	Es	%change demand	ΔMC	Es	%change demand
0.01	0.001	9.3762	0.07	0.001	65.6334	0.20	0.001	187.5241
	0.005	0.3953		0.005	2.7674		0.005	7.9067
	0.01	0.1132		0.01	0.7926		0.01	2.2646
	0.05	0.0133		0.05	0.0931		0.05	0.2659
	0.1	0.0064		0.1	0.0451		0.1	0.1290
	0.2	0.0032		0.2	0.0224		0.2	0.0640
	0.3	0.0021		0.3	0.0149		0.3	0.0426
	0.4	0.0016		0.4	0.0112		0.4	0.0319
	0.5	0.0013		0.5	0.0089		0.5	0.0255
	0.6	0.0011		0.6	0.0074		0.6	0.0213
	0.7	0.0009		0.7	0.0064		0.7	0.0182
	0.8	0.0008		0.8	0.0056		0.8	0.0160
	0.9	0.0007		0.9	0.0050		0.9	0.0142
	1	0.0006		1	0.0045		1	0.0128
	3	0.0002		3	0.0015		3	0.0043
	5	0.0001		5	0.0009		5	0.0026
	10	0.0001		10	0.0004		10	0.0013
0.03	0.001	28.1286	0.10	0.001	93.7621			
	0.005	1.1860		0.005	3.9534			
	0.01	0.3397		0.01	1.1323			
	0.05	0.0399		0.05	0.1330			
	0.1	0.0193		0.1	0.0645			
	0.2	0.0096		0.2	0.0320			
	0.3	0.0064		0.3	0.0213			
	0.4	0.0048		0.4	0.0160			
	0.5	0.0038		0.5	0.0128			
	0.6	0.0032		0.6	0.0106			
	0.7	0.0027		0.7	0.0091			
	0.8	0.0024		0.8	0.0080			
	0.9	0.0021		0.9	0.0071			
	1	0.0019		1	0.0064			
	3	0.0006		3	0.0021			
	5	0.0004		5	0.0013			
	10	0.0002		10	0.0006			
0.05	0.001	46.8810	0.15	0.001	140.6431			
	0.005	1.9767		0.005	5.9300			
	0.01	0.5661		0.01	1.6984			
	0.05	0.0665		0.05	0.1995			
	0.1	0.0322		0.1	0.0967			
	0.2	0.0160		0.2	0.0480			
	0.3	0.0106		0.3	0.0319			
	0.4	0.0080		0.4	0.0239			
	0.5	0.0064		0.5	0.0191			

(*continued on next page*)

Table 16.1 Percentage change in demand associated with various measures of marginal cost and own-price elasticities of supply—cont'd

ΔMC	Es	%change demand	ΔMC	Es	%change demand	ΔMC	Es	%change demand
	0.6	0.0053		0.6	0.0160			
	0.7	0.0046		0.7	0.0137			
	0.8	0.0040		0.8	0.0120			
	0.9	0.0035		0.9	0.0106			
	1	0.0032		1	0.0096			
	3	0.0011		3	0.0032			
	5	0.0006		5	0.0019			
	10	0.0003		10	0.0010			

Source: Calculation by the authors.
Es represents the own-price elasticity of supply. Details of the calculations are available from Bouhlal (2012). ΔMC represents the change in marginal costs due to fortification.
%change demand refers to the percentage change in demand necessary in order to maintain producer profitability.

Increasing the change in marginal cost due to fortification leads to increases in the percentage shift in demand needed to make the fortification strategy viable for producers so as to maintain profitability. We observe that when ΔMC= \$0.03/ounce instead of the initial value of \$0.01/ounce, all values associated with the percentage changes in demand corresponding to different own-price elasticities of supply rose considerably. Nevertheless, the demand shift values are still very small for ΔMC = \$0.03/ounce, namely 0.34% when the own-price elasticity of supply is equal to 0.01, 0.005% when the own-price elasticity of supply is equal to 0.4, and 0.002% when the own-price elasticity of supply is equal to 1. Small own-price elasticities of supply are indicative of short-run effects, while larger own-price elasticities of supply are indicative of long-run effects. Hence, fortification is feasible for manufacturers not only in the short-run but also in the long-run. Importantly, the percentage change in demand needed to maintain producer profitability decreases with increases in the own-price elasticity of supply, making the fortification process more attractive in the long-run.

The same results are obtained even when choosing very high changes in marginal cost due to fortification. In the extreme case where ΔMC = \$0.20/ounce, we still observe the same pattern, namely that the percentage change in demand needed to maintain producer profitability decreases with increases in the own-price elasticity of supply. In this case, the change is marginal cost incidentally is tantamount to the average price of a slice of processed cheese (Bouhlal, 2012). Under this assumption then, fortification induces a 100% increase or doubling in the price of processed cheese. Even with this considerably high incremental or marginal cost, we can still say that fortification is economically feasible to producers since at non-extreme values of the own-price elasticity of supply, the demand shift needed is still manageable.

16.5.7 Changes in the own-price elasticity of demand

The previously discussed simulation analysis hinges on the estimated own-price elasticity of demand to be −00.68 from Bouhlal (2012). To push our analysis further, we

Table 16.2 Percentage change in demand due to various own-price elasticities of demand and supply for the case where $\Delta MC = \$0.03$/ounce.

Es	Ed	%change demand
0.005	−00.4	1.8160
	−00.68	1.1860
	−00.8	0.6813
	−01.2	0.4543
0.1	−00.4	0.0296
	−00.68	0.0193
	−00.8	0.0111
	−01.2	0.0074
0.3	−00.4	0.0098
	−00.68	0.0064
	−00.8	0.0037
	−01.2	0.0024
0.6	−00.4	0.0049
	−00.68	0.0032
	−00.8	0.0018
	−01.2	0.0012
1	−00.4	0.0029
	−00.68	0.0019
	−00.8	0.0011
	−01.2	0.0007

Source: Calculation by the authors.
Es and Ed refer to the own-price elasticity of supply and the own-price elasticity of demand for processed cheese, respectively.
%change demand refers to the percentage change in demand necessary in order to maintain producer profitability.

investigate the impact of fortification with omega-3 fatty acids on producer surplus or profitability for own-price elasticity of demand values other than −00.68. Recall that from the economic literature, the own-price elasticity for processed cheese has ranged from −00.36 to −00.99. Subsequently, we allow the own-price elasticity of demand for processed cheese to vary from −00.4 to −01.2, and we compute the percentage change in demand needed to maintain producer profitability not only for different values of the own-price elasticity of supply but also for different values of marginal costs. To conform to space limitations, as exhibited in Table 16.2, we consider the situation where $\Delta MC = \$0.03$/ounce, and we allow the own-price elasticities of supply to vary from 0.005 to 1. Results for other levels of marginal costs from \$0.01/ounce to \$0.20/ounce with various assumptions about the own-price elasticities of supply are available upon request from the authors.

As exhibited in Table 16.2, increases in the value of the own-price elasticity of demand (in absolute value) produce decreases in the percentage change in demand needed to offset the change in marginal cost of production due to fortification. When the $\Delta MC = \$0.03$/ounce and coupled with the own-price elasticity of supply equal to 0.005 for example, the shift in demand needed is 1.82% when the own-price elasticity of demand is −00.4; however, the demand shift needed drops to 0.68% when the own-price elasticity of demand is −00.8 and drops further to 0.45% when the own-price elasticity of demand of is −01.2. For the range of own-price elasticities of demand investigated, the shift in demand needed to offset the incremental or marginal cost due to fortification never exceeded 2%. This finding supports the contention that fortification of processed cheese is economically feasible for manufacturers. In doing so, manufacturers may diversify their product line, maintain profitability, and provide a healthier product to consumers.

16.6 Conclusion

Given the current market situation concerning processed cheese in the United States, we focus on the economic feasibility of fortifying processed cheese with omega-3 fatty acids. Heretofore, studies have been conducted in the extant literature dealing with the fortification of reduced-fat cheddar cheese with n-3 fatty acids, processed cheese fortified with fish oil emulsion, selected cheeses fortified with vegetable and animal sources of omega-3 fatty acids and the fortification of processed cheese spread with omega-3 fatty acids. But no studies currently exist that deal with the economic feasibility of the fortification of processed cheese.

This ex-ante analysis takes into account simulated market conditions, via demand and supply curves, and evaluates the percentage increase in the demand for processed cheese needed to offset the incremental costs of fortification in order to maintain producer profitability. The methodology involves the use of a partial equilibrium framework, often referred to as EDM. To eliminate concerns about uncertainty of parameter values with the EDM approach, our work utilizes not only various values of own-price elasticities of demand and supply but also a range of the marginal costs associated with the fortification process. In this way we provide a check on the robustness of the empirical results.

Bottom line, with only minimal shifts in the demand for processed cheese, we demonstrate that fortification with omega-3 fatty acids can occur without any loss in producer profits. This finding supports the contention that fortification of processed cheese indeed is economically feasible for manufacturers. The additional important by-products for manufacturers in doing so are the diversification of their product line and the provision of a healthier product to consumers. In addition, the potential increases in the demand for processed cheese due to the fortification with omega-3 fatty acids also will be beneficial to dairy farmers, to retailers, and to consumers.

Next steps include extending the EDM analysis to consider multiple products, conventional processed cheese and processed cheese fortified with omega-3 fatty acids. As well, with the introduction of processed cheese fortified with omega-3 fatty acids, we are in position to track the demand for this new product using scanner data from vendors such as Nielsen or from Information Resources, Inc. (IRI). With this tracking, we can empirically discern market shares between conventional processed cheese and processed cheese fortified with omega-3 fatty acids as well as discern the own-price elasticity of demand for the respective products. In this way, we may statistically test whether or not the own-price elasticities for the multiple products are the same.

As well, it would be advantageous to survey manufacturers of processed cheese in order to obtain industry estimates of the marginal costs associated with fortification rather than rely on a broad range of \$0.01/ounce to \$0.20/ounce. Finally, determining whether or not cannibalization effects are evident for firms that offer multiple products for processed cheese would be of interest. The identification and assessment of cannibalization are integral factors when making strategic decisions about new product introductions such as processed cheese fortified with omega-3 fatty acids.

References

Allen, L., de Benoist, B., Dary, O., Hurrell, R., 2006. Guidelines of Food Fortification with Micronutrients. World Health Organization and Food and Agricultural Organization of the United Nations.

Beery, D. 2018. Fortification Driving Innovation in Dairy. Food Business News. Available online at: https://www.foodbusinessnews.net/articles/12713-fortification-driving-innovation-in-dairy. Accessed October 25, 2018.

Bermudez-Aguirre, D., Barbosa-Cánovas, G.V., 2011. Quality of selected cheeses fortified with vegetable and animal sources of omega-3. Food Sci. Technol. 44 (7), 1577–1584.

Bouckley, B. 2017. "Scientists Claim High Omega-3 Retention Rates in Popular Cheeses." DairyReporter.com. Available online at: https://www.DairyReporter.com/article/202/02/22/scientists-claim-high-omega-3-retention-rates-in-popular-cheeses.

Bouhlal, Y., 2012. A Retrospective and Prospective Analysis of the Demand for Cheese Varieties in the Unites States. Separtment of Agricultural Economics, Texas A&M University. Ph.D. Dissertation.

Capps Jr., O.G.W., Williams, V.S.S, Brown, D.S., 2020. Quantitative Evaluation of the Effectiveness of Activities by the National Dairy Promotion and Research Program and the National Fluid Milk Processor Promotion Program-1995 to 2018. Agribusiness, Food, and Consumer Economics Research Center, Texas A&M University Technical Report.

Davis, C., Dong, D., Blayney, D.P., Owens, A., 2010. An Analysis of U.S. Household Dairy Demand. USDA Economic Research Service. Technical Bulletin Number 1928.

Davis, G.C., Espinoza, M.C., 1998. A unified approach to sensitivity analysis in equilibrium displacement models. Am. J. Agric. Econ. 80 (4), 868–879.

Ganesan, B., Brothersen, C., McMahon, D.J., 2014. Fortification of foods with omega-3 polyunsaturated fatty acids. Crit. Rev. Food Sci. Nutr. 54 (1), 98–114.

Garg, M.L., Wood, L.G., Singh, H., Moughan, P.J., 2006. Means of delivering recommended levels of long chain n-3 polyunsaturated fatty acids in human diets. J. Food Sci. 71 (5), R66–R71.

Gebauer, S.K., Psota, T.L., Harris, W.S., Kris-Etherton PM, P.M., 2006. n− 3 fatty acid dietary recommendations and food sources to achieve essentiality and cardiovascular benefits. Am. J. Clin. Nutr. 83 (6), 1526S–1535S.

Global Organization for EPA and DHA Omega-3s (GOED). 2020. Available online at: www.goedomega3.com.

Goldberg, R.J., Katz, J., 2007. A meta-analysis of the analgesic effects of omega-3 polyunsaturated fatty acid supplementation for inflammatory joint pain. Pain 129 (1-2), 210–223.

Gogus, U., Smith, C., 2010. n-3 omega fatty acids: a review of current knowledge. Int. J. Food Sci. Technol. 45 (3), 417–436.

Gould, B.W., Lin, H.C., 1994. The demand for cheese in the United States: the role of household composition. Agribusiness 10 (1), 43–57.

Institute of Medicine (US) Committee on Use of Dietary Reference Intakes in Nutrition Labeling, 2003. Dietary Reference Intakes: Guiding Principles for Nutrition Labeling and Fortification. National Academies Press, Washington, DC.

Kolanowski, W., 2005. Bioavailability of Omega-3 PUFA from Foods Enriched with Fish Oil. Polish J. Food Nutr. Sci. 55 (4), 335–340.

Kolanowski, W., Weißbrodt, J., 2007. Sensory quality of dairy products fortified with fish oil. Int. Dairy J. 17 (10), 1248–1253.

Kris-Etherton, P.M., Harris, W.S., Appel, L.J., 2002. AHA scientific statement, fish consumption, fish oil, omega-3 fatty acids, and cardiovascular disease. Circulation 106 (21), 2747–2757.

Lombardi, F., Terranova, P., 2007. Anti-arrhythmic properties of N-3 poly-unsaturated fatty acids (n-3 PUFA). Curr. Med. Chem. 14 (19), 2070–2080.

MacLean, C.H., Newberry, S.J., Mojica, W.A., Khanna, P., Issa, A.M., Suttorp, M.J., Lim, Y.W., Traina, S.B., Hilton, L., Garland, R., Morton, S.C., 2006. Effects of omega-3 fatty acids on cancer risk: a systematic review. J. Am. Med. Assoc. 295 (4), 403–415.

Martini, S., Thurgood, J.E., Brothersen, C., Ward, R., McMahon, D.J., 2009. Fortification of reduced-fat cheddar cheese with n-3 fatty acids: effect on off-flavor generation. J. Dairy Sci. 92 (5), 1876–1884.

Renuka, V., Ramasamy, D., Dhineshkumar, V., 2016. Fortification of omega-3 fatty acids in processed cheese spread. Int. J. Sci. Environ. Technol. 5 (4), 2557–2565.

Report of the Joint FAO/WHO Expert Consultation on the Risks and Benefits of Fish Consumption. 2010. FAO Fisheries and Aquaculture Report, No. 979, January.

Robbins, O. 2019. Is cheese healthy? See what the research says. https://foodrevolution.org/blog/is-cheese-good-for-you/. Accessed on January 16, 2020.

Schmit, T.M., Gould, B.W., Dong, D., Kaiser, H.M., Chung, C., 2003. The impact of generic advertising on U.S. household cheese purchases: a censored autocorrelated regression approach. Can. J. Agric. Econ. 51 (1), 15–37.

Shoup, M.E. 2019. US consumers continue to eat more cheese. https://www.foodnavigator-usa.com/Article/2019/01/30/US-consumers-continue-to-eat-more-natural-and-processed-cheeses#. Accessed on Febraury 13, 2020.

Simopoulos, A.P., 2002. Omega-3 fatty acids in inflammation and autoimmune diseases. J. Am. Coll. Nutr. 21 (6), 495–505.

United States Department of Agriculture, Economic Research Service, 2021 https://www.ers.usda.gov/data-products/dairy-data/. Accessed on March 7, 2021.

Yokoyama, M., Origasa, H., Matsuzaki, M., Matsuzawa, Y., Saito, Y., Ishikawa, Y., Oikawa, S., Sasaki, J., Hishida, H., Itakura, H., Kita, T., Kitabatake, A., Nakaya, N., Sakata, T., Shimada, K., Shirato, K., 2007. Effects of eicosapentaenoic acid on major coronary events in hypercholesterolaemic patients (JELIS): a randomized open-label, blinded endpoint analysis. Lancet North Am. Ed. 369 (9567), 1090–1098.

Ye, A., Cui, J., Taneja, A., Zhu, X., Singh, H., 2009. Evaluation of processed cheese fortified with fish oil emulsion. Food Res. Int. 42 (8), 1093–1098.

Index

Page numbers followed by "*f*" and "*t*" indicate, figures and tables respectively.

FSC
www.fsc.org
MIX
Papier | Fördert gute Waldnutzung
FSC® C083411

Kolibri 360 GmbH
Ferdinand-Jühlke-Straße 7
99095 Erfurt, Deutschland
produktsicherheit@kolibri360.de